Studies in Big Data

Volume 34

Series editor

Janusz Kacprzyk, Polish Academy of Sciences, Warsaw, Poland
e-mail: kacprzyk@ibspan.waw.pl

The series "Studies in Big Data" (SBD) publishes new developments and advances in the various areas of Big Data- quickly and with a high quality. The intent is to cover the theory, research, development, and applications of Big Data, as embedded in the fields of engineering, computer science, physics, economics and life sciences. The books of the series refer to the analysis and understanding of large, complex, and/or distributed data sets generated from recent digital sources coming from sensors or other physical instruments as well as simulations, crowd sourcing, social networks or other internet transactions, such as emails or video click streams and other. The series contains monographs, lecture notes and edited volumes in Big Data spanning the areas of computational intelligence incl. neural networks, evolutionary computation, soft computing, fuzzy systems, as well as artificial intelligence, data mining, modern statistics and Operations research, as well as self-organizing systems. Of particular value to both the contributors and the readership are the short publication timeframe and the world-wide distribution, which enable both wide and rapid dissemination of research output.

More information about this series at http://www.springer.com/series/11970

Sławomir T. Wierzchoń · Mieczysław A. Kłopotek

Modern Algorithms
of Cluster Analysis

 Springer

Sławomir T. Wierzchoń
Institute of Computer Science
Polish Academy of Sciences
Warsaw
Poland

Mieczysław A. Kłopotek
Institute of Computer Science
Polish Academy of Sciences
Warsaw
Poland

ISSN 2197-6503   ISSN 2197-6511   (electronic)
Studies in Big Data
ISBN 978-3-319-88752-4   ISBN 978-3-319-69308-8   (eBook)
https://doi.org/10.1007/978-3-319-69308-8

Printed on acid-free paper

This Springer imprint is published by Springer Nature
The registered company is Springer International Publishing AG
The registered company address is: Gewerbestrasse 11, 6330 Cham, Switzerland

Contents

Symbols

$\mathfrak{X}$	Set of objects (observations)
x	An element of the set $\mathfrak{X}$
$X \in \mathbb{R}^{m \times n}$	Matrix, representing the set of objects, $X = (\mathbf{x}_1 \ \cdots \ \mathbf{x}_m)^{\mathrm{T}}$
m	Cardinality of the set X
n	Dimensionality (number of coordinates) of the vector of features of $\mathbf{x} \in X$
C	Set of clusters
k	Number of clusters
$diag(\mathbf{v})$	Diagonal matrix having the main diagonal equal to $\mathbf{v}$
$G = (V, E)$	Graph spanned over the set of vertices V linked with the edges indicated in the set E
$N(v)$	Set of neighbours of the vertex v in graph G
$N_k(v)$	Set of k nearest neighbours of the vertex v in graph G
$\partial(C)$	Edge separator in graph G: $\partial(C) = \{\{u, v\} \in E : u \in C, v \notin C\}$
A	Neighbourhood matrix having elements $a_{ij} \in \{0, 1\}$, $A = [a_{ij}]_{m \times m}$
S	Similarity matrix having elements $s_{ij} \in [0, 1]$, $S = [s_{ij}]_{m \times m}$
$\mathfrak{m}$	Number of edges in graph G
d_i	Degree (weight) of the i-th object, $d_i = \sum_{j=1}^{m} s_{ij}$
D	Matrix of degrees, $D = diag([d_1, \ldots, d_m])$
L	Combinatorial Laplacian, $L = D - S$
$\mathfrak{L}$	Normalised Laplacian, $\mathfrak{L} = D^{-1/2}LD^{-1/2}$
$\mathbb{L}$	A variant of the Laplacian, $\mathbb{L} = \mathbb{I} - P$
P	Column-stochastic transition matrix, describing walk on graph G, $P = SD^{-1}$
$\widetilde{P}$	Row-stochastic transition matrix, describing walk on graph G, $\widetilde{P} = D^{-1}S$
$\widehat{P}$	Column-stochastic transition matrix in a lazy walk on graph G, $\widehat{P} = \frac{1}{2}(\mathbb{I} + SD^{-1})$
π	Stationary distribution of the Markov chain having transition matrix P, that is, a vector such that $\pi = P\pi$

$\rho(s, \alpha)$ Global PageRank with positive preference vector s and the coefficient of teleportation α: $\rho(s, \alpha) = \alpha s + (1 - \alpha) P \rho(s, \alpha)$

$\mathfrak{p}(s, \alpha)$ Personalised PageRank vector with nonnegative preference vector s and the coefficient of teleportation α

$\mathbf{e}$ Column vector having elements equal 1

$\mathbb{I}$ Unit matrix

List of Figures

List of Tables

List of Algorithms

Chapter 1
Introduction

Abstract This chapter characterises the scope of this book. It explains the reasons why one should be interested in cluster analysis, lists major application areas, basic theoretical and practical problems, and highlights open research issues and challenges faced by people applying clustering methods.

The methods of data analysis can be roughly classified into two categories[1]: (a) the descriptive (exploratory) ones, which are recommended when, having no initial models or hypotheses, we try to understand the general nature and structure of the high-dimensional data, and (b) the confirmatory (inferential) methods, applied in order to confirm the correctness of the model or of the working hypotheses, concerning the data collected. In this context, a particular role is played by various statistical methods, such as, for instance, analysis of variance, linear regression, discriminant analysis, multidimensional scaling, factor analysis, or, finally, the subject of our considerations here—*cluster analysis*[2] [237, 303].

It is really difficult to list all the domains of theoretical and practical applications of cluster analysis. Jain [272] mentions three basic domains: image segmentation [273], information retrieval [347], and bio-informatics [46].

The task of cluster analysis or grouping is to divide the set of objects into homogeneous groups: two arbitrary objects belonging to the same group are more similar to each other than two arbitrary objects belonging to different groups. If we wish to apply this recipe in practice, we must find the answers to two basic questions: (a) how to define the similarity between the objects, and (b) in what manner should one make use of the thus defined similarity in the process of grouping. The fact that the answers to these questions are provided independently one from another results in the multiplicity of algorithms.

A number of natural processes can give rise to clusters in the data [195]. One may have collected data from multiple experiments on same or intersecting sets of objects, or one obtained longitudinal data (same objects measured at different time

---

[1]See, e.g., J. W. Tukey, *Exploratory Data Analysis*. Addison-Wesley, 1977.

[2]This area is also referred to as Q-analysis, typology, grouping, partitioning, clumping, and taxonomy, [273], depending upon the domain of application.

© Springer International Publishing AG 2018
S. T. Wierzchoń and M. A. Kłopotek, *Modern Algorithms of Cluster Analysis*,
Studies in Big Data 34, https://doi.org/10.1007/978-3-319-69308-8_1

points), data may have been collected at different experimental sites, data may stem from subjects (human or animal) related to one another like in families etc. Seasonality, climate, geography, measurement equipment manufacturer are further possible hidden variables, inducing cluster formation among objects of the study. If one is unaware of such subtleties of the data collection process (no explicit information is available) but one is suspicious of such a situation, then one is advised to apply a clustering method prior to further analysis because the existence of clusters may impact the results of further analyses as an intra-cluster correlation may exist so that the frequent independence assumptions are violated. For example, if one intends to perform regression analysis on the data, and there exist clusters, then one is advised to add a factor for the cluster effect into the standard regression model (e.g. ANOVA analysis). A number of statistical tests, like χ^2, t-test, Wilcoxon test have also been adjusted for clusters. In image analysis, clustering may be performed prior to applying more advanced methods, like object recognition.

Division of data into clusters may also by itself be a subject of interest.

An interesting example of application of cluster analysis in modern information retrieval is constituted by the CLUSTY search engine (clusty.com). The methods of cluster analysis play an important role in the development of the recommender systems [258, 416], in various kinds of economic, medical etc. analyses. Application of these methods results from at least three reasons:

(a) To gain insight into the nature of the data, and, first of all, to indicate the typical and the atypical (*outlying*) data items, to uncover the potential anomalies, to find hidden features, or, finally—to be able to formulate and verify the hypotheses concerning the relations between the observations;
(b) To obtain a compact description of data, to select the most representative objects; here, the classical application is image compression;
(c) To obtain a natural classification of data, for instance—by determining the similarities between pairs of objects to establish the respective hierarchical structures.

Methods of cluster analysis are being applied there, where we wish to understand the nature of the phenomenon, represented by the set of observations, to get a sort of summary of the content of large data sets, and to process such sets effectively.

Jain and Dubes [273] list the following challenges, linked with the practical use of clustering:

(a) What is a cluster (group, module)?
(b) What features ought to be used to analyse the data collected?
(c) Should the data be normalised?
(d) Are there outliers in the analysed dataset, and if so—how should they be treated?
(e) How to define the similarity between the pairs of objects?
(f) How many clusters are really there in the data set? Do we deal, at all, with clusters?
(g) What method should be used in a given situation?
(h) Is the obtained partition of data justified?

Some of these questions find at least partial answers in a multiplicity of books, devoted to various aspects of cluster analysis, like [12, 19, 70, 164, 273], or [173].

The recent years, though, brought a number of new challenges. First, cluster analysis is being applied nowadays to processing of huge sets of data, which makes it necessary to develop specialised algorithms. Second, the necessity of analysing the data sets having complex topologies (data being situated in certain submanifolds) leads to the requirement of applying more refined tools. We mean here the spectral methods, the methods making use of kernel functions, and the relations between these two groups of methods.

The starting point for these methods is constituted by the matrix, describing the strength of interrelations between the objects making up the data set. In the case of the kernel-based methods, considerations are being transferred to the highly dimensional and nonlinear space of features for the purpose of strengthening of the separability of data. Due to application of the so-called *kernel trick* to determine the distance between objects in this new space, the knowledge of a certain kernel function suffices. The elements of the matrix, mentioned before, are the values of the kernel function, calculated for the respective pairs of objects. On the other hand, in the case of spectral methods, the elements of the matrix correspond to the values of similarity of the pairs of objects. Eigenvalues and eigenvectors of this matrix, or a matrix, being its certain transformation (usually a form of the Laplacian of the graph, corresponding to the similarity matrix), provide important information on relations between objects.

An interesting example is constituted by the problem of ordering of documents, collected by the crawlers, cooperating with a search engine. We deal in such cases with enormous collections of objects (of the order of 10^9), with the issue of their effective representation, and, finally, with the task of indicating the important documents. The common sense postulate that the important websites (documents) are the ones that are referred to (linked to) by other important websites allows for the reformulating of the problem of assigning ranks in the ordering into the problem of determining the dominating eigenvector of the stochastic matrix P, being a slight modification of the original matrix A, representing connections between the websites, that is $a_{ij} = 1$ when on the ith website there is a link to the jth one. Vector $\mathbf{r}$, which satisfies the matrix equation $\mathbf{r} = P^\mathsf{T}\mathbf{r}$, is called the PageRank vector [379]. Both this vector and its diverse variants (TotalRank, HillTop, TrustRank, GeneRank, IsoRank, etc.) are the examples of the so-called spectral ranking, which find application not only in the ordering of documents (or, more generally, in various aspects of bibliometrics), but also in graph theory (and, consequently, also in the analysis of social networks), in bio-informatics, and so on.

A development over the latter idea is constituted by the methods based on random walk on graphs. Namely, the task of grouping is here considered as an attempt of finding such a division of the graph that a randomly moving wanderer shall remain for a long time in a single cluster (subgraph), only rarely jumping between the clusters. An interesting work on this subject is presented in, e.g., [211], while a number of theoretical results have been shown by Chung [23, 115, 116], and by Meila [352].

The transformation of the matrices, representing relations (e.g. similarity) between the elements of the set of objects, into the stochastic Markovian matrices leads to yet another interesting formalism. As we treat the eigenvectors of these matrices as forming a new system of coordinates, we transform the multidimensional data into a cloud of points in a space with a much lower number of dimensions. This new representation reflects the essential properties of the original multidimensional structure. In this manner we enter the domain of the so-called diffusion maps [125], which have two significant features, distinguishing them from the classical methods: they are nonlinear, and, besides this, they do faithfully map the local structure of data. From the point of view of, for instance, processing of text documents, an essential advantage of such an approach is a relatively straightforward adaptation to the semi-supervised clustering problems and the *coclustering* problems—the important, quite recently developing, directions of machine learning.

These most up-to-date trends in grouping have not been, until now, systematically presented nor considered in an integrated manner against the background of the classical methods of cluster analysis. Such a survey would be very welcome indeed from the point of view of students with an interest in the techniques of knowledge extraction from data and computer-based modelling, in the context of their application in pattern recognition, signal analysis, pattern and interdependence identification, and establishment of other interesting characteristics in the large and very large collections of data, especially those put together automatically in real time.

That is why we prepared this monograph, which:

(a) Contains the survey of the basic algorithms of cluster analysis, along with presentation of their diverse modifications;
(b) Makes apparent the issue of evaluation and proper selection of the number of clusters;
(c) Considers the specialisation of selected algorithms regarding the processing of enormous data sets;
(d) Presents in a unified form a homogeneous material, constituting the basis for developing new algorithms of spectral and diffusion data analysis;
(e) Comments upon the selected solutions for the semi-supervised learning;
(f) Provides a rich bibliography, which might also be a starting point to various individual research undertakings.

This book has been divided into 7 chapters. Chapter 2 outlines the major steps of cluster analysis. Chapters 3 and 5 are devoted to two distinct brands of clustering algorithms. While Chap. 3 focusses on clustering of objects that are embedded in some metric space, Chap. 5 concentrates on ones abandoning the embedding space and concentrating on the similarities between objects alone. The former are referred to as combinatorial algorithms, the latter are called spectral algorithms. Chapter 4 addresses the issues of parameter tuning for clustering algorithms and the assessment of the resultant partitions. Chapter 6 discusses a specific, yet quite important application area of cluster analysis that is of community detection in large empirical graphs. Finally, in Chap. 7 we present the data sets used throughout this book in illustrative examples.

From the point of view of knowledge discovery task the clustering can be split into the following stages:

- problem formulation,
- collecting data,
- choice and application of a clustering algorithm,
- evaluation of its results.

While the problem of clustering seems to be obvious in general terms, its formalisation is not as obvious in concrete case. It requires first a formal framework for the data to be learned from, as outlined in Sect. 2.1. Furthermore, the user is requested to define similarity between the objects (the geometry of the sample space) as well as a cost function (a measure of value of the obtained clusters). The definition of the geometry must be application-specific, tuned to the specific needs of the user. Section 2.2 provides with a guidance in this process, listing many possibilities of measuring similarity both in numerical and non-numerical data and explaining some special properties of these measures. Section 2.4.1 provides with a set of possible cost functions to be used in the clustering process. Note, however, that each concrete algorithm targets at some specific cost function.

Advices on the choice of data for cluster analysis can be found in Sect. 4.1.

To enable the choice of an appropriate clustering algorithm, Sect. 2.3 provides with an overview of hierarchical clustering methods, Sect. 2.4 presents partitional (called also combinatorial) clustering methods, Sect. 2.5 hints at other categories of clustering procedures like relational, spectral, density, kernel and ensemble methods. Some of these types of algorithms are reflected on more deeply in Chaps. 3, 5 and 6, as already mentioned.

The results of the clustering may be looked at from two perspectives: whether or not the choice of parameters of the algorithm was appropriate (see Sect. 4.2 for the choice of the number of clusters), and if the results meet the quality expectations (Sect. 4.3ff). These quality expectations may encompass:

- the cost function reached sufficiently low value,
- the homogeneity of clusters and their separation are sufficient (aesthetically or from business point of view),
- the geometry of the clusters corresponds to what the algorithm is expected to provide,
- the clusters correspond to some external criteria, like manually clustered subsets etc.

User's intuition and/or expectations with respect to the geometry of clusters is essential for the choice of the clustering algorithm. One shall therefore always inspect the description of the respective algorithm if it can properly respond to our expectations. In fact the abundance of clustering algorithms is a derivative of the fact that these expectations may be different, depending on actual application.

So if the geometry of the data makes them suitable for representation in a feature space, then visit first the methods described in Chap. 3. If they are conveniently represented as a graph, then either Chap. 5 or 6 are worth visiting.

If in the feature space, you may expect that ideally your clusters should form clearly separated (hyper)balls. In this case k-means algorithm (Sect. 3.1) may be most appropriate. If the shape should turn out to be more ellipsoidal, see Sect. 3.3.5.4. Should the clusters take forms like line segments, circles etc., visit Sect. 3.3.5.3. If you need to cluster text documents in vector space representation (which lie essentially on a sphere in this representation), then you may prefer spherical-k-means algorithm (Sect. 3.1.5.3). It may happen that you do not expect any particular shape but rather oddly shaped clouds of points of constant density, you may prefer DBSCAN or HDBSCAN algorithm (Sect. 2.5.4). If you expect that the border line between the clusters is not expressible in simple terms, you may try kernel methods (Sects. 2.5.7 and 3.3.5.6). Last not least you may seek a sharp split of objects into clusters or a more fuzzy one, using e.g. Fuzzy-C-Means algorithm.

There exist also multiple geometry choices for graph-based data representation. You may seek dense subgraphs, subgraphs, where a random walker would be trapped, subgraphs separable by removal of a low number of edges or subgraphs where connections evidently are different from random.

The main topic of this book is to provide with a basic understanding of the formal concepts of the cluster, clustering, partition, cluster analysis etc. Such an understanding is necessary in the epoch of Big Data, because due to the data volume and characteristics it is no more feasible to rely solely on viewing the data when facing the clustering problem. Automated processing to the largest possible extent is required and it can only be achieved if the concepts are defined in a clear-cut way. But this is not the only contribution of this book to the solving problems in the realm of Big Data. *Big Data* is usually understood as large amounts of data in terms of overall volume measured in (peta)bytes, large number of objects described by the data and lengthy object descriptions in a structured, unstructured or semi-structured way. Therefore a user interested in Big Data issues may find it helpful to get acquainted with various measures of object similarity as described in Sect. 2.2, where measures based on quantitative (like numerical measurement results) and on qualitative features (like text) as well as on their mixtures are described. As Big Data encompass hypertext data or other linked objects, graph-based similarity measures described in Sects. 6.2 and 6.4.4 may be also of interest. For the discussion of other complex structures in the data, like multilayered graphs, see Sect. 6.10. Various variants, how such similarity measures can be exploited when defining clustering cost functions are described in Sect. 2.4.1. Alternatively, they can be used in spectral analysis of graphs, described in Chap. 5.

Last not least we present a number of approaches to handle large number of objects in reasonable time. The diversity of such algorithms encompasses so-called grid-based methods, see Sect. 2.5.5, methods based on sampling, see Sect. 3.10, parallelisation via map-reduce, usage of tree-structures, see Sect. 3.1.4, to various heuristic approaches like those used for community detection, see Sect. 6.7. For particular algorithms we provide also with references to publications of variants devoted to huge data sets. Note that in some cases, like spectral clustering, accommodation to big data sets requires some new theoretical developments, enabling for example to sample linked data in a reasonable way, see Sect. 5.5.

Alternatively, especially with large bodies of textual data, data projection onto a smaller feature space can accelerate processing, like e.g. random projection, see Sect. 3.9.

In order to concentrate the narrative of the book on the essentials of the clustering problem, some more general considerations have been moved to Appendices A to E.

The limited volume of this book did not allow to cover all the developments in the field. On the one hand, we do not discuss algorithms elaborated for theoretical purposes of clusterability theory. On the other hand, we do not consider algorithms developed for particular applications only, like medical imaging. We concentrated on the clustering paradigm of a cluster as group of objects that are more similar to one another than to objects from other clusters. But other concepts are possible, based e.g. on classification/regression capabilities. Some researchers consider clusters as sets of objects where the attributes are (more) easily predictable from one another than in general population (mutual prediction). We mention only in passing that one may seek clusters either in a given space or in its subspaces (projections of the original space), but this is by itself a separate research area, answering questions what is the optimal number of features and which subspace (possibly linearly transformed) should be used. Still another research area concentrates around the so-called optimal modularity, where we seek clusters in which similarity pattern does not occur to be random. Still another area of investigation are special forms of optimised cluster quality function which ensure better convergence, like sub-modular functions.

Chapter 2
Cluster Analysis

Abstract This chapter outlines the major steps of cluster analysis. It starts with an informal introduction to clustering, its tools, methodology and applications. Then it proceeds with formalising the problem of data clustering. Diverse measures of object similarity, based on quantitative (like numerical measurement results) and on qualitative features (like text) as well as on their mixtures, are described. Various variants, how such similarity measures can be exploited when defining clustering cost functions are presented. Afterwards, major brands of clustering algorithms are explained, including hierarchical, partitional, relational, graph-based, density-based and kernel methods. The underlying data representations and interrelationships between various methodologies are discussed. Also possibilities of combining several algorithms for analysis of the same data (ensemble clustering) are presented. Finally, the issues related to easiness/hardness of the data clustering tasks are recalled.

Cluster analysis consists in distinguishing, in the set of analysed data, the groups, called clusters. These groups are disjoint[1] subsets of the data set, having such a property that data belonging to different clusters differ among themselves much more than the data, belonging to the same cluster. The role of cluster analysis is, therefore, to uncover a certain kind of natural structure in the data set. The means enabling performing that task is constituted usually by a certain measure of similarity or dissimilarity—the issue is discussed further on in Sect. 2.2. Cluster analysis is not only an important cognitive tool, but, as well, a method for reducing large sets of data, since it allows for the replacement of a group of data by its compact characterisation, like, e.g. the centre of gravity of the given group.

The task of cluster analysis can be perceived as a problem of grouping of objects according to their mutual similarity. Objects, which are mutually similar in a sufficiently high degree, form a homogeneous group (a cluster). It is also possible to consider the similarity of objects to certain characteristic entities (called prototypes) of the classes. In this case we deal more with the problem of classification, that is—of finding the model patterns—see [164]. Yet, if the characteristics corresponding to

---

[1] The requirement of disjoint subsets is used in the classical data analysis. In the general case, the groups distinguished might constitute the *coverage* of the data set. This is the case, for instance, in the fuzzy data analysis.

© Springer International Publishing AG 2018
S. T. Wierzchoń and M. A. Kłopotek, *Modern Algorithms of Cluster Analysis*,
Studies in Big Data 34, https://doi.org/10.1007/978-3-319-69308-8_2

classes are not given a priori, they should be established. And this is exactly what cluster analysis is about.

The term *data clustering* (meaning grouping of data) appeared for the first time in 1954 in the title of a paper concerning the analysis of anthropological data [272, p. 653]. Other equivalent names, given cluster analysis, are *Q-analysis*, *typology*, *clumping*, and *taxonomy* [273], depending on the domain, in which clustering is applied. There is a number of very good books, devoted to cluster analysis. The classical ones include[2]: [19, 164, 173, 273, 284, 436, 463]. Among the more recent and specialised monographs we can mention:[9, 70, 73, 75, 140, 297, 387, 402, 453]. A reader, interested in survey works may wish to have a look at, e.g., [66, 272, 274, 515]. More advanced techniques of cluster analysis are considered, in particular, in [176].

It is worth noting that both in common language and in computer science there exists a bunch of synonyms or closely related concepts for the term *cluster analysis* which reveal various aspects of this term. One of them is the *unsupervised learning* or *learning without a teacher*. This term suggests that we are looking for a hidden structure behind the data that we want to reveal. It suggests also that it must pay off to have recovered such a structure that is there must exist a criterion saying whether or not a useful structure was discovered. Last not least the hidden structure has to be learnable from the data in the sense of learnability theory. One speaks frequently about *client segmentation, image segmentation* etc. Image segmentation means that we look for a particular structure behind the image data—a trace of one or more physical objects, to be separated from the background. Criteria of belonging to the same object may include local properties, like continuity of colour, shading, texture, or global ones, like alignment along a line of limited curvature etc. Client segmentation on the other hand tries to split the population into homogeneous sections concentrated around some centric concepts so that predominantly global quality criteria will be of interest. And of course one asks the representativeness of the client population via the split into categories (generalization capability), easiness of assignment of new client to an existent category (mutual prediction of attributes within a category) and about profitability of cluster assignment from the business point of view (a kind of the cluster separation criterion). The frequently used term *taxonomy formation* points at the need of not only splitting the objects into classes or categories, but also of providing with a simple, compact description. Hence it is important to note that we are not talking just about clustering or grouping of objects, but rather the term *analysis* should be stressed because various cluster analysis methods reveal in the data structures of different types and the user of such methods either should be aware of what kind of hidden information he is looking for so that he chooses appropriate methods or he should be aware of what type of output he got if he applied a bunch of various methods for exploratory purposes.

For these reasons, the authors of this book do not try to point out the best method for clustering, neither explicitly nor implicitly, but rather concentrate on pointing

---

[2]Many of those listed have been modified several times over and successive editions have been published.

at particular advantages of individual methods and methodologies while covering a broad spectrum of potential and actual applications of cluster analysis.

It is the researcher himself who needs to know his goals and to choose the appropriate clustering approach that fits best his purposes. In order to help to understand the importance of user's goals, let us recall the once famous "Ugly Duckling Theorem" of Watanabe[3] Assume objects in a collection are described by binary attributes taking on values "true" and "false". Assume that two objects are more similar if they share more "true"-valued attributes. Furthermore, let us include in object description not only the original attributes but also their derivatives (obtained via logical operations "and", "or", "not"). Under these circumstances any object is similar to any other so that no clustering makes sense. This fact has a number of other consequences, which are frequently ignored: dropping/selecting/weighting appropriate subset of these attributes may lead to any similarity structure we wish to have. It means also that the very same set of objects may have different meaningful clusterings depending on the purpose of clustering and that application of an irrelevant clustering algorithm may lead to completely useless cluster sets. Therefore before starting cluster analysis, we need to have clear criteria on how we evaluate the quality of a single cluster (homogeneity) and/or the quality of the interrelationships between clusters (separation), and of course, what similarity/dissimilarity of objects means in our application area.

So e.g. when applying cluster analysis for purposes of delimitation of species of plants or animals in biology, the choice of algorithms and attributes will be driven by the assumed definition of species, e.g. via the requirement of (direct or indirect) interbreeding capability. While there is a limited amount of experimental data on interbreeding, the clustering performed on other available attributes is considered correct if it matches the known interbreeding possibilities and/or restrictions. Medical classification of diseases would be validated on/driven by a compromise between common symptoms, known causes and treatment possibilities. Image segmentation will understand clusters of pixels as homogeneous if they belong to the same organ, blood vessel, building, connected area covered with same plants etc. Market segmentation will be successful if we discover groups of clients worthy/not worthy addressing in marketing campaigns. Geographical client segmentation for purposes of store chains may be evaluated on the discovery of new profitable locations of stores etc. Clustering in databases aims at efficient organization of data for query processing.

Whatever the domain-specific expectations, one may apply cluster analysis with diverse goals of later usage of the results. One may perform exploratory data analysis seeking interesting patterns just to identify hypotheses worthy further investigation. One may be interested in data reduction (compression) and structuring for efficient storage and access. Or one may look for a clustering that aligns with other groupings to investigate their pertinence or universality.

---

[3] Watanabe, Satosi (1969). Knowing and Guessing: A Quantitative Study of Inference and Information. New York: Wiley, pp. 376–377.

Note also that depending on the application, the criteria for homogeneity and separation may be diverse. Let us just mention a few:

- homogeneity criteria

  - small within-cluster dissimilarities,
  - clusters fitting some homogeneous probability models like the Gaussian or a uniform distribution on a convex set, or some functional, time series or spatial process models,
  - members of a cluster well represented by its centroid,
  - clusters corresponding to connected areas in data space with high density,
  - clusters fitting into convex shapes (e.g. balls)
  - clusters characterisable by a small number of variables.
  - features approximately independent within clusters;

- separation criteria

  - large between-cluster dissimilarities
  - dissimilarity matrix of the data reflected by the clustering
  - constraints on co-occurrence/not co-occurrence in the same cluster matched
  - cluster stability (e.g. under resampling, bootstrapping)
  - low number of clusters
  - clusters of roughly the same size
  - balanced distances between cluster centres along a minimum weight spanning tree
  - clusters corresponding well to an externally given partition or values of one or more variables that were not used for computing the clustering
  - clusters fitting partial labelling
  - learnability of cluster membership from data
  - for hierarchical clusters the distance between cluster members reflected by the hierarchy level when they join the same cluster

The above criteria serve only as examples of possible cluster/clustering quality and whether or not any, some or large portion of them is actually used in a concrete application case will depend on the interests of the researcher.

We would further need to specify, in what sense a given criterion should be satisfied, e.g. by just passing a threshold or by reaching a local/global optimum or deviate only by some percentage from such an optimum. If we define multiple optimality criteria, we would need to reconcile them by stating a strategy of compromise between them. These and various other aspects of clustering philosophy are discussed in-depth in the paper [250] by Hennig.

2.1 Formalising the Problem

It is usually assumed in cluster analysis, that a set of m objects $\mathfrak{X} = \{\mathfrak{x}_1, \ldots, \mathfrak{x}_m\}$ is given.

For purposes of analysis the set of objects may be characterised by an embedded or relational representation.

In the *embedded* representation, every object is described by an n-dimensional vector $\mathbf{x}_i = (x_{i1} \ldots x_{in})^{\mathrm{T}}$, where x_{ij} denotes the value of the j-th feature of the object $\mathfrak{x}_i$. The vector $\mathbf{x}_i$ is being called feature vector or image.[4]

The subject of cluster analysis is constituted, therefore, not so much by the original set of objects $\mathfrak{X}$, as by its representation, given through the matrix $X = (\mathbf{x}_1 \ldots \mathbf{x}_m)^{\mathrm{T}}$, whose i-th row is a vector of features, describing the i-th object. In view of the fact that to object $\mathfrak{x}_i$ corresponds the i-th row of matrix X, the term "object" shall be used to denote both the element $\mathfrak{x}_i \in \mathfrak{X}$ and the vector of feature values $\mathbf{x}_i$, characterising this object. In statistics, vector $\mathbf{x}_i$ is called (n-dimensional) observation. Even though the values of individual measurements of feature values might be expressed in different scales (nominal, ordinal or quotient), the majority of practical and theoretical results have been obtained under the assumption that the components of vectors $\mathbf{x}_i$ are real numbers. Thus, for a vast majority of cases, considered in the present book, we shall be assuming that the observations are given by the vectors $\mathbf{x}_i \in \mathbb{R}^n$.

This kind of data representation will be explored in the current chapter as well as in Chaps. 3 and 4.

It is alternatively assumed (in the *relational* representation) that information on the data set is provided in the form of the matrix S of dimension $m \times m$, the elements of this matrix s_{ij} represent similarities (or dissimilarities) for the pairs of objects $\mathfrak{x}_i$, $\mathfrak{x}_j$. When making use of such a representation, we can give up the requirement of having the features, describing the objects, measured on the quantitative scales. The pioneer of such a perspective was Polish anthropologist, ethnographer, demographer and statistician—Jan Czekanowski.[5] The method, developed by Czekanowski, and presented in the first handbook of modern methods of data analysis and interpretation of its results [129], consists in replacing numbers in the matrix S by the appropriately selected graphical symbols. In this manner, an unordered diagram (called Czekanowski's diagram) arises, which, after an adequate reordering of rows and columns of the matrix, makes apparent the existence of groups of objects mutually similar. Extensive description of his idea provide Graham and Hell in [213]. To illustrate this approach consider a small matrix

---

[4]The latter term is particularly justified, when we treat the measurements as mappings $\mathbf{f} \colon \mathfrak{X} \to \mathbb{R}^n$ of the set of objects into a certain set of values. Then, $\mathbf{x}_i = \mathbf{f}(\mathfrak{x}_i)$, and in the mathematical nomenclature $\mathbf{x}_i$ is the image of the object $\mathfrak{x}_i$.

[5]See J. Gajek. Jan Czekanowski. Sylwetka uczonego. *Nauka Polska*, 6(2), 1958, 118–127.

$$
\nabla = \begin{array}{c} {} \\ 1 \\ 2 \\ 3 \\ 4 \\ 5 \\ 6 \end{array}
\begin{array}{cccccc}
1 & 2 & 3 & 4 & 5 & 6 \\
\end{array}
\left[
\begin{array}{cccccc}
1.0 & 0.2 & 0.1 & 0.3 & 0.8 & 0.4 \\
0.2 & 1.0 & 0.1 & 0.9 & 0.3 & 0.3 \\
0.1 & 0.1 & 1.0 & 0.2 & 0.2 & 0.7 \\
0.3 & 0.9 & 0.2 & 1.0 & 0.4 & 0.1 \\
0.8 & 0.3 & 0.2 & 0.4 & 1.0 & 0.2 \\
0.4 & 0.3 & 0.7 & 0.1 & 0.2 & 1.0 \\
\end{array}
\right]
$$

and let us denote by □ the degrees smaller than 0.5 and by ■ the degrees greater than 0.5. Then the above matrix can be represented graphically as in left part of Eq. (2.1). After reordering its rows and columns we obtain the matrix depicted in right part of the Eq. (2.1) revelling three clusters: $\{1, 5\}$, $\{2, 4\}$ and $\{3\}$.

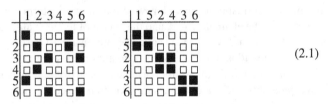 (2.1)

Although the method was developed originally more than 100 years ago, it is still in use, mainly in archaeology,[6] in economic sciences,[7] and even in musicology. Using the method of Czekanowski, the mathematicians from Wrocław: K. Florek, J. Łukaszewicz, J. Perkal, H. Steinhaus and S. Zubrzycki, elaborated the so-called "Wrocław taxonomy", which they presented in 1957 in 17th issue of the journal *Przegląd Antropologiczny* (see also [515]). In further parts of the book we present other methods of analysing the matrix of similarities. More advanced considerations of application of the similarity matrix in cluster analysis have been presented in the references [44, 45]. This type of representation is addressed primarily in Chaps. 5 and 6.

The role of the classical cluster analysis is to split the set of objects (observations) into $k < m$ groups $\mathcal{C} = \{C_1, \ldots, C_k\}$, where each i-th group C_i is called cluster. Such a division fulfils three natural requirements:

(i) Each cluster ought to contain at least one object, $C_j \neq \emptyset$, $j = 1, \ldots, k$.
(ii) Each object ought to belong to a certain cluster, $\bigcup_{j=1}^{k} C_j = \mathfrak{X}$.
(iii) Each object ought to belong to exactly one cluster, $C_{j_1} \cap C_{j_2} = \emptyset$, $j_1 \neq j_2$.

---

[6]See, e.g., A. Sołtysiak, and P. Jaskulski. Czekanowski's Diagram: A method of multidimensional clustering. In: J.A. Barceló, I. Briz and A. Vila (eds.) *New Techniques for Old Times. CAA98. Computer Applications and Quantitative Methods in Archaeology.* Proc. of the 26th Conf., Barcelona, March 1998 (BAR International Series 757). Archaeopress, Oxford 1999, pp. 175–184.

[7]See, e.g., A. Wójcik. Zastosowanie diagramu Czekanowskiego do badania podobieństwa krajów Unii Europejskiej pod względem pozyskiwania energii ze źródeł odnawialnych. *Zarządzanie i Finanse (J. of Management and Finance),* 11(4/4), 353–365, 2013, http://zif.wzr.pl/pim/2013_4_4_25.pdf.

In particular, when $k = m$, each cluster contains exactly one element from the set $\mathfrak{X}$. This partition is trivial, and so we shall be considering the cases, in which k is much smaller than m.

An exemplary illustration of the problem that we deal with in cluster analysis, is provided in the Fig. 2.1. The "clouds" of objects, situated in the lower left and upper right corners constitute distinctly separated clusters. The remaining objects form three, two, or one cluster, depending on how we define the notion of similarity between the objects.

Cluster analysis is also referred to as unsupervised learning. We lack here, namely, the information on the membership of the objects in classes, and it is not known, how many classes there should really be. Even though the mechanical application of the algorithms, which are presented in the further parts of the book allows for the division of any arbitrary set into a given number of classes, the partition thus obtained may not have any sense. Assume that $\mathfrak{X}$ is a set of points selected conform to the uniform distribution from the set $[0, 1] \times [0, 1]$—see Fig. 2.2a. Mechanical application of an algorithm of grouping, with the predefined parameter $k = 3$ leads to the result, shown in Fig. 2.2b. It is obvious that although the partition obtained fulfils the conditions set before, it has no sense.

In informal terms, the presence of a structure in a data set is manifested through the existence of separate areas, and hence of clusters, enjoying such a property that any two objects, belonging to the common cluster C_i are more mutually similar than any two objects, picked from two different clusters, i.e.

Fig. 2.1 The task of cluster analysis: to break down the set of observations into k disjoint subsets, composed of similar elements

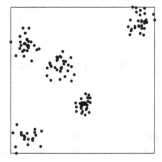

Fig. 2.2 Grouping of the set of randomly generated objects. **a** 100 points randomly selected from the set $[0, 1] \times [0, 1]$. **b** Division of the objects into three groups using the k-means algorithm. Bigger marks indicate the geometrical centers of groups

(a)   (b)

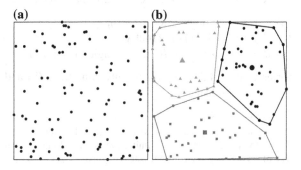

$$s(\mathfrak{x}', \mathfrak{x}'') > s(\mathfrak{y}', \mathfrak{y}'')$$

if only $\mathfrak{x}', \mathfrak{x}'' \in C_i$, $\mathfrak{y}' \in C_{j_1}$, $\mathfrak{y}'' \in C_{j_2}$ and $j_1 \neq j_2$. Symbol s denotes here a certain measure of similarity, that is, a mapping $s: \mathfrak{X} \times \mathfrak{X} \to \mathbb{R}$. In many situations it is more convenient to make use of the notion of dissimilarity (like, e.g., distance) and require the objects, belonging to different clusters, to be more distant than the objects, belonging to the same cluster. Various definitions of the measures of similarity or dissimilarity are considered in the subsequent Sect. 2.2. The choice of the appropriate measure constitutes an additional factor of complexity of the data analysis task. Yet another such factor is the choice of an adequate criterion for determining the partition of the set $\mathfrak{X}$, that is stating whether or not a given partition would be satisfactory or not. For this purpose qualitative criteria listed on p. 12 need to be chosen and formalised for the purposes of a particular clustering task.

The majority of the methods of grouping consists in an "intelligent" extraction of information from the matrix S, with elements representing similarity or dissimilarity of the pairs of objects. An excellent example of this kind of procedure is provided by the hierarchical methods, shortly presented in Sect. 2.3 or by the spectral methods, being the primary subject of Chap. 5.

The most popular methods of grouping include the hierarchical methods, the combinatorial methods (referred also to as relocation-based or partitional), the density-based methods, the grid methods (being a brand of density based ones) and the methods based on models. The descriptions of these methods and comments on them can be found in numerous survey studies, like, e.g. [272, 274, 515]. In the further part of the chapter we shall present their short characteristics. The rapidly developing spectral methods will be presented in Chap. 5.

2.2 Measures of Similarity/Dissimilarity

In order to be able to quantify associations between pairs of objects, a measure of similarity $s: \mathfrak{X} \times \mathfrak{X} \to \mathbb{R}$ or of dissimilarity is introduced. The two measures are, in principle, dual, that is—the lower the value of dissimilarity, the more similar the two compared objects. A particular example of dissimilarity is distance (metric), that is, a function $d: \mathfrak{X} \times \mathfrak{X} \to \mathbb{R}_+ \cup \{0\}$ fulfilling three conditions:

(a) $d(\mathfrak{x}, \mathfrak{y}) = 0$ if and only if $\mathfrak{x} \equiv \mathfrak{y}$,
(b) $d(\mathfrak{x}, \mathfrak{y}) = d(\mathfrak{y}, \mathfrak{x})$ (symmetry),
(c) $d(\mathfrak{x}, \mathfrak{y}) \leq d(\mathfrak{x}, \mathfrak{z}) + d(\mathfrak{z}, \mathfrak{y})$ (triangle inequality),

for arbitrary $\mathfrak{x}, \mathfrak{y}, \mathfrak{z} \in \mathfrak{X}$. When only conditions (b) and (c) are satisfied, then d is called pseudo-distance.[8]

---

[8]Note that, as Gower et al. [209] states, see their Theorem 1, any non-metric dissimilarity measure $d(\mathfrak{z}, \mathfrak{y})$ for $\mathfrak{z}, \mathfrak{y} \in \mathfrak{X}$ where $\mathfrak{X}$ is finite, can be turned into a (metric) distance function $d'(\mathfrak{z}, \mathfrak{y}) = d(\mathfrak{z}, \mathfrak{y}) + c$ where c is a constant where $c \geq \max_{\mathfrak{x}, \mathfrak{y}, \mathfrak{z} \in \mathfrak{X}} \|d(\mathfrak{x}, \mathfrak{y}) + d(\mathfrak{y}, \mathfrak{z}) - d(\mathfrak{z}, \mathfrak{x})\|$.

For instance, if d_{max} denotes the maximum value of distance between the pairs of objects from the set $\mathfrak{X}$, then distance can be transformed into a measure of similarity (proximity) $s(\mathfrak{x}_i, \mathfrak{x}_j) = d_{max} - d(\mathfrak{x}_i, \mathfrak{x}_j)$. The thus obtained measure of proximity attains the maximum values, when $i = j$ (object $\mathfrak{x}_i$ is identical with itself), and the lower the value of this measure, the less mutually similar (more dissimilar) the objects compared are.

In the above example, the maximum value of similarity is the number $d_{max} = s(\mathfrak{x}_i, \mathfrak{x}_i)$. It is more convenient to operate with the normalised similarity $s(\mathfrak{x}_i, \mathfrak{x}_j) = 1 - d(\mathfrak{x}_i, \mathfrak{x}_j)/d_{max}$.

Another, frequently applied example of the measure of similarity is provided by the transformation $s(\mathfrak{x}_i, \mathfrak{x}_j) = \exp[-d^2(\mathfrak{x}_i, \mathfrak{x}_j)/\sigma^2]$, where $\sigma > 0$ is a parameter, or, yet, $s(\mathfrak{x}_i, \mathfrak{x}_j) = 1/[d(\mathfrak{x}_i, \mathfrak{x}_j) + \epsilon]$, where $\epsilon > 0$ is a small number.[9]

Generally, if $f : \mathbb{R} \to \mathbb{R}$ is a monotonously decreasing function, such that $f(0) > 0$ and $\lim_{\xi \to \infty} f(\xi) = a \geq 0$, then

$$s(\mathfrak{x}_i, \mathfrak{x}_j) = f(d(\mathfrak{x}_i, \mathfrak{x}_j)) \tag{2.2}$$

is a measure of similarity, induced by the distance d.

The condition $a \geq 0$, given above, is, in principle, not necessary; it is introduced in order to preserve the symmetry with the non-negative values of the distance measure d. Note also that if d is a distance, then the measure of similarity, defined through the transformation f, referred to above, fulfils the triangle condition in the form—see [110]:

$$s(\mathfrak{x}, \mathfrak{y}) + s(\mathfrak{y}, \mathfrak{z}) \leq s(\mathfrak{x}, \mathfrak{z}) + s(\mathfrak{y}, \mathfrak{y})$$

if function f is *additionally* convex.[10]

The result can be derived as follows: Either (1) $d(\mathfrak{x}, \mathfrak{z}) \leq d(\mathfrak{x}, \mathfrak{y})$ or (2) this is not true but $d(\mathfrak{x}, \mathfrak{z}) \leq d(\mathfrak{y}, \mathfrak{z})$ or (3) $d(\mathfrak{x}, \mathfrak{z}) > d(\mathfrak{x}, \mathfrak{y})$ & $d(\mathfrak{x}, \mathfrak{z}) > d(\mathfrak{y}, \mathfrak{z})$ holds. In the first case $f(d(\mathfrak{x}, \mathfrak{z})) \geq f(d(\mathfrak{x}, \mathfrak{y}))$. As $0 \leq d(\mathfrak{y}, \mathfrak{z})$, we have $f(0) \geq f(d(\mathfrak{y}, \mathfrak{z}))$. But summing up both we get $f(d(\mathfrak{y}, \mathfrak{y})) + f(d(\mathfrak{x}, \mathfrak{z})) \geq f(d(\mathfrak{x}, \mathfrak{y})) + f(d(\mathfrak{y}, \mathfrak{z}))$. So the claim is proven. In the second case the reasoning is analogous (just flip $\mathfrak{x}$ and $\mathfrak{z}$).

So let us consider the third case when $d(\mathfrak{x}, \mathfrak{z}) > d(\mathfrak{x}, \mathfrak{y})$&$d(\mathfrak{x}, \mathfrak{z}) > d(\mathfrak{y}, \mathfrak{z})$. As metric is assumed, $d(\mathfrak{x}, \mathfrak{z}) \leq d(\mathfrak{x}, \mathfrak{y}) + d(\mathfrak{y}, \mathfrak{z})$. Hence $1 \leq \frac{d(\mathfrak{y}, \mathfrak{z})}{d(\mathfrak{x}, \mathfrak{z})} + \frac{d(\mathfrak{x}, \mathfrak{y})}{d(\mathfrak{x}, \mathfrak{z})}$. Therefore $0 \leq 1 - \frac{d(\mathfrak{y}, \mathfrak{z})}{d(\mathfrak{x}, \mathfrak{z})} \leq \frac{d(\mathfrak{x}, \mathfrak{y})}{d(\mathfrak{x}, \mathfrak{z})} \leq 1$. Let us pick any $\lambda \in \left[0, \frac{d(\mathfrak{x}, \mathfrak{y})}{d(\mathfrak{x}, \mathfrak{z})}\right]$ such that $1 - \frac{d(\mathfrak{y}, \mathfrak{z})}{d(\mathfrak{x}, \mathfrak{z})} \leq \lambda$. Apparently $0 \leq \lambda \leq 1$. We see immediately that $\lambda d(\mathfrak{x}, \mathfrak{z}) \leq d(\mathfrak{x}, \mathfrak{y})$ and $(1 - \lambda)d(\mathfrak{x}, \mathfrak{z}) \leq d(\mathfrak{y}, \mathfrak{z})$ From the convexity definition we have that $(1 - \lambda)f(0) + \lambda f(d(\mathfrak{x}, \mathfrak{z})) \geq f((1 - \lambda) \cdot 0 + \lambda d(\mathfrak{x}, \mathfrak{z})) = f(\lambda d(\mathfrak{x}, \mathfrak{z})) \geq f(d(\mathfrak{x}, \mathfrak{y}))$, with the last inequality being due to the decreasing monotonicity of f.

---

[9]In practice, a small number means one that is small compared to distances but still numerically significant under the available machine precision.

[10]A real-valued function $f(x)$ is said to be convex over the interval $[a, b] \subset \mathbb{R}$ if for any $x_1, x_2 \in [a, b]$ and any $\lambda \in [0, 1]$ we have $\lambda f(x_1) + (1 - \lambda)f(x_2) \geq f(\lambda x_1 + (1 - \lambda)x_2)$. All three mentioned examples of transformation of a distance into similarity measure are in fact convex functions.

Similarly $\lambda \cdot f(0) + (1 - \lambda) f(d(\mathfrak{x}, \mathfrak{z})) \geq f(\lambda \cdot 0 + (1 - \lambda) d(\mathfrak{x}, \mathfrak{z})) = f((1 - \lambda)$ $d(\mathfrak{x}, \mathfrak{z})) \geq f(d(\mathfrak{y}, \mathfrak{z}))$

By summing these two inequalities we get $f(0) + f(d(\mathfrak{x}, \mathfrak{z})) \geq f(d(\mathfrak{x}, \mathfrak{y}) + f(d(\mathfrak{y}, \mathfrak{z})$ so obviously the triangle inequality holds here too.

This derivation by the way differs a bit from [110] but follows the general idea contained therein.

Since the measures of similarity and dissimilarity are dual notions, see, e.g., [110], we shall be dealing in further course primarily with various measures of dissimilarity, and in particular—with distances.[11] Making use of distances requires having the possibility of assigning to each object $\mathfrak{x}_i$ its representation $\mathbf{x}_i$. An interesting and exhaustive survey of various measures of similarity/dissimilarity can be found, for instance, in [100].[12]

2.2.1   Comparing the Objects Having Quantitative Features

When all the features, which are used to describe objects from the set $\mathfrak{X}$ are quantitative, then every object $\mathfrak{x}_i \in \mathfrak{X}$ is identified with an n-dimensional vector $\mathbf{x}_i = (x_{i1}, x_{i2}, \ldots, x_{in})^{\mathrm{T}}$. The most popular measure of dissimilarity is the Euclidean distance

$$d(\mathfrak{x}_i, \mathfrak{x}_j) = \|\mathbf{x}_i - \mathbf{x}_j\| = \sqrt{\sum_{l=1}^{n} (x_{il} - x_{jl})^2}$$

or, more generally, the norm defined by the square form

$$d_W(\mathfrak{x}_i, \mathfrak{x}_j) = \|\mathbf{x}_i - \mathbf{x}_j\|_W = \sqrt{(\mathbf{x}_i - \mathbf{x}_j)^{\mathrm{T}} W (\mathbf{x}_i - \mathbf{x}_j)} \qquad (2.3)$$

where W is a positive definite matrix of the dimensions $n \times n$.

If W is a unit matrix, then Eq. (2.3) defines the Euclidean distance. If, on the other hand, W is a diagonal matrix having the elements

$$w_{ij} = \begin{cases} \omega_i & \text{if } i = j \\ 0 & \text{otherwise} \end{cases}$$

then (2.3) defines the weighted Euclidean distance, i.e.

$$d_W(\mathfrak{x}_i, \mathfrak{x}_j) = \sqrt{\sum_{l=1}^{n} \omega_l (x_{il} - x_{jl})^2} = \sqrt{\sum_{l=1}^{n} (y_{il} - y_{jl})^2}$$

---

[11] In Chaps. 5 and 6 we will predominantly concentrate on similarity measure based clustering methods.

[12] In Sect. 6.2 we discuss some similarity measures defined for data in form of graphs/networks.

where $y_{il} = \sqrt{\omega_l} x_{il}$ is the weighted value of the feature l, measured for the i-th object.

The Euclidean distance is being generalized in various manners. These most commonly used are commented upon below.[13]

2.2.1.1 Minkowski Distance

Minkowski distance (norm) is defined as follows

$$d_p(\mathbf{x}_i, \mathbf{x}_j) = \|\mathbf{x}_i - \mathbf{x}_j\|_p = \left[\sum_{l=1}^{n} |x_{il} - x_{jl}|^p \right]^{1/p}, \quad p \geq 1, p \in \mathbb{R} \qquad (2.4)$$

When we take $p = 1$, we obtain the city block distance (called also taxicab or Manhattan distance)

$$d_1(\mathbf{x}_i, \mathbf{x}_j) = \|\mathbf{x}_i - \mathbf{x}_j\|_1 = \sum_{l=1}^{n} |x_{il} - x_{jl}| \qquad (2.5)$$

For $p = 2$, Eq. (2.4) defines the Euclidean distance. In view of the popularity of this distance definition, we shall be writing $\|\mathbf{x}_i - \mathbf{x}_j\|$ instead of $\|\mathbf{x}_i - \mathbf{x}_j\|_2$.

Finally, when $p = \infty$, we get Chebyshev distance

$$d_\infty(\mathbf{x}_i, \mathbf{x}_j) = \|\mathbf{x}_i - \mathbf{x}_j\|_\infty = \max_{l=1,\ldots,n} |x_{il} - x_{jl}| \qquad (2.6)$$

Minkowski distance is used not only in exact sciences, but also in psychology,[14] industrial design and generally in designing. Unit circles are described in Minkowski metric by the equation

$$|x|^p + |y|^p = 1$$

which is also called the curve (or oval) of Lamé. The respective shapes for three values of the parameter p are presented in Fig. 2.3a. Danish mathematician Piet Hein[15] concluded that the case of $p = 2.5$ leads to the shape, featuring *high aesthetic qualities*, see Fig. 2.3b, this fact having been made use of in designing Sergels roundabout in Stockholm.

---

[13]Note that the Euclidean distance has been investigated itself to a great depth, see e.g. [209]. See Sect. B.5 for a discussion of criteria of a dissimilarity matrix being Euclidean distance matrix. Gower et al. [209] points out that any dissimilarity matrix D may be turned to an Euclidean distance matrix, see their Theorem 7, by adding an appropriate constant, e.g. $d'(\mathfrak{z}, \mathfrak{y}) = \sqrt{d(\mathfrak{z}, \mathfrak{y})^2 + h}$ where h is a constant such that $h \geq -\lambda_m$, λ_m being the smallest eigenvalue of $(\mathbf{I} - \mathbf{1}\mathbf{1}^T/m)(-1/2D_{sq})(\mathbf{I} - \mathbf{1}\mathbf{1}^T/m)$, D_{sq} is the matrix of squared values of elements of D, m is the number of rows/columns in D.

[14]See Chap. 3 in: C.H. Coombs, R.M. Dawes, A. Tversky. *Mathematical Psychology: An Elementary Introduction*. Prentice Hall, Englewood Cliffs, NJ 1970.

[15]His profile can be found on the website http://www.piethein.com/.

Fig. 2.3 Unit circles for the selected values of the parameter p: (a) $p = 1, 2, \infty$, (b) $p = 2.5$ (Hein's super-ellipse)

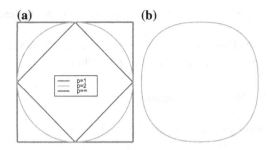

Distances, deriving from the Minkowski metric, have two important shortcomings. First, as the dimensionality of the problem increases, the difference between the close and the far points in the space $\mathbb{R}^n$ disappears, this being the effect of summing of the differences in the locations of objects in the particular dimensions.[16]

More precisely, the situation is as follows. Denote by $d_{p,n}^{min}, d_{p,n}^{max}$, respectively, the minimum and the maximum values of distance (measured with distance d_p) between two arbitrary points, selected from the set of m randomly generated points in n-dimensional space. Then, see [8]

$$C_p \leq \lim_{n \to \infty} \mathbb{E}\left[\frac{d_{p,n}^{max} - d_{p,n}^{min}}{n^{1/p-1/2}}\right] \leq (m-1)C_p \qquad (2.7)$$

where C_p is a constant, depending upon the value of p, and $\mathbb{E}$ denotes the expected value. This inequality implies that in a high-dimensional space the difference $d_{p,n}^{max} - d_{p,n}^{min}$ increases proportionally to $n^{1/p-1/2}$ irrespective of the distribution of data [8]. This property plays a dominating role, when $n \geq 15$. In particular (see Fig. 2.4a)

$$d_{p,n}^{max} - d_{p,n}^{min} \rightarrow \begin{cases} C_1\sqrt{n} & \text{if } p = 1 \\ C_2 & \text{if } p = 2 \\ 0 & \text{if } p \geq 3 \end{cases}$$

In order to prevent this phenomenon, Aggarwal, Hinnenburg and Keim proposed in [8] application of the fractional Minkowski distances with the parameter $p \in (0, 1]$—see Fig. 2.4b. Yet, in this case (2.4) is no longer a distance, since the triangle condition is not satisfied. If, for instance, $\mathbf{x} = (0, 0)$, $\mathbf{y} = (1, 1)$ and $\mathbf{z} = (1, 0)$, then

$$d(\mathbf{x}, \mathbf{y}) = 2^{1/p} > d(\mathbf{x}, \mathbf{z}) + d(\mathbf{y}, \mathbf{z}) = 1 + 1$$

A subsequent, but not so critical issue is constituted by the fact that the values of the Minkowski metric are dominated by these features, whose values are measured on the scales with the biggest ranges. This issue can be relatively easily resolved by introducing weighted distance, that is—by replacing each component of the

---

[16]Equivalently, one can say that two arbitrary vectors in $\mathbb{R}^n$ are orthogonal [402, p. 7.1.3].

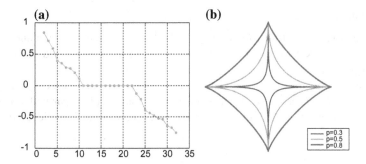

Fig. 2.4 The influence of the parameter p on the properties of Minkowski distance. **a** Average values of the difference between the most distant points from a 100-point set in dependence upon the number of dimensions, n, and the value of the exponent, p. **b** Unit circles for $p < 1$

Eq. (2.4) by the expression $\omega_l(x_{il} - x_{jl})^p$, where w_l is the weight equal, e.g., the inverse of the standard deviation of the l-th feature, or the inverse of the range of variability of the l-th feature. The counterpart to the second variant is constituted by the initial normalization of data, ensuring that $x_{il} \in [0, 1]$ for each of the features $l = 1, \ldots, n$. This is, usually, a routine procedure, preceding the proper data analysis. In some cases, instead of a simple normalization, standardization is applied, that is—the original value x_{il} is replaced by the quotient $(x_{il} - \mu_l)/\sigma_l$, where μ_l, σ_l are the average value and the standard deviation of the l-th feature.

2.2.1.2 Mahalanobis Distance

When defining distance (2.4) it is by default assumed that features are not mutually correlated. When this assumption is not satisfied, Mahalanobis distance is usually applied,

$$d_\Sigma(\mathbf{x}_i, \mathbf{x}_j) = \sqrt{(\mathbf{x}_i - \mathbf{x}_j)^\mathsf{T} \Sigma^{-1} (\mathbf{x}_i - \mathbf{x}_j)} \tag{2.8}$$

which is a variant of the distance (2.3), where W is equal the inverse of the covariance matrix. Covariance matrix is calculated in the following manner

$$\Sigma = \frac{1}{m} \sum_{i=1}^{m} (\mathbf{x}_i - \overline{\boldsymbol{\mu}})(\mathbf{x}_i - \overline{\boldsymbol{\mu}})^\mathsf{T} \tag{2.9}$$

with

$$\overline{\boldsymbol{\mu}} = \frac{1}{m} \sum_{i=1}^{m} \mathbf{x}_i \tag{2.10}$$

being the vector of average values.

Note that:

(a) By applying the transformation $\mathbf{y}_i = \Sigma^{-1/2}\mathbf{x}_i$ we reduce Mahalanobis distance $d_\Sigma(\mathbf{x}_i, \mathbf{x}_j)$ to Euclidean distance between the transformed vectors, that is $d_\Sigma(\mathbf{x}_i, \mathbf{x}_j) = \|\mathbf{y}_i - \mathbf{y}_j\|$.

(b) When the features are independent, then the covariance matrix is a diagonal matrix: the nonzero elements are equal to the variances of the particular features. In such a case Mahalanobis distance becomes the weighted Euclidean distance of the form

$$d_S(\mathbf{x}_i, \mathbf{x}_j) = \sqrt{\sum_{l=1}^{n} \left(\frac{x_{il} - x_{jl}}{\sigma_l}\right)^2} \tag{2.11}$$

Mahalanobis distance is useful in identification of the *outliers* (atypical observations). A number of properties of this distance are provided in the tutorial [341].

2.2.1.3 Bregman Divergence

The distances, that is—the measures of dissimilarity—considered up till now, have a differential character: $d(\mathbf{x}, \mathbf{y}) = \varphi(\mathbf{x} - \mathbf{y})$, where $\varphi: \mathbb{R}^n \to \mathbb{R}$ is an appropriately selected function. In certain situations, e.g. in problems concerning signal compression, measures are needed that would account for more complex relations between the vectors compared. An exhaustive survey thereof is given in [54]. An instance is represented in this context by the measure, introduced for the systems of speech compression by Chaffee[17]

$$d_{Ch}(\mathbf{x}, \mathbf{y}) = (\mathbf{x} - \mathbf{y})^T R(\mathbf{x})(\mathbf{x} - \mathbf{y}) \tag{2.12}$$

While in the case of Mahalanobis distance (2.8) the covariance matrix, which appears there, is established a priori, the matrix of weights, which is used in the Definition (2.12) depends upon the currently considered object $\mathbf{x}$. Another example is provided by the measure of Itakura-Saito.[18]

All these measures are generalized by the so-called Bregman divergence. It is defined as follows [50].

Definition 2.2.1 Let $\phi: S \to \mathbb{R}$ be a strictly convex function, defined on a convex set $S \subset \mathbb{R}^n$. Besides, we assume that the relative interior, $rint(S)$, of the set S is not

---

[17]D.L. Chaffee, Applications of rate distortion theory to the bandwidth compression of speech, Ph.D. dissertation, Univ. California, Los Angeles, 1975. See also R.M. Gray, et al., Distortion measures for speech processing, *IEEE Trans. on Acoustics, Speech and Signal Processing*, **28**(4), 367–376, Aug. 1980.

[18]See F. Itakura, S. Saito, Analysis synthesis telephony based upon maximum likelihood method, *Repts. of the 6th Intl. Cong. Acoust.* Tokyo, C-5-5, C-17-20, 1968.

Table 2.1 Bregman divergences generated by various convex functions [50]

Domain	$\phi(\mathbf{x})$	$d_\phi(\mathbf{x}, \mathbf{y})$	Divergence
$\mathbb{R}$	$\mathbf{x}^2$	$(\mathbf{x} - \mathbf{y})^2$	Quadratic loss function
$\mathbb{R}_+$	$\mathbf{x} \log \mathbf{x}$	$\mathbf{x} \log(\frac{\mathbf{x}}{\mathbf{y}}) - (\mathbf{x} - \mathbf{y})$	
$[0,1]$	$\mathbf{x} \log \mathbf{x} + (1 - \mathbf{x}) \log(1 - \mathbf{x})$	$\mathbf{x} \log(\frac{\mathbf{x}}{\mathbf{y}}) + (1 - \mathbf{x}) \log(\frac{1-\mathbf{x}}{1-\mathbf{y}})$	Logistic loss function
$\mathbb{R}_{++}$	$-\log \mathbf{x}$	$\frac{\mathbf{x}}{\mathbf{y}} - \log(\frac{\mathbf{x}}{\mathbf{y}}) - 1$	Itakura-Saito distance
$\mathbb{R}_+^n$	$\sum_{j=1}^n (\mathbf{x}_j \log \mathbf{x}_j - \mathbf{x}_j)$	$\sum_{j=1}^n \left(\mathbf{x}_j \log(\frac{\mathbf{x}_j}{\mathbf{y}_j}) - (\mathbf{x}_j - \mathbf{y}_j)\right)$	Generalized I-divergence
$\mathbb{R}^n$	$\|\mathbf{x}\|^2$	$\|\mathbf{x} - \mathbf{y}\|^2$	Squared Euclidean distance
$\mathbb{R}^n$	$\mathbf{x}^\mathsf{T} W \mathbf{x}$	$(\mathbf{x} - \mathbf{y})^\mathsf{T} W (\mathbf{x} - \mathbf{y})$	Mahalanobis distance
n-simplex	$\sum_{j=1}^n \mathbf{x}_j \log_2 \mathbf{x}_j$	$\sum_{j=1}^n \mathbf{x}_j \log_2(\frac{x_j}{y_j})$	KL-divergence

empty, and ϕ is a function differentiable on $rint(S)$. Bregman divergence is then such a function $d_\phi \colon S \times rint(S) \to [0, \infty)$ that for the arbitrary $\mathbf{x}, \mathbf{y} \in \mathbb{R}^d$

$$d_\phi(\mathbf{x}, \mathbf{y}) = \phi(\mathbf{x}) - \phi(\mathbf{y}) - (\mathbf{x}-\mathbf{y})^\mathsf{T} \nabla \phi(\mathbf{y}) \tag{2.13}$$

Symbol $\nabla \phi(\mathbf{y})$ denotes the gradient of the function $\phi(\mathbf{y})$. □

Examples of Bregman divergence are shown in Table 2.1. Special attention ought to be paid to the last three examples. If we take for $\phi(\mathbf{x})$ the squared length of vector $\mathbf{x}$, then $d_\phi(\mathbf{x}, \mathbf{y})$ is equal squared Euclidean distance between the points, represented by the vectors $\mathbf{x}$ and $\mathbf{y}$. If the mapping ϕ is defined as $\mathbf{x}^\mathsf{T} W \mathbf{x}$, where W is a positive definite matrix, then we obtain the squared distance (2.3), in particular—defined in Sect. 2.2.1.2 Mahalanobis distance. Finally, if $\mathbf{x}$ is a stochastic vector,[19] while ϕ is a negative value of entropy, $\phi(\mathbf{x}) = \sum_{j=1}^n x_j \log_2 x_j$, then we obtain Kullback-Leibler divergence, referred to frequently as KL-divergence. This notion plays an important role in information theory, machine learning, and in information retrieval.

A closely related concept is the *Bregman information* of a set X of data points $\phi(\mathbf{x}_j)$, drawn from a probability distribution π.

$$I_\phi(X) = \min_{\mathbf{y} \in \mathbb{R}^n} \sum_{j=1}^m \pi_j d_\phi(\mathbf{x}_j, \mathbf{y})$$

The most interesting property of Bregman information is that the $\mathbf{y}$ minimising $I_\phi(X)$ does not depend on the particular Bregman divergence ϕ and is always equal

---

[19]That is—all of its components are non-negative and $\sum_{j=1}^n x_j = 1$.

to $\mu = \sum_{j=1}^{m} \pi_j \mathbf{x}_j$ that is the mean value vector of the data set. As we will see, this can be used in construction of clustering algorithms.[20]

Define $J_\phi(\mathbf{y}) = \sum_{j=1}^{m} \pi_j d_\phi(\mathbf{x}_j, \mathbf{y})$. Then $I_\phi = \min_{\mathbf{y}} J_\phi(\mathbf{y})$. Let us now compute

$$
\begin{aligned}
J_\phi(\mathbf{y}) - J_\phi(\mu) &= \sum_{j=1}^{m} \pi_j d_\phi(\mathbf{x}_j, \mathbf{y}) - \sum_{j=1}^{m} \pi_j d_\phi(\mathbf{x}_j, \mu) \\
&= \sum_{j=1}^{m} \left(\pi_j \left(\phi(\mathbf{x}_j) - \phi(\mathbf{y}) - (\mathbf{x}_j - \mathbf{y})^{\mathrm{T}} \nabla \phi(\mathbf{y}) \right) \right) \\
&\quad - \sum_{j=1}^{m} \left(\pi_j \left(\phi(\mathbf{x}_j) - \phi(\mu) - (\mathbf{x}_j - \mu)^{\mathrm{T}} \nabla \phi(\mu) \right) \right) \\
&= \sum_{j=1}^{m} \pi_j \phi(\mathbf{x}_j) - \sum_{j=1}^{m} \pi_j \phi(\mathbf{y}) - \sum_{j=1}^{m} \pi_j \mathbf{x}_j^{\mathrm{T}} \nabla \phi(\mathbf{y}) + \sum_{j=1}^{m} \pi_j \mathbf{y}^{\mathrm{T}} \nabla \phi(\mathbf{y}) \\
&\quad - \sum_{j=1}^{m} \pi_j \phi(\mathbf{x}_j) + \sum_{j=1}^{m} \pi_j \phi(\mu) + \sum_{j=1}^{m} \pi_j \mathbf{x}_j^{\mathrm{T}} \nabla \phi(\mu) - \sum_{j=1}^{m} \pi_j \mu^{\mathrm{T}} \nabla \phi(\mu) \\
&= \sum_{j=1}^{m} \pi_j \phi(\mathbf{x}_j) - \phi(\mathbf{y}) - \sum_{j=1}^{m} \pi_j \mathbf{x}_j^{\mathrm{T}} \nabla \phi(\mathbf{y}) + \mathbf{y}^{\mathrm{T}} \nabla \phi(\mathbf{y}) - \sum_{j=1}^{m} \pi_j \phi(\mathbf{x}_j) + \phi(\mu) \\
&\quad + \sum_{j=1}^{m} \pi_j \mathbf{x}_j^{\mathrm{T}} \nabla \phi(\mu) - \mu^{\mathrm{T}} \nabla \phi(\mu) \\
&= -\mu^{\mathrm{T}} \nabla \phi(\mu) + \mathbf{y}^{\mathrm{T}} \nabla \phi(\mathbf{y}) - \phi(\mathbf{y}) + \sum_{j=1}^{m} \pi_j \phi(\mathbf{x}_j) - \sum_{j=1}^{m} \pi_j \mathbf{x}_j^{\mathrm{T}} \nabla \phi(\mathbf{y}) \\
&\quad - \sum_{j=1}^{m} \pi_j \phi(\mathbf{x}_j) + \phi(\mu) + \sum_{j=1}^{m} \pi_j \mathbf{x}_j^{\mathrm{T}} \nabla \phi(\mu) \\
&= +\phi(\mu) - \phi(\mathbf{y}) + \sum_{j=1}^{m} \pi_j \mathbf{x}_j^{\mathrm{T}} \nabla \phi(\mu) - \mu^{\mathrm{T}} \nabla \phi(\mu) \\
&\quad + \mathbf{y}^{\mathrm{T}} \nabla \phi(\mathbf{y}) - \sum_{j=1}^{m} \pi_j \mathbf{x}_j^{\mathrm{T}} \nabla \phi(\mathbf{y}) + \sum_{j=1}^{m} \pi_j \phi(\mathbf{x}_j) - \sum_{j=1}^{m} \pi_j \phi(\mathbf{x}_j) \\
&= +\phi(\mu) - \phi(\mathbf{y}) + \left(\left(\sum_{j=1}^{m} \pi_j \mathbf{x}_j \right) - \mu \right)^{\mathrm{T}} \nabla \phi(\mu) \\
&\quad + \left(\mathbf{y} - \left(\sum_{j=1}^{m} \pi_j \mathbf{x}_j \right) \right)^{\mathrm{T}} \nabla \phi(\mathbf{y})
\end{aligned}
$$

[20]Note that Bregman divergence is in general neither symmetric nor fits the triangle condition hence it is not a metric distance. However, there exist some families of Bregman distances, like Generalized Symmetrized Bregman and Jensen-Bregman divergence, which under some conditiomns are a square of a metric distance, and may be even embedded in Euclidean space. For details see Acharyya, S., Banerjee, A., and Boley, D. (2013). Bregman Divergences and Triangle Inequality. In SDM'13 (SIAM International Conference on Data Mining), pp. 476–484.

$$\begin{aligned} &= +\phi(\boldsymbol{\mu}) - \phi(\mathbf{y}) + (\boldsymbol{\mu} - \boldsymbol{\mu})^T \nabla \phi(\boldsymbol{\mu}) + (\mathbf{y} - \boldsymbol{\mu})^T \nabla \phi(\mathbf{y}) \\ &= +\phi(\boldsymbol{\mu}) - \phi(\mathbf{y}) - (\boldsymbol{\mu} - \mathbf{y})^T \nabla \phi(\mathbf{y}) \\ &= d_\phi(\boldsymbol{\mu}, \mathbf{y}) \geq 0 \end{aligned} \tag{2.14}$$

So we see that $J_\phi(\mathbf{y}) \geq J_\phi(\boldsymbol{\mu})$ hence $\boldsymbol{\mu}$ minimises J_ϕ.

2.2.1.4 Cosine Distance

Another manner of coping with the "curse of dimensionality", pointed at in Sect. 2.2.1.1, is suggested by, for instance, Hamerly [229], namely by introducing distance $d_{cos}(\mathbf{x}_i, \mathbf{x}_j)$, defined as 1 minus cosine of the angle between the vectors $\mathbf{x}_i, \mathbf{x}_j$,

$$d_{cos}(\mathbf{x}_i, \mathbf{x}_j) = 1 - \cos(\mathbf{x}_i, \mathbf{x}_j) = 1 - \frac{\mathbf{x}_i^T \mathbf{x}_j}{\|\mathbf{x}_i\| \|\mathbf{x}_j\|} = 1 - \frac{\sum_{l=1}^n x_{il} x_{jl}}{\|\mathbf{x}_i\| \|\mathbf{x}_j\|} \tag{2.15}$$

The value of cosine of an angle, appearing in the above formula, constitutes an example of a measure of similarity, which is the basic measure, applied in the information retrieval systems for measuring the similarity between the documents [39]. In this context the components x_{il} represent the frequency of appearance of a keyword indexed l in the i-th document. Since frequencies are non-negative, then, for two arbitrary vectors, representing documents, $0 \leq \cos(\mathbf{x}_i, \mathbf{x}_j) \leq 1$ and $0 \leq d_{cos}(\mathbf{x}_i, \mathbf{x}_j) \leq 1$. Other measures, which quantify the similarity of documents, are considered in [259].

Note that $d_{cos}(\mathbf{x}_i, \mathbf{x}_j)$ is not metric (the triangle condition does not hold). In order to obtain a metric distance measure, the $d_{arccos}(\mathbf{x}_i, \mathbf{x}_j) = \arccos(\cos(\mathbf{x}_i, \mathbf{x}_j))$ distance is introduced, which is simply the angle between the respective vectors. Other metric possibility is the sine distance, $d_{sin}(\mathbf{x}_i, \mathbf{x}_j) = \sin(\mathbf{x}_i, \mathbf{x}_j) = \sqrt{1 - \cos^2(\mathbf{x}_i, \mathbf{x}_j)}$.

Still another cosine similarity based metric distance is $d_{sqrtcos}(\mathbf{x}_i, \mathbf{x}_j) = \sqrt{1 - \cos(\mathbf{x}_i, \mathbf{x}_j)}$.

2.2.1.5 Power Distance

When we wish to increase or decrease the growing weight, assigned to these dimensions, for which the objects considered differ very much, we can apply the so-called power distance[21]:

$$d_{p,r}(\mathbf{x}_i, \mathbf{x}_j) = \left(\sum_{l=1}^n |x_{il} - x_{jl}|^p \right)^{1/r} \tag{2.16}$$

---

[21] See e.g. *Web-based handbook of statistics. Cluster analysis: Agglomeration.* http://www.statsoft.pl/textbook/stathome.html.

where p and r are parameters. The parameter p controls the increasing weight, which is assigned to the differences for the particular dimensions, while parameter r controls the increasing weight, which is assigned to the bigger differences between objects. Of course, when $p = r$, this distance is equivalent to Minkowski distance.

In general, power distance is not metric.

2.2.2   Comparing the Objects Having Qualitative Features

Similarly as in the preceding point, we assume here, that for reasons related to the facility of processing, each object is represented by the vector $\mathbf{x}$, but now its components are interpreted more like labels. If, for instance, the feature of interest for us is *eye color*, then this feature may take on such values as 1–blue, 2–green, etc. It is essential that for such labels there may not exist a natural order of such label "values", proper for a given phenomenon under consideration. In such situations one can use as the measure of dissimilarity the generalized Hamming distance: $d_H(\mathbf{x}_i, \mathbf{x}_j)$, equal the number of these features, whose values for the compared objects are different.

When we deal with mixed data, that is—a part of features have a qualitative character, and a part—quantitative character, then we can apply the so-called Gower coefficient [303], which is the weighted sum of the partial coefficients of divergence $\delta(i, j, l)$, determined for each feature of the objects $\mathbf{x}_i$ i $\mathbf{x}_j$. For a nominal (qualitative) feature we take $\delta(i, j, l) = 1$, when the value of this feature in both objects is different and $\delta(i, j, l) = 0$ in the opposite case. If, on the other hand, we deal with the quantitative feature having values from the interval $[x_l^{min}, x_l^{max}]$, then we replace $\delta(i, j, l)$ by

$$\delta(i, j, l) = \frac{|x_{il} - x_{jl}|}{x_l^{max} - x_l^{min}}$$

Ultimately, we take as the distance between two objects

$$d_m(\mathbf{x}_i, \mathbf{x}_j) = \frac{\sum_{l=1}^n w(i, j, l)\delta(i, j, l)}{\sum_{l=1}^n w(i, j, l)} \tag{2.17}$$

where $w(i, j, l)$ is the weight equal zero when either one of the values x_{il}, x_{jl} was not observed, or we deal with a so-called asymmetric binary feature[22] and for one of the compared objects the value of this feature is equal zero. In the remaining cases we have $w(i, j, l) = 1$.

The recommender systems [89], whose purpose is to suggest to the users the selection of broadly understood goods (books, movies, discs, information, etc.) matching in a possibly best manner the tastes and the preferences of the users, make use of

---

[22]E.g. in medical tests it is often assumed that lack of a given feature for a patient is denoted by symbol 0, while its presence—by the symbol 1. In such situations it is better not to account in the comparisons for the number of zeroes.

the similarity measure $s(\mathbf{x}_i, \mathbf{x}_j)$ between the preferences of the given user, $\mathbf{x}_i$, and the preferences of other users, $\mathbf{x}_j$, where $j = 1, \ldots, m$, $j \neq i$. Here, the l-th component of the preference vector corresponds to the evaluation of the l-th good. One of the most often applied measures of similarity is, in this case, the modified Pearson correlation coefficient

$$r(\mathbf{x}_i, \mathbf{x}_j) = \frac{(\mathbf{x}_i - \bar{\mathbf{x}}_i) \cdot (\mathbf{x}_j - \bar{\mathbf{x}}_j)}{\|\mathbf{x}_i - \bar{\mathbf{x}}_i\| \|\mathbf{x}_j - \bar{\mathbf{x}}_j\|} \tag{2.18}$$

where $\bar{\mathbf{x}}_i$ denotes the average of the vector $\mathbf{x}_i$. Modification concerns the numerator of the above expression: when summing up the respective products one accounts for only those components of the vectors, which represent the common evaluations of the users compared. Instead of Pearson correlation one can apply, of course, other measures of correlation, adapted to the character of the features used in describing objects. The most popular variants applied for the qualitative features are Spearman or Kendall correlations.

Just like in the case of the cosine similarity measure, also here one can introduce the correlation-based distance $d_{r1}(\mathbf{x}_i, \mathbf{x}_j) = 1 - r(\mathbf{x}_i, \mathbf{x}_j)$, taking values from the interval $[0, 2]$. Another variant was proposed in [467]: $d_{r2}(\mathbf{x}_i, \mathbf{x}_j) = \sqrt{2[1 - r(\mathbf{x}_i, \mathbf{x}_j)]}$. This distance also takes values from the interval $[0, 2]$; the positively and strongly correlated variables correspond to small distances, while negatively and strongly correlated variables correspond to large distances, the weakly correlated variables being situated midway. One should also note that when we deal with the centered variables (i.e. $\bar{\mathbf{x}}_i = \bar{\mathbf{x}}_j = 0$), then the value of the correlation coefficient is identical with the value of cosine of the angle between the two vectors. This fact was taken advantage of by Trosset [467] to formulate the angular representation of correlation, used then to construct a new algorithm of cluster analysis. Finally, in [203], side by side with d_{r2}, the (pseudo-)distance

$$d_{r3}(\mathbf{x}_i, \mathbf{x}_j) = \left(\frac{1 - r(\mathbf{x}_i, \mathbf{x}_j)}{1 + r(\mathbf{x}_i, \mathbf{x}_j)} \right)^\beta$$

where $\beta > 0$ is a parameter, was introduced. This distance was applied in the FCM algorithm (which is presented in Sect. 3.3). The coefficient β controls, in this case, the degree of fuzziness of the resulting partition. When $r(\mathbf{x}_i, \mathbf{x}_j) = -1$, then the distance is undefined.

Since the value of r represents the cosine of the angle between the (centered) vectors $\mathbf{x}_i, \mathbf{x}_j$, then

$$d_{\tan}(\mathbf{x}_i, \mathbf{x}_j) = \sqrt{\frac{1 - r^2(\mathbf{x}_i, \mathbf{x}_j)}{r^2(\mathbf{x}_i, \mathbf{x}_j)}} \tag{2.19}$$

can be treated as tangent distance. The tangent distance measure, $d_{\tan}$, is based on the volume of joint information, carried by the features analysed. The parallel vectors ($r = 1$), as well as the anti-parallel ones ($r = -1$) carry the very same information, and hence their distance is equal 0, while similarity is the highest and equal 1. On the other hand, the orthogonal vectors are infinitely distant and have zero similarity, since each of them carries entirely different information. Vectors, having correlation coefficients different from 0 and ± 1 contain partly a specific information, and partly common information. The volume of the common information constitutes the measure of their similarity. The tangent measure of distance has nowadays a wide application in the analysis of similarity.[23] Another, normalized variant of the correlation-based similarity measure is

$$r_n(\mathbf{x}_i, \mathbf{x}_j) = \frac{1}{2}\left(r(\mathbf{x}_i, \mathbf{x}_j) + 1\right) \tag{2.20}$$

which guarantees that $r_n(\mathbf{x}_i, \mathbf{x}_j) \in [0, 1]$. In bioinformatics the so-called squared correlation distance is being applied $d_b(\mathbf{x}_i, \mathbf{x}_j) = 1 - r^2(\mathbf{x}_i, r_j)$ when comparing genetic profiles.[24] One can find a number of interesting comments on the properties of the correlation coefficient in [407].

Pearson correlation is an adequate yardstick when the variables compared are normally distributed. When this is not the case, other measures of similarity ought to be applied.

Let us also mention the chi-square distance, which is defined as follows

$$d_{\chi^2}(\mathbf{x}, \mathbf{y}) = \frac{1}{2}\sum_{i=1}^{n} \frac{(x_i - y_i)^2}{x_i + y_i} \tag{2.21}$$

and which is used in comparing histograms.[25] This distance finds application in correspondence analysis, as well as in the analysis of textures of digital images. Another measure, which is used in this context, is the Bhattacharyya distance, which measures the separability of classes; this distance is defined as follows:

$$d_B(\mathbf{x}, \mathbf{y}) = \left(1 - BC(\mathbf{x}, \mathbf{y})\right)^{1/2} \tag{2.22}$$

where $BC(\mathbf{x}, \mathbf{y}) = \sum_{i=1}^{n} \sqrt{x_i y_i}$ is the so-called Bhattacharyya coefficient. Sometimes, the following definition is used: $d_B(\mathbf{x}, \mathbf{y}) = -\ln BC(\mathbf{x}, \mathbf{y})$.

---

[23] see, e.g., J. Mazerski. *Podstawy chemometrii*, Gdańsk 2004. Electronic edition available at http://www.pg.gda.pl/chem/Katedry/Leki_Biochemia/dydaktyka/chemometria/podstawy_chemometrii.zip.

[24] See http://www.improvedoutcomes.com/docs/WebSiteDocs/Clustering/Clustering_Parameters/Pearson_Correlation_and_Pearson_Squared_Distance_Metric.htm.

[25] See V. Asha, N.U. Bhajantri, and P. Nagabhushan: GLCM-based chi-square histogram distance for automatic detection of defects on patterned textures. *Int. J. of Computational Vision and Robotics*, **2**(4), 302–313, 2011.

A survey on the measures of similarity/dissimilarity, used in grouping of the time series is provided, for instance, in [328].[26]

2.3 Hierarchical Methods of Cluster Analysis

Hierarchical methods are among the traditional techniques of cluster analysis. They consist in successive aggregation or division of the observations and their subsets. Resulting from this kind of procedure there is a tree-like structure, which is referred to as dendrogram.

The agglomerative techniques start from the set of observations, each of which is treated as a separate cluster. Clusters are aggregated in accordance with the decreasing degree of similarity (or the increasing degree of dissimilarity) until one, single cluster is established. The manner of proceeding is represented by the following pseudo-code 2.1:

Algorithm 2.1 Algorithm of agglomerative cluster analysis

Require: Data $X = (\mathbf{x}_1, \ldots, \mathbf{x}_m)^{\mathrm{T}}$.
Ensure: Dendrogram $\mathcal{C} = \{C_1, \ldots, C_{2m-1}\}$.
1: *Initialization.* Establish m single-element clusters and calculate distance for each pair of such clusters. Memorize the distances calculated in the symmetric square matrix $D = [d_{ij}]$.
2: Find a pair C_i, C_j of the clusters that are the closest to each other.
3: Form a new cluster $C_k = C_i \cup C_j$. In the generated dendrogram this corresponds to introducing a new node and connecting it with the nodes, corresponding to the clusters C_i, C_j.
4: Update the distance matrix, i.e. calculate the distance between the cluster C_k and the remaining clusters, except for C_i and C_j.
5: Remove from the matrix D rows and columns, corresponding to the aggregated clusters C_i, C_j and add a row and a column, for the new cluster C_k.
6: Repeat steps (2)–(5) until only one, single cluster is created.

In step 3 of the above algorithm we join together two closest clusters. By defining more precisely the notion, related to the new distances from the cluster thus created to the remaining ones, one obtains seven different variants of the agglomerative algorithms (abbreviations in brackets correspond to the names, introduced in [434]):

(a) Single linkage method or nearest neighbour method (*single linkage*): Distance between two clusters is equal to the distance between two closest elements belonging to different clusters. The resulting clusters form, in this case, long

---

[26] An extensive overview of distance measures is provided in the *Encyclopedia of Distances* by Michel Marie Deza and Elena Deza, Springer 2009, http://www.uco.es/users/ma1fegan/Comunes/asignaturas/vision/Encyclopedia-of-distances-2009.pdf.

"chains". In order to find the optimum solution to the task, involving the method specified, the algorithms are used, referring to the minimum spanning tree.[27]

(b) Complete linkage method or the farthest neighbour method (*complete linkage*): Distance between two clusters is equal to the distance between two farthest objects, belonging to different clusters. This method is most appropriate, when the real objects form well separated and compact clusters.

(c) Average linkage method (*unweighted pair-group average*, UPGA): Distance between two clusters is equal the average distance between all pairs of objects belonging to both clusters considered.

(d) Weighted pair-group average method (*weighted pair-group average*, WPGA): This method is similar to the preceding one, but calculations are carried out with the weights, equal the numbers of objects in the two clusters considered. This method is advised in cases, when we deal with clusters having distinctly different numbers of objects.

(e) Centroid method (*unweighted pair-group centroid*, UPGC): Distance between two clusters is equal to distance between their centroids (gravity centers).

(f) Method of weighted centroids or of the median (*weighted pair-group centroid*, WPGC): Distance between two clusters is calculated as in the previous method, but with introduction of weights, which are equal the numbers of objects in the clusters considered.

(g) Ward method of minimum variance. In this method the sum of squares of distances between objects and the center of the cluster, to which the objects belong, is minimized. This method, even though considered to be very effective, tends to form clusters having similar (low) cardinalities.

The opposition to the agglomerative techniques is constituted by the divisive techniques. Here, analysis starts from a single all-encompassing cluster, which is subject to successive divisions, according to the increasing degree of similarity. These techniques, even though apparently symmetric to the agglomerative ones, are used in practice much less frequently.

Since for a given set of observations one can obtain multiple different hierarchies, the question arises: "To what extent a dendrogram reflects the distances between the particular pairs from the set X?" One of the popular means for assessing the quality of grouping was introduced by Sokal and Rohlf,[28] namely the *cophenetic correlation coefficient*. Given a dendrogram, the matrix D_T is formed, presenting the levels of aggregation, at which the pairs of objects appeared for the first time in the same cluster. Let further E be a vector (variable) formed of the elements located above the main diagonal of the distance matrix D and let T be the vector (variable)

---

[27]M. Delattre and P. Hansen. "Bicriterion cluster analysis", *IEEE Trans. on Pattern Analysis and Machine Intelligence*, Vol-2, No. 4, pp. 277–291, 1980.

[28]R.R. Sokal, F.J. Rohlf. 1962. The comparison of dendrograms by objective methods. *Taxon*, 11(2), 1962, 33–40.

formed out of the elements situated above the main diagonal[29] of the D_T matrix. The cophenetic correlation coefficient is the Pearson correlation coefficient between these two variables. The computational details are presented in the example 2.3.1. Another coefficient, which allows for assessing the degree of matching between the dendrogram and the matrix of distances (similarities) is the Goodman-Kruskal coefficient[30] (*Goodman-Kruskal gamma coefficient, gamma index*). It was introduced with the intention of assessing the concordance of orderings of features expressed on the ordinal scale.

The hierarchical methods have some important shortcomings. We mention below some of the most important ones:

(i) They lose their clarity with the increase of the number of analysed objects.
(ii) There is no way to shift objects from one cluster to another, even if they had been wrongly classified at the initial stages of the procedure.
(iii) The results reflect the degree, to which the data match the structure implied by the algorithm selected ("chain" or a compact "cloud").

Example 2.3.1 Consider the data from the Fig. 2.5a. The matrix of distances between the objects is of the form

$$
D = \begin{array}{c|cccccc}
 & 1 & 2 & 3 & 4 & 5 & 6 \\
\hline
1 & 0 & 4.4721 & 4.2426 & 2.2361 & 2.8284 & 3.1623 \\
2 & & 0 & 1.4142 & 3.0000 & 2.0000 & 3.1623 \\
3 & & & 0 & 2.2361 & 1.4142 & 2.0000 \\
4 & & & & 0 & 1.0000 & 1.0000 \\
5 & & & & & 0 & 1.4142 \\
6 & & & & & & 0
\end{array}
$$

By applying the complete link (farthest neighbour) method, we construct the corresponding dendrogram. In the first step we aggregate the objects with numbers 4 and 6, between which distance is equal 1 unit. In the next step we add object number 5, situated at the distance $d(\{r_4, r_6\}, r_5) = \max\left(d(r_4, r_5), d(r_6, r_5)\right) = 1.4142$. In the third step we glue together objects having numbers 2 and 3, between which distance is equal 1.4142. Then, in the fourth step, we aggregate the clusters formed until now into one cluster $\{\{\{r_4, r_6\}, r_5\}, \{r_2, r_3\}\}$, the distance between these clusters being equal 3.1623. Finally, in the last step, we add the element r_1, which is situated at the distance of 4.4721 from the cluster already established. The dendrogram obtained therefrom is shown in Fig. 2.5b. The matrix of the dendritic distances, determined on the basis of the calculations performed, is as follows:

---

[29]Both distance matrices are symmetric, it suffices, therefore, to consider their elements situated above (or below) the main diagonal.

[30]L.A. Goodman, W.H. Kruskal. Measures of association for cross classifications. *J. of the American Statistical Association*, 49(268), 1954, 732–764.

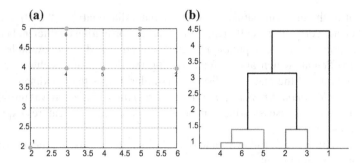

Fig. 2.5 The exemplary data set **a** and the corresponding dendrogram, **b** obtained from the complete link method

$$
D_T = \begin{bmatrix}
0 & 4.4721 & 4.4721 & 4.4721 & 4.4721 & 4.4721 \\
 & 0 & 1.4142 & 3.1623 & 3.1623 & 3.1623 \\
 & & 0 & 3.1623 & 3.1623 & 3.1623 \\
 & & & 0 & 1.4142 & 1.0000 \\
 & & & & 0 & 1.4142
\end{bmatrix}
$$

Hence, vectors E and T are equal

$$
\begin{array}{l|llllllllllllll}
E & 4.47 & 4.24 & 2.23 & 2.83 & 3.16 & 1.41 & 3.00 & 2.00 & 3.16 & 2.24 & 1.41 & 2.00 & 1.00 & 1.00 & 1.41 \\
T & 4.47 & 4.47 & 4.47 & 4.47 & 4.47 & 1.41 & 3.16 & 3.16 & 3.16 & 3.16 & 3.16 & 3.16 & 1.41 & 1.00 & 1.41
\end{array}
$$

The cophenetic correlation coefficient, expressing the degree of match between the dendrogram from Fig. 2.5b and the matrix of distances D is equal the coefficient of Pearson correlation $c(E, T) = 0.7977$. ☐

Interestingly, there exist solid theoretical foundations for in-the-limit behaviour of hierarchical clustering. Imagine, following [105], a (continuous) subset $\mathcal{X}$ of an n-dimensional real-valued space $\mathbb{R}^n$ with a probabilistic density function $f : \mathcal{X} \to \mathbb{R}$. Let an Euclidean distance function be defined in this space and $B(\mathbf{x}, r)$ be a ball of radius r around $\mathbf{x}$ in this space. Furthermore let S be a subset of $\mathcal{X}$. We say that $\mathbf{x}, \mathbf{y}$ are connected in S if there exists a continuous function (called "path") $P : [0, 1] \to S$ such that $P(0) = \mathbf{x}$ and $P(1) = \mathbf{y}$. If there exists a path between two points in S, then we say that they are connected in S. The relation "be connected in S" is an equivalence relation in S, hence it splits S into disjoint sets of points where the points are connected within the sets but not between them.

Definition 2.3.1 Let us define, for a real value λ, the set $S = \{\mathbf{x} \in \mathcal{X}; f(\mathbf{x}) >\geq \lambda\}$. Then we will say that the set $\mathbb{C}_{f,\lambda}$ of disjoint subsets of S is a *clustering* of $\mathcal{X}$, and each of these subsets will be called a *cluster*.

A cluster is intuitively a "high density connected subregion of $\mathcal{X}$". the smaller λ, the bigger will be the clusters.

Definition 2.3.2 The function $\mathbb{C}_f$ assigning each real value λ a clustering $\mathbb{C}_{f,\lambda}$ would be called a *cluster tree* of $\mathcal{X}$. $\mathbb{C}_f(\lambda) = \mathbb{C}_{f,\lambda}$

This function is called "a cluster tree" because for any two $\lambda' \leq \lambda$ it has the following properties:

- If $C \in \mathbb{C}_f(\lambda)$, then there exists such a cluster $C' \in \mathbb{C}_f(\lambda')$ that $C \subseteq C'$
- If $C \in \mathbb{C}_f(\lambda)$ and $C' \in \mathbb{C}_f(\lambda')$ then either $C \subseteq C'$ or $C \cap C' = \emptyset$.

Let us mention also their notion of σ, ϵ-separation under the density function f.

Definition 2.3.3 Two sets $A, A' \subseteq \mathcal{X}$ are said to be σ, ϵ-*separated*, if there exists a set $S \subseteq \mathcal{X}$

- Any path in $\mathcal{X}$ from A to A' intersects with S
- $\sup_{x \in S_\sigma} f(x) < (1 - \epsilon) \inf_{x \in A_\sigma \cup A'_\sigma} f(x)$

Hereby for any set Y the notation Y_σ means all points not more distant from Y than σ.

Under this definition, $A_\sigma \cup A'_\sigma$ must lie within $\mathcal{X}$ (so that the density is non-zero), but S_σ does not need to.

From this point of view, an actual data set $\mathfrak{X}$ can be considered as a sample from the set $\mathcal{X}$ and the result of hierarchical clustering of $\mathfrak{X}$ may be deemed of as an approximation of the (intrinsic) cluster tree of $\mathcal{X}$. We may ask how well and under what circumstances this approximation is good. Hartigan [236] introduced the following notion of consistency of a clustering $\mathbb{C}_\mathfrak{X}$, where $\mathfrak{X}$ is a sample containing m elements, with cluster tree of $\mathcal{X}$. For any two sets $A, A' \subset \mathcal{X}$, let A_*, A'_* resp. denote the smallest cluster of $\mathbb{C}_\mathfrak{X}$ containing $A \cap \mathcal{X}$, $A' \cap \mathcal{X}$ resp. We say that $\mathbb{C}_\mathfrak{X}$ is consistent, if for a λ and A, A' being disjoint elements of $\mathbb{C}_f(\lambda)$ probability that A_*, A'_* are disjoint equals 1 in the limit when the sample size m grows to infinity. Hartigan [236] showed that the single link algorithm is consistent only in a one dimensional space.

Chaudhuri and Dasgupta [105] proposed therefore the "robust" single link Algorithm 2.2. This algorithm reduces to single link algorithm in case of $\alpha = 1$ and $k = 1$, or to Wishart algorithm for $\alpha = 1$ and k larger. In order to broaden the class of consistent hierarchical clustering algorithms, Chaudhuri and Dasgupta suggest to use (end of Sect. 3.2.) $\alpha \in [\sqrt{2}, 2]$ and k_m equal to closest bigger integer to $n(\ln m)^2$ where n is the dimensionality and m is the sample size.

Please note in passing that conceptually the idea of a "cluster tree" refers to some probability density distribution. Nonetheless a distance-based clustering algorithm could be analysed. It can be stated generally, that the clustering concepts based on distance, similarity graphs and density can be considered as strongly related because it is always possible for a data set, for which we know only distances, or similarities or linkage graphs, to embed it in an Euclidean space reflecting faithfully the distances/similarities so that the respective clusters can be analysed in a density-based manner.

Algorithm 2.2 "Robust" Single Link

Require: Data $X = (\mathbf{x}_1, \ldots, \mathbf{x}_m)^\mathrm{T}$.
Ensure: Dendrogram $\mathcal{C} = \{C_1, \ldots, C_{2m-1}\}$.
1: *Initialization.* Natural number $k \geq 1$ and a real number $\alpha \in [1, 2]$ are user-defined parameters. Establish m single-element clusters and calculate distance for each pair of such clusters. Memorize the distances calculated in the symmetric square matrix $D = [d_{ij}]$. For each data element $\mathbf{x}$ calculate $r_k(\mathbf{x})$ as the diameter of the smallest ball around $\mathbf{x}$ containing k elements (including $\mathbf{x}$).
2: **for** r taking on distinct values of $r_k(\mathbf{x})$ computed above, from the lowest to the largest **do**
3:    Find a pair C_i, C_j of the clusters that contain the closest elements to each other among those pairs for which for each $\mathbf{x} \in C_i \cup C_j$ $r_k(\mathbf{x}) \leq r$ and there exist $\mathbf{x} \in C_i$, $\mathbf{y} \in C_j$ such that $\|\mathbf{x}, \mathbf{y}\| \leq \alpha r$.
4:    Form a new cluster $C_k = C_i \cup C_j$. In the generated dendrogram this corresponds to introducing a new node and connecting it with the nodes, corresponding to the clusters C_i, C_j.

5:    Update the distance matrix, i.e. calculate the distance between the cluster C_k and the remaining clusters, except for C_i and C_j.
6:    Remove from the matrix D rows and columns, corresponding to the aggregated clusters C_i, C_j and add a row and a column, for the new cluster C_k.
7:    Repeat steps (2)–(5) until no more link can be added.
8:    Connected components at this point are regarded as clusters belonging to one level of the cluster tree
9: **end for**

2.4   Partitional Clustering

Let $U = [u_{ij}]_{m \times k}$ denote the matrix with elements indicating the fact of assignment of i-th object to the j-th class. When $u_{ij} \in \{0, 1\}$, then we speak of a "crisp" partition of the set $\mathfrak{X}$, while for $u_{ij} \in [0, 1]$—we deal with the "fuzzy" partition. The latter case is considered in Sect. 3.3. Here, we concentrate on the "crisp" partitions.

If matrix U is supposed to represent the partition of the set of objects (see conditions, mentioned in Sect. 2.1), then this matrix has to satisfy the following conditions: (a) each object has to belong to exactly one cluster, that is $\sum_{j=1}^{k} u_{ij} = 1$ and (b) each cluster must contain at least one object, but cannot contain all the objects, i.e. $1 \leq \sum_{i=1}^{m} u_{ij} < m$. Denote by $\mathcal{U}_{m \times k}$ the set of all the matrices, representing the possible partitions of the set of m objects into k disjoint classes. It turns out, see, e.g., [19, 303], that for the given values of m and k the cardinality of the set $\mathcal{U}_{m \times k}$ is defined by the formula

$$\vartheta(m, k) = \frac{1}{k!} \sum_{j=1}^{k} (-1)^{k-j} \binom{k}{j} j^m \tag{2.23}$$

For instance, $\vartheta(3, 2) = 3$, but already $\vartheta(100, 5)$, is the number of the order of 10^{68}. Generally, the number of ways, in which m observations can be divided among k clusters is approximately equal $k^m/k!$, meaning that it is of the order of $O(k^m)$. In particular, when $k = 2$, then $\vartheta(m, 2) = 2^{m-1} - 1$. The problem of selection of the proper partition is, therefore, an $\mathcal{NP}$-complete task of combinatorial optimisation.

An effective navigation over the sea of the admissible partitions is secured by the criteria of grouping and the methods of optimisation, coupled with them.

2.4.1 Criteria of Grouping Based on Dissimilarity

The fundamental criteria, applied in the partitional clustering are homogeneity and separation. Homogeneity of a cluster means that two arbitrary objects, which belong to it, are sufficiently similar, while separation of clusters means that two arbitrary objects, belonging to different clusters are sufficiently different. In other words, the partition, induced by the clustering algorithm should contain homogeneous and well separated groups of objects.

Let $D = [d_{ij}]_{m \times m}$ denote the matrix with elements d_{ij} corresponding to the dissimilarities of objects i and j. We assume that D is a symmetric and non-negative matrix, having zeroes on the diagonal. The homogeneity of a cluster C_l composed of n_l objects can be measured with the use of one of four indicators,[31] see, e.g., [233]:

$$
\begin{aligned}
h_1(C_l) &= \sum_{\mathbf{x}_i, \mathbf{x}_j \in C_l} d_{ij} \\
h_2(C_l) &= \max_{\mathbf{x}_i, \mathbf{x}_j \in C_l} d_{ij} \\
h_3(C_l) &= \min_{\mathbf{x}_i \in C_l} \max_{\mathbf{x}_j \in C_l} d_{ij} \\
h_4(C_l) &= \min_{\mathbf{x}_i \in C_l} \sum_{\mathbf{x}_j \in C_l, j \neq i} d_{ij}
\end{aligned}
\tag{2.24}
$$

The first of these indicators is the sum of dissimilarities between the pairs of objects belonging to the cluster C_l. If the elements of the set C_l are mutually sufficiently similar, the set can be treated as a clique,[32] and so h_1 represents the weight of a clique. The second indicator is the maximum value of dissimilarity in the group C_l; it corresponds to the diameter of the set C_l. The third indicator is being referred to as the radius of the set C_l, while the fourth one is defined as a minimum sum of dissimilarities between the objects from the set C_l and its representative. This latter indicator is called *star index* or medoid.

Separation is quantified with the use of the following indicators

---

[31]For computational reasons, instead of distance, its squared value is often used, allowing for omitting the square root operation.

[32]In graph theory, a clique is such a subgraph, in which any two vertices are connected by an edge. Admitting that we connect with an edge the vertices that are little dissimilar, we treat a clique as a set of mutually similar vertices.

$$s_1(C_l) = \sum_{x_i \in C_l} \sum_{x_j \notin C_l} d_{ij}$$

$$s_2(C_l) = \min_{x_i \in C_l, x_j \notin C_l} d_{ij} \qquad (2.25)$$

The first of them, $s_1(C_l)$, called the cutting cost, is the sum of distances between the objects from the set C_l and the objects from outside of this set. The second one, $s_2(C_l)$ is equal the minimum dissimilarity between the elements of the set C_l and the remaining elements of the set $\mathfrak{X}$.

Based on the measures as defined above, we can quantify the quality $J(m, k)$ of the partition of the set of m elements into k disjoint groups with the use of the indicators given below:

$$J_1(m, k) = \frac{1}{k} \sum_{i=1}^{k} \mathfrak{w}_i$$

$$J_2(m, k) = \max_{i=1,\dots,k} \mathfrak{w}_i \qquad (2.26)$$

$$J_3(m, k) = \min_{i=1,\dots,k} \mathfrak{w}_i$$

Symbol $\mathfrak{w}_i$ denotes one of the previously defined indicators of homogeneity/ separability, assigned to the i-th group. If $\mathfrak{w}$ represents homogeneity, then the indicator $J_1(n, k)$ corresponds to the average homogeneity inside groups, $J_2(n, k)$—the maximum homogeneity, and $J_3(n, k)$—the minimum homogeneity in the partition produced. A good partition ought to—in the case of homogeneity—be characterised by the possibly low values of these indicators, while in the case of separability—by their possibly high values.

2.4.2   The Task of Cluster Analysis in Euclidean Space

Conform to the convention adopted, each object $x_i \in \mathfrak{X}$ is described by the n-dimensional vector of features, so that the set $\mathfrak{X}$ is identified with the set of m points in the n-dimensional Euclidean space. Let

$$\bar{\mu} = \frac{1}{m} \sum_{i=1}^{m} x_i \qquad (2.27)$$

be the gravity centre of the set of m objects and let

$$\mu_j = \frac{1}{|C_j|} \sum_{x_i \in C_j} x_i = \frac{1}{\sum_{i=1}^{m} u_{ij}} \sum_{i=1}^{m} u_{ij} x_i = \frac{1}{n_j} \sum_{i=1}^{m} u_{ij} x_i \qquad (2.28)$$

denote the gravity centre of the j-th cluster, where $n_j = |C_j| = \sum_{i=1}^{m} u_{ij}$ is the cardinality of the j-th cluster.

We define two matrices (see, e.g., [173, 303]):

$$W = \sum_{i=1}^{m} \sum_{j=1}^{k} u_{ij} (\mathbf{x}_i - \boldsymbol{\mu}_j)(\mathbf{x}_i - \boldsymbol{\mu}_j)^{\mathrm{T}} \tag{2.29}$$

$$B = \sum_{j=1}^{k} \left(\sum_{i=1}^{m} u_{ij} \right) (\boldsymbol{\mu}_j - \overline{\boldsymbol{\mu}})(\boldsymbol{\mu}_j - \overline{\boldsymbol{\mu}})^{\mathrm{T}} \tag{2.30}$$

Matrix W is the in-group covariance matrix, while B is the inter-group covariance matrix. Both matrices together constitute a decomposition of the dispersion matrix, or variance-covariance matrix

$$T = \sum_{i=1}^{m} (\mathbf{x}_i - \overline{\boldsymbol{\mu}})(\mathbf{x}_i - \overline{\boldsymbol{\mu}})^{\mathrm{T}} \tag{2.31}$$

i.e.: $T = W + B$.

The typical objective functions, which are used as the criteria of selection of the proper partition, are [173]:

(a) Minimisation of the trace of matrix W. This criterion is equivalent to minimisation of the sum of squares of the Euclidean distances between the objects and the centres of clusters, to which these objects belong, that is

$$\begin{aligned} J_1(m, k) &= \sum_{j=1}^{k} \sum_{i=1}^{m} u_{ij} \|\mathbf{x}_i - \boldsymbol{\mu}_j\|^2 \\ &= \sum_{j=1}^{k} \frac{1}{n_j} \sum_{\mathbf{x}_i, \mathbf{x}_l \in C_j} \|\mathbf{x}_i - \mathbf{x}_l\|^2 \end{aligned} \tag{2.32}$$

In other words, minimisation of the indicator J_1 is equivalent to minimisation of the criterion of homogeneity, $h_1(C_j)/n_j$. Such approach favours spherical clusters.

(b) Minimisation of the determinant of matrix W. This criterion is useful in the situations, when the natural clusters are not spherical.

(c) Maximisation of the trace of matrix BW^{-1}. This is a generalisation of the Mahalanobis distance (2.8) for the case of more than two objects. The shortcoming of this criterion is its sensitivity to scale. Grouping obtained from the raw data may drastically differ from the one obtained after the data are rescaled (e.g. through standardisation or normalisation).

2.4.2.1 Minimising the Trace of In-Group Covariance

Despite the limitations, mentioned before, the criterion (2.32) is among the most willingly and most frequently used in practice. We shall soon see that this is not so much the effect of its simplicity, as—surprisingly—its relatively high degree of universality.

Minimisation of the quality index (2.32) leads to the mathematical programming problem of the form

$$\min_{u_{ij}\in\{0,1\}} \sum_{i=1}^{m}\sum_{j=1}^{k} \left\| \mathbf{x}_i - \frac{\sum_{l=1}^{m} u_{lj}\mathbf{x}_l}{\sum_{l=1}^{m} u_{lj}} \right\|^2$$

$$\text{subject to } \sum_{i=1}^{m} u_{ij} > 1, \ j = 1, \ldots, k \qquad (2.33)$$

$$\sum_{j=1}^{k} u_{ij} = 1, \ i = 1, \ldots, m$$

It is a 0/1 (binary) programming problem with a nonlinear objective function, [16, 233]. The integer constraints, along with the nonlinear and non-convex objective function make the problem (2.33) $\mathcal{NP}$-hard. For this reason the determination of the minimum of the function J_1 is usually performed with the use of the heuristic methods. Yet, attempts have been and are being made, aiming at a satisfactory solution to the problem (2.33). A survey of these attempts can be found, for instance, in [16, 30, 390]. The goal of such studies is not only to find the optimum solution, but also to gain a deeper insight into the very nature of the task of grouping the objects. We provide below several equivalent forms of the problem (2.33), enabling its generalisation in various interesting ways.

Let $F \in \mathbb{R}^{m \times m}$ be a matrix having the elements

$$f_{ij} = \begin{cases} \frac{1}{n_j} & \text{if } (\mathbf{x}_i, \mathbf{x}_j) \in C_j \\ 0 & \text{otherwise} \end{cases} \qquad (2.34)$$

where, as before, n_j denotes the number of objects in the group C_j.

If we number the objects from the set $\mathfrak{X}$ in such a manner that the first n_1 objects belong to group C_1, the successive n_2 of objects belong to group C_2, etc., then F is a block-diagonal matrix, $F = \text{diag}(F_1, \ldots, F_k)$. Each block F_j is an $n_j \times n_j$ matrix having elements $1/n_j$, i.e. $F_j = (1/n_j)\mathbf{e}\mathbf{e}^{\mathsf{T}}, \ j = 1, \ldots, k$.

Lemma 2.4.1 *Matrix F of dimensions $m \times m$, having elements defined as in Eq. (2.34), displays the following properties:*

(a) it is a non-negative symmetric matrix, that is, $f_{ij} = f_{ji} \geq 0$, for $i, j = 1, \ldots, m$,
(b) it is a doubly stochastic matrix, that is, $F\mathbf{e} = F^{\mathsf{T}}\mathbf{e} = \mathbf{e}$,
(c) $FF = F$ (idempotency)

(d) $\mathrm{tr}\,(F) = k$,

(e) spectrum of the matrix F, $\sigma(F) = \{0, 1\}$, and there exist exactly k eigenvalues equal 1.

Proof Properties (a)–(d) are obvious. We shall be demonstrating only the property (e). Since every block has the form $F_j = \mathbf{e}\mathbf{e}^{\mathrm{T}}/n_j$, then exactly one eigenvalue of this submatrix is equal 1, while the remaining $n_j - 1$ eigenvalues are equal zero. The spectrum of the matrix F, $\sigma(F) = \bigcup_{j=1}^{k} \sigma(F_j)$, hence F has exactly k eigenvalues equal 1. □

If $X = (\mathbf{x}_1 \dots \mathbf{x}_m)^{\mathrm{T}}$ is the matrix of observations, then

$$M = FX$$

is the matrix, whose i-th row represents the gravity centre of the group, to which the i-th object belongs.

Given the above notation, the quality index (2.32) can be written down in the equivalent matrix form

$$J_1(C_1, \dots, C_k) = \sum_{j=1}^{k} \sum_{x_i \in C_j} \|\mathbf{x}_i - \boldsymbol{\mu}_j\|^2$$
$$= \mathrm{tr}\left((X - M)^{\mathrm{T}}(X - M)\right)$$
$$= \mathrm{tr}\left(X^{\mathrm{T}}X + M^{\mathrm{T}}M - 2X^{\mathrm{T}}M\right)$$

Taking advantage of the additivity and commutativity of the matrix trace, see properties (b) and (c) in p. 321, as well as symmetry and idempotence of matrix F, we transform the above expression to the form

$$J_1(C_1, \dots, C_k) = \mathrm{tr}\,(X^{\mathrm{T}}X + X^{\mathrm{T}}F^{\mathrm{T}}FX - 2X^{\mathrm{T}}FX)$$
$$= \mathrm{tr}\,(X^{\mathrm{T}}X + X^{\mathrm{T}}FX - 2X^{\mathrm{T}}FX)$$
$$= \mathrm{tr}\,(X^{\mathrm{T}}X - X^{\mathrm{T}}FX)$$
$$= \mathrm{tr}\,(X^{\mathrm{T}}X - X^{\mathrm{T}}XF)$$

Let

$$K = X^{\mathrm{T}}X$$

It is a symmetric and non-negative matrix. The expression $\mathrm{tr}\,(X^{\mathrm{T}}XF) = \mathrm{tr}\,(KF) = \sum_{ij} k_{ij} f_{ij}$ is a linear combination of the elements of matrix K, hence, in case of the nonlinear objective function J_1 we obtain a linear function! Finally, the task of grouping of objects in the Euclidean space reduces to either *minimisation* of the indicator

$$J_1(C_1, \dots, C_k) = \mathrm{tr}\left(K(\mathbb{I} - F)\right) \tag{2.35}$$

or, equivalently, if we ignore the constant component K, to *maximisation* of the indicator

$$J_1'(C_1, \ldots, C_k) = \text{tr}\,(KF) \tag{2.36}$$

The first of these forms is used, in particular, by Peng and Wei in [390], while the second, by, for instance, Zass and Shashua in [527]. We concentrate on the task of maximisation[33]

$$\max_{F \in \mathbb{R}^{m \times m}} \text{tr}\,(KF)$$
$$\text{subject to } F \geq 0,\, F^{\text{T}} = F,\, F\mathbf{e} = \mathbf{e} \tag{2.37}$$
$$F^2 = F,\, \text{tr}\,(F) = k$$

The above formulation allows for the generalisation of the task of grouping in a variety of manners:

(a) Minimisation of the indicator (2.32) assumed that the observations are points in n-dimensional Euclidean space. In such a case the prototypes are sought, being the gravity centres of the possibly compact groups. The quality index of the form of $\text{tr}\,(KF)$ makes it possible to replace the distances between the points by a more general kernel function.[34] This shifts our interest in the direction of the so-called relational grouping, where, instead of the Euclidean distance between the pairs of points, a more general measure of similarity is being used, $k_{ij} = s(\mathbf{x}_i, \mathbf{x}_j)$, e.g. the Gaussian kernel function $k_{ij} = \exp\left(-\gamma \|\mathbf{x}_i - \mathbf{x}_j\|^2\right)$, where $\gamma > 0$ is a parameter. Thereby we give up the assumption that the set $\mathcal{X}$ has to have the representation $X \in \mathbb{R}^{m \times n}$.

(b) The objective function, appearing in the problem (2.37) can be replaced by the Bregman divergence (see Definition 2.2.1 on p. 22) D_ϕ, this leading to the problem, see [492]

$$\max_{F \in \mathbb{R}^{m \times m}} D_\phi(K, F)$$
$$\text{subject to } F \geq 0,\, F^{\text{T}} = F,\, F\mathbf{e} = \mathbf{e} \tag{2.38}$$
$$F^2 = F,\, \text{tr}\,(F) = k$$

When $\phi = r^2$, problem (2.38) is reduced to problem (2.37). The generalised variant of the k-means algorithm for such a case is considered in the study [50]. It is shown there that, in particular, the prototypes of groups are determined as the (weighted) gravity centres of these groups, and, even more importantly, that such a definition of the prototype is correct only if the dissimilarity of the objects is measured through some Bregman divergence.

---

[33]Peng and Wei note in [390] that the idea of representing the objective function appearing in the problem (2.33) in the form of minimisation of the trace of an appropriate matrix was forwarded first by A.D. Gordon and J.T. Henderson in their paper, entitled "An algorithm for Euclidean sum of squares", which appeared in the 33rd volume of the journal *Biometrica* (year 1977, pp. 355–362). Gordon and Henderson, though, wrote that this idea had been suggested to them by the anonymous referee!.

[34]See Sect. 2.5.7.

Let us also note here that formulation (2.37) turns our attention towards the doubly stochastic matrices, that is—such symmetric and non-negative matrices that the sum of every row and every column is equal 1. One can find interesting remarks on applications of such matrices in, for instance [296].

In order to enhance the flexibility of the formulation (2.37), let us replace the matrix F by the product GG^T, where G is the matrix of the dimensions $m \times k$, having the elements

$$g_{ij} = \begin{cases} \frac{1}{\sqrt{n_j}} & \text{if } (\mathbf{x}_i, \mathbf{x}_j) \in C_j \\ 0 & \text{otherwise} \end{cases}$$

Matrix G fulfils the following conditions: (a) it is non-negative, $g_{ij} \geq 0$, $i = 1, \ldots, m$, $j = 1, \ldots, k$, (b) $G^T G = \mathbb{I}$, (c) $GG^T \mathbf{e} = \mathbf{e}$. The non-negativity of the matrix G means that F is a completely positive matrix, and its cp-rank[35] equals k.

By referring to the fact that $\text{tr}(AB) = \text{tr}(BA)$, provided both products do exist, we can turn the task of maximisation (2.37) into the one of the form

$$\max_{G \in \mathbb{R}^{m \times k}} \text{tr}(G^T K G)$$
$$\text{subject to } G \geq 0,$$
$$G^T G = \mathbb{I},$$
$$GG^T \mathbf{e} = \mathbf{e}$$

(2.39)

Taking into account the formulation (2.39) we can treat the problem of grouping as a search for such a matrix F, for which $\text{tr}(KF)$ attains the maximum in the set of all matrices $\mathbb{R}^{m \times m}$ fulfilling two additional conditions: (a) F is a doubly stochastic matrix, (b) F is a completely positive matrix, and its cp-rank $= k$, that is: $F = GG^T$, where G is a non-negative matrix of the dimensions $m \times k$. Zass and Shashua propose in [527] a two-stage procedure:

(i) The given matrix K is replaced by the doubly stochastic matrix $\widetilde{F}$. In order to do this, one can use the Sinkhorn-Knopp method, which consists in the intermittent normalisation of rows and columns of matrix K. Other, more effective methods of turning a matrix into the doubly stochastic form are commented upon by Knight in [296].

(ii) In the set of the non-negative matrices of dimensions $m \times k$ such a matrix G is sought, which minimises the error $\|\widetilde{F} - GG^T\|_F$, where $\| A \|_F = \sqrt{\text{tr} A^T A} = \sqrt{\sum_{i=1}^m \sum_{j=1}^m a_{ij}^2}$ denotes the Frobenius norm.

The concepts here barely outlined have been developed into the advanced methods of cluster analysis. A part of them makes use of the so-called semi-definite programming; we can mention here the studies, reported in [107, 308, 390], or in [512]. They concern not only the typical problems of minimising the trace of an appropriate matrix, but also more advanced methods of grouping, which are considered in the further parts of this book.

---

[35] See Definition B.2.4 in p. 320.

2.4.2.2   Approximating the Data Matrix

The quality index (2.32) can be transformed to

$$J_1 = \|X - UM\|_F^2 \tag{2.40}$$

where $M \in \mathbb{R}^{k \times n}$ is the matrix with group centroids being its rows, $M = (\boldsymbol{\mu}_1, \ldots, \boldsymbol{\mu}_k)^{\mathrm{T}}$, while $U \in \mathbb{R}^{m \times k}$ is the matrix indicating the assignment of the i-th object to the j-th group, $U = (\mathbf{u}_1, \ldots, \mathbf{u}_m)^{\mathrm{T}}$.

Minimisation of the indicator (2.40) allows for taking a different perspective on the task of grouping: we look for a possibly good approximation of the data matrix by the product of two matrices, U and M. If $u_{ij} \in \{0, 1\}$, then this task is being carried out with the help of the following procedure:

(a) If $\widehat{M} = (\widehat{\boldsymbol{\mu}}_1, \ldots, \widehat{\boldsymbol{\mu}}_k)^{\mathrm{T}}$ is the current approximation of the matrix M, then the elements $\widehat{u}_{ij}$ of the matrix $\widehat{U}$, constituting the approximation of U, have the form

$$\widehat{u}_{ij} = \begin{cases} 1 \text{ if } j = \underset{1 \le t \le k}{\arg\min} \|\mathbf{x}_i - \widehat{\boldsymbol{\mu}}_t\|_F^2 \\ 0 \text{ otherwise} \end{cases} \tag{2.41}$$

(b) If $\widehat{U}$ is the current approximation of the matrix U, then determination of the matrix $\widehat{M}$ minimising the indicator J_1 is a classical problem of regression. From the condition[36] $\partial J_1 / \partial \widehat{M} = \widehat{U}^{\mathrm{T}} (\widehat{U} \widehat{M} - X) = 0$ we get

$$\widehat{M} = (\widehat{U}^{\mathrm{T}} \widehat{U})^{-1} \widehat{U}^{\mathrm{T}} X \tag{2.42}$$

(c) Steps (a) and (b) are repeated until the terminal condition is fulfilled, this condition consisting in the performance of a given number of repetitions, or in stabilisation of the elements of the matrix $\widehat{U}$.

The algorithm is initiated by specifying either an approximation $\widehat{U}$, or the matrix $\widehat{M}$. A better variant, ensuring faster convergence, is to start from the matrix $\widehat{M}$. The methods of its initialisation are considered in Sect. 3.1.3.

Minimisation of the indicator (2.40) constitutes, in its essence, the problem of the so-called non-negative factorisation, playing an important role in machine learning [497], bioinformatics [145], text analysis [67, 516], or in recommender systems [301]. Yet, the problem (2.40) differs somewhat from the classical formulation [319], where it is required to have both matrices non-negative. In the case of grouping, matrix M does not have to be non-negative, while matrix U must satisfy certain additional constraints, like, e.g., $\sum_{j=1}^{k} u_{ij} = 1$. The thus formulated task of factorisation of matrix X is a subject of intensive studies, e.g. [151, 153, 264, 326, 327].

---

[36]We take advantage here of the fact that $\|A\|_F^2 = \mathrm{tr}(A'A)$.

By generalising the indicator (2.40) to the form

$$J_1 = \|X - U^\alpha M^\mathsf{T}\|_F^2 \tag{2.43}$$

where $\alpha > 1$, we obtain the formulation leading to fuzzy grouping that we consider in Sect. 3.3. An example of application of this technique in bioinformatics shall be presented in Sect. 2.5.8.

2.4.2.3 Iterative Algorithm of Finding Clusters

The algorithms of determination of the partition of objects into k classes are usually iterative procedures, which converge at a local optimum [230]. An instance thereof has been presented in the preceding section. Its weak point is the necessity of performing operations on matrices in step (b). Note, though, that due to a special structure of the matrix of assignments U:

(a) Product $\widehat{U}^\mathsf{T}\widehat{U} = \widetilde{U}$ is a diagonal matrix having elements

$$\widetilde{u}_{jj} = \sum_{i=1}^{m} \widehat{u}_{ij} = n_j, \quad j = 1, \ldots, k$$

where n_j denotes the number of elements of the j-th group. Hence, $\widetilde{U}^{-1}$ is a diagonal matrix, as well, having elements $\widetilde{u}_{jj}^{-1} = 1/n_j$.

(b) Matrix $\widetilde{M} = \widehat{U}^\mathsf{T} X$ has the dimensions $k \times n$, and its i-th row is the sum of rows of the matrix X, corresponding to the elements of the i-th cluster. So, the i-th row of the matrix $\widetilde{U}^{-1}\widetilde{M}$ is the arithmetic mean of the coordinates of objects, assigned to the i-th group.

The general form of the iterative procedure of assigning objects to clusters is shown in the pseudocode 2.3. Here, the weights are additionally used, indicating the contribution of a given object to the relocation of the gravity centres. This mechanism was introduced by Zhang [530], and was applied, in particular, by Hamerly and Elkan, [230]. When all weights are equal 1, the Algorithm 2.3 corresponds to the algorithm from the preceding section and represents the classical Lloyd's heuristics [334].

The essence of the algorithm is the iterative modification of the assignment of objects to clusters. It is most common to assign an object to the cluster with the closest gravity centre, i.e.

$$u_{ij} = \begin{cases} 1 & \text{if } j = \underset{1 \le t \le k}{\arg\min} \|\mathbf{x}_j - \boldsymbol{\mu}_t\| \\ 0 & \text{otherwise} \end{cases} \tag{2.45}$$

Algorithm 2.3 Iterative algorithm of cluster analysis (generalised Lloyd's heuristics)

Require: Data set X and number of groups k.

Ensure: Gravity centres of classes $\{\boldsymbol{\mu}_1, \ldots, \boldsymbol{\mu}_k\}$ along with the assignment of objects to classes $U = [u_{ij}]_{m \times k}$.

1: *Initialisation.* Select the gravity centres of clusters and assign weights to objects $w(\mathbf{x}_i)$.

2: Updating of assignments: for each object determine its assignment to a cluster and, possibly, also its weight.

3: Updating of the gravity centres of clusters:

$$\boldsymbol{\mu}_j = \frac{\sum_{i=1}^{m} u_{ij} w(\mathbf{x}_i) \mathbf{x}_i}{\sum_{i=1}^{m} u_{ij} w(\mathbf{x}_i)} \qquad (2.44)$$

4: Repeat steps 2 and 3 until the stopping condition is fulfilled, usually assumed to be the lack of changes in the assignment of objects to clusters.

---

where $\boldsymbol{\mu}_t$ denotes the gravity centre of cluster C_t (this rule was applied in step (a) of the algorithm from the preceding section). It is the rule *the winner takes all*—known also from the theory of competitive supervised learning.[37] The clusters, determined in this manner, are called Voronoi clusters.[38] Formally, if $\boldsymbol{\mu}_1, \ldots, \boldsymbol{\mu}_k$ is a set of prototypes, then the Voronoi cluster W_j is the set of points, for which $\boldsymbol{\mu}_j$ is the closest prototype, that is:

$$W_j = \{\mathbf{x} \in \mathbb{R}^n \mid j = \arg\min_{1 \le l \le k} \|\mathbf{x} - \boldsymbol{\mu}_l\|\} \qquad (2.46)$$

These clusters are convex sets, i.e.

$$[(\mathbf{x}' \in W_j) \wedge (\mathbf{x}^* \in W_j)] \Rightarrow [\mathbf{x}' + \alpha(\mathbf{x}^* - \mathbf{x}')] \in W_j, \quad 0 \le \alpha \le 1$$

The division of the space $\mathbb{R}^n$ into Voronoi sets is called Voronoi tessellation (tiling) or Dirichlet tessellation. Their nature is illustrated in Fig. 2.6. In the two-dimensional case the lines, separating the regions, belonging to different clusters, are the lines of symmetry of the segments, linking the neighbouring gravity centres. Although effective algorithms for tiling are known for the two-dimensional case [397], the very notion remains useful also in the n-dimensional case.

---

[37] See, e.g., J. Hertz, A. Krogh, R.G. Palmer: *Introduction to the Theory of Neural Computation. Santa Fe Institute Series*, Addison-Wesley, 1991.

[38] Georgiy Fedosiyevich Voronoi, whose name appears in the Voronoi diagrams, was a Russian mathematician of Ukrainian extraction. He lived in the years 1868–1908. Interesting information on this subject can be found in 17th chapter of popular book by Ian Stewart, entitled *Cows in the Maze. And other mathematical explorations*, published in 2010 by OUP Oxford.

Fig. 2.6 Voronoi
tessellation, that is, the
boundaries of clusters
determined by the gravity
centres marked by dark dots

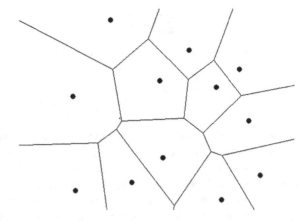

2.4.3 Grouping According to Cluster Volume

Minimisation of the trace or the determinant of the matrix W leads to clusters having similar numbers of elements. Besides, minimisation of tr (W) favours spherical clusters.

Report [429] presents the MVE algorithm (*Minimum Volume Ellipsoids*), in which clusters are represented by the hyper-ellipsoids having minimum volume. In distinction from the algorithms outlined in the preceding section, the MVE algorithm is independent of scale and allows for the determination of clusters having different numbers of elements. The dissimilarity measure, applied in this algorithm, is related to Mahalanobis distance. A similar issue is also discussed by Kumar and Orlin in [310].

The quality criterion adopted has the following form:

$$\min \sum_{j=1}^{k} \text{vol}(C_j)$$
$$\text{subject to } \sum_{j=1}^{k} |C_j| \geq (1-\alpha)m, \ 0 \leq \alpha < 1$$
$$C_j \subset X$$

The first limiting condition means that existence of at most αm outliers in the data set is allowed, while the remaining observations have to be assigned to k clusters.

The hyper-ellipsoid E_j, containing the objects from the group C_j is defined by its centre $\mathbf{c}_j$ and the symmetric and positive definite matrix Q_j, e.g. the covariance matrix, characterising this group of data. So,

$$E_j = \{\mathbf{x} \in \mathbb{R}^n : (\mathbf{x} - \mathbf{c}_j)^{\mathrm{T}} Q_j^{-1} (\mathbf{x} - \mathbf{c}_j) \leq 1\} \tag{2.47}$$

Its volume is equal $\sqrt{\det(Q_j)}$. Taking advantage of this fact, we can reduce the task of partitioning the set X into k groups to the problem of semi-definite programming of the form (see, e.g., [390])

$$\min \sum_{j=1}^{k} \sqrt{\det(Q_j)}$$

$$\text{subject the } (\mathbf{x}_i - \mathbf{c}_j)^{\mathrm{T}} Q_j^{-1}(\mathbf{x}_i - \mathbf{c}_j), \forall(\mathbf{x}_i \in C_j), j = 1, \ldots, k \quad (2.48)$$

$$\sum_{j=1}^{k} |C_j| \geq (1 - \alpha)m, \ 0 \leq \alpha < 1$$

$$C_j \subset X$$

$$C_j \succ 0, \ j = 1, \ldots, k$$

Symbol $C_j \succ 0$ means that C_j is a symmetric and positive definite matrix.

Publication [310] presents two algorithms that solve the problem formulated above: (1) kVolume, an iterative algorithm, which divides up the set of observations into the predefined number of groups, and (2) hVolume, that is—a hierarchical grouping algorithm, which generates a monotonic family of clusters. In both cases a fast algorithm of calculating volumes is made use of, as presented in [450].

2.4.4 Generalisations of the Task of Grouping

In many instances, such as: grouping of documents, social network analysis, or bioinformatics, particular objects may belong simultaneously to several groups. In other words, instead of partitioning the data set, we look for the covering of this set. One of the ways to deal with such situations is to apply the algorithms of fuzzy clustering, for instance the FCM algorithm from Sect. 3.3. Yet, in recent years, emphasis is being placed on the algorithms allowing not only for an explicit reference to simultaneous assignment of objects to various groups, but also making it possible to identify the optimum coverings of the given set of objects. Their detailed consideration exceeds the assumed framework of this book. We shall only mention a couple of interesting solutions, encouraging the reader to an own study of this matter:

– Banerjee et al. presented in [49] MOC–*Model based Overlapping Clustering*, which can be seen as the first algorithm, which produces the optimum covering of the data set. The authors mentioned make use of the probabilistic relational model,[39] proposed for purposes of analysis of the microarrays. While the original solution concentrates on the normal distributions, MOC operates on arbitrary distributions from the exponential family. Besides, by introducing Bregman divergence, one can apply here any of the distances, discussed in Sect. 2.2.1. In the opinion of the respective authors, this new algorithm can be applied in document analysis, recommender systems, and in all situations, in which we deal with highly dimensional and sparse data.

---

[39]E. Segal, A. Battle, and D. Koller. Decomposing gene expression into cellular processes. In: *Proc. of the 8th Pacific Symp. on Biocomputing* (PSB), 2003, pp. 89–100.

- Cleuziou[40] proposed OKM–*Overlappingk-Means*, which is a generalisation of the k-means algorithm. This idea was then broadened to encompass the generalised k-medoids algorithm.
- In other approaches to the search for the coverings, graph theory and neural networks are being applied. In the case of the graph theory methods, first the similarity graph is constructed (analogously as in the spectral data analysis, considered in Sect. 5), and then all the cliques contained in it are sought.[41]

A survey of other algorithms is provided also in [391].

Grouping of objects in highly dimensional spaces on the basis of distances between these objects gives rise to problems discussed in Sect. 2.2.1.1. A classical solution consists in projecting the entire data set on a low dimensional space (by applying, for instance, multidimensional scaling, random projections, or principal component analysis) and using some selected algorithm to cluster the thus obtained transformed data. In practice, though, it often turns out that various subsets of data ("clusters") may be located in different subspaces.

That is why the task of grouping is defined somewhat differently. Namely, such a breakdown $C_1, \ldots, C_k$ of the data set is sought, along with the corresponding subsets of features, $F_i \subset \{1, \ldots, n\}$, that points assigned to C_j be sufficiently close one to another in the F_j dimensional space. It can be said that C_j is composed of points which, after projection on the F_j dimensional space, constitute a separate cluster. A survey of methods meant to solve the thus formulated task is provided in [304, 386].

2.4.5   Relationship Between Partitional and Hierarchical Clustering

Similarly as for hierarchical clustering, we may consider a theoretical partitioning of a probability distribution underlying the actual samples. But contrary to the concept of clustering three (see Definition 2.3.2), we need a more elaborate concept of a clustering. It is generally agreed that a clustering can be viewed as a cut of the clustering tree that is a subset of the clusters of clustering tree (that is for each cluster of the cluster tree there exists either a subset or superset of it as a cluster of our clustering) such that no two clusters intersect. This particular view has been adopted e.g. in the HDBSCAN algorithm in which the dendrogram is the starting point of flat partitioning the data set.[42]

---

[40] G. Cleuziou. A generalization of k-means for overlapping clustering. Université d'Orléans, LIFO, Rapport No RR-2007-15.

[41] See, e.g., W. Didimo, F. Giordano, G. Liotta. Overlapping cluster planarity. In: Proc. APVIS 2007, pp. 73–80, M. Fellows, J. Guo, C. Komusiewicz, R. Niedermeier, J. Uhlmann. Graph-based data clustering with overlaps, *COCOON*, 2009, pp. 516–526.

[42] Also one can proceed in the reverse direction: One may use a partitional algorithm to create a hierarchical clustering of data, simply by applying recursively the partitional algorithm to clusters obtained previously, like in case of bi-sectional k-means algorithm in Sect. 3.1.5.2.

In particular, a clustering may be the set of clusters emerging for a given value of λ. This would correspond to a cut on the same level of the cluster tree. But for various practical reasons, also other criteria are used, either jointly or separately, like balanced size (in terms of enclosing tight ball), balanced probability mass, sufficient separation for a given sample size etc. Hence each cluster may be obtained for a different density threshold value λ_j. In this case the cluster tree would be cut at different levels at different branches.

Like in case of hierarchical clustering, upon obtaining a clustering, its consistency with the clustering tree, σ, ϵ-separation and other statistical properties with respect to the underlying distribution should be checked. Such checks may fail due to at least one of the following problems (1) no structure in the data exists, (2) no structure compatible with the class of structures sought by the applied algorithm exists, (3) the sample size is too small to reveal the underlying structure.

2.5 Other Methods of Cluster Analysis

Methods, which have been outlined in the two preceding sections belong to two essential streams developing within cluster analysis. In both cases a cluster is understood as a set of objects that are mutually much more similar than any two objects selected from two different clusters.

2.5.1 Relational Methods

It has been assumed till now that each object is represented by a vector of feature values. An alternative description is constituted by the relation of similarity or dissimilarity for the pairs of objects. We encounter such situation in social sciences or in management [138, 242]. By operating on the relation of similarity / dissimilarity we can conceal the values of the attributes, which may be of significance in such domains as, for instance, banking. It also is simpler to deal with mixed attribute types (both quantitative and qualitative). The sole problem to be solved at this level is the choice of the function measuring similarity / dissimilarity of objects.

The notion of "relational methods" is sometimes used in a wider context. Thus, for instance, a set of relations may be given, S_i, defined on different subsets $\mathfrak{X}_i$ of objects [27], or relations may have a more complex character. In particular, when grouping documents, it is worthwhile to consider relations between subject groups and keywords.

Graph-based methods mentioned below can be viewed as special cases of relational methods understood in this way. Notably, the discussion of graph clustering in Chap. 5 is not restricted to classical graphs alone, but addresses at least partially also weighted graphs that can well represent the problems of relational clustering.

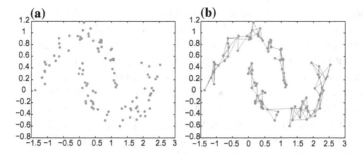

Fig. 2.7 Identification of clusters by examining mutual similarity between objects: **a** a set of points. In illustrations like this we follow the intuitive convention: The points are meant to lie in a plane, the horizontal axis reflects the first coordinate and the vertical one the second coordinate, **b** a graph obtained by joining five nearest neighbors of each node

2.5.2   Graph and Spectral Methods

The set X is often identified with the set of vertices of a certain graph Γ, whose edges represent the connections between the objects; e.g. the pair $\{x_i, x_j\}$ is an edge in Γ, if the two objects are similar in the degree not lower than s_τ. In such a context a cluster becomes equivalent to, for instance, a clique, that is, a connected subgraph of the graph Γ, i.e. such a one that every two vertices of the subgraph are connected by an edge. In another definition it is assumed that a cluster is such a subgraph Γ_i, whose vertices communicate exclusively among themselves, and do not communicate with other vertices, outside this subgraph. This means that when extracting clusters we take into account their connectivity and not only their compactness. This is illustrated on Fig. 2.7. Using pairwise distances between the points from the left panel we construct a graph shown on the right panel. Here, each node is linking with other five nearest nodes.

The progenitor of graph cut clustering is Zahn's[43] approach, consisting of two steps. First, using similarity matrix describing relationships among the objects, a maximum spanning tree is constructed, and then the edges with small weights are removed from this tree to get a set of connected components. This method is successful in detecting clearly separated clusters, but if the density of nodes is changed, its performance will deteriorate. Another disadvantage is that the cluster structure must be known in advance.

The understanding of clusters, as described above, places the problem of their extraction in the context of graph cutting. In particular, when we assign to every edge $\{v_i, v_j\}$ in the graph Γ a similarity degree s_{ij}, the problem boils down to removing the edges with small weights in order to decompose Γ into connected components. Such a procedure ensures that the sum of weights of the edges, connecting vertices from different groups is small in comparison with the sum of weights linking the vertices,

---

[43]C.T. Zahn, Graph-theoretic methods for detecting and describing gestalt clusters. *IEEE Trans. Comput.*, 20(1):68–86, 1971.

belonging to the same subgraph.[44] An exhaustive survey of techniques, which are applied in the partitioning of graphs, can be found e.g. in [179, 418].

A particularly important domain of application of this group of algorithms is parallel computing. Assume that solving a certain problem requires performing of m tasks, each of which constitutes a separate process, program or thread, realised by one of c processors. When the processors are identical, and all tasks are of similar complexity, then we can assign to every processor the same number of tasks. Usually, though, realisation of a concrete task requires the knowledge of partial results, produced by other tasks, which implies the necessity of communication between these tasks. In this manner we obtain a graph, with vertices corresponding to individual tasks, while edges—correspond to their communication needs. Communication between the processors, which form a parallel computing machine is much slower than data movement within one single processor. In order to minimise communication to a necessary level, the processes (vertices of the graph) ought to be partitioned into groups (i.e. assigned to processors) in such a way that the number of edges, linking different groups, be minimal. This is the formulation of a problem, whose solutions are considered, in particular, in [179].

Another group of problems is constituted by image segmentation. Here, an image is modelled as an undirected weighted graph, its vertices being pixels, or groups of pixels, while the weights of edges correspond to the degree of similarity (or dissimilarity) between the neighbouring pixels. This graph (image) is to be segmented conform to a certain criterion, defining "good" clusters.

An interesting offer, allowing for identification of complex cluster structures is spectral graph theory—its application in clustering is considered in detail in this book in Chap. 5. Various methods of spectral cluster analysis are reviewed in [176, 418, 482]. It is worth noting that spectral methods play nowadays a significant role in practical tasks of data analysis and mining, such as information retrieval [68], bioinformatics [251], or recommender systems [1, 311].

2.5.3   Relationship Between Clustering for Embedded and Relational Data Representations

Let us recall once again the cluster tree concept representing the dense and less dense regions of the data. If we sample objects from a space with various densities, then it is obvious that more objects will be clustered from dense regions and if we construct a similarity graph, then nodes corresponding to denser regions will be more strongly interconnected. Hence finding denser regions will correspond to cutting a graph in such a way as to remove the lowest number of edges (a kind of minimum cut). Due to these analogies, as we will see later in this book, various algorithmic ideas are adopted across the boundary of data representation.

---

[44]which means that objects, belonging to the same class are mutually sufficiently similar, while objects, belonging to different groups are sufficiently mutually dissimilar.

2.5.4 Density-Based Methods

In a different perspective, a cluster is conceived as a dense area in the space of objects, surrounded by low density areas. This kind of definition is convenient in situations, when clusters have irregular shapes, and the data set contains outliers and noisy observations—see Fig. 2.8. A typical representative of this group of algorithms is DBSCAN [171] and its later modifications [26, 170, 414].

Before presenting the essential idea of this algorithm, we shall introduce the necessary definitions.

(a) Let $d(\mathbf{x}, \mathbf{y})$ denote distance between any two points $\mathbf{x}, \mathbf{y} \in X$ and let $\epsilon > 0$ be a parameter. The ϵ-neighbourhood of the object $\mathbf{x}$ is the set

$$N_\epsilon(\mathbf{x}) = \{\mathbf{x}' \in X : d(\mathbf{x}, \mathbf{x}') \le \epsilon\}$$

(b) An object $\mathbf{x} \in X$ is called the internal point of a cluster, if its ϵ-neighbourhood contains at least $minPts$ objects, i.e. when $|N_\epsilon(\mathbf{x})| \ge minPts$, where $minPts$ is a parameter.
(c) An object $\mathbf{x} \in X$ is called a border point, if $|N_\epsilon(\mathbf{x})| < minPts$, but this neighbourhood contains at least one internal point.
(d) If $\mathbf{x} \in X$ is neither an internal point, nor a border point, then it is treated as a disturbance (*outlier*).

In construction of the particular clusters use is made of the notion of density-reachability. Namely, a point $\mathbf{y} \in X$ is *directly* density-reachable from the point $\mathbf{x} \in X$ if $\mathbf{y} \in N_\epsilon(\mathbf{x})$, and, besides, $\mathbf{x}$ is an internal point, that is, it is surrounded by a sufficiently high number of other points. Note that the relationship of direct density-reachability is asymmetric. Then, $\mathbf{y}$ is called density-reachable from the point $\mathbf{x}$ if there exists a sequence $\mathbf{x}_1, \ldots, \mathbf{x}_n$ of points such that $\mathbf{x}_1 = \mathbf{x}$, $\mathbf{x}_n = \mathbf{y}$, and each

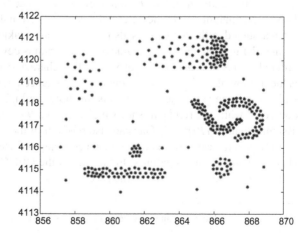

Fig. 2.8 An example of clusters with irregular shapes and various densities. The data set contains also *outliers*

point $\mathbf{x}_{i+1}$ is directly density-reachable from $\mathbf{x}_i$, $i = 1, \ldots, n-1$. Note that the thus defined relation of reachability is asymmetric ($\mathbf{y}$ may be a border point). That is why the notion of density-connectedness is introduced: points $\mathbf{x}, \mathbf{y} \in X$ are density-connected, if there exists such a point $\mathbf{z} \in X$ that both $\mathbf{x}$ and $\mathbf{y}$ are density-reachable from $\mathbf{z}$. Clusters, generated by the DBSCAN algorithm have the following properties:

(i) All points, belonging to a cluster are mutually density-connected.
(ii) If an internal point is density-connected with another point of a cluster, then it is also an element of this cluster.

Generation of clusters, having such properties, is outlined here through the pseudocode 2.4. Note that we make use here only of the internal and border points. Each two internal points, whose mutual distance does not exceed the value of ϵ are put in the same cluster. On the other hand, the border points are classified in any cluster, provided a neighbour of this point is an internal point of the cluster.

Algorithm 2.4 DBSCAN algorithm

Require: Data $X = (\mathbf{x}_1, \ldots, \mathbf{x}_m)^{\mathrm{T}}$, maximal distance between internal points ϵ.
Ensure: A partition $C = \{C_1, \ldots, C_k\}$, the number of clusters k.
1: *Initialisation.* Mark the points from the set X as internal, border or noisy points.
2: Remove the noisy points.
3: Connect with an edge the neighbouring internal points (i.e. the internal points situated at a distance not bigger than ϵ).
4: Form a cluster out of the neighbouring internal points.
5: Assign the border points to one of the clusters, upon which they neighbour.

Choice of the appropriate value of the radius ϵ influences significantly the results of the algorithm. If ϵ is too big—density of each point is identical and equal m (that is—the cardinality of the set X). If, however, the value of the radius is too small, then $|N_\epsilon(\mathbf{x})| = 1$ for any $\mathbf{x} \in X$. The value of the parameter ϵ is often assumed as the so-called k-distance, k-$dist(\mathbf{x})$, namely the distance between the point $\mathbf{x}$ and its k-th nearest neighbour. In the case of the two-dimensional sets the value of $k = 4$ is often assumed, although there is also a suggestion of taking $k = n + 1$.

In order to select the proper value of ϵ a diagram is developed of the increasingly ordered values of k-$dist(\mathbf{x})$ for all $\mathbf{x} \in X$. It is characterised by the appearance of a value, following which an abrupt increase of distance takes place. This is the value, which is chosen as ϵ. The procedure is illustrated in Fig. 2.9. Its left part shows the data set,[45] while the right part shows the diagram of k-$dist(\mathbf{x})$ for $k = 4$. It is usually assumed that $min Pts = k$. One can also read out of the diagram that $\epsilon \approx 1.0$.

In a general case, instead of k-distance, one can use a certain function $g(\mathbf{x})$, characterising density at point $\mathbf{x}$. In the case of the DENCLUE algorithm [252] this is the sum of the components of the function

---

[45]It arose from adding 14 randomly generated points to the set, described in Chap. 7, namely
`data3_2`.

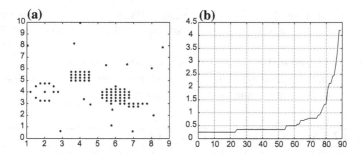

Fig. 2.9 Diagram of k-distances between the points of an exemplary data set $\mathfrak{X}$: **a** An exemplary data set composed of 90 points, **b** increasingly ordered values of the average distance between each of the 90 points of the data set $\mathfrak{X}$ and their $k = 4$ nearest neighbours

$$g(\mathbf{x}) = \sum_{\mathbf{x}' \in X} f(\mathbf{x}, \mathbf{x}')$$

where $f(\mathbf{x}, \mathbf{x}')$ is, in principle, an arbitrary function, which describes the influence exerted by the object $\mathbf{x}'$ on the object $\mathbf{x}$, e.g.

$$f(\mathbf{x}, \mathbf{x}') = \exp\left(-\frac{d(\mathbf{x}, \mathbf{x}')}{2\sigma^2}\right)$$

with $\sigma > 0$ being a parameter. In this context, clusters are described as local maxima of the function $g(\mathbf{x})$.

The fundamental qualities of the DBSCAN algorithm are as follows:

1. Knowledge of the number of clusters existing in the set X is not required.
2. Clusters may have arbitrary shapes. Owing to the parameter *minPts* the effect of a single connection is reduced, as manifested by the appearance of thin lines, composed of points, belonging to different clusters.
3. Data are allowed to contain noise and outliers.
4. The algorithm requires specification of just two parameters: the radius ϵ, and the number *minPts*, used in classification of points.
5. The algorithm is only slightly sensitive to the order, in which the individual objects from the set X are considered.

Yet, there is a significant requirement that clusters have similar densities. This shortcoming is done with in the variants of the algorithm—GDBSCAN [414] and LDB-SCAN [162]. Another essential weak point is the fact that the quality of the algorithm strongly depends upon the definition of distance $d(\mathbf{x}, \mathbf{y})$. We already know that, for instance, Euclidean distance loses on its usefulness in the case of analysis of highly dimensional data.

Moore [361] proposed the so-called anchor algorithm to analyse highly dimensional data—see the pseudocode 2.5. This algorithm belongs to the class of algorithms grouping on the basis of the minimum class diameter.

Algorithm 2.5 Anchor grouping algorithm, [361]

Require: Data set X, number of clusters K, minimum number of objects in a cluster n_{min}.
Ensure: A partition $\mathcal{C} = \{C_1, \ldots, C_K\}$.
1: Select randomly a point. Initiate $k = 1$, $k' = 0$. Take as the anchor a^1 the point from the set X which is most distant from the initial point.
2: Assign to the group C_k, initiated by the anchor a^k, these objects, which are closer to a^k than to other anchors. The objects form a list, ordered decreasingly with respect to their distance from the anchor.
3: Check, whether the anchor a^k has the sufficient number of objects assigned to it. If $|C_k| < n_{min}$, then $k' = k' + 1$.
4: Substitute $k = k + 1$. For a^k substitute the point that is most distant from the remaining anchors.
5: If $k - k' < K$, then go to step 2.
6: **return** partition of the set X into disjoint groups.

It should be noted that the algorithm may produce more than K groups. It may also happen that the algorithm shall not produce sets containing more than n_{min} elements.

Still another path for resolving DBSCAN shortcomings was the HDBSCAN algorithm [98]. It produces a flat set of clusters by first creating a hierarchy of clusters (like in the agglomerative methods) and then cutting this hierarchy at varying levels of hierarchy optimising some quality criteria. In this way clusters of varying density may be extracted. As a first step the minimum spanning tree is constructed where however the distances between elements are not the original ones but are modified. First a neighbourhood size parameter k is defined (like in robust single link on p. 33). Then the distances are updated in such a way that it is the maximum of the distance between elements, the original distance to its k-th closest neighbour of the first element and of the second element. When building the tree, a dendrogram is created as in agglomerative methods. Once the dendrogram is finished, it is "pruned" to mark as "non-clusters" those clusters that were too small. For this purpose the user has to define the minimum cluster size. All clusters that are smaller in the dendrogram, are marked as "non-clusters". Afterwards the persistence of clusters is computed. In the dendrogram, if a cluster A splits into a cluster B and a non-cluster C, then the cluster A is deemed to be the same as B. When the cluster identities are fixed in this way, the birth of a cluster is defined as the first occurrence of this cluster from the top of the dendrogram, and its death—the last occurrence. For each cluster λ_{birth} and λ_{death} are computed as inverses of the edge lengths added to dendrogram upon birth and death resp. Additionally, for each data point x that ever belonged to the cluster its $\lambda_{x,left}$ is computed as the inverse of the distance when it left the cluster. Cluster stability is defined as the sum of $(\lambda_{x,left} - \lambda_{birth})$ over all data points ever belonging to it. We first select all leaf nodes as "candidate flat clusters". Then working from the bottom of the tree, we check if the sum of stabilities of child clusters of a node is lower or equal the stability of a given node, then we deselect its children and select it as a candidate flat cluster. Otherwise we substitute the original stability of the node with the sum of stabilities of its child nodes. Upon reaching the top of the dendrogram the current candidate flat clusters become our actual set of clusters, being the result of HDBSCAN.

Summarizing, HDBSCAN has two parameters that need to be set by the user: the number of neighbours k and the minimal cluster size. They substitute the ϵ radius of the neighbourhood that needed to contain at least $minPts$ objects in DBSCAN. By replacing ϵ with k, the flexibility of having clusters of varying density is achieved. Setting of $minPts$ in DBSCAN was a bit artificial from the point of view of the user, as he had to estimate cluster density in advance, prior to looking into the data. Whereas the choice of minimal cluster size in advance is a bit easier because it may be derived from some business requirements and is easy to adjust upon look in the final clustering. HDBSCAN provides also with the insight into degree of obscureness of some points in the clusters, if one looks through the dendrogram and analyses the "non-clusters".

2.5.5 Grid-Based Clustering Algorithms

Grid-based approaches are recommended for large multidimensional spaces where clusters are regarded as regions that are denser than their surroundings. In this sense they can be deemed as an extension of density-based approaches.

The grid based methods handle the complexity of huge data sets by not dealing with the data points themselves but rather with the value space surrounding data points. A typical grid-based clustering algorithm proceeds as follows [210]

1. Creating the grid, that is partitioning the data space into a finite number of cells
2. Calculating the cell density for each cell
3. Sorting of the cells according to their densities.
4. Identifying cluster centres.
5. Merging information from neighbouring cells.

One example of such an algorithm, STatistical INformation Grid-based clustering method (STING), was proposed by Wang et al. [496] for clustering spatial databases. The goal of the algorithm was to enable spatial queries in two-dimensional space. The data is organised hierarchically: four lower level cells are combined into one higher level cell.

2.5.6   Model-Based Clustering

Model-based clustering assumes that a parametric model is underlying the cluster structure of the data. Typical approach here is to consider data as being generated by a mixture of Gaussian models. Clustering is reduced to identifying these models by detecting the parameters, e.g. mean, standard deviations, covariance matrices of the elements of the mixture etc.

2.5.7   Potential (Kernel) Function Methods

These methods originate from the work of Ajzerman, Braverman and Rozonoer [12], where these authors used the term of potential function. A short introduction to this approach can be found in Sect. 5.6 of the monograph [463]. Along with the development of the theory of support vector machines (SVM) [477] the term "kernel function" replaced the earlier term of the "potential function". We present below, following [176], the fundamental assumptions of cluster analysis methods using the notion of kernel function.

We assume (for simplicity and in view of practical applications), that we deal with n-dimensional vectors having real-valued components (and not the complex numbers, as this is assumed in the general theory). Hence, as until now, $X = \{\mathbf{x}_1, \dots, \mathbf{x}_m\}$ denotes the non-empty set of objects, with $\mathbf{x}_i \in \mathbb{R}^n$.

Definition 2.5.1 A function $\mathsf{K} \colon X \times X \to \mathbb{R}$ is called the positive definite kernel function (Mercer kernel or simply kernel) if: (i) $\mathsf{K}(\mathbf{x}_i, \mathbf{x}_j)$ is a symmetric function, and (ii) for any vectors $\mathbf{x}_i, \mathbf{x}_j \in X$ and any real-valued constants $c_1, \dots, c_m$ the following inequality holds: [46]

---

[46]An introduction of Kernel functions should start with definition of a *vector space*. Let V be a set, $\oplus \colon V \times V \to V$ be so-called inner operator (or vector addition) and $\odot \colon \mathbb{R} \times V \to V$ be so-called outer operator (or scalar vector multiplication). Then $(V, \oplus, \odot)$ is called vector space over the real numbers if the following properties hold: $u \oplus (v \oplus w) = (u \oplus v) \oplus w$, there exists $0_v \in V$ such that $v \oplus 0_V = 0_V \oplus v = v$, $v \oplus u = u \oplus v$, $\alpha \odot (u \oplus v) = (\alpha \odot u) \oplus (\alpha \odot v)$, $(\alpha + \beta) \odot v = (\alpha \odot v) \oplus (\beta \odot v)$, $(\alpha \cdot \beta) \odot v = \alpha \odot (\beta \odot v)$, $1 \odot v = v$.

Given the vector space, one can define the inner product space as a vector space V over real numbers in which the scalar product $\langle \cdot, \cdot \rangle \colon V \times V \to \mathbb{R}$ $\langle v, v \rangle \geq 0$, $\langle v, v \rangle = 0 \Leftrightarrow v = 0$, $\langle u, v \rangle = \langle v, u \rangle$, $\langle u, \lambda \odot v \rangle = \lambda \cdot \langle u, v \rangle$, $\langle u, v + w \rangle = \langle u, v \rangle + \langle u, w \rangle$.

Now let X be a the space of our input data. A mapping $K \colon X \times X \to \mathbb{R}$ is called a *kernel*, if there exists an inner product space $(F, \langle \cdot, \cdot \rangle)$ (F being called the feature space) and a mapping $\Phi \colon X \to F$ such that in this inner product space $K(x, y) = \langle \Phi(x), \Phi(y) \rangle$ for all $x, y \in X$. As for inner product $\langle \Phi(x), \Phi(y) \rangle = \langle \Phi(y), \Phi(x) \rangle$ holds, obviously $K(x, y) = K(y, x)$.

Mercel has shown that if $\int_{x \in X} \int_{y \in X} K^2(x, y) dx dy < +\infty$ (compactness of K) and for each function $f \colon X \to \mathbb{R}$ $\int_{x \in X} \int_{y \in X} K(x, y) f(x) f(y) dx dy \geq 0$ (semipositive-definiteness of K) then there exists a sequence of non-negative real numbers (eigenvalues) $\lambda_1, \lambda_2, \dots$ and a sequence of functions $\phi_1, \phi_2, \dots \colon X \to \mathbb{R}$ such that $K(x, y) = \sum_{i=1}^{\infty} \lambda_i \phi_i(x) \phi_i(y)$, where the sum on the right side is absolute convergent. Moreover $\int_{x \in X} \phi_i(x) \phi_j(x) dx$ is equal 1 if $i = j$ and equal 0 otherwise.

$$\sum_{i=1}^{m}\sum_{j=1}^{m} c_i c_j \mathsf{K}(\mathbf{x}_i, \mathbf{x}_j) \geq 0$$

□

If $\mathsf{K}_1(\mathbf{x}, \mathbf{y})$, $\mathsf{K}_2(\mathbf{x}, \mathbf{y})$ are kernel functions, then their sum, product, and $a\mathsf{K}(\mathbf{x}, \mathbf{y})$, where $a > 0$, are also kernel functions. The typical kernel functions, which are used in machine learning are:

(a) Linear kernel $\mathsf{K}_l(\mathbf{x}, \mathbf{y}) = \mathbf{x}^T\mathbf{y} + c$. In the majority of cases, the algorithms, which use linear kernel functions, are close to equivalence with their "non-kernel" counterparts (thus, e.g., the kernel-based variant of the principal component analysis with the linear kernel is equivalent to the classical PCA algorithm).

(b) Polynomial kernel $\mathsf{K}(\mathbf{x}, \mathbf{y}) = (\alpha\mathbf{x}^T\mathbf{y} + c)^d$, where α, c and the degree of the polynomial, d, are parameters. The polynomial kernel functions are applied, first of all, in the situations, in which normalised data are used.

(c) Gaussian kernel $\mathsf{K}(\mathbf{x}, \mathbf{y}) = \exp\left(-\|\mathbf{x} - \mathbf{y}\|^2/(2\sigma^2)\right)$, where $\sigma > 0$ is a parameter, whose choice requires special care. If its value is overestimated, the exponent will behave almost linearly, and the very nonlinear projection will lose its properties. In the case of underestimation, function K loses the regularisation capacities and the borders of the decision area become sensitive to the noisy data. The Gaussian kernel functions belong among the so-called radial basis functions of the form

$$\mathsf{K}(\mathbf{x}, \mathbf{y}) = \exp\left(-\frac{\sum_{j=1}^{n} |x_j^a - y_j^a|^b}{2\sigma^2}\right), \ b \leq 2$$

The strong point of the Gaussian kernel is that (for a correctly chosen value of the parameter σ) it filters out effectively the noisy data and the outliers.

Every Mercer kernel function can be represented as a scalar product

$$\mathsf{K}(\mathbf{x}_i, \mathbf{x}_j) = \Phi(\mathbf{x}_i)^T\Phi(\mathbf{x}_j) \tag{2.49}$$

where $\Phi: X \to \mathcal{F}$ is a nonlinear mapping of the space of objects into a highly dimensional space of features $\mathcal{F}$. An important consequence of this representation is the possibility of calculating the Euclidean distance in the space $\mathcal{F}$ without knowledge of the explicit form of the function Φ. In fact,

---

(Footnote 46 continued)
Obviously, the function ϕ may be of the form of an infinite vector $\Phi = (\sqrt{\lambda_1}\phi_1, \sqrt{\lambda_2}\phi_2, \ldots)$.

The above kernel definition constitutes a special application of this general formula to the case of a finite set X. The function K over a finite set X can be represented in such a case as a matrix, which must be therefore semipositive definite and the function ϕ can be expressed for each $x \in X$ as the vector of corresponding eigenvector components multiplied with square root of the respective eigenvalues.

$$\|\Phi(\mathbf{x}_i) - \Phi(\mathbf{x}_j)\|^2 = \left(\Phi(\mathbf{x}_i) - \Phi(\mathbf{x}_j)\right)^{\mathrm{T}}\left(\Phi(\mathbf{x}_i) - \Phi(\mathbf{x}_j)\right)$$
$$= \Phi(\mathbf{x}_i)^{\mathrm{T}}\Phi(\mathbf{x}_i) + \Phi(\mathbf{x}_j)^{\mathrm{T}}\Phi(\mathbf{x}_j) - 2\Phi(\mathbf{x}_i)^{\mathrm{T}}\Phi(\mathbf{x}_j) \quad (2.50)$$
$$= K(\mathbf{x}_i, \mathbf{x}_i) + K(\mathbf{x}_j, \mathbf{x}_j) - 2K(\mathbf{x}_i, \mathbf{x}_j)$$

In view of the finiteness of the set X it is convenient to form a matrix K having elements $k_{ij} = K(\mathbf{x}_i, \mathbf{x}_j)$. Since $k_{ij} = \Phi(\mathbf{x}_i)^{\mathrm{T}}\Phi(\mathbf{x}_j)$, then, from the formal point of view, K is a Gram matrix, see the Definition B.2.3. Given this notation, we can write down the last equality in the form

$$\|\Phi(\mathbf{x}_i) - \Phi(\mathbf{x}_j)\|^2 = k_{ii} + k_{jj} - 2k_{ij} \quad (2.51)$$

The kernel functions are being used in cluster analysis in three ways, referred to through the following terms, see [176, 515]:

(a) kernelisation of the metrics,
(b) clustering in feature space $\mathcal{F}$,
(c) description via support vectors.

In the first case we look for the prototypes in the space X, but the distance between objects and prototypes is calculated in the space of features, with the use of Eq. (2.50). The counterpart to the criterion function (2.32) is now constituted by

$$J_1^{\Phi} = \sum_{j=1}^{k}\sum_{i=1}^{m} u_{ij}\|\Phi(\mathbf{x}_i) - \Phi(\boldsymbol{\mu}_j)\|^2$$
$$= \sum_{j=1}^{k}\sum_{i=1}^{m} u_{ij}\left(K(\mathbf{x}_i, \mathbf{x}_i) + K(\boldsymbol{\mu}_j, \boldsymbol{\mu}_j) - 2K(\mathbf{x}_i, \boldsymbol{\mu}_j)\right) \quad (2.52)$$

If, in addition, $K(\mathbf{x}_i, \mathbf{x}_i) = 1$, e.g. K is a Gaussian kernel, the above function simplifies to the form

$$J_1^{\Phi} = 2\sum_{j=1}^{k}\sum_{i=1}^{m} u_{ij}\left(1 - K(\mathbf{x}_i, \boldsymbol{\mu}_j)\right) \quad (2.53)$$

In effect, in this case the function $d(\mathbf{x}, \mathbf{y}) = \sqrt{1 - K(\mathbf{x}, \mathbf{y})}$ is a distance, and if, in addition, K is a Gaussian kernel, then $d(\mathbf{x}, \mathbf{y}) \to \|\mathbf{x} - \mathbf{y})\|$ when $\sigma \to \infty$.

An example of such an algorithm is considered in deeper details in Sect. 3.3.5.6. The idea of calculating distances in the space of features was also made use of in the kernelised and effective algorithm of hierarchical grouping, as well as in the kernelised version of the mountain algorithm,[47] see also [288].

---

[47]The mountain algorithm is a fast algorithm for determining approximate locations of centroids. See R.R. Yager and D.P. Filev. Approximate clustering via the mountain method. *IEEE Trans. on Systems, Man and Cybernetics*, 24(1994), 1279–1284.

In the second case we operate with the images $\Phi(\mathbf{x}_i)$ of the objects and we look for the prototypes $\boldsymbol{\mu}_j^\Phi$ in the space of features. The criterion function (2.32) takes now on the form

$$J_2^\Phi = \sum_{j=1}^{k} \sum_{i=1}^{m} u_{ij} \| \Phi(\mathbf{x}_i) - \boldsymbol{\mu}_j^\Phi \|^2 \tag{2.54}$$

where $\boldsymbol{\mu}_j^\Phi \in \mathcal{F}$. In Sect. 3.1.5.5 we show how this concept is applied to the classical k-means algorithm, and in Sect. 3.3.5.6.2.—to the k-fuzzy-means algorithm (FCM).

Finally, the description based on the support vectors refers to the single-class variant of the support vector machine (SVM), making it possible to find in the space of features the sphere of minimum radius, containing *almost* all data, that is—the data with exclusion of the *outliers* [63]. By denoting the centre of the sphere with the symbol $\mathbf{v}$, and its radius with the symbol R, we obtain the constraint of the form

$$\| \Phi(\mathbf{x}_i) - \mathbf{v} \|^2 \le R^2 + \xi_i, \ i = 1, \ldots, m \tag{2.55}$$

where ξ_i are artificial variables. More extensive treatment of this subject is presented in Sect. 3.5 of the book [176].

The basic characteristics of the kernel-based clustering algorithms are as follows:

(a) They enable formation and description of clusters having shapes different from spherical or ellipsoidal.
(b) They are well adapted to analysing the incomplete data and the data containing *outliers* as well as disturbances (noise).
(c) Their shortcoming consists in the necessity of estimating additional parameters, e.g. the value of σ in the case of the Gaussian kernel.

Even though the characteristic (a) sounds highly encouraging, it turns out that the classical partitional algorithms from Sect. 2.4 may also be applied in such situations. We deal with this subject at greater length below.

2.5.8 Cluster Ensembles

Similarly as in machine learning, where the so-called families of classifiers are used in classification (see Sects. 4.5 and 4.6 in [303]), in data grouping attempts are made to enhance the effectiveness of grouping by applying the families of groupings (*cluster ensembles* [257]). This kind of approach is also referred to as aggregation of clusterings or consensus partitioning . The data set X is analysed from various points of view, and the conclusions, resulting therefrom, are used in the construction of the final partition. As noted by Strehl and Ghosh [447] it is, in a way, the problem of the so-called *knowledge reuse*, with which we deal in, for instance, marketing or banking. Thus, for instance, a company disposes of various profiles, describing

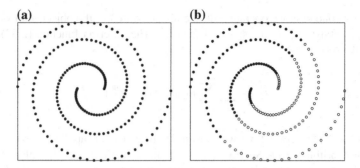

Fig. 2.10 a The set 2spirals is composed of two spirals, situated one inside the other. **b** Grouping produced by the k-means algorithm

the behaviour of customers in terms of demographic and geographical aspects, the history of purchases done, etc. Aggregation of such descriptions allows for formulating of composite judgements, which support the design of effective trade strategies, addressed at well selected groups of customers. It is essential that in formulation of such judgements the entire analysis does not have to be repeated from scratch, but knowledge, originating from various sources is used and creatively processed.

Example 2.5.1 Consider a simple problem, represented by the data set,[48] which is shown in Fig. 2.10a. Data points are here located along two spirals, of which one is situated inside the other one. The classical k-means algorithm produces the output, which is shown in Fig. 2.10b.

In order to obtain the correct partition, Fred and Jain [187] ran N times the k-means algorithm, assuming a different number of classes at each time. They aggregated the partial results, establishing the so-called co-association matrix, composed of the elements $w_{ij} = r_{ij}/N$, where r_{ij} is the number of cases, in which the pair of objects (i, j) was assigned to the same class. In order to determine the ultimate partition on the basis of this new matrix, the single link hierarchical algorithm was used, i.e. variant (a) from Sect. 2.3.   □

The above way of proceeding can be formalised as follows: Let $\mathfrak{C} = \{\mathcal{C}^1, \ldots, \mathcal{C}^N\}$ be a family of partitions of the data set X, with $\mathcal{C}^i = \{C_1^i, \ldots, C_{k_i}^i\}$, $i = 1, \ldots, N$, where k_i is the number of groups, proposed in the i-th partition. The problem consists in finding such a partition C^* of the set X, which has the following properties [188]

(i) Conformity with the family of partitions $\mathfrak{C}$, i.e. the partition C^* ought to reflect the essential features of each partition $C^i \in \mathfrak{C}$.

(ii) Robustness with respect to small disturbances in $\mathfrak{C}$, namely the number of clusters and their content ought not undergo drastic changes under the influence of slight disturbances of the partitions, forming the family $\mathfrak{C}$.

---

[48]Information on this data set is provided in Chap. 7.

(iii) Conformity with the additional information on the elements of the set X, provided this information is available. Thus, e.g., if the assignment of (all or some) objects to classes is known, the partition C^* ought to be to the maximum degree in agreement with this assignment.

To measure the degree of agreement between the partitions, forming the family $\mathfrak{C}$, Fred and Jain applied in [188], similarly, anyway, as Strehl and Ghosh in [447], the normalised measure of mutual information $NMI(C^\alpha, C^\beta)$, where C^α, C^β denote the partitions compared. The form and the properties of this measure are considered in detail in Sect. 4.4.3. We only note here that $NMI(C^\alpha, C^\beta)$ is a number from the interval [0, 1].

The degree of agreement of the partition C^* with the partitions from the family $\mathfrak{C}$ is calculated as

$$NMI(C^*, \mathfrak{C}) = \frac{1}{N} \sum_{i=1}^{N} NMI(C^*, C^i) \tag{2.56}$$

Let, further on, $\mathfrak{P}(k) = \{\mathcal{P}_1(k), \ldots, \mathcal{P}_{\vartheta(m,k)}(k)\}$ denote all the possible partitions of the set X into k disjoint classes. It should be remembered that $\vartheta(m, k)$ is the number, defined by the Eq. (2.23), of all the possible partitions of an m-element set X into k disjoint classes. Hence, as C^* we can take the partition

$$C^* = \underset{1 \leq and \leq \vartheta(m,k)}{\arg \max} \ NMI(\mathcal{P}_i(k), \mathfrak{C}) \tag{2.57}$$

This partition satisfies the first of the postulates, formulated before. In order to account for the requirements of flexibility, the authors quoted here form with a bootstrap method an M-element family of partitions $\mathbb{B} = \{\mathfrak{B}_1, \ldots, \mathfrak{B}_M\}$, randomly assigning objects from the set X, with repetitions, to the appropriate sets from the family $\mathfrak{C}$. A more extensive treatment of the subject, along with a description of the performed experiments, is provided in [188, 189].

Alternative methods of aggregating multiple partitions of the data set are considered by Strehl and Ghosh [447]. Hore, Hall and Goldgof [257] formulate the procedure that ensures scalability of aggregation of partitions (represented by the gravity centres of classes), corresponding to a sparse data set. These latter authors consider two situations: (a) in each portion of data the same number of classes is distinguished, or (b) the numbers of classes are different. In the first case these authors obtain the so-called BM (*Bipartire Merger*) algorithm, and in the second case—the MM (*Metis Merger*) algorithm. An additional strong point of this work is the rich bibliography, concerning the families of partitions.

Then, Thangavel and Visalakshi [458] describe the application of the families of partitions in the k-harmonic means algorithm.[49] Finally, Kuncheva and Vetrov wonder in [313] whether the outputs from cluster ensembles are more stable than partitions obtained from the a single clustering algorithm. They understand stability as

---

[49]This algorithm is presented in Sect. 3.1.5.4 of this book.

sensitivity (or, more precisely, lack of sensitivity) with respect to small disturbances in the data or in the parameters of the grouping algorithm. The considerations therein allowed for the formulation of certain recommendations, concerning the selection of the number of clusters, resulting from the analysed family of partitions.

Let us mention, at the end, one more method of aggregating the partial groupings, which is used in microarray analysis.

Example 2.5.2 One of the most popular applications of clustering in bioinformatics is microarray analysis. Suppose we treat X as a matrix with rows corresponding to genes, and columns—to experiments or samples. The value of x_{ij} corresponds to the level of expression of the i-th gene in the sample (experiment) j. In typical applications from this domain the matrix X is exceptionally "slender": it has thousands of rows and not more than 100 columns.

The study [92] presents the problem of formation of meta-genes, being the linear combinations of n genes. In solving this problem, the non-negative factorisation of matrices is used, mentioned here in Sect. 2.4.2.1, i.e. finding of such matrices W and H, whose product WH^{T} is an approximation of the matrix X. In this concrete case columns of the matrix W correspond to meta-genes (or to diagnostic classes), while the number w_{ij} defines the value of the coefficient of contribution from the i-th gene in the j-th meta-gene. Then, the elements h_{ij} of the matrix H indicate the levels of expression of the meta-gene j in sample i. Matrix H is made use of for grouping of samples: i-th sample is assigned to this meta-gene j^*, which corresponds to the maximum value of h_{ij}.

In the general case, by performing the decomposition of the matrix X many times over, we obtain the set of matrices (W^t, H^t), where $t = 1, \ldots, t_{max}$ denotes the successive number of the NMF decomposition, while t_{max} is the total number of the decompositions performed.

In the study, reported in [92], the following manner of aggregating the partial results was applied. Let C^t denote the concordance matrix, obtained in experiment t, having dimensions $m \times m$. Its element $c_{ij}^t = 1$ if genes i and j belong to the same class, and $c_{ij}^t = 0$ in the opposite case. Let, further, $\overline{C} = (C_1 + \cdots + C_{t_{max}})/t_{max}$ be the aggregate (averaged) concordance matrix. The numbers $\overline{c}_{ij}$ can be treated as the degrees of similarity of the gene pairs (i, j). By turning similarities into distances, that is—by forming the matrix D, having elements $d_{ij} = 1 - \overline{c}_{ij}$, we construct the dendrogram and calculate the cophenetic correlation coefficient (see Example 2.3.1 in p. 31), indicating the degree of agreement between the distances, contained in matrix D, and the distances, resulting from the dendrogram developed. If the results of grouping, obtained in each run of the algorithm are similar (meaning that the grouping obtained has a stable character), then the elements of matrix $\overline{C}$ (and of matrix D) will have values close to 0 or 1, and the calculated correlation coefficient will be close to 1. In case, when the data set analysed does not represent a clear k-group structure, the correlation coefficient shall have value well below 1. In addition, the resulting dendrogram serves in the ordering of column and rows of the concordance matrix. The use is made here of the order, in which the leaves of the dendrogram are marked. In the left hand part of Fig. 2.11 the unordered matrices of concordance,

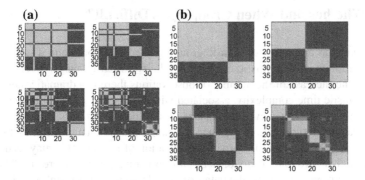

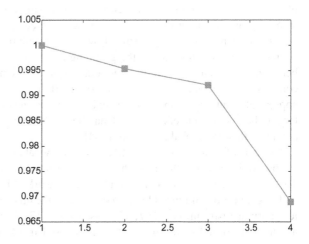

Fig. 2.11 Unordered (**a**) and ordered (**b**) concordance matrices, generated with the use of the method, described in the text. Experiments were carried out for $k = 2, 3, 4, 5$ groups

Fig. 2.12 Values of the cophenetic correlation coefficient for the partitions obtained. The horizontal axis represents the number of groups, the vertical axis shows the cophenetic correlation coefficient

$\overline{C}$, are shown, as obtained for $k = 2, 3, 4, 5$, while in the right-had part—the same matrices with the appropriately ordered rows and columns. Matrix X represents, in this case, the levels of expression of 5000 genes, registered in 38 samples, taken from the bone marrow.[50] Conform to the claim from the authors of [92], the method here outlined allows for a precise distinction between two types of leukaemia (myeloid and lymphoblastic leukaemia), this being indicated in the upper matrix in part (b) of Fig. 2.11. A reader, interested in the interpretation of the remaining figures is kindly referred to the publication [92].

It should be noted that the quality of partitions, as measured by the cophenetic correlation coefficient, decreases with the number of groups—see Fig. 2.12. This corresponds to the existence of less distinct structures in the data set, so that the algorithm is not capable of determining them sufficiently precisely. □

---

[50]The data, as well as the MATLAB code, are available at http://www.broadinstitute.org/cgi-bin/cancer/publications/pub_paper.cgi?mode=view\&paper_id=89.

2.6   Whether and When Grouping Is Difficult?

When treated as an optimisation task, grouping is a "hard" problem, if we consider it in the context of pessimistic complexity (the worst case analysis). This means that with the increase of the number of observations there is a dramatic increase in the pessimistic time complexity, associated with finding the global optimum of the criterion function. If, however, the data represent the true clusters, and the number of observations is sufficiently high, then use can be made of a number of methods of local search, allowing for the correct identification of groups. This may lead to the conviction that "grouping is not difficult; it is either easy or not interesting" [441].

It turns out, in fact, that if the data originate from a (well separable) mixture of normal distributions, and the number of observations is sufficiently high, then the task of grouping is easy. There exists an algorithm, having polynomial time complexity, which identifies—with high probability—the correct division into groups. In particular, this algorithm locates sufficiently precisely (with an assumed error) the gravity centres of groups. This allows for formulating the upper bound on the computational conditioning of the grouping algorithm, that is—the minimum gap between groups and the minimum number of observations in the sample, ensuring correct classification. So, for instance, for an arbitrary Gaussian mixture, if only an appropriate number of observations exist, then the maximum likelihood estimates tend to the true parameters, provided that the local maxima of the likelihood function are lower than the global maximum [404].

One can, of course—and, in fact, should—ask what the "sufficiently high" number of observations means, that is—what is the information limit for the task of grouping. This issue is discussed also in the already cited work [441].

Dasgupta and Schulman [134] concentrate on the mixture of n-dimensional spherical normal distributions $N(\boldsymbol{\mu}, \sigma \mathbb{I}_n)$—see Fig. 2.13. The data, which come from a spherical normal distribution, can be enclosed inside the hypersphere with the

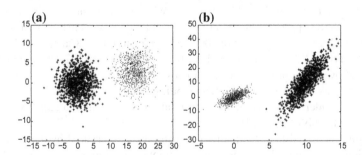

Fig. 2.13 Exemplary data having: **a** spherical and **b** non-spherical normal distribution

radius[51] $r = \sigma \sqrt{n}$. Therefore, it can be assumed that the data, originating from two distributions, $N(\boldsymbol{\mu}_1, \sigma_1 \mathbb{I}_n)$ and $N(\boldsymbol{\mu}_2, \sigma_2 \mathbb{I}_n)$ are c-separable, if [134]

$$\|\boldsymbol{\mu}_1 - \boldsymbol{\mu}_2\| \geq c \max(\sigma_1, \sigma_2)\sqrt{n} = c \max(r_1, r_2) \tag{2.58}$$

We have endowed, in this manner, the notion of separability of distributions, with a precise meaning.[52] In particular, when we deal with anisotropic multi-normal distributions $N(\boldsymbol{\mu}, \Sigma)$, where Σ denotes the covariance matrix, then $r = \sqrt{tr(\Sigma)}$. In the case of a mixture of k distributions, we denote by c_{ij} the separability of the i-th and jth distribution, and by $c = \min_{i \neq j} c_{ij}$ the separability of the mixture. So, e.g., the 2-separable mixture of distributions represents almost exclusively the disjoint clusters of n dimensional points, and with the increasing n lower and lower value of c is required to secure the disjointness of clusters, this being due to the specific properties of the Euclidean distance (see Sect. 2.2).

If the mixture of k spherical normal distributions is sufficiently separable (at the order of $\Omega(n^{1/4})$), and the sample contains $O(k)$ observations, then it suffices to perform two iterations of the EM algorithm.[53] Further advance is achieved by the application of the spectral projection methods (considered in Chap. 5). Vempala and Wang [479] show that if c is a constant of the order $\Omega(n^{1/4} \log^{1/4} nk)$ and the dimension of the sample is of the order $\Omega(n^3 k^2 \log ckn/\delta)$, then the k-dimensional spectral projection allows for identification of the group centres with probability $1 - \delta$.

Kanungo et al. adopt, as the measure of separation of clusters, in [281], the quotient

$$sep = \frac{r_{min}}{\sigma_{max}} \tag{2.59}$$

where r_{min} is half of distance between the closest centres of classes, while σ_{max} denotes the maximum of the standard deviations, characterising clusters. The authors quoted show, see Theorem 1 in [281], that if the class centres are sufficiently close to the gravity centres of clusters, then as the value of sep increases, the time of execution of the appropriately implemented k-means algorithm improves.

A similar conclusion was formulated by Zhang in the report [531]. Call *clusterability* a measure characterising the partition $\mathcal{C}$ of the set X. For a given partition $\mathcal{C} = \{X_1, \dots, X_k\}$ it is, for instance, possible to define the within-group variance, $W_\mathcal{C}(X) = \sum_{j=1}^{k} p_j \sigma^2(X_i)$, and the between-group variance, $B_\mathcal{C}(X) = \sum_{j=1}^{k} p_j \|\boldsymbol{\mu}_j - \boldsymbol{\mu}\|$, where $p_i = |X_i|/m$, $\boldsymbol{\mu}_j$ is the gravity centre of the j-th group,

---

[51] It is, actually, the approximate average length of the random vector, having exactly this distribution. If $\mathbf{x}$ is a vector, having as coordinates random numbers distributed according to $N(0, \sigma)$, then its expected length is $\mathbb{E}(\|\mathbf{x}\|) = \sigma\sqrt{2}\Gamma((n+1)/2)/\Gamma(n/2)$. In an approximation, $\mathbb{E}(\|\mathbf{x}\|) \approx \sigma\sqrt{2}[1 - 1/(4n) + 1/(21n^2)]$, and so $\mathbb{E}(\|\mathbf{x}\|) \to \sigma\sqrt{2}$ when $n \to \infty$.

[52] Recall, that on p. 12 we have listed a number of requirements for the separation notion in general, and on pp. 35, 28, and 33 we have already mentioned a couple of other separation measures, intended for more general types of probability distributions.

[53] This algorithm is commented upon in Sect. 3.2.

and μ is the gravity centre of the entire set X. Then, the measure of clusterability is the quotient [531]

$$C(X) = \max_{C \in \mathfrak{C}} \frac{B_C(X)}{W_C(X)}$$

where $\mathfrak{C}$ is the set of all possible clusterings of the set X. The higher the value of $C(X)$, the more separate are individual groups. One of the results, presented in the report quoted, proposes that the higher the clusterability (corresponding to the existence of natural clusters), the easier it is to find the appropriate partition [2].

For other notions of clusterability, less dependent on a particular form of probability distribution, see Definition D.1.5 and its subsequent discussion on p. 348.

Last not least, when choosing an algorithm and evaluating its results, we shall keep in mind the ultimate goal—the consumption of clustering results. If we seek a way to create a taxonomy, a catalogue of e.g. a large collection of products, we would apply a hierarchical clustering algorithm and test how well the hierarchy represents distances between products. If we look for covering an area with points of sales, we will apply a partitioning algorithm and verify whether or not our consumers have a reasonable distance to the closest POS and weight the loss of consumers against our investment budget. If we want to develop effective marketing strategies, we will probably seek well separated clusters.

Let us terminate this section with the following statement. The assessment of quality of a concrete tool, in this case—an algorithm of grouping—remains, actually, in the competence of the person, using the tool. Thus, instead of looking for the "best" tool, one should rather consider the available algorithms in the categories of their *complementarity*: the capacity of compensating for the weak points of one algorithm with the qualities of the other, or the capacity of strengthening the positive qualities of an algorithm by some other one.

Chapter 3
Algorithms of Combinatorial Cluster Analysis

Abstract While Chap. 2 presented the broadness of the spectrum of clustering methods, this chapter focusses on clustering of objects that are embedded in some metric space and gives an insight into the potential variations of clustering algorithms driven by peculiarities of data at hand. It concentrates essentially around k-means clustering algorithm and its variants, which serve as an illustration on the kinds of issues the algorithms need to accommodate to. After presenting the basic variants of k-means, implementational issues are discussed, including initialisation and efficiency (e.g. k-means++). Then accommodations to special forms of data, e.g. text data (spherical k-means), or data with clusters separated by complicated boundary lines (kernel k-means), or ones residing in non-Euclidean space (k-medoids, k-modes) are presented. Subsequently non-sharp separation areas between clusters are discussed. First the EM algorithm is presented, then fuzzy k-means, with fuzzified variants of the aforementioned algorithms. k-means application presumes apriorical knowledge of the number of clusters. To overcome this limitation, so-called affinity propagation algorithm is presented. Finally approaches to handling data of non-typical relation between the number of objects and the number of dimensions, that is residing (predominantly) in a narrow section of the feature space are discussed (projective clustering, random projection, k q-flats, manifold clustering), or with too large data sets (subsampling), or clustering changing over time. The latter issue is closely connected to clustering under partial pre-labelling of objects. Also methods for simultaneous clustering of both objects and their features (co-clustering) are briefly outlined.

Combinatorial cluster analysis may be viewed as a method to investigate (in a heuristic way) the space of all possible arrangements of data into groups and picking up the one that fits an optimised target function. It differs for example from graph data analysis where there exist a priori ties between objects (links in the graph) that one seeks to preserve during clustering.

Let us have a look at basic algorithms of combinatorial cluster construction via target function optimisation.

© Springer International Publishing AG 2018   67
S. T. Wierzchoń and M. A. Kłopotek, *Modern Algorithms of Cluster Analysis*,
Studies in Big Data 34, https://doi.org/10.1007/978-3-319-69308-8_3

3.1 k-means Algorithm

k-means is considered to be the most typical representative of this group of algorithms. Early outlines of this algorithm can be found in papers of Steinhaus [444] (published in 1956), Lloyd [334] (1957), Ball and Hall [47] (1965) and MacQueena [338] (1967). The authors of [509] place it among the ten most important data mining algorithms. The 50-years-long history of the algorithm is exposed by Jain in [272].

As a matter of fact, the modern version on the k-means algorithm is an adaptation of Lloyd heuristic,[1] [334]. Originally, the heuristic was proposed for scalar quantisation in PCM (*Pulse Code Modulation*) systems. Its essence lies in alternating assignment of objects to clusters (based on known centroid location) and prototype updates (based on known cluster membership of objects). It is assumed that the signal, subject to quantisation, is in general a random vector X with a probability density function f_X. Actually, the density function f_X is not known and only a sample $X = \{\mathbf{x}_1, \ldots, \mathbf{x}_m\}$ is available. An adaptation of Lloyd algorithm to such conditions is called LBG algorithm [330], which is nearly identical with the k-means algorithm presented below. Like any other heuristic, this algorithm does not guarantee finding the optimal solution. Yet, it was observed that the algorithm behaves very well if the elements of the data set form natural groups with different properties. The more the properties of these groups differ, the fewer iterations are needed to discover the internal structure of the data set.[2]

In spite of more than half century that passed since invention of the k-means algorithm, its properties and convergence conditions have not been sufficiently investigated. A reader interested in these aspects is advised to have a look at papers[3] [32, 378] and the bibliography contained therein. We recommend also the paper [31], the authors of which exploit the so-called *smoothed analysis*.[4]

In case of k-means algorithm the target is assumed to be the minimisation of the trace of the matrix W, defined by Eq. (2.29).[5] Let U denote once more the

---

[1]Though the algorithm was presented in a report dated July 31st, 1957, it was not officially published till 1982. A similar algorithm was published earlier in a paper by J. Max, Quantizing for minimum distortion. *IEEE Trans. on Info. Th.*, **50**(5), 937–994, 1960. For this reason it is called also Lloyd-Max algorithm, or Max-Lloyd algorithm.

[2]We mentioned this property in Sect. 2.6.

[3]See also Q. Du, M. Emelianenko, and L. Ju, Convergence of the Lloyd algorithm for computing centroidal Voronoi tessellations, *SIAM J. on Numerical Analysis*, **44**(1), 102–119, 2006; M. Emelianenko, L. Ju, and A. Rand, Nondegeneracy and weak global convergence of the Lloyd algorithm in $\mathbb{R}^d$, *SIAM J. on Numerical Analysis*, **46**(3), 1423–1441, 2009.

[4]It is a mixture (a hybrid) of worst-case and average-case analyses. It allows to estimate realistically the algorithm complexity in practical settings. The research in this area was initiated by D. Spielman and S.-H. Teng with the paper "Smoothed analysis of algorithms: why the simplex algorithm usually takes polynomial time", *Proc. of the 33-rd Annual ACM Symposium on Theory of Computing*, ACM 2001, pp. 296–305. We recommend "Mini course on smoothed analysis" available at the Web site http://roeglin.org/publications/SmoothedAnalysis.pdf.

[5]Pollard in [395] considers a bit different perspective of the k-means algorithm. He assumes that a probability density function f is defined over $\mathbb{R}^n$ and a set C of k or fewer points $\mu_i, i = 1, \ldots, k$ is sought the minimises the function $\Phi(C, f) = \int \min_{\mu \in C} ||\mathbf{x} - \mu||_2^2 f(\mathbf{x})d\mathbf{x}$. The k-means algorithm

assignment matrix, telling which object belongs to which cluster (cluster assignment to or split/partition into clusters), and M—the matrix with rows representing gravity centres of the clusters.[6] Then the optimisation criterion function, called also the partition cost function, is of the form

$$J(U, M) = \sum_{i=1}^{m} \sum_{j=1}^{k} u_{ij} \|\mathbf{x}_i - \boldsymbol{\mu}_j\|^2 \tag{3.1}$$

Its minimisation is equivalent to minimisation of the sum of error squares. Here, an "error" is understood as the distance of the i-th observation from the centre of the cluster to which it belongs.

In general, the Euclidean distance $\|\mathbf{x}_i - \boldsymbol{\mu}_j\|$ can be replaced by the Minkowski distance $d_p(\mathbf{x}_i - \boldsymbol{\mu}_j)$ and the criterion function (3.1) will then have the form

$$J_p(U, M) = \sum_{i=1}^{m} \sum_{j=1}^{k} u_{ij} d_p^p(\mathbf{x}_i, \boldsymbol{\mu}_j) = \sum_{i=1}^{m} \sum_{j=1}^{k} u_{ij} |\mathbf{x}_i - \boldsymbol{\mu}_j|^p \tag{3.2}$$

The points, at space in which the partial derivatives[7] are equal zero, are good candidates for minima of the partition cost function:

$$\frac{\partial}{\partial \boldsymbol{\mu}_t} J_p(U, M) = \sum_{i=1}^{m} \sum_{j=1}^{k} u_{ij} \frac{\partial}{\partial \boldsymbol{\mu}_t} |\mathbf{x}_i - \boldsymbol{\mu}_j|^p$$

$$= \sum_{i=1}^{m} u_{it} \frac{\partial}{\partial \boldsymbol{\mu}_t} |\mathbf{x}_i - \boldsymbol{\mu}_t|^p$$

$$= \sum_{\mathbf{x}_i \in C_t} p |\mathbf{x}_i - \boldsymbol{\mu}_t|^{p-1}$$

Components of the gradient vector are of the form

$$\frac{\partial}{\partial \mu_{tl}} |x_{il} - \mu_{tl}|^p = p \cdot \operatorname{sgn}(x_{il} - \mu_{tl}) \cdot |x_{il} - \mu_{tl}|^{p-1}$$

In particular, when $p = 2$, we get

(Footnote 5 continued)
serves then as a means of clustering a finite sample of m elements from the distribution f. He asks the question whether or not the sequence of sets C_m of cluster centres of samples of size m for increasing m to infinity would converge to the set C of "true" cluster centres. He proves that it would so if there existed a unique C and if we had at our disposal an algorithm computing the true optimum of Eq. (3.1) for any finite sample. Such an algorithm was shown to be NP-hard, however.

[6]Cluster gravity centres are called also cluster prototypes, or class centres or just prototypes.

[7]We treat here J_p as a *continuous* function of cluster centre coordinates.

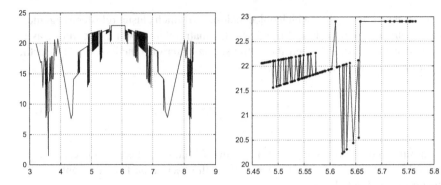

Fig. 3.1 Local optima in the problem of partitioning of the (one dimensional) set X, consisting of 10 points. Figure to the left: values of the indicator J (vertical axis) for the gravity centre of one cluster placed at locations from the interval $[3, 9]$. Figure to the right: a detailed look at these values when the centre belongs to the set $[5.4686, 5.7651]$

$$\sum_{\mathbf{x}_i \in C_t} 2(x_{il} - \mu_{tl}) = 0 \Rightarrow \mu_{tl} = \frac{1}{n_t} \sum_{\mathbf{x}_i \in C_t} x_{il} \tag{3.3}$$

where n_t means the cardinality of the class C_t.

It must be stressed, however, that, for a fixed set X with cardinality m, the cost of partitioning this set into k clusters via gravity centres determined by the above equation is a function which: (a) takes on at most $\vartheta(m, k)$ distinct values[8] and (b) possesses many local minima. For example, in Fig. 3.1, a diagram is shown of the partition cost as a function of gravity centre of *one* of the clusters.[9] The set X consists here of 10 points. Five of them were drawn randomly according to the rule[10] $3 + \mathtt{rand}$, and the remaining five—according to the rule $8 + \mathtt{rand}$.

The above remarks indicate that the task of determining an optimal split is $\mathcal{NP}$-hard. It is that hard even if $k = 2$, and also even when $n = 2$ (see [30, p. 4] and the bibliography cited therein).[11] This fact justifies application of

---

[8]The number $\vartheta(m, k)$ is defined in p. 34.

[9]Note that the gravity centre of the second cluster would make the diagram more complex.

[10]$\mathtt{rand}$ means a random number generator sampling with uniform distribution from the interval $[0, 1]$.

[11]Let us illustrate the complexity issue following the ideas of Inaba et al. [269]. The k cluster centres generated by the k-means algorithm create a Voronoi tessellation of the n-dimensional space into k regions each of which is separated from each of its neighbours by a $n - 1$ dimensional hyperplane passing through the middle of the line segment connecting the two centres and orthogonal to this line segment. As there are k regions, each has at most $k - 1$ neighbours, so at most $k(k - 1)/2$ such separating hyperplanes exist. Out of the m data points, assume that no $n + 1$ of them are hypercoplanar. In $\mathbb{R}^n$ n non-hypercoplanar points are needed to define a hyperplane π (splitting the space into two halves). It assigns the remaining points to either side of this hyperplane, marking them say with + and −. The set of hyperplanes preserving the "side" of these other points shall be called "associated" with the n points selected. The "associated" hyperplanes may be deemed of as the "slight" change of position of the hyperplane π, containing the selected points. It is easily

various heuristics.[12] However, a common drawback of the heuristics is the lack of
warranties on achieving an optimal solution. For example, in case of the Lloyd heuristic, that is, the variant of the Algorithm 2.3 with weights $w(\mathbf{x}_i) \equiv 1, i = 1, \ldots, m$,
the ratio of the cost returned by the algorithm to the optimal cost can be any large
number. It is not hard to encounter such situations, as illustrated by the example
below:

Example 3.1.1 [30, p. 89] Assume that the set $\mathfrak{X}$ consists of four points with coor-
dinates: $\mathbf{x}_1 = (-a, -1)$, $\mathbf{x}_2 = (a, -1)$, $\mathbf{x}_3 = (-a, 1)$ and $\mathbf{x}_4 = (a, 1)$, where $a \geq 1$.
If $k = 2$, and initial cluster centres are located at points $\mathbf{x}_1$ and $\mathbf{x}_3$, then the *k*-means
algorithm returns clusters $C_1 = \{\mathbf{x}_1, \mathbf{x}_2\}$, $C_2 = \{\mathbf{x}_3, \mathbf{x}_4\}$ with centres $\boldsymbol{\mu}_1 = (0, -1)$,
$\boldsymbol{\mu}_2 = (0, 1)$ and with cost $J = 4a^2$. But if the initial centres were placed at points
$\mathbf{x}_1$ i $\mathbf{x}_4$ then we get the optimal split $C'_1 = \{\mathbf{x}_1, \mathbf{x}_3\}$, $C'_2 = \{\mathbf{x}_2, \mathbf{x}_4\}$ with cost $J_{opt} = 4$
and centres $\boldsymbol{\mu}'_1 = (-a, 0)$, $\boldsymbol{\mu}'_2 = (a, 0)$. This means that $J/J_{opt} = a^2 \rightarrow \infty$ when
$a \rightarrow \infty$.

Let us note, in passing, that for $a < 1$ the optimal split is of the form $C_1 = \{\mathbf{x}_1, \mathbf{x}_2\}$,
$C_2 = \{\mathbf{x}_3, \mathbf{x}_4\}$. Also in this case the quotient $J/J_{opt} = 1/a^2 \rightarrow \infty$ when $a \rightarrow 0$. $\square$

For this reason there emerges a growing interest in approximation algorithms, that
is, algorithms returning a partition with cost $J = (1+\varepsilon)J_{opt}$, as mentioned by Arthur
in page 4 of his thesis [30]. Regrettably, in the majority of cases these algorithms
have exponential complexity in the number of clusters and the amount of data. This
fact renders them next to useless in the analysis of big data sets. An exception here
is constituted by the papers [280, 309].

Subsequently, we describe the base variant of the *k*-means algorithm and a number
of its modifications.

(Footnote 11 continued)
checked that these "associated" hyperplanes can assign the set of the n points any $+/-$ labelling,
that is 2^n different labellings of the n selected points such that they agree with the hyperplane π
labelling of the remaining points. As there are m data points, we can select at most $\binom{m}{n}$ distinct
hyperplanes, containing exactly n of the points, yielding at most $2^n \binom{m}{n}$ different labellings. Out of
them you will pick up the mentioned $k(k-1)/2$ hyperplanes separating the Voronoi regions. So
the choice is restricted to $\binom{2^n \binom{m}{n}}{k(k-1)/2} \ll (2m)^{nk(k-1)/2}$ and this is the upper bound on the number of
iterations of a *k*-means algorithm or its variants. The number is prohibitively big, but for low k and
large m it is still orders of magnitude lower than k^{m-k}, that is checking any split of m data points
into k subsets. Actually the number of distinct splits of m data points into k subsets is much larger,
but we use this expression as an illustrative lower bound.

However, the expected run time has been shown to be polynomial in the number of elements in
the sample, see D. Arthur, B. Manthey, and H. Röglin. *Smoothed Analysis of the k-Means Method.*
J. ACM **58** (5), 2011, Article 19 (October 2011), 31 pages.

[12]See Hruschka, E.R., et al. A survey of evolutionary algorithms for clustering. *IEEE Trans. on*
Systems Man, and Cybernetics, Part C: Applications and Reviews, **39**(2), 2009, 133–155; A. Ajith,
S. Das, S. Roy. Swarm intelligence algorithms for data clustering. *Soft Computing for Knowledge*
Discovery and Data Mining. Springer US, 2008. 279–313.

3.1.1 The Batch Variant of the k-means Algorithm

The control flow of the base algorithm is depicted in pseudo-code 2.3 on p. 44. It is assumed that $w(\mathbf{x}_i) = 1, i = 1, \ldots, m$. The algorithm starts with determination of gravity centres of the clusters. The simplest, though least efficient, initialisation method, namely a random split of the set of objects into k clusters and computation of the mean value in each cluster, is applied. Then, repeatedly, two steps are executed: (a) each object is moved to the cluster with the closest centre to it, and subsequently (b) in each cluster its centre is updated given the new object assignment. The procedure is terminated when the cluster assignment stabilizes.[13]

Such a *batch* version of the algorithm is called also Forgy algorithm [181] or H-MEANS heuristic [436]. It resembles the EM algorithm, described in Sect. 3.2. Selim and Ismail have shown in [424] that the algorithm converges in a finite number of steps and have defined conditions for which the obtained solution is a local minimum of the function (3.1).

3.1.2 The Incremental Variant of the k-means Algorithm

The "classic version" (from the point of view of code optimisation) of the algorithm passes iteratively through *each* object and checks if its relocation to some other cluster would improve the target function. In case such an improvement is observed, the object changes its cluster membership. Let us assume we want to move a certain object $\mathbf{x}^*$ from cluster C_j to cluster C_l. Then, the new coordinates of the gravity centre of this latter cluster would be computed from the formula [164]:

$$\boldsymbol{\mu}_l^* = \frac{n_l \cdot \boldsymbol{\mu}_l + \mathbf{x}^*}{n_l + 1} = \boldsymbol{\mu}_l + \frac{\mathbf{x}^* - \boldsymbol{\mu}_l}{n_l + 1}$$

The coordinates of the gravity centre of the cluster C_j, after removal of the object $\mathbf{x}^*$ from it, undergo a similar transformation:

$$\boldsymbol{\mu}_j^* = \boldsymbol{\mu}_j - \frac{\mathbf{x}^* - \boldsymbol{\mu}_j}{n_j - 1}$$

Let us denote by $V(C_j)$ the sum of squares of distances of all objects of the cluster C_j from its gravity centre. It can be easily verified that (see also [164]):

$$V(C_j - \{\mathbf{x}^*\}) = V(C_j) - \frac{n_j}{n_j - 1}\|\mathbf{x}^* - \boldsymbol{\mu}_j\|^2 = V(C_j) - \delta_j(\mathbf{x}^*)$$

$$V(C_l \cup \{\mathbf{x}^*\}) = V(C_l) + \frac{n_l}{n_l + 1}\|\mathbf{x}^* - \boldsymbol{\mu}_l\|^2 = V(C_l) + \delta_l(\mathbf{x}^*)$$

---

[13]If U^t denotes the split matrix obtained in the t-th iteration, then the split corresponding to it is called stabilised one if $U^{t+1} = U^t$.

So, we see that the relocation of the object $\mathbf{x}^*$ would be advantageous if the condition $\delta_j(\mathbf{x}^*) > \delta_l(\mathbf{x}^*)$ holds, that is:

$$\frac{n_j}{n_j - 1} \|\mathbf{x}^* - \boldsymbol{\mu}_j\|^2 > \frac{n_l}{n_l + 1} \|\mathbf{x}^* - \boldsymbol{\mu}_l\|^2$$

We obtained in this way an algorithm called in [164] BIMSEC (*Basic Iterative Minimum-Squared-Error Clustering*) or K-MEANS (see e.g. [436]). Pseudo-code 3.1 describes the basic steps of the algorithm.

Algorithm 3.1 Incremental clustering algorithm BIMSEC

Input: Data set X, number of clusters k.
Output: A partition defined by the set of gravity centres $\{\boldsymbol{\mu}_1, \ldots, \boldsymbol{\mu}_k\}$.
1: Determine the initial split of the data set into k non-empty clusters and compute their gravity centres.
2: Choose a (next) element $\mathbf{x}^*$ from a cluster, say cluster C_j.
3: If $n_j = 1$ – go to step 6. Otherwise compute the increments

$$\delta_l(\mathbf{x}^*) = \begin{cases} \dfrac{n_l}{n_l + 1} \|\mathbf{x}^* - \boldsymbol{\mu}_l\|^2 & \text{if } l \neq j \\ \dfrac{n_j}{n_j - 1} \|\mathbf{x}^* - \boldsymbol{\mu}_j\| & \text{if } l = j \end{cases}$$

4: Move object $\mathbf{x}^*$ to the cluster C_{j^*} if $\delta_{j^*} \leq \delta_l, l = 1, \ldots, k$.
5: Compute new coordinates of gravity centres $\boldsymbol{\mu}_j, \boldsymbol{\mu}_{j^*}$ and the value of the indicator $J(U, M)$.
6: If the indicator value did not change after testing m objects – halt. Otherwise go back to step 2.

Though the "classic" incremental version is more expensive than the "batch" version (H-MEANS algorithm), it turns out to be more successful [164]. One can say that while a solution returned by K-MEANS cannot be improved by H-MEANS, it is nonetheless possible to improve the result of H-MEANS by subsequent application of K-MEANS—see, e.g., [234]. A compromise between computational efficiency and accuracy can be achieved via alternating application of both algorithms. For example, we may repeat in a loop many times the steps (2) and (3) of the Algorithm 2.3 and thereafter only several times (due to higher computational costs) we would carry out object moves according to the K-MEANS heuristic.

3.1.3 Initialisation Methods for the k-means Algorithm

It is beyond doubt that the big advantage of the k-means algorithm for big data analysis is its simplicity and linear time complexity with respect to the number of objects. It has, however, also a number of disadvantages, the most important of them being the following:

(a)   **(b)**

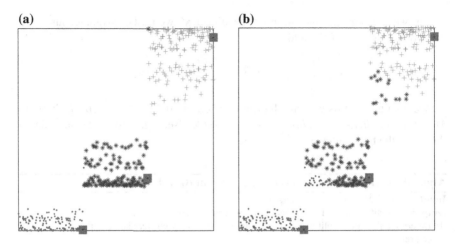

Fig. 3.2 The influence of initialisation and scaling on the results of clustering of a set of 450 observations. Initial location of cluster centres is marked with red squares, and respective symbols denote the assignment of objects to clusters. In case **a** $x_{i1} \in \{1, \ldots, 450\}$, and in case **b** $x_{i1} \in 2.222185 \cdot 10^{-3} + 2.034705 \cdot 10^{-10} \cdot \{1, \ldots, 450\}, i = 1, \ldots, 450$

(a) The result depends on the order, in which the data is processed and on the value range (scale) of individual components of the vector, describing the objects. Initial normalisation allows to eliminate the impact of value ranges.
(b) It is a greedy algorithm the result of which depends on initial conditions, see Fig. 3.2.
(c) It is sensitive to the presence of abnormal observations (*outliers*).
(d) The number of clusters must be known in advance.
(e) It can be used only for analysis of numerical data. Simple generalisations for the case of mixed (qualitative and quantitative) data are discussed in Sect. 3.1.5.7.

Furthermore, application of the H-MEANS heuristic leads frequently to obtaining several empty clusters [436]. In such cases the H-MEANS+ heuristic, introduced by Hansen and Mladenovic in [234], may prove helpful. If one got $k - k_1$ non-degenerate (i.e. non-empty) clusters, then k_1 objects are selected that are most distant from the centres of clusters, to which they belong. These objects are treated as new, additional cluster centres and all the objects are re-classified according to the rule "the winner takes all", (2.45).

In order to reduce the impact of initialisation on the final result, a number of "clever" initialisation methods were proposed. Let us present a couple of the most frequently used methods.

(a) k objects randomly selected from the set X are treated as cluster centres. The remaining objects are assigned to respective clusters according to the clustering update rule, that is, they are assigned to the closest gravity centre. This method was proposed by Forgy in [181].

(b) As previously, we start with a random choice of k gravity centres and an order is imposed on the set of objects. The objects (visited according to the imposed order) are assigned to the cluster with the closest gravity centre and after this assignment the coordinates of gravity centre of this cluster are updated. This method was suggested by MacQueen in 1967.

(c) The next method was proposed by many authors. Among the first ones we shall mention Goznalez [206] and Hochbaum and Shmoys [253]; it was also suggested, in particular, by Katsavounidis, Kuo and Zhang [283]. The crucial idea of the method is to choose from the data set k objects most distant one from another. In the first step, the μ_1 is set to the coordinates of the object $\mathbf{x} \in X$ of maximal length, that is $\mu^1 = \arg\max_{1 \le i \le m} \|\mathbf{x}_i\|$. It is the seed of the set of centroids M, that is: $M = \{\mu^1\}$. Subsequently, the object $\mathbf{x} \in X$ is identified that is most distant from μ^1. This is the second centroid μ_2 inserted into the set M. To determine the j-th candidate, for each $\mathbf{x} \in X \backslash M$ the distance from all elements from the set M is computed and the smallest one is assumed to be the distance $d(\mathbf{x}, M)$. The object most distant from the set M is selected as μ_j, that is

$$\mu_j = \arg\max_{\mathbf{x}_i \in X} \left(\min_{\mu \in M} \|\mathbf{x}_i - \mu\| \right) \tag{3.4}$$

The process of choosing the object most distant from M is terminated, when we get k candidates. A quick algorithm, implementing this method, is presented in Sect. 3.1.4, and its probabilistic variant—in Sect. 3.1.3.1.

(d) Ng, Jordan and Weiss proposed in [374] the Orthogonal Initialization: normalize each data item to have unit, and then select a data item at random, and successively select $k - 1$ data whose directions are most orthogonal to the previously chosen points—see footnote 3 in [374, Sect. 4] for details.

(e) Another, most elaborate method, is co-authored by Kaufman and Rousseeuw, [284]. The cluster centres are chosen iteratively till we get k of them. The first centroid μ_1 is the most central object of the whole data set. Assume we chose already $s < k$ gravity centres. For each pair of objects, $\mathbf{x}_i, \mathbf{x}_j$, not selected into the set of centroids, the value $\beta_{ij} = \max[B_j - d(\mathbf{x}_i, \mathbf{x}_j), 0]$ is computed, where $B_j = \min_{l=1,\dots,s} d(\mathbf{x}_j, \mu_l)$. Thereafter, the gain from selecting the location $\mathbf{x}_i$ equal $\sum_j \beta_{ij}$, is computed and the location $\mathbf{x}_i$, is selected as the next gravity centre, which maximizes this gain value.

(f) An initial clustering of the set of objects is obtained using some other method, like the Ward algorithm.

Experiments described in [389] indicate quite good properties of random cluster initialisation and of Kaufman/Rousseeuw initialisation. Ng et al. claim that their initialisation method ensures a quick convergence of the clustering algorithm, [374]. Other initialisation methods were elaborated by Su and Dy in [448], though their proposals, we think, seem to be quite complex algorithmically. A number of bibliographic positions devoted to this topic are mentioned by Kuncheva and Vetrow [313]. A careful comparison of various initialisation methods can be found in [345].

(i)   **(ii)**

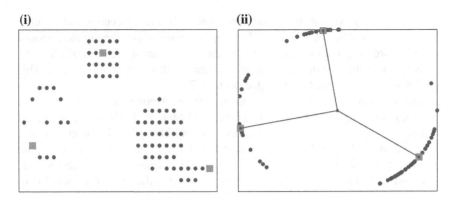

Fig. 3.3 The distribution of the gravity centres (denoted with squares) in two methods of initialisation. (i)—initialisation described in point (c), (ii)—initialisation described in point (d). In this last case the coordinates were initially normalised in such a way that $\mathbf{x}_{ij} \in [-1, 1]$, $i = 1, \ldots, m$, $j = 1, \ldots, n$, and afterwards—as required by the method—each row of the matrix X was so normalized that $\|\mathbf{x}_i\| = 1$. The lines connecting the coordinate system origin (0, 0) with gravity centres play here an explanatory role only—they allow to imagine the angles between the centres

Figure 3.3 illustrates the qualitative difference between methods (c) and (d). Note that in both cases each candidate centre belongs to a different cluster.

Still another idea is the random initialisation of the assignment matrix. However, it is a much worse solution. For example, for the data set `data3_2.txt`, in the case of random initialisation of the class centres, we got 72 correct solutions in 100 runs, while in the case of random initialisation of the U matrix—only 38 correct solutions in 100 runs.

A thorough analysis of the impact of initialisation on the stability of k-means algorithm was carried out in [93]. An advanced initialisation method was also proposed there. It is a two-stage procedure. In the first stage, $k' > k$ gravity centres are picked randomly, objects are assigned to corresponding clusters and the centre coordinates are updated. Thereafter, most "valuable" centres are selected, and among them, using the variant (c)—k centres most distant from one another are chosen.

3.1.3.1 k-means++ Algorithm

An interesting initialisation method was presented in [32] (compare also [378]). It is, in fact, a variant of the (c) method from the previous section. Its authors did notice that if the data set contains at least k outliers, then the choice of the most distant observations will result in malformed initial centres. Hence, they propose a probabilistic variant of the selection rule. Let $u(\mathbf{x})$ denote the distance of the object

x from the set C consisting of $j < k$ centres. The next centre is picked according to the probability distribution[14]

$$p(\mathbf{x}) = \frac{u^2(\mathbf{x})}{\sum_{\mathbf{x}' \in X} u^2(\mathbf{x}')} \tag{3.5}$$

The pseudo-code 3.2 describes the steps of the algorithm. The procedure of probabilistic selection of candidate centres is in fact very quick. One can apply the method presented in the subsequent section or the method proposed in [281]. Both methods allow to avoid unnecessary distance computations. In the first case, the triangle inequality is exploited, and in the second one—a special data structure, the so-called kd-tree, is used.

Algorithm 3.2 *k*-means++ algorithm, [32]

Input: Data set X, number of clusters k.
Output: A partition defined by the set of gravity centres $\{\mu_1, \ldots, \mu_k\}$.
1: Pick randomly the object $\mathbf{x} \in X$ and assume $\mu_1 = \mathbf{x}$. Let $C = \{\mu_1\}$.
2: **for** $j = 2$ to k **do**
3:    For each object $\mathbf{x}$ compute the distance to the nearest centre from the set C, that is $u(\mathbf{x}) = \min_{\mu \in C} \|\mathbf{x} - \mu\|^2$.
4:    Substitute μ_j with the object $\mathbf{x}$ picked randomly according to the distribution (3.5).
5:    Add μ_j to the set of centres, $C \leftarrow C \cup \{\mu_j\}$.
6: **end for**
7: Run the *k*-means algorithm.

Arthur and Vassilvitskii [32] demonstrate that under some conditions the algorithm converges in super-polynomial time. Beside this, they prove the following property.

Theorem 3.1.1 (*[32]*) *The expected value of the partition cost, computed according to Eq. (3.1), is delimited in case of k-means++ algorithm by the inequality*

$$\mathbb{E}(J) \le 8(\ln k + 2) J_{opt} \tag{3.6}$$

where J_{opt} denotes the optimal value of the partition cost. ☐

Furthermore, it can be shown that this property holds in the limit, when the sample size grows to infinity, see Appendix E.

The sequential nature is the main disadvantage of the algorithm. The proper initiation requires k-fold visiting of each element of the data set in order to choose the candidate cluster centres. The paper [40] contains a modification addressing this

---

[14]If we replace the Euclidean distance with Minkowski d_l then the elements of the set X should be picked with probability $p(\mathbf{x}) = u^l(\mathbf{x}) / \sum_{\mathbf{x}' \in X} u^l(\mathbf{x}')$. Consult [30, p. 98].

issue—the k-means|| algorithm, specially designed for very big data analysis. An additional advantage of the proposal is the possibility of paralleled implementation under the `MapReduce` paradigm.[15]

There exists a number of theoretical papers studying the possibilities of other initialisation of k-means algorithm in order to ensure either an approximation of the optimal cost function value, or reducing the computational time complexity to linear or sublinear, either without assumptions about the clustered set or requiring a kind of clusterability (good separation of clusters). Regrettably, these efforts are of no practical value so far, as the probabilistic guaranties are too loose (below 50%), the scaling factors in the time-linear versions are too large (thousands), or the required separation of clusters turns out to be extremely large (16 times the standard deviation of points within a cluster or more).

Let us illustrate the last problem in more detail. Assume that the data is well clusterable in the following sense: 99% of data in each cluster lies within a ball of radius r, the distance between cluster centres is at least $d = 4r$, and the cardinality of each cluster is about the same. Under these circumstances, k-means would for sure find the optimal clusters, given that within the initialisation phase a point is picked from each cluster. Under k-means with random initialisation this probability amounts to $\frac{k!}{k^k}$. For $k = 2$ this amounts to 0.5, hence by repeating the clustering $r = 7$ times, we have the acceptable success probability of $1 - \left(\frac{k!}{k^k}\right)^r$ of over 99%. In case of $k = 10$, the success rate of a single proper initialisation drops down to 0.0003, so that the number of repetitions needed increases to $r = 10,000$.[16] On the other hand, if we use k-means++ initialisation, for $k = 2$ the success probability in a single try is about 0.8, so that with 3 repetitions, the guarantees of success are sufficient. However, with $k = 10$, the probability of hitting the proper clustering in a single initialisation step decreases to 0.04 which is much better than for k-means, but still a significant number of repetitions is needed, about $r = 100$.

So, let us propose a modification of the distribution used by k-means++ initialisation (3.5) to the form

$$p(\mathbf{x}) = \frac{u^s(\mathbf{x})}{\sum_{\mathbf{x}' \in X} u^s(\mathbf{x}')} \tag{3.7}$$

where s is a parameter (equal 2 in (3.5)). Under this modification, for $s = 10$ and $k = 10$ you have 98% chance of successful initialisation in a single pass, so that $r = 1$ suffices. The negative effect of course is the increased susceptibility to outliers.

---

[15] A reader interested in this type of solutions is encouraged to consult the tutorial https://hadoop. apache.org/docs/r1.2.1/mapred_tutorial.html.

[16] Of course bad initialisations will be corrected by proper k-means to a reasonable extent.

3.1.4 Enhancing the Efficiency of the k-means Algorithm

Practical applications of k-means algorithm encounter the obstacle of the computational burden of multiple calculations of the distances $\|\mathbf{x}_i - \boldsymbol{\mu}_j\|$ for $i = 1, \ldots, m$, $j = 1, \ldots, k$. It turns out to be particularly expensive in case of large data sets where usually we have also a large number of clusters.

The simplest remedy, improving time complexity, is to compute the distance between a pair of objects by using the following identity

$$\|\mathbf{x}_i - \boldsymbol{\mu}_j\|^2 = \|\mathbf{x}\|^2 + \|\boldsymbol{\mu}_j\|^2 - 2\mathbf{x}_i^{\mathrm{T}} \boldsymbol{\mu}_j$$

The first and the second elements of this sum can be computed once for each $i = 1, \ldots, m$, $j = 1, \ldots, k$, while all the scalar products can be computed as $X M^{\mathrm{T}}$, where M stands for the $k \times n$ matrix of centroids. The product of two matrices can be computed using BLAS package.[17]

The classical approach to reduce the computational burden is to use special data structures, for example, k-dimensional binary search trees, so-called k-d trees [281], their variant called BBD-trees [33], metric trees [472], or the so-called R-trees [219]. An extensive review of methods of identification of the closest neighbour is contained in chapter [120], and a rich bibliography on the subject can be found on the Web page [329].

Another method consists in intelligent exploitation of the triangle inequality to eliminate superfluous computations of the distances [169]. One can achieve implementationally cheap algorithms with time complexity comparable to the algorithms of the former group.

The k-means algorithm may be parallelised quite easily. Assume that the data set has been split into P disjoint subsets $Y_p \subset X$, each of which is allocated to one of the processors. Assume further that global information is available to each of the processors on the identifiers of cluster centres and their coordinates and it has the form of a list of <key, value> pairs. Given this information, each processor performs locally two basic operations:

(a) It assigns elements of the set Y_p, allocated to it, to the clusters following the rule *winner takes all* (2.45), that is, it partitions Y_p into

$$Y_p = Y_{p,1} \cup \cdots \cup Y_{p,k}$$

and thereafter
(b) it computes the sums $s_{p,j} = \sum_{\mathbf{x} \in Y_{p,j}} \mathbf{x}$ and cardinalities of the "subclusters" $m_{p,j} = |Y_{p,j}|, j = 1, \ldots, k, p = 1, \ldots, P$.

---

[17]Its source codes are available e.g. from http://people.sc.fsu.edu/~jburkardt/c_src/blas3/blas3.html.

The local lists of the form $<j, (s_{p,j}, m_{p,j})>$ are aggregated to the global list (by one of the processors), whereby the coordinates of the j-th centre are computed as follows:

$$\mu_j = \frac{1}{m_j} \sum_{p=1}^{P} s_{p,j} \tag{3.8}$$

where $m_j = \sum_{p=1}^{P} m_{p,j}$. Both summations can be carried out in common **for** the loop, and thereafter the resulting sums can be divided by one another.

The described steps are easy to implement within the MapReduce framework, see e.g. [536]. The step (a) fits the specification of the map function, the step (b)—that of the combine function, whereas the Eq. (3.8) may be embedded in the body of the reduce function. Under practical settings, the functions map and combine are integrated into one function. By extending this function according to the description provided in Sect. 3.1.4 one can significantly reduce the number of calls of the record-group distance by computing the function in step (a), thereby significantly reducing the execution time.

k-means algorithm possesses also hardware implementations, which are particularly useful for huge data analysis. A short review of such approaches is contained in [268].

An interesting attempt to improve the properties of the standard k-means algorithm was made with the development of ISODATA algorithm (*Iterative Self-Organizing Data Analysis Technique*). Its authors, Ball i Hall, enriched the base algorithm with the possibility of combining similar low cardinality clusters and with the possibility of splitting high cardinality clusters. Hence, ISODATA, features adaptive choice of the number of clusters. The details of the algorithm can be found in the monograph [463] and in [47].

Bejarano et al. [58] propose a sampling version of k-means algorithm that shall accelerate it for large data samples. It is claimed that the largest sample size needed, m^* amounts to

$$m^* = k \left(\frac{1}{m} + \left(\frac{w}{z^*} \right)^2 \right)^{-1}$$

where m is the size of the (large) data set to be sub-sampled, w is the (user-defined) width of the confidence intervals for cluster means, and z^* is the percentage point of the normal distribution $\mathcal{N}(0, 1)$ for user-specified confidence level. The algorithm is further refined so that not m^*, but even fewer elements of the sample are used in subsequent iteration steps, based on the variances within each cluster. So for the next iteration, the iteration sample size m^{**} equals the sum of sample sizes needed to estimate properly each cluster mean:

$$m^{**} = \sum_{j=1}^{k} m_j^{**}$$

where the sample size needed for cluster j is estimated from the previous step as

$$m_j^{**} = \left(\frac{1}{m_j^{estim}} + \left(\frac{w}{2z^* \sigma_j} \right)^2 \right)^{-1}$$

where σ_j is the (estimate of) variance of cluster j from the previous step, and m_j^{estim} is the estimate of the cluster size in the database, computed from the previous iteration step as

$$m_j^{estim} = m \frac{m_{cj}^{**prev}}{m^{**prev}}$$

where m is the number of elements in the database, m^{**prev} is the sample size from the previous step, m_{cj}^{**prev} is the number of elements of the sample assigned in the previous step to cluster j.

Though different sample sizes may be used in each step, no resampling is performed, but rather first m^{**} elements of the initial sample of size m^* are used.

Bradley, Benett and Bemiriz introduced in [84] the concept of a balanced k-means algorithm in which they require each cluster to contain at least $\tilde{n}$ objects. Their numerical experiments show that their algorithm avoids poor local minima.

Wagstaff et al. proposed in [488] a semi-supervised variant of the classical k-means algorithm. It exploits the background knowledge on which data elements must belong to the same cluster and which cannot be members of the same cluster.[18]

3.1.5 Variants of the k-means Algorithm

3.1.5.1 On Line Variant of the k-means Algorithm

For big data analysis, the so-called *on line* version of k-means algorithm has been developed. Its idea is outlined in the pseudo-code 3.3.

Algorithm 3.3 *On line* version of the k-means algorithm

---
1. Initialise (e.g. randomly) k prototypes.
2. Select (according to the assumed probability distribution $p(\mathbf{x})$) an element $\mathbf{x} \in X$.
3. Determine the winner, that is, the prototype $\mu_{s(\mathbf{x})}$ that is the closest to the considered object $\mathbf{x}$.
4. Modify the coordinates of the winning prototype, i.e.

$$\mu_{s(\mathbf{x})} \leftarrow \mu_{s(\mathbf{x})} + \alpha(\mathbf{x} - \mu_{s(\mathbf{x})}) \qquad (3.9)$$

   where $\alpha \in (0, 1]$ is called the learning coefficient
5. Go to step (2) if the termination condition is not met

---

[18]Read more in Sect. 3.15.2 about semi-supervised clustering.

The terminal condition of this algorithm is defined as stabilisation of prototypes, meaning that when each coordinate of the prototype before and after the modification does not differ by more than a pre-defined parameter ϵ.

The learning coefficient α can be either a constant or a function of (execution) time. It is known that in the first case the prototype coordinates $\boldsymbol{\mu}_j(t)$ at iteration t behave like the exponentially falling mean of the signals $\mathbf{x}^{(j)}(t)$, for which the prototype $\boldsymbol{\mu}_j(t)$ was a winner, that is

$$\boldsymbol{\mu}_j(t) = (1 - \alpha)^t \boldsymbol{\mu}_j(0) + \alpha \sum_{i=1}^{t} (1 - \alpha)^{t-i} \mathbf{x}^{(j)}(t)$$

The above equation implies that the current value of the prototype is most strongly influenced by the current observation, while the impact of the preceding observations decays exponentially with time. This property is the source of instability of the algorithm. Even after presenting a large number of observations, the next observation can radically impact the coordinates of the winning prototype.

MacQueen proposed in [338] to reduce the α coefficient hyperbolically with time, i.e.

$$\alpha(t) = \frac{\alpha_0}{t}, \quad t \geq 1 \tag{3.10}$$

In this case, the vector $\boldsymbol{\mu}_j(t)$ is the arithmetic mean of the presented observations, that is:

$$\boldsymbol{\mu}_j(t) = \boldsymbol{\mu}_j(t - 1) + \alpha(t)\left(\mathbf{x}(t)^{(t)} - \boldsymbol{\mu}_j(t - 1)\right) = \frac{1}{t}(\mathbf{x}^{(j)}(1) + \cdots + \mathbf{x}^{(j)}(t))$$

In fact,

$$\boldsymbol{\mu}_j(1) = \boldsymbol{\mu}_j(0) + \alpha(1)(\mathbf{x}^{(j)}(1) - \boldsymbol{\mu}_j(0)) = \mathbf{x}^{(j)}(1)$$

$$\boldsymbol{\mu}_j(2) = \boldsymbol{\mu}_j(1) + \alpha(2)(\mathbf{x}^{(j)}(2) - \boldsymbol{\mu}_j(1)) = \tfrac{1}{2}(\mathbf{x}^{(j)}(1) + \mathbf{x}^{(j)}(2))$$

etc.

Another scenario of reducing the learning coefficient was suggested in [192]:

$$\alpha(t) = \alpha_p \left(\frac{\alpha_k}{\alpha_p}\right)^{\frac{t}{t_{max}}} \tag{3.11}$$

where α_p, α_k mean the initial and the final value of the coefficient with $\alpha_p > \alpha_k$, and t_{max} is the maximal number of iterations. Further variants are discussed by Barbakh and Fyfe in [52].

3.1.5.2 Bisection Variant of the *k*-means Algorithm

Practical experience indicates that clusters obtained via agglomerative data analysis are generally of higher quality that those generated by the *k*-means algorithm. To increase the quality of the latter, Steinbach, Karypis and Kumar [443] proposed the so-called bisection variant of the *k*-means algorithm. Its idea (see pseudo-code 3.4) consists in initial division of the data set X into two clusters. Subsequently, one of the clusters is selected to be partitioned (again using *k*-means algorithm) into two new clusters. The process is repeated till a satisfactory number of clusters is obtained. Though diverse criteria may be applied to choose a cluster for partitioning, but, as the authors claim, [443], good results are obtained if one takes the largest cardinality cluster.

Algorithm 3.4 Bisectional *k*-means algorithm

Input: Data set X, number of clusters k.
Output: A partition defined by the set of gravity centres $\{\mu_1, \ldots, \mu_k\}$.
1: *Initialisation*. Split the set X into two clusters.
2: Choose a cluster C for further partitioning (bisection).
3: Split the set C into two subsets using *k*-means algorithm.
4: Repeat step (3) a fixed number of times (parameter ITER) till a partition of the highest quality (homogeneity) is obtained.
5: Repeat steps 2–3 till a proper number of clusters is obtained.

It is crucial to update cluster centres sequentially in the algorithm used in step (3), which leads to better results. Expressing it differently, if the cluster C_j of cardinality n_j and centre vector μ_j is extended by the element $\mathbf{x}^*$, then the coordinates of the centre vector are updated according to the equation

$$\mu_{jl} = \frac{n_j \cdot \mu_{jl} + \mathbf{x}_l^*}{n_j + 1}, \quad l = 1, \ldots, n$$

Though *k*-means algorithm guarantees reaching the local minimum of the function (3.1), it is not more the case in the incremental version where it is applied "locally" to a selected cluster. Nonetheless the bisectional variant produces clusters of similar size and of high quality. Here, the quality is measured as the averaged entropy

$$E = \frac{1}{m} \sum_{j=1}^{k} n_j E_j$$

where n_j is the cardinality of the j-th cluster, E_j is the entropy of this cluster

$$E_j = -\sum_{t=1}^{k} p_{tj} \log p_{tj}$$

and p_{tj} denotes the probability that an element of, j-th cluster will be classified into the cluster C_t.

3.1.5.3   Spherical k-means Algorithm

This is a variant of the k-means algorithm, developed for clustering of text documents, [147, 148, 152]. It is assumed that documents are represented by vectors with components reflecting the frequencies of word occurrences in respective documents. These vectors are deemed to be normalized, that is $\|\mathbf{x}\| = 1, \mathbf{x} \in X$, hence they belong to the surface of a unit sphere—which implies the name of the algorithm. In such a case, the squared distance between the i-th document and the j-th prototype is equal

$$d^2(\mathbf{x}_i, \boldsymbol{\mu}_j) = \|\mathbf{x}_i - \boldsymbol{\mu}_j\|^2 = \|\mathbf{x}\|^2 - 2\mathbf{x}_i^T \boldsymbol{\mu}_j + \|\boldsymbol{\mu}_j\|^2 = 2(1 - \mathbf{x}_i^T \boldsymbol{\mu}_j) \quad (3.12)$$

where the product $\mathbf{x}_i^T \boldsymbol{\mu}_j$ equals to the cosine of the angle between the (normalised) vectors representing both objects. Hence, the minimisation of the error (3.1) corresponds to maximisation of the expression

$$J_s(U, M) = \sum_{i=1}^m \sum_{j=1}^k u_{ij} \mathbf{x}_i^T \boldsymbol{\mu}_j \quad (3.13)$$

The search for the optimal partition of the set of documents could follow the k-means algorithm, i.e. the prototype $\boldsymbol{\mu}_j$ components could be determined according to Eq. (3.3), and the assignment of objects to clusters could be governed by the "winner-takes-all" rule. However, the vector $\boldsymbol{\mu}_j$, determined in this way, would not be a unit vector, as required. Therefore, it is replaced with the direction vector $\mathbf{c}_j$, as defined by Eq. (3.14):

$$\mathbf{c}_j = \frac{\boldsymbol{\mu}_j}{\|\boldsymbol{\mu}_j\|} = \frac{\sum_{\mathbf{x}_i \in C_j} \mathbf{x}_i}{\|\sum_{\mathbf{x}_i \in C_j} \mathbf{x}_i\|} \quad (3.14)$$

The Cauchy -Schwartz -Buniakowski inequality implies for each unit vector $\mathbf{z}$:

$$\sum_{\mathbf{x} \in C_j} \mathbf{x}^T \mathbf{z} \leq \sum_{\mathbf{x} \in C_j} \mathbf{x}^T \mathbf{c}_j \quad (3.15)$$

so that $\mathbf{c}_j$ is the most similar vector (with respect to the cosine similarity measure) to the vectors forming the cluster C_j. This means that $\mathbf{c}_j$ represents the leading *topic* in the cluster C_j.

Knowing the main (leading) topic of cluster C_j, one can define the *consistency* of this cluster, that is, its quality, measured as the sum of similarities of the collected documents on this topic:

$$q(C_j) = \begin{cases} \sum_{\mathbf{x}_i \in C_j} \mathbf{x}_i^T \mathbf{c}_j & \text{if } C_j \neq \emptyset \\ 0 & \text{otherwise} \end{cases} \tag{3.16}$$

The property $\mu_j = \frac{1}{n_j} \sum_{\mathbf{x} \in C_j} \mathbf{x}^T$, combined with the Eq. (3.14), implies:

$$q(C_j) = \sum_{\mathbf{x}_i \in C_j} \mathbf{x}_i^T \mathbf{c}_j = n_j \mu^T \mathbf{c}_j = n_j \|\mu_j\| \|\mathbf{c}_j^T \mathbf{c}_j = n_j \mu_j = \| \sum_{\mathbf{x} \in C_j} \mathbf{x}\| \tag{3.17}$$

which points at an additional interpretation of the cluster quality index, [147, 148]. The quality of the partitioning $C = \{C_1, \ldots, C_k\}$ is now defined by the formula:

$$Q(\{C_1, \ldots, C_k\}) = \sum_{j=1}^{k} q(C_j) = \sum_{j=1}^{k} \sum_{\mathbf{x}_i \in C_j} \mathbf{x}_i^T \mathbf{c}_j \tag{3.18}$$

constituting an analogue of quality index (3.1), but with the amendment that we seek now such a partition $\{C_1, \ldots, C_k\}$ that maximises the indicator (3.18). An efficient heuristic has been proposed to solve this problem. It is presented as the pseudo-code 3.5.

Algorithm 3.5 Spherical *k*-means algorithm

Input: Data set X, number of clusters k.
Output: A partition defined by the set of gravity centres $\{\mu_1, \ldots, \mu_k\}$.
1: *Initialisation.* Create any partition of the set of documents $C^0 = \{C_1^0, \ldots, C_k^0\}$ and determine the topics $\mathbf{c}_j^0$, $j = 1, \ldots, k$. Set the iteration counter $t = 0$.
2: For each document $\mathbf{x}_i$ find the closest topic (with respect to the cosine measure). Create the new partition

$$C_j^{t+1} = \{\mathbf{x} \in X : \mathbf{x}^T \mathbf{c}_j^t \geq \mathbf{x}^T \mathbf{c}_l^t, l = 1, \ldots, m\}, j = 1, \ldots, k$$

   i.e. C_j^{t+1} contains the documents that are most similar to the topic $\mathbf{c}_j$. If a document happens to be equally similar to several topics, assign it to any of them randomly.
3: Update the characteristics of the topics $\mathbf{c}_j^{t+1}$ as indicated by the Eq. (3.14), substituting C_j with C_j^{t+1}.
4: If the stop condition does not hold, set the counter t to $t + 1$ and return to step 2.

It has been demonstrated in [152] that for each $t \geq 0$, the Algorithm 3.5 preserves the following property

$$Q(\{C_j^{t+1}\}_{j=1}^k) \geq Q(\{C_j^t\}_{j=1}^k) \tag{3.19}$$

The paper [146] presents several algorithms for improving the performance for big document collections.

The spherical algorithm exhibits the properties similar to those of the previously discussed *k*-means algorithm. In particular, it is sensitive to initialisation and the optimisation process can get stuck at a local optimum.

The significant difference between the spherical and non-spherical k-means algorithms is related to the shape of the clusters [152]. In case of spherical algorithm, the border between the clusters C_j and C_l, represented by topics c_j, c_l, is described by the equation

$$\mathbf{x}^\mathrm{T}(\mathbf{c}_j - \mathbf{c}_l) = 0$$

This is an equation of a hyperplane, passing through the origin of the coordinate system. Such a hyperplane cuts the unit sphere at the equator. In case of standard k-means algorithm, the distance between i-th point and j-th cluster is equal

$$\|\mathbf{x}_i - \boldsymbol{\mu}_j\|^2 = (\mathbf{x}_i - \boldsymbol{\mu}_j)^\mathrm{T}(\mathbf{x}_i - \boldsymbol{\mu}_j) = \mathbf{x}_i^\mathrm{T}\mathbf{x}_i - 2\mathbf{x}_i^\mathrm{T}\boldsymbol{\mu}_j + \boldsymbol{\mu}_j^\mathrm{T}\boldsymbol{\mu}_j = \mathbf{x}_i^\mathrm{T}\mathbf{x}_i - 2(\mathbf{x}_i^\mathrm{T}\boldsymbol{\mu}_j - \frac{1}{2}\boldsymbol{\mu}_j^\mathrm{T}\boldsymbol{\mu}_j)$$

Hence, we have the final form of the border equation:

$$\|\mathbf{x}_i - \boldsymbol{\mu}_j\|^2 - \|\mathbf{x}_i - \boldsymbol{\mu}_l\|^2 = 0 \equiv \mathbf{x}^\mathrm{T}(\boldsymbol{\mu}_j - \boldsymbol{\mu}_l) = \frac{1}{2}(\boldsymbol{\mu}_j^\mathrm{T}\boldsymbol{\mu}_j - \boldsymbol{\mu}_l^\mathrm{T}\boldsymbol{\mu}_l)$$

It is also an equation of a hyperplane, but it intersects the sphere at any place.

More remarks about the algorithm and its various implementations (batch and incremental ones) can be found in the fourth chapter of the book [297]. The paper [537] presents an adaptation of the algorithm for the processing of large data collections.

3.1.5.4   KHM: The Harmonic k-means Algorithm

Let us rewrite the Eq. (3.1) in the equivalent form

$$J(U, M) = \sum_{i=1}^{m} \min_{j=1,\ldots,k} , \|\mathbf{x}_i - \boldsymbol{\mu}_j\|^2 \tag{3.20}$$

Zhang [530] found that the properties of the function $\min(a_1, \ldots, a_k)$, where a_j, $j = 1, \ldots, k$ are positive real numbers can be quite faithfully represented by the harmonic mean of these numbers

$$m_h(a_1, \ldots, a_k) = \frac{k}{\sum_{j=1}^{k} \frac{1}{a_j}}$$

Contour diagrams of both functions are presented in Fig. 3.4.

Similarities between both functions can be strengthened or weakened by introduction of an additional parameter—the power exponent, to which the arguments of harmonic mean are raised. Finally, by replacing in Eq. (3.20) the minimum operator by the harmonic mean operator, we obtain a new criterion function of the form

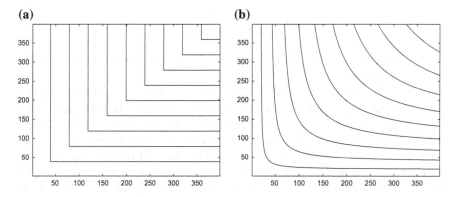

Fig. 3.4 Contour diagrams of the function: **a** $\min(a_1, a_2)$, and **b** $m_h(a_1, a_2)$. It was assumed in both cases that $a_1, a_2 \in (0, 400]$. Horizontal axis: a_1 value, vertical axis: a_2 value, the contour lines were drawn for equidistant values

$$J_h(U, M) = \sum_{i=1}^{m} \frac{k}{\sum_{j=1}^{k} \|\mathbf{x}_i - \boldsymbol{\mu}_j\|^{-p}} \tag{3.21}$$

where $p \geq 2$ is a parameter. It is suggested to set $p = 3.5$, [530].

So far, we have been dealing with a sharp (crisp) partition. But the function (3.21) induces a fuzzy partition, in which the membership of i-th object in j-th cluster is a number $u_{ij} \in [0, 1]$. It is determined from the equation, [530]

$$u_{ij} = \frac{\|\mathbf{x}_i - \boldsymbol{\mu}_j\|^{-2-p}}{\sum_{j=1}^{k} \|\mathbf{x}_i - \boldsymbol{\mu}_j\|^{-2-p}} \tag{3.22}$$

To compute the coordinates of the cluster centres, the weight $w(\mathbf{x}_i)$, occurring in the third step of the Algorithm 2.3, presented on p. 44, needs to be determined first[19]:

$$w(\mathbf{x}_i) = \frac{\sum_{j=1}^{k} \|\mathbf{x}_i - \boldsymbol{\mu}_j\|^{-2-p}}{\left(\sum_{j=1}^{k} \|\mathbf{x}_i - \boldsymbol{\mu}_j\|^{-p}\right)^2} \tag{3.23}$$

This weight promotes the points that are distant from all the centres, which enforces relocation of these centres in such a way as to cover all data.

If $\mathbf{x}_i \approx \boldsymbol{\mu}_j$, then the original distance is replaced by the value $\max\{\|\mathbf{x}_i - \boldsymbol{\mu}_j\|, \epsilon\}$, where ϵ is a "sufficiently small number", as suggested by [530].

Fuzzy cluster membership of objects and weights of these objects allow for the computation of the centre coordinates—see the third step of the Algorithm 2.3

---

[19]The plain *k*-means algorithm assumes $w(\mathbf{x}_i) = 1$.

$$\mu_j = \frac{\sum_{i=1}^{m} u_{ij} w(\mathbf{x}_i) \mathbf{x}_i}{\sum_{i=1}^{m} u_{ij} w(\mathbf{x}_i)} \qquad (3.24)$$

This algorithm is less sensitive to initialisation [530]. However, it can still converge to a local minimum.

Hamerly and Elkan proposed in [230] two modifications, called Hybrid-1 i Hybrid-2. The first hybrid is a combination of the k-means algorithm with KHM. The split into clusters is crisp (an object belongs to the cluster with the closest centre), but each object is assigned a weight according to the formula (3.23). The membership degree of an object to a cluster in the second hybrid is determined by the Eq. (3.22), but the weights are constant, $w(\mathbf{x}_i) = 1, i = 1, \ldots, m$. None of these hybrids considered has a quality superior to KHM algorithm. However, their analysis indicates that the most important factor is the fuzzy membership u_{ij}. Hence, Hybrid-2 proved to be only a little bit inferior to the original KHM algorithm. The introduction of weights, on the other hand, improves the properties of the k-means algorithm.

Experiments presented in papers [230, 530], show that KHM provides with higher quality results than not only the k-means algorithm but also the fuzzy k-means algorithm, presented in Sect. 3.3. Further improvements of the algorithm, based on meta-heuristics, are presented in [519] and in the literature cited therein.

3.1.5.5 Kernel Based k-means Algorithm

The idea of the algorithm is to switch to a multidimensional feature space $\mathcal{F}$ and to search therein for prototypes $\boldsymbol{\mu}_j^{\Phi}$ minimizing the error

$$J_2^{\Phi}(U) = \sum_{i=1}^{m} \min_{1 \le j \le k} \| \Phi(\mathbf{x}_i) - \boldsymbol{\mu}_j^{\Phi} \|^2 \qquad (3.25)$$

where $\Phi \colon \mathbb{R}^n \to \mathcal{F}$ is a non-linear mapping of the space X into the feature space. This is the second variant of application of the kernel functions in cluster analysis, according to classification from the Sect. 2.5.7. In many cases, switching to the space $\mathcal{F}$ allows to reveal clusters that are not linearly separable in the original space.[20]

---

[20]Dhillon et al. [149] suggest for example to use a weighted kernel k-means to discover clusters in graphs analogous to ones obtained from a graph by normalised cuts. They optimise a weighted version of the cost function, that is

$$J_2^{\Phi}(U) = \sum_{i=1}^{m} \min_{1 \le j \le k} w(\mathbf{x}_i) \| \Phi(\mathbf{x}_i) - \boldsymbol{\mu}_j^{\Phi} \|^2$$

where

$$\boldsymbol{\mu}_j^{\Phi} = \frac{1}{\sum_{\mathbf{x}_i \in C_j} w(\mathbf{x}_i)} \sum_{\mathbf{x}_i \in C_j} w(\mathbf{x}_i) \Phi(\mathbf{x}_i)$$

In analogy to the classical k-means algorithm, the prototype vectors are updated according to the equation

$$\boldsymbol{\mu}_j^{\Phi} = \frac{1}{n_j} \sum_{\mathbf{x}_i \in C_j} \Phi(\mathbf{x}_i) = \frac{1}{n_j} \sum_{i=1}^{m} u_{ij} \Phi(\mathbf{x}_i) \tag{3.26}$$

where $n_j = \sum_{i=1}^{m} u_{ij}$ is the cardinality of the j-th cluster, and u_{ij} is the function, allocating objects to clusters, i.e. $u_{ij} = 1$ when $\mathbf{x}_i$ is an element of the j-th cluster and $u_{ij} = 0$ otherwise. A direct application of this equation is not possible, because the function Φ is not known. In spite of this, it is possible to compute the distances between the object images and prototypes in the feature space, making use of Eq. (2.50). The reasoning runs as follows:

$$\begin{aligned}
\|\Phi(\mathbf{x}_i) - \boldsymbol{\mu}_j^{\Phi}\|^2 &= \left(\Phi(\mathbf{x}_i) - \boldsymbol{\mu}_j^{\Phi}\right)^{\mathrm{T}} \left(\Phi(\mathbf{x}_i) - \boldsymbol{\mu}_j^{\Phi}\right) \\
&= \Phi(\mathbf{x}_i)^{\mathrm{T}} \Phi(\mathbf{x}_i) - 2\Phi(\mathbf{x}_i)^{\mathrm{T}} \boldsymbol{\mu}_j^{\Phi} + (\boldsymbol{\mu}_j^{\Phi})^{\mathrm{T}} \boldsymbol{\mu}_j^{\Phi} \\
&= \Phi(\mathbf{x}_i)^{\mathrm{T}} \Phi(\mathbf{x}_i) - \frac{2}{n_j} \sum_{h=1}^{m} u_{hj} \Phi(\mathbf{x}_i)^{\mathrm{T}} \Phi(\mathbf{x}_h) + \\
&\quad + \frac{1}{n_j^2} \sum_{r=1}^{m} \sum_{s=1}^{m} u_{rj} u_{sj} \Phi(\mathbf{x}_r)^{\mathrm{T}} \Phi(\mathbf{x}_s) \\
&= k_{ii} - \frac{2}{n_j} \sum_{h=1}^{m} u_{hj} k_{hi} + \frac{1}{n_j^2} \sum_{r=1}^{m} \sum_{s=1}^{m} u_{rj} u_{sj} k_{rs}
\end{aligned} \tag{3.27}$$

where, like in Sect. 2.5.7, $k_{ij} = \Phi(\mathbf{x}_i)^{\mathrm{T}} \Phi(\mathbf{x}_j) = K(\mathbf{x}_i, \mathbf{x}_j)$. If we denote with $\mathbf{u}_j$ the j-th column of the matrix U and substitute $\widetilde{\mathbf{u}}_j = \mathbf{u}_j / \|\mathbf{u}_j\|_1$, then the above equation can be rewritten in a closed form:

$$\|\Phi(\mathbf{x}_i) - \boldsymbol{\mu}_j\|^2 = k_{ii} - 2(\widetilde{\mathbf{u}}_j^{\mathrm{T}} K)_i + \widetilde{\mathbf{u}}_j^{\mathrm{T}} K \widetilde{\mathbf{u}}_j \tag{3.28}$$

where K is a Gram matrix with elements k_{ij}, and the scalar $(\widetilde{\mathbf{u}}_j^{\mathrm{T}} K)_i$ denotes the i-th component of the vector $(\widetilde{\mathbf{u}}_j^{\mathrm{T}} K)$.

In this way, one can update the elements of the matrix U without determining the prototypes explicitly. This is possible, because $\mathbf{x}_i$ is assigned to the cluster minimising the above-mentioned distance, i.e.

$$u_{ij} = \begin{cases} 1 & \text{if } \|\Phi(\mathbf{x}_i) - \boldsymbol{\mu}_j^{\Phi}\|^2 = \min_{1 \leq t \leq k} \|\Phi(\mathbf{x}_i) - \boldsymbol{\mu}_t^{\Phi}\|^2 \\ 0 & \text{otherwise} \end{cases} \tag{3.29}$$

(Footnote 20 continued)
The application of kernel k-means would liberate from the necessity of computing eigenvalues and eigenvectors needed in spectral clustering of graphs.

Though it is not possible to use the Eq. (3.26) directly, one can determine (in the original feature space X) the approximate cluster prototypes by assuming μ_j to be the object $\mathbf{x}^j$ matching the condition [533]

$$\mathbf{x}_i^j = \arg\min_{\mathbf{x}_i \in C_j} \| \Phi(\mathbf{x}_i) - \boldsymbol{\mu}_j^\Phi \|^2 \tag{3.30}$$

Prototypes, defined in this way, are in fact medoids (see the next section). Another method of determining the prototypes is presented in Sect. "**KFCM-$\mathcal{F}$** Algorithm" on page 121.

The algorithm is summarized in the form of the pseudo-code 3.6. It needs to be stressed that, like the k-means algorithm, its kernel based variant is also sensitive to initialisation, meaning the 2nd step of the algorithm. The simplest way to initialise it is to assign $k-1$ (randomly selected) objects to $k-1$ different clusters and to allocate the remaining $m-k+1$ objects to the k-th cluster. Another method is to assume the partition returned by the classical k-means algorithm. It guarantees that at least a part of the objects will belong to proper clusters and the Algorithm 3.6 will only modify erroneous assignments.

Algorithm 3.6 Kernel-based k-means algorithm

Input: Data set X, number of clusters k.
Output: A partition defined by the set of gravity centres $\{\boldsymbol{\mu}_1, \ldots, \boldsymbol{\mu}_k\}$.
1: Choose a kernel function K and compute the elements of the Gram matrix $k_{ij} = \mathsf{K}(\mathbf{x}_i, \mathbf{x}_j)$ for the set of objects $\{\mathbf{x}_1, \ldots, \mathbf{x}_m\}$.
2: Compute the initial allocation of elements among clusters.
3: For each object $\mathbf{x}_i \in X$ compute its distance from each cluster by applying Eq. (3.27), and assign $\mathbf{x}_i$ to the closest cluster
4: Repeat step (3) till none of u_{ij} values changes.
5: Determine the approximate prototypes according to Eq. (3.30).

The presented algorithm clusters correctly the data similar to the ones in Fig. 3.5, and, as reported by authors of [288], even in case of such sets as iris one gets better clustering performance (compared to k-means algorithm). An *on line* version of this algorithm was presented by Schölkopf, Smola and Müller in [419], and its modifications are presented in [194].

3.1.5.6 *k*-medoids Algorithm

The k-means algorithm is formally applicable only if the dissimilarity between pairs of objects is equal to the square of their Euclidean distance. This means that the features describing the object properties must be measured on quantitative scale, e.g. on the ratio scale. In such a case, the objects can be represented as n-dimensional vectors with real-valued components. Beside this, the partition cost, given by Eq. (3.1) depends on the maximal distances between the prototype μ_i and the object of the

Fig. 3.5 Results of application of the kernel-based variant of the k-means algorithm. Object-to-cluster allocation is indicated by different colours. $\sigma = 1$ was assumed

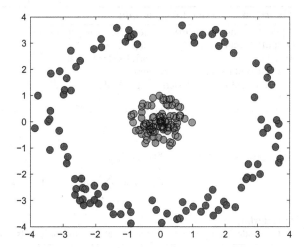

cluster represented by this prototype. Hence, the k-means algorithm is not resistant to the presence of outliers. Furthermore, the centres $\boldsymbol{\mu}_i$ are abstract objects not belonging to the set $\mathfrak{X}$. To avoid these disadvantages, one may measure the partition cost as follows:

$$J_{med}(\mathfrak{p}_1, \ldots, \mathfrak{p}_k) = \min_{1 \le j \le k} \sum_{i \in C_j} d(\mathfrak{x}_i, \mathfrak{p}_j) \tag{3.31}$$

where $\mathfrak{p}_1, \ldots, \mathfrak{p}_k \in \mathfrak{X}$ are cluster centres, called also prototypes, examples, exemplars, medoids or just centres, and $d: \mathfrak{X} \times \mathfrak{X} \to \mathbb{R}$ is a dissimilarity measure. There is no need for this measure to be symmetric or even metric (as required for Euclidean distance or its generalisations).

The k-medoids (or k-centres) algorithm aims at such a choice of centres $\{\mathfrak{p}_1, \ldots, \mathfrak{p}_k\} \subset \mathfrak{X}$ for which the index J_{med} reaches its minimum. By introducing the dissimilarity measure $d_{ij} = d(\mathfrak{x}_i, \mathfrak{p}_j)$ we broaden the applicability of the algorithm requiring only that for each pair of objects from the set $\mathfrak{X}$ it is possible to compute their dissimilarity. Working with dissimilarity matrix has an additional advantage: its dimension depends on the number of objects only.

As the object representation stops playing any role here, the practical implementations of the algorithm makes use only of the vector $\mathbf{c}$ having the elements c_i, indicating to which example (and in this way to which cluster) the i-th object belongs. More precisely:

$$c_i = \begin{cases} j & \text{if } \mathfrak{x}_i \in C_j \\ i & \text{if } \mathfrak{x}_i \text{ is an exemplar} \end{cases} \tag{3.32}$$

An elementary implementation of the k-medoids algorithm is demonstrated by the pseudo-code 3.7, see e.g. [237, sec. 14.3.10].

Algorithm 3.7 k-medoids algorithm

Input: Dissimilarity matrix $D = [d_{ij}]_{m \times m}$, number of clusters k
Output: Partition $C = \{C_1, \ldots, C_k\}$.
1: Select the subset $\mathcal{K} \subset \{1, \ldots, m\}$. Its elements are pointers to examples (prototypes)
2: **while (not** termination condition) **do**
3:   Assign objects to clusters using the rule

$$c_i = \begin{cases} \underset{j \in \mathcal{K}}{\arg\min} \ d_{ij} & \text{if } i \notin \mathcal{K} \\ i & \text{otherwise} \end{cases} , \ i = 1, \ldots, m \qquad (3.33)$$

4:   Update the examples, that is

$$j_r^* = \arg\min_{t:\, c_t = r} \sum_{t':\, c_{t'} = r} d_{tt'}, \ r = 1, \ldots, k \qquad (3.34)$$

5: **end while**
6: If the index value remained unchanged after testing m objects—Stop. Otherwise return to step 2.

In the simplest case the examples are picked at random, but very good results can be achieved by adapting the methods from Sect. 3.1.3. In particular, one can start with selection of k most dissimilar objects.

The Eq. (3.33) tells us that the object i is assigned to the least dissimilar example from the set $\mathcal{K}$. On the other hand, the Eq. (3.34) states that for a set of objects sharing a common example we select as the new example such an object for which the sum of dissimilarities to other objects of the cluster is the lowest. Like the k-means algorithm, the k-medoids algorithm stops when the new examples are identical with those of the previous iteration.

In spite of its superficial simplicity, the algorithm is much more time-consuming than k-means algorithm: determining a new example requires $O(|C_r|)$ operations, and the cluster allocation updates—$O(mk)$ comparisons.

Kaufman and Rousseeuw initiated research on elaboration of efficient method of determination of examples. In [284] they proposed two algorithms: PAM (*Partitioning Around Medoids*) and CLARA (*Clustering LARge Applications*). PAM follows the just described principle, that is- it seeks in $\mathfrak{X}$ such *medoid* objects which minimise the index (3.1). This approach is quite expensive. The big data analysis algorithm CLARA uses several (usually five) samples consisting of $40 + 2k$ objects and applies the PAM algorithm to each of the samples to obtain a set of proposed sets of medoids. Each proposal is evaluated using the index (3.1); the set yielding the lowest value of the criterion function is chosen.

The next improvement was the CLARANS (*Clustering Large Applications based upon RANdomized Search*), [375]—an algorithm with quadratic complexity in the number of objects m. The algorithm construes a graph of sets of k medoids; two graph nodes are connected by an edge if the assigned sets differ by exactly one element. The neighbours are generated using random search techniques. Another variant of the k-centres algorithm was proposed in [385].

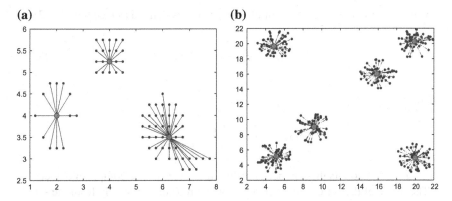

Fig. 3.6 Clustering of separable data sets with the *k*-medoids algorithm: **a** the set `data3_2`, **b** the set `data6_2`. Cluster examples are marked with red dots

In Fig. 3.6 allocations of objects to examples for the data sets `data3_2` and `data6_2` are presented.

Remark 3.1.1 The *k*-medoids algorithm should not be confused with the *k*-medians algorithm, proposed to make the *k*-means algorithm resistant against *outliers*. In the *k*-medians algorithm the Euclidean distance is replaced with the Manhattan distance, assuming $p = 1$ in the Eq. (3.2), [84, 273]. In such a case the cluster centres are determined from the equation

$$\sum_{\mathbf{x}_i \in C_j} \frac{\partial}{\partial \mu_{jl}} |x_{il} - \mu_{jl}| = 0 \Rightarrow \sum_{\mathbf{x}_i \in C_j} \text{sgn}(x_{il} - \mu_{jl}) = 0$$

i.e. $\mu_j l$ are the medians of the respective components of the vectors $\mathbf{x}_i$ assigned to the cluster C_j.

This algorithm plays an important role in operations research, in particular in the choice of location of service centres.[21]   □

3.1.5.7 *k*-modes Algorithm

Huang presented in [263] an adaptation of the *k*-means algorithm for cases where the features describing the objects are measured on nominal scale. Examples of such features are: sex, eye colour, hair colour, nationality etc. Each feature takes usually a value from a small value set; for example the value of the feature "eye colour" can be: "blue", "brown", "black" etc. By replacing these values with consecutive integers we can assign each real object $\mathfrak{x}_i$ a concise representation in terms of a vector $\mathbf{x}_i$, elements of which x_{il} are integers *pointing at* the proper value of the *i*-th feature.

---

[21] See e.g. N. Mladenović, J. Brimberg, P. Hansen, J.A. Moreno-Pérez. The p-median problem: A survey of meta-heuristic approaches. *European J. of Op. Res.*, 179(3), 2007, 927–939.

Now the dissimilarity of the pair of objects i and j is defined as (see Sect. 2.2.2):

$$d_{ij} = d(\mathbf{x}_i, \mathbf{x}_j) = \sum_{l=1}^{n} \delta(x_{il}, x_{jl}) \tag{3.35}$$

where

$$\delta(x_{il}, x_{jl}) = \begin{cases} 1 \text{ if } x_{il} = x_{jl} \\ 0 \text{ otherwise} \end{cases}$$

hence $d(\mathbf{x}_i, \mathbf{x}_j)$ counts the number of differences between the compared objects.

The mode of the set C is defined as an object $\mathbf{m}$ (not necessarily from the set $\mathfrak{X}$) minimizing the rating:

$$d(C, \mathbf{m}) = \sum_{\mathbf{x} \in C} d(\mathbf{x}, \mathbf{m}) \tag{3.36}$$

Huang noted in [263] the following property that allows to identify the mode of the set C quickly:

Lemma 3.1.1 *Let $Dom(A_l)$ denote the set of values of the feature A_l and let c_{ls} denote the number of objects in the set C in which the feature l takes on the value s. Vector $\mathbf{m}$ with components matching the condition*

$$m_l = \arg\max_{s \in Dom(A_l)} c_{ls}, \ l = 1, \ldots, n \tag{3.37}$$

is the mode of the set C.   □

Let $\mathbf{m}_j$ denote the mode of the j-th cluster. The problem of k-modes consists in finding such modes $\mathbf{m}_1, \ldots, \mathbf{m}_k$ that the index

$$J_{mod}(\mathbf{m}_1, \ldots, \mathbf{m}_k) = \sum_{j=1}^{k} d(C_j, \mathbf{m}_j) = \sum_{j=1}^{k} \sum_{\mathbf{x}_i \in C_i} d(\mathbf{x}_i, \mathbf{m}_j) \tag{3.38}$$

is minimised. A method to solve this problem is presented in the form of the pseudo-code 3.8.

The subsequent theorem points at the difference between the k-medoids and the k-modes algorithms

Theorem 3.1.2 *Let $\mathfrak{p}_1, \ldots, \mathfrak{p}_k$ be a set of medoids and let $\mathbf{m}_1, \ldots, \mathbf{m}_k$ be the set of modes. Let both be computed using respective algorithms. Then*

$$J_{med}(\mathfrak{p}_1, \ldots, \mathfrak{p}_k) \leq 2J_{mod}(\mathbf{m}_1, \ldots, \mathbf{m}_k) \tag{3.39}$$

□

Algorithm 3.8 *k*-modes algorithm, [263]

Input: Matrix X representing the set of objects, number of clusters k.
Output: Partition $\mathcal{C} = \{C_1, \ldots, C_k\}$.
1: Determine the dissimilarities d_{ij} for each pair of objects.
2: determine in any way k modes.
3: **while** (**not** termination condition) **do**
4: Using the dissimilarity measure 3.35 assign objects to the closest clusters
5: Update mode of each cluster applying the Lemma 3.1.1.
6: **end while**
7: Return a partition into k clusters.

Let us note however that the execution of the *k*-medoids algorithm requires only the dissimilarity matrix $D = [d_{ij}]$ while the *k*-modes algorithm insists on access to the matrix X with rows of which containing the characteristics of individual objects.

Huang considers in [263] the general case when the features describing objects are measured both on ratio scale and on nominal scale. Let us impose such an order on the features that the first n_1 features are measured on ratio scale while the remaining ones—on nominal scale. In such a case the cost of partition represented by the matrix $U = [u_{ij}]$ can be determined as follows:

$$J_p(U, M) = \sum_{j=1}^{k} \sum_{i=1}^{m} \left[u_{ij} \sum_{l=1}^{n_1} (x_{il} - m_{jl})^2 + \gamma u_{ij} \sum_{l=n_1+1}^{n} \delta(x_{il}, m_{jl}) \right] \quad (3.40)$$

where $\gamma > 0$ is a coefficient balancing both types of dissimilarities.

By substituting

$$P_j^r = \sum_{i=1}^{m} u_{ij} \sum_{l=1}^{n_1} (x_{il} - m_{jl})^2, \quad P_j^n = \gamma \sum_{i=1}^{m} u_{ij} \sum_{l=n_1+1}^{n} \delta(x_{il}, m_{jl})$$

we rewrite the Eq. (3.40) in the form

$$J_p(U, M) = \sum_{j=1}^{k} (P_j^r + P_j^n) \quad (3.41)$$

Optimisation of this index is performed iteratively, namely, starting with the initial centroid matrix M objects are assigned to clusters with the least differing centroids. The dissimilarity measure of the object $\mathbf{x}_i$ with respect to centroid $\mathbf{m}_i$ equals to

$$d(\mathbf{x}_i, \mathbf{m}_i) = \sum_{l=1}^{n_1} (x_{il} - m_{jl})^2 + \gamma \sum_{l=N_1+1}^{n} \delta(x_{ij}, m_{jl}) \quad (3.42)$$

Subsequently new cluster centres are determined. Their components being real numbers are computed as in the classic k-means algorithm, see Eq. (3.3), while for the nominal valued components the Lemma 3.1.1 is applied.

According to Huang [263], there are three prevailing differences between this algorithm (called by the author k-*prototypes*) and the algorithm CLARA, mentioned in the previous section:

- The k-prototypes algorithm processes the entire data set, while CLARA exploits sampling when applied to large data sets.
- The k-prototypes algorithm optimises the cost function on the entire data set so that at least a local optimum for the entire data collection is reached. CLARA performs optimization on samples only and hence runs at risk of missing an optimum, especially when for some reason the samples are biased.
- Sample sizes required by CLARA grow with increase of the size of the entire data set and the complexity of interrelationships within it. At the same time efficiency of CLARA decreases with the sample size: it cannot handle more than dozens of thousands of objects in a sample. On the contrary, the k-prototypes algorithm has no such limitations.

It is worth mentioning that the dissimilarity measure used here fulfils the triangle inequality, which implies that accelerations mentioned in Sect. 3.1.4 are applicable here.

3.2 EM Algorithm

Instead of concentrating on a given set of objects, a clustering process may target a population from which the actually obtained set has been sampled. In this case one assumes that the population can be described as a sum of statistical models, where each of the models describes a cluster of objects. A statistical model may be a probability density function of occurrence of objects of the given cluster in the feature space.

Frequently, as a matter of convenience, it is assumed that the probability density distribution function that describes the set of observations X is a mixture of Gaussian distributions,

$$p(\mathbf{x}) = \sum_{j=1}^{k} p(C_j) p(\mathbf{x}|C_j; \boldsymbol{\mu}_j, \Sigma_j) \qquad (3.43)$$

where $p(C_j)$ denotes the a priori probability of the object to belong to the j-th cluster, while $p(\mathbf{x}|C_j; \boldsymbol{\mu}_j, \Sigma_j)$ is the probability density of the n-dimensional Gaussian distribution with the vector of expected values $\boldsymbol{\mu}_j$ and covariance matrix Σ_j according to which the observations belonging to the j-th cluster are generated. Let us recall the formula for the probability density function of multidimensional (multivariate)

Gaussian distribution:

$$p(\mathbf{x}|C_j; \boldsymbol{\mu}_j, \Sigma_j) = \frac{1}{(2\pi)^{n/2}\sqrt{|\Sigma_j|}} \exp\left[-\frac{1}{2}(\mathbf{x} - \boldsymbol{\mu}_j)^{\mathsf{T}}\Sigma^{-1}(\mathbf{x} - \boldsymbol{\mu}_j)\right] \quad (3.44)$$

The assumption of Gaussian (or normal) distributions of elements in a cluster is a "classical" variant of the discussed approach but it is nonetheless sufficiently general. It is possible to demonstrate that every continuous and constrained probability density distribution can be approximated with any required precision by a mixture of Gaussian distributions [303]. Sample sets that were generated from a mixture of two normal distributions were presented already in Fig. 2.13 on p. 64. But, in general, mixtures of any type of distributions can be considered. A rich bibliography on this subject can be found on the Web page of David Dowe http://www.csse.monash.edu. au/~dld/mixture.modelling.page.html.

Let us reformulate the task of clustering as follows: We will treat the vectors of expected values as cluster centres (rows of the M matrix). Instead of the cluster membership matrix U, we will consider the matrix $P = [p_{ij}]_{m \times k}$, where $p_{ij} = p(C_j|\mathbf{x}_i)$ means the probability that the i-th object belongs to the j-th cluster. Now let us define the target function for clustering as follows:

$$J_{EM}(P, M) = -\sum_{i=1}^{n} \log\left(\sum_{j=1}^{k} p(\mathbf{x}_i|C_j)p(C_j)\right) \quad (3.45)$$

The function $\log(\cdot)$ was introduced to simplify the computations. The minus sign allows to treat the problem of parameter estimation as the task of minimization of the expression (3.45).

Assuming that the set of observations consists really of k clusters we can imagine that each observation $\mathbf{x}_i, i = 1, \ldots, m$ was generated by exactly one of the k distributions, though we do not know by which. Hence, the feature that we shall subsequently call "*class*" is a feature with unknown values.[22] We can say that we have incomplete data of the form $\mathbf{y} = (\mathbf{x}, class)$. Further, we do not know the parameters of the constituent distributions of the mixture (3.43). In order to estimate them, we can use the EM (*Expectation Maximization*) method, elaborated by Dempster, Laird and Rubin [144], which is, in fact, a maximum likelihood method, adapted to the situation when data is incomplete. The method consists in repeating the sequence of two steps (see [75, 348]):

- Step **E** (*Expectation*): known feature values of objects, together with estimated parameters of models are used to compute for these objects the expected values of the unknown ones.
- Step **M** (*Maximization*): both known (observed) and estimated (unobserved) feature values of objects are used to estimate the model parameters using the model likelihood maximisation given the data.

---

[22]More precisely, it is a so called "hidden feature" or "hidden variable".

The EM algorithm is one more example of the broad class of hill-climbing algorithms. Its execution starts with a choice of mixture parameters and proceeds by stepwise parameter re-estimation until the function (3.45) reaches a local optimum.

Step **E** consists in estimation of the probability that the i-th object belongs to the class j. If we assume that we know the parameters of distributions belonging to the mixture (3.43) then we can estimate this probability using the Bayes theorem:

$$p(C_j|\mathbf{x}_i) = \frac{p(\mathbf{x}_i|C_j)p(C_j)}{p(\mathbf{x}_i)} = \frac{p(\mathbf{x}_i|C_j)p(C_j)}{\sum_{l=1}^{k} p(\mathbf{x}_i|C_l)p(C_l)} \tag{3.46}$$

$p(\mathbf{x}_i|C_j)$ is the value of the density function $\phi(\mathbf{x}_i; \boldsymbol{\mu}, \Sigma)$, having mean vector $\boldsymbol{\mu}$ and covariance matrix Σ, at the point $\mathbf{x}_i$, i.e. $p(\mathbf{x}_i|C_j) = \phi(\mathbf{x}_i; \boldsymbol{\mu}_j, \Sigma_j)$. The value $p(\mathbf{x}_i)$ is computed according to the total probability theorem.

To execute step **M**, we assume that we know the class membership of all objects, which is described by the aposteriorical distribution $p(C_j|\mathbf{x}_i)$. Under these circumstances the method of likelihood maximisation can be used to estimate the distribution parameters. The estimator of the vector of expected values of the j-th cluster is of the form:

$$\boldsymbol{\mu}_j = \frac{1}{mp(C_i)} \sum_{i=1}^{m} p(C_j|\mathbf{x}_i)\mathbf{x}_i \tag{3.47}$$

Note that this equation has the same form as the equation (2.44) given that the weights $w(\mathbf{x}_i)$ are constant, e.g. $w(\mathbf{x}_i) = 1$. If we assume, additionally, the preferential class membership (i.e. $p(C_j|\mathbf{x}_i) \in \{0, 1\}$) then we will discover that the formula (3.47) can be reduced to the classical formula of computation of gravity centre of a class.

Similarly, we compute the covariance matrix from the formula:

$$\Sigma_j = \frac{1}{mnp(C_j)} \sum_{i=1}^{m} (\mathbf{x}_i - \boldsymbol{\mu}_j)(\mathbf{x}_i - \boldsymbol{\mu}_j)^{\mathrm{T}} \tag{3.48}$$

and the apriorical class membership probability from the equation:

$$p(C_j) = \frac{1}{m} \sum_{i=1}^{m} p(C_j|\mathbf{x}_i) \tag{3.49}$$

We can summarize these remarks with the Algorithm 3.9.

If we speak in the language of the k-means algorithm, then we will say that the step **E** is equivalent to an update of the assignment of points to clusters, and the step **M** can be deemed as determining characteristics of the new partition of the set X. A convenient approach (see e.g. [134]) to implement the above-mentioned pseudo-code consists in updating apriorical probabilities right after step **E**, thereafter determining centres $\boldsymbol{\mu}_i^{t+1}$, and lastly computing the matrix Σ_j^{t+1}.

Algorithm 3.9 EM algorithm for cluster analysis

Input: Matrix X representing the set of objects, number of clusters k.

Output: A set of pairs $(\mu_1, \Sigma_1), \ldots, (\mu_k, \Sigma_k)$, representing the components C_j of a mixture of Gaussian distributions fitting the data, for each object from X the probability of belonging to each of the components C_j.

1: Initialisation. Set the iteration counter to $t = 0$. Give the initial estimates of the parameters of the distributions μ_j^t, Σ_j^t and of apriorical values of probabilities $p(C_j)$, $j = 1, \ldots, k$.

2: (Step E): Using Bayes Theorem compute the membership of i-th object in j-th class, $p^{t+1}(C_j|\mathbf{x}_i), i = 1, \ldots, m, j = 1, \ldots, k$ – compare equation (3.46).

3: (Step M): Knowing class memberships of objects update the distribution parameters μ_i^{t+1}, Σ_j^{t+1}, and apriorical probabilities $p^{t+1}(C_j)$, $j = 1, \ldots, k$ – equations (3.47) – (3.49).

4: $t = t + 1$

5: Repeat steps 2 – 4 till the estimate values stabilise.

System WEKA [505] implements the EM algorithm for cluster analysis. Many other systems exploit mixtures of distributions, e.g. Snob [491], AutoClass [106] and MClust [186].

Experiments presented in [229] show that the EM algorithm is, with respect to quality, comparable to k-means algorithm. Just like in the case of the latter, also performance of the EM algorithm depends on initialisation.[23] However, as noticed in [15], the EM algorithm is not suitable for processing of high-dimensional data due to losses in computational precision. Let us remember, too, that determining the covariance matrix requires computation of $\frac{1}{2}(n^2 + n)$ elements, which becomes quite expensive with the growth of the number of dimensions. Finally, the disadvantage of the EM algorithm formulated in the here presented matter is its slow convergence, [144]. But, on the other hand, it is much more universal than the k-means algorithm: it can be applied in the analysis of non-numerical data. The quick algorithm developed by Moore in [360] allows to reduce, at least partially, the deficiencies listed.

Structural similarity between the EM and k-means algorithms permits to state that if the distribution $p(C_j|\mathbf{x})$ is approximated by the rule "the winner takes all", then the EM algorithm with a mixture of Gaussian distributions with covariance matrix $\Sigma_j = \epsilon \mathbb{I}$, $j = 1, \ldots, k$, where $\mathbb{I}$ means a unit matrix, becomes the k-means algorithm when $\epsilon \to 0$ [285].

Dasgupta and Schulman [134] analyse deeply the case of a mixture of spherical normal distributions, that is—those with $\Sigma_j = \sigma_j \mathbb{I}$. They prove that if the clusters are sufficiently well separable, see Eq. (2.58), then the estimation of mixture parameters requires only two iterations of a slightly modified Algorithm 3.9. This is particularly important in case of high-dimensional data ($n \gg \ln k$). They developed also a simple procedure, applicable in the cases when the exact value of k is not known in advance. They propose an initialisation method, which is a variant of the method (c) from Sect. 3.1.3.

In recent years, we have been witnessing a significant advancement of theoretical description and understanding of the problems related to learning of parameters of

---

[23]One of initialisation methods for the EM algorithm, estimating mixture parameters, consists in applying the k-means algorithm.

a mixture of distributions. A reader interested in these aspects is recommended to study the paper [278].

Slow convergence of the standard version of the EM algorithm urged the development of its various modifications, such as greedy EM,[24] stochastic EM,[25] or random swap EM. Their brief review can be found in sections 4.4. i 4.5 of the Ph.D. thesis [535] and in paper [392], where a genetic variant of the EM algorithm has been proposed. It allows not only to discover the mixture parameters but also the number of constituent components.

3.3 FCM: Fuzzy c-means Algorithm

3.3.1 Basic Formulation

Ruspini [412] was the first to propose to represent clusters as fuzzy sets. If $\widetilde{\chi}_j$ denotes the membership function representing j-th cluster then the fuzzy partition is the set of fuzzy subsets $\widetilde{F}_1, \ldots, \widetilde{F}_k$. Their membership functions are defined as follows:

$$\sum_{j=1}^{k} \widetilde{\chi}_j(\mathbf{x}_i) = 1, i = 1, \ldots, m \qquad (3.50)$$

This definition is intended to take into account *outliers* and other irregularities in the induced partition. Additionally, to be able to determine representatives of individual clusters (called prototypes or—as in the case of crisp sets—gravity centres), Dunn [166] introduced the concept of compact and well separated (CWS) clusters. Let $C = \{C_1, \ldots, C_k\}$ mean a crisp[26] partition of a set of objects. Let $(\mathbf{x}, \mathbf{y}) \in X$ be two such points that $\mathbf{x} \in C_i$, $\mathbf{y} \in conv(C_i)$ where $conv(C_i)$ is a convex hull C_i, and let $i, j, k, j \neq k$ be any indexes. The set C_i is a CWS-cluster if $(\mathbf{x}, \mathbf{y})$ are much closer to one another in the sense of some assumed distance d than any two points $(\mathbf{u}, \mathbf{v}) \in X$ such that $\mathbf{u} \in C_j$, $\mathbf{v} \in conv(C_k)$. This property is described by the index

$$\beta(k, C) = \left(\min_{j=1,\ldots,k} \; \min_{l=1,\ldots,m, l \neq j} d(C_j, conv(C_l)) \right) \Big/ \max_{j=1,\ldots,k} diam(C_j) \qquad (3.51)$$

where $diam(C_j)$ is the diameter of the set C_j, and $d(A, B)$ is the distance between the sets A and B. It turns out that the set X can be divided into k CWS-clusters (when the distance d is known), if [166]

---

[24]See J.J. Verbeek, N. Vlassis, B. Kröse: "Efficient greedy learning of Gaussian mixture models", *Neural computation*, **15**(2), 2003, pp. 469–485.

[25]G. Celeux, D. Chauveau, J. Diebolt: Stochastic versions of the EM algorithm: an experimental study in the mixture case. *J. of Stat. Computation and Simulation*, **55**(4), 1996, pp. 287–314.

[26]a "crisp" partition is the traditional view of splitting a set: each element is assigned to exactly one cluster. A "fuzzy" partition allows each element to be assigned to "some degree" to several clusters.

$$\bar{\beta}(k) = \max_{\mathcal{C}} \beta(k, \mathcal{C}) > 1 \tag{3.52}$$

The problem of finding a partition, for which the index $\beta(k, \mathcal{C}) = \bar{\beta}(k)$, is very difficult. In particular, Dunn [166] has shown that the algorithm ISODATA, introduced in Sect. 3.1.4, finds the CWS-partition also in cases when it does not exist. For this reason the criterion (3.1) was weakened allowing for fuzzy partitions of the set of objects. Let $U = [u_{ij}]_{m \times k}$ be a matrix fulfilling the following conditions

$$(a) \ 0 \le u_{ij} \le 1, i = 1, \ldots, m, j = 1, \ldots, k$$

$$(b) \ \sum_{j=1}^{k} u_{ij} = 1, i = 1, \ldots, m \tag{3.53}$$

$$(c) \ 0 < \sum_{i=1}^{m} u_{ij} < m, j = 1, \ldots, k$$

Element u_{ij} determines the membership degree of the object $\mathbf{x}_i$ in the class C_j. The condition (a) means a partial membership of objects the classes, and condition (b) enforces the full membership of objects in a distinguished set of classes.[27] Finally, the condition (c) does not permit to create empty classes. The set of all matrices fulfilling the conditions (3.53) is called the set of fuzzy partitions and we denote it with the symbol $\mathcal{U}_{fk}$. Let us make the remark that the dimension of the space of fuzzy clusterings significantly exceeds the dimension of the space of crisp partitions [73].

Prototypes of classes are represented by the rows of the matrix $M = (\boldsymbol{\mu}_1, \ldots, \boldsymbol{\mu}_k)^T$ having the dimensions $k \times n$.

The search for a fuzzy partition fulfilling the conditions mentioned is carried out via minimisation of the quality index of the form:

$$J_\alpha(U, M) = \sum_{i=1}^{m} \sum_{j=1}^{k} u_{ij}^\alpha \|\mathbf{x}_i - \boldsymbol{\mu}_j\|^2 \tag{3.54}$$

where $\alpha > 1$ is a parameter (so-called fuzziness exponent). It has been shown [73] that when $\alpha \to 1$ then the algorithm generates clusters identical with ones obtained from the algorithm ISODATA (which justifies the name *Fuzzy ISODATA* for the algorithm described here), and when $\alpha \to \infty$ then the values $u_{ij} \to 1/k$, i.e. we obtain a completely fuzzy partition. Typical values of the parameter α are 1.25 and 2.

Introduction of partial object membership allows to distinguish between the "typical" group representatives and the objects that are less likely characterised as belonging to a given group. The idea has been illustrated in Fig. 3.7.

---

[27] In other words, we do not allow for the existence of other (unknown) classes, to which objects could partially or completely belong.

(a)   **(b)**

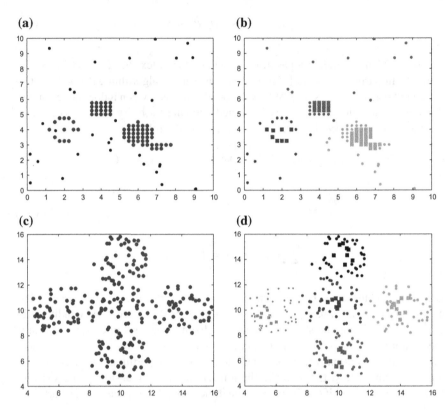

(c)   **(d)**

Fig. 3.7 Clustering of data with the use of the FCM algorithm: **a** Set data3_2 (blue points) with added noise (black points). **b** Assignments to groups generated by the FCM algorithm. The bigger squares denote typical objects, smaller circles—the other objects; the colour indicates the class membership. Same convention applies to Figure (**d**) representing the results obtained for the set data5_2 from Figure (**c**). $\alpha = 2$ was assumed in computations. Object $\mathbf{x}_i$ is treated as typical if $\max_j u_{ij} \geq 0.9$

Remark 3.3.1 Due to similarity of the function (3.54) to (3.1), the subsequently discussed algorithm could be called fuzzy k-means algorithm. Its authors denoted the number of clusters with the letter c, hence they coined the name *Fuzzy c-Means*. We will, however, continue to denote the number of clusters with the letter k, while keeping the abbreviation FCM when referring to this algorithm.   □

Remark 3.3.2 The Euclidean distance occurring in the Eq. (3.54) can be replaced with any norm $\|\mathbf{x}\|_W = (\mathbf{x}^T W \mathbf{x})^{1/2}$, where W is a positive definite matrix of dimension $n \times n$. This more general formulation has been used, e.g., in the book [70]. Subsequent formulas for the membership degrees u_{ij} remain valid if the Euclidean distance used therein is replaced by the norm $\| \cdot \|_W$.   □

3.3.2 Basic FCM Algorithm

The indicator (3.54), depending on the partition U and the prototypes M, is not a convex function of its arguments. Nonetheless, the problem of its minimisation can be simplified by fixing the values of either matrix U or M, that is, by assuming $U = \tilde{U}$, $M = \tilde{M}$, respectively. In such a case the simplified functions $\tilde{J}_m(U) = J_m(U, \widehat{M})$ and $\tilde{J}_m(M) = J_m(\widehat{U}, M)$ are convex functions of their arguments—consult [69, 271]. Hence, the classical optimisation methods are applicable, i.e. to determine the vector of prototypes, the equation system of k equations of the form given below, is solved

$$\frac{\partial}{\partial \mu_j} J_\alpha(U, M) = 0, \quad j = 1, \ldots, k \tag{3.55}$$

and solving for the matrix U relies on creating the Lagrange function

$$L(J_\alpha, \lambda) = \sum_{i=1}^{m} \sum_{j=1}^{k} u_{ij}^\alpha \|\mathbf{x}_i, \mu_j\|^2 - \sum_{i=1}^{m} \lambda_i \left(\sum_{j=1}^{k} u_{ij} - 1 \right) \tag{3.56}$$

and computing values u_{ij} from the equation system

$$\begin{cases} \dfrac{\partial}{\partial u_{ij}} L(J_\alpha, \lambda) = 0 \\[2mm] \dfrac{\partial}{\partial \lambda_i} L(J_\alpha, \lambda) = 0 \end{cases} \quad i = 1, \ldots, m, \quad j = 1, \ldots, k \tag{3.57}$$

Solution of the task of minimisation of function (3.54) in the set of fuzzy partitions $\mathcal{U}_{fk}$ is obtained iteratively by computing, for a fixed partition U, the prototypes (see Appendix A)

$$\mu_{jl} = \frac{\sum_{i=1}^{m} u_{ij}^\alpha x_{il}}{\sum_{l=1}^{m} u_{lj}^\alpha}, \quad j = 1, \ldots, k, l = 1, \ldots, n \tag{3.58}$$

and then new assignments to clusters

$$u_{ij} = \begin{cases} \left[\displaystyle\sum_{l=1}^{k} \left(\frac{\|\mathbf{x}_i - \mu_j\|}{\|\mathbf{x}_i - \mu_l\|} \right)^{\frac{2}{\alpha-1}} \right]^{-1} & \text{if } Z_i = \emptyset \\[4mm] \epsilon_{ij} & \text{if } j \in Z_i \neq \emptyset \\[2mm] 0 & \text{if } j \notin Z_i \neq \emptyset \end{cases} \tag{3.59}$$

where $Z_i = \{j : 1 \leq j \leq k, \|\mathbf{x}_i - \mu_j\| = 0\}$, and values ϵ_{ij} are chosen in such a way that $\sum_{j \in Z_i} \epsilon_{ij} = 1$. Usually, the non-empty set Z_i contains one element, hence $\epsilon_{ij} = 1$. if we substitute $d(\mathbf{x}_i, \mu_j) = \|\mathbf{x}_i - \mu_j\| + \varepsilon$, where ε is a number close to zero, e.g. $\varepsilon = 10^{-10}$, then the above equation can be simplified to

$$u_{ij} = \left[\sum_{l=1}^{k} \left(\frac{d(\mathbf{x}_i, \boldsymbol{\mu}_j)}{d(\mathbf{x}_i, \boldsymbol{\mu}_l)} \right)^{\frac{2}{\alpha-1}} \right]^{-1} \tag{3.60}$$

This last equation can be rewritten in an equivalent form that requires a lower number of divisions

$$u_{ij} = \frac{d^{\frac{2}{1-\alpha}}(\mathbf{x}_i, \boldsymbol{\mu}_j)}{\sum_{l=1}^{k} d^{\frac{2}{1-\alpha}}(\mathbf{x}_i, \boldsymbol{\mu}_l)} \tag{3.61}$$

Note that the membership degree u_{ij} depends not only on the distances of the i-th object from the centre of the j-th cluster but also on its distance to centres of other clusters. Furthermore, when $\alpha = 2$, the denominator of the expression (3.61) is the harmonic average of the squares of distance of the object from the cluster centres. In this case, the FCM resembles a little bit the KHM algorithm from Sect. 3.1.5.4.

The FCM algorithm termination condition is usually defined as stabilisation of the partition matrix. If U^t, U^{t+1} denote the matrices obtained in subsequent iterations, then the computations are terminated as soon as $\max_{i,j} |u_{ij}^{t+1} - u_{ij}^{t}| \leq \epsilon$, where ϵ is a predefined precision, e.g. $\epsilon = 0.0001$. Of course, one can change the order of steps, i.e. first initiate the prototype matrix M and determine the corresponding cluster assignments u_{ij}, and subsequently update the prototypes. The last two steps are repeated till the vectors stabilise $\boldsymbol{\mu}_j$, i.e. till $|\mu_{jl}^{t+1} - \mu_{jl}^{t}| < \epsilon$, where $j = 1, \ldots, k, l = 1, \ldots, n$. Note that the number of comparisons required in deciding on computation termination is in the second case usually lower than in the first one: matrix M contains kn elements, while matrix U has only mk elements.

The fuzziness coefficient α is an important parameter of the algorithm, because its properties heavily depend on the value of this coefficient. If this value is close to one, the algorithm behaves like the classic k-means algorithm. If α grows without bounds, then prototypes converge to the gravity centre of the object set X. Several heuristic methods for selection of this coefficient have been proposed in the literature [64, 380]. The best recommendation is to choose it from the interval [1.5, 2.5]. One has, however, to remember that the mentioned heuristics result from empirical investigations and may not reflect all the issues present in real world data.[28] The paper [522] suggests some rules for the choice of the value of α, pointing at the fact that coefficient choice depends on the data themselves. The impact of parameter choice on the behaviour of the FCM algorithm was investigated by Choe and Jordan in [114]. The influence of parameter α on the number of iterations until stabilisation of the matrix U and on the distance of returned prototypes $\boldsymbol{\mu}_j$ from the intrinsic ones $\boldsymbol{\mu}_j^*$ is depicted in Fig. 3.8. The average distance has been computed as

---

[28]Dunn [166] recommended $\alpha = 2$ on the grounds of the capability of the algorithm to reconstruct well separated clusters.

$$d_{avg} = \frac{1}{k} \sum_{j=1}^{k} \|\boldsymbol{\mu}_j^* - \boldsymbol{\mu}_j\| \tag{3.62}$$

We chose the set iris.data stemming from the repository [34] to illustrate the comparison. Intrinsic centres of the groups are presented in the table below.

$$\mathbf{M}^* = \begin{bmatrix} 5.00\ 3.42\ 1.46\ 0.24 \\ 5.93\ 2.77\ 4.26\ 1.32 \\ 6.58\ 2.97\ 5.55\ 2.02 \end{bmatrix} \tag{3.63}$$

We can deduce from the figure that the increase of the value of the coefficient α causes the increase in the number of iterations. Interestingly, for α values close to 1 the error measured with the quantity d_{avg} initially decreases and then, after exceeding the optimal value (here $\alpha^* = 1.5$), it grows. The standard deviation is close to zero for low α values, which means a high repeatability of the results (Fig. 3.8).

Remark 3.3.3 Schwämmle and Jensen [422] investigated randomly generated data sets and concluded that the best value of the α parameter is a function of the dimension n and the amount of data m. Based on their experiments, they proposed the following analytical form of this function

$$\begin{aligned} \alpha(m, n) = 1 &+ \left(\frac{1418}{m} + 22.05\right)n^{-2} \\ &+ \left(\frac{12.33}{m} + 0.243\right)n^{-0.0406\ln(m)-0.1134} \end{aligned} \tag{3.64}$$

This result contradicts the common practice to choose $\alpha = 2$.

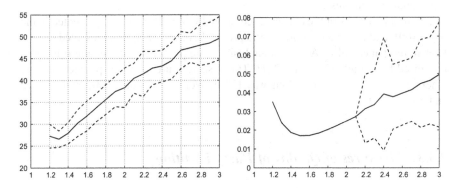

Fig. 3.8 Influence of the value of fuzziness exponent α on the number of iterations (picture to the left) and the quality of prototypes (picture to the right), expressed in terms of the average distance d_{avg} according to the Eq. (3.62). Solid line indicates the mean value $\overline{m}$ of the respective quantity (averaged over 100 runs), and the dotted lines mark the values $\overline{m} \pm \overline{s}$, with $\overline{s}$ being the standard deviation. $\epsilon = 10^{-8}$ was assumed. Experiments were performed for the data set iris.txt

It has been also observed that investigation of the minimal distance between the cluster centres, V_{MCD}, is a good hint for choice of the proper number of clusters k. It is recommended to choose as a potential candidate such a value k^* for which an abrupt decrease of the value of V_{MCD} occurs.

One should, however, keep in mind that these results were obtained mainly for the Gaussian type of clusters. Both authors suggest great care when applying the derived conclusions to the real data.   □

The FCM algorithm converges usually quickly to a stationary point. Slow convergence should be rather treated as an indication of a poor starting point. Hathaway and Bezdek cite in [239, p. 243] experimental results on splitting a mixture of normal distributions into constituent sets. An FCM algorithm with parameter $\alpha = 2$ needed 10–20 iterations to complete the task, while the EM algorithm required hundreds, and in some cases thousands of iterations.

Encouraged by this discovery, the authors posed the question: Let $p(\mathbf{y}; \alpha_0, \alpha_1) = \alpha_0 p_0(\mathbf{y}) + \alpha_1 p_1(\mathbf{y})$, where p_0 i p_1 are symmetric density functions with mean values 0 and 1, respectively, and the expected values (with respect to components) of the variable $|Y|^2$ being finite. Can these clusters be identified? Regrettably, it turns out that (for $\alpha = 2$) there exist such values of α_0, α_1 that the FCM algorithm erroneously identifies the means of both sub-populations. This implies that even if one could observe an infinite number of objects the algorithm possesses only finite precision (with respect to estimation of prototype location). This result is far from being a surprise. FCM is an example of non-parametric algorithm and the quality index J_α does not refer to any statistical properties of the population. Hathaway and Bezdek conclude [239]: if population components (components of a mixture of distributions) are sufficiently separable, i.e. each sub-population is related to a clear "peak" in the density function, then the FCM algorithm can be expected to identify the characteristic prototypes at least as well as the maximum likelihood method (and for sure much quicker than this).

An initial analysis of the convergence of the FCM algorithm was presented in the paper [69], and a correct version of the proof of convergence was presented 6 years later in the paper [243]. A careful analysis of the properties of this algorithm was initialised by Ismail and Selim [271]. This research direction was pursued, among others, by Kim, Bezdek and Hathaway [290], Wei and Mendel [500], and also Yu and Yang [523].

3.3.3   Measures of Quality of Fuzzy Partition

Though the FCM algorithm requires only a small number of iteration till a stable partition of objects is obtained, the partition quality assessment requires the introduction of appropriate quality measures that would depend on the parameters k and α. The reader will find a thorough discussion of clustering quality evaluation methods in

Chap. 4. Nonetheless, we will present here a couple of measures that play a special role in the evaluation of fuzzy clustering algorithms.

If the intrinsic group membership of objects is known, $\mathcal{P}^t = \{C_1^t, \ldots, C_k^t\}$, then the so-called purity index is applied. It reflects the agreement between the intrinsic partition and the one found by the algorithm. First, the fuzzy assignment matrix is transformed to a Boolean group membership matrix U^b with entries

$$u_{ij}^b = \begin{cases} 1 \text{ if } j = \displaystyle\arg\max_{1 \le t \le k} u_{it} \\ 0 \text{ otherwise} \end{cases} \qquad (3.65)$$

A partition $\mathcal{P}^f = \{C_1^f, \ldots, C_k^f\}$ is obtained in this way, with $C_j^f = \{i : u_{ij}^b = 1\}$. Subsequently, we construct the contingency table with entries $m_{ij} = |C_i^t \cap C_j^f|$. Finally, the agreement of the two partitions is calculated as

$$\mathfrak{P}(\mathcal{P}^1, \mathcal{P}^2) = \frac{1}{m} \sum_{i=1}^{k_1} \max_{1 \le j \le k_2} m_{ij} \qquad (3.66)$$

While the purity is a general purpose measure, the so called reconstruction error is a measure designed exclusively for algorithms producing fuzzy partitions [387]. It is defined as the average distance between the original object and the reconstructed one, i.e.

$$e_r = \frac{1}{m} \sum_{i=1}^{m} \| \mathbf{x}_i - \widetilde{\mathbf{x}}_i \|^2 \qquad (3.67)$$

where the reconstruction $\widetilde{\mathbf{x}}_i$ is performed according to the formula

$$\widetilde{\mathbf{x}}_i = \frac{\sum_{j=1}^{k} u_{ij}^\alpha \mu_j}{\sum_{j=1}^{k} u_{ij}^\alpha}, \quad i = 1, \ldots, m \qquad (3.68)$$

The lower the reconstruction error the better the algorithm performance. One should, however, remember that low values of the k parameter induce rather high error values ($e_r \rightarrow 0$ when $k \rightarrow m$). The purity index evaluates, therefore, the algorithm precision, while the reconstruction error describes the quality of encoding/decoding of objects by the prototypes and the assignment matrix. One can say that e_r is a measure of dispersion of prototypes in the feature space. In particular, the reconstruction error decreases when the prototypes are moved towards the centres of dense areas of feature space [214]. Hence, e_r measures the capability of prototypes to represent individual clusters. The dependence of the reconstruction error on the fuzziness exponent α is illustrated in Fig. 3.9.

Other measures of quality of fuzzy partitions are partition coefficient $F_k(U)$ and fuzzy partition entropy $H_k(U)$. They are defined as follows, see e.g. [70]:

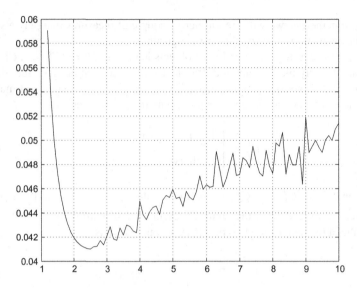

Fig. 3.9 Influence of the α exponent (abscissa axis) on the reconstruction error (ordinate axis). Test data: file `iris.txt`

$$F_k(U) = \operatorname{tr}(UU^{\mathsf{T}})/m = \frac{1}{m} \sum_{i=1}^{m} \sum_{j=1}^{k} u_{ij}^2 \tag{3.69}$$

$$H_k(U) = -\frac{1}{m} \sum_{i=1}^{m} \sum_{j=1}^{k} u_{ij} \log_a u_{ij}, \quad a > 1 \tag{3.70}$$

They exhibit the following properties

$$F_k(u) = 1 \Leftrightarrow H_k(U) = 0 \Leftrightarrow U \text{ is a crisp partition}$$

$$F_k(U) = \tfrac{1}{k} \Leftrightarrow H_k(U) = \log_a(k) \Leftrightarrow U = [\tfrac{1}{k}] \tag{3.71}$$

$$\tfrac{1}{k} \le F_k(U) \le 1; \quad 0 \le H_k(U) \le \log_a(k)$$

The entropy H_k is more sensitive to local changes in partition quality than the coefficient F.

When data tend to concentrate into a low number of well separable groups, then these indicators constitute a good hint for the proper selection of the number of clusters.

Remark 3.3.4 The quantity $H_k(U)$ is a measure indicating the degree of fuzziness of a partition. If U is a crisp partition, then $H_k(U) = 0$ for any matrix U with elements $u_{ij} \in \{0, 1\}$. The entropy $H(U)$, defined later in Eq. (4.31) in Sect. 4.4.2, allows to explore deeper the nature of a crisp partition. One should not confuse these two measures. □

Example 3.3.1 To illustrate the above-mentioned thesis, consider two sets presented in Fig. 3.10. The first set data_6_2 contains two dimensional data forming six clear-cut clusters. The second set data_4_2 was obtained by randomly picking same number of points from a two dimensional normal distribution $N(m_i, \mathbb{I}), i = 1, \ldots, 4$, where $m_1 = (3, 0)^T$, $m_2 = (0, 3)^T$, $m_3 = (3, 3)^T$, $m_4 = (0, 6)^T$, and $\mathbb{I}$ means a unit covariance matrix. In both cases $F_k(U)$ and $H_k(U)$ for various values of k were computed, see Fig. 3.11. In the first case of a clear structure, we observe clear cut optima reached by both indexes. In the second case both indices behave monotonically. □

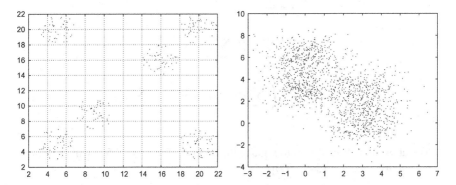

Fig. 3.10 Test sets data_6_2 (Figure to the left) and data_4_2 (Figure to the right)

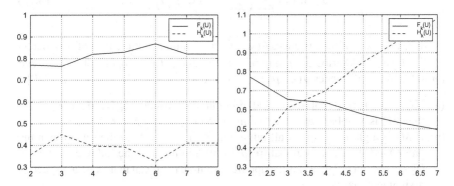

Fig. 3.11 Dependence of the values of quality criteria for the sets data_6_2 (to the left) and data_4_2 (to the right) on the assumed number of classes

3.3.4 An Alternative Formulation

The method of determining the membership degree via Eq. (3.60) does not depend on the definition of distance, [70]. By replacing the Euclidean distance $\|\mathbf{x}_i - \boldsymbol{\mu}_j\|$ used there with some distance $d(\mathbf{x}_i, \boldsymbol{\mu}_j)$, we can state that

$$
u_{ij}^{\alpha-1} d^2(\mathbf{x}_i, \boldsymbol{\mu}_j) = \left(\frac{1}{\sum_{l=1}^{k} d^{\frac{2}{1-\alpha}}(\mathbf{x}_i, \boldsymbol{\mu}_l)} \right)^{\alpha-1} = \left(\sum_{l=1}^{k} d^{\frac{2}{1-\alpha}}(\mathbf{x}_i, \boldsymbol{\mu}_l) \right)^{1-\alpha}
$$

Hence

$$
\begin{aligned}
\tilde{J}_\alpha(M) &= \sum_{i=1}^{m} \sum_{j=1}^{k} u_{ij} u_{ij}^{\alpha-1} d^2(\mathbf{x}_i, \boldsymbol{\mu}_j) \\
&= \sum_{i=1}^{m} \sum_{j=1}^{k} u_{ij} \left(\sum_{l=1}^{k} d^{\frac{2}{1-\alpha}}(\mathbf{x}_i, \boldsymbol{\mu}_l) \right)^{1-\alpha} \\
&= \sum_{i=1}^{m} \left(\sum_{l=1}^{k} d^{\frac{2}{1-\alpha}}(\mathbf{x}_i, \boldsymbol{\mu}_l) \right)^{\alpha-1} \cdot \sum_{j=1}^{k} u_{ij}
\end{aligned}
$$

The above equation allows to replace the criterion function (3.54) with an equivalent function (see e.g. [240, 500])

$$
\tilde{J}_\alpha(M) = \sum_{i=1}^{m} \left(\sum_{j=1}^{k} d^{\frac{2}{1-\alpha}}(\mathbf{x}_i, \boldsymbol{\mu}_j) \right)^{1-\alpha} \tag{3.72}
$$

More precisely, if the distances $d(\mathbf{x}_i, \boldsymbol{\mu}_j)$ are continuous functions of parameters $M \in \mathcal{M}$, which describe the group prototypes, and $\mathcal{M}$ is an open subset $\mathbb{R}^{kn}$, and for $M^* \in \mathcal{M}$ all distances $d(\mathbf{x}_i, \boldsymbol{\mu}_j) > 0$, $i = 1, \ldots, m$, $j = 1, \ldots, k$, then, [240]

(a) If (U^*, M^*) is a global (resp. local) minimum of the index $J_\alpha(U, M)$ in $\mathcal{U}_{kc} \times \mathcal{M}$ then M^* is a global (resp. local) minimum of index $\tilde{J}_\alpha(M)$ in the set $\mathcal{M}$.
(b) If M^* is a global (resp. local) minimum of the index $\tilde{J}_\alpha$ in the set $\mathcal{M}$ then the pair $(\Phi(M^*), M^*)$ is a global (resp. local) minimum of the index J_α in the set $\mathcal{U}_{fk} \times \mathcal{M}$, where $\Phi(M)$ is a mapping assigning the given matrix M a respective assignment matrix U.

The difference between the formulations (3.54) and (3.72) is not only formal. Let us notice that in the first case the number of parameters to be identified is $k(m+n)$, and in the second case there are kn of them.

More importantly, in the first case we deal with an optimisation task with constraints (3.61). The solution of the problem stated in this way is the alternating algorithm from Sect. 3.3.2.

In the second case, the determination of the allocations from the family $\mathcal{U}_{fk}$ is ignored, with concentration on finding such prototypes that minimise the indicator (3.72). Here we deal with a much simpler task of unconditional optimisation. Class allocations for objects are computed from Eq. (3.61) only if needed.

To illustrate more convincingly the difference between these approaches, let us assume that we replace the Euclidean distance used so far with a distance induced by Minkowski metric.[29] This requires a modification of the iterative algorithm in the formulation (3.54). Such changes are unnecessary in the optimisation driven formulation (3.72). Only the value of the indicator $\widetilde{J}_\alpha(M)$ is computed in a slightly different manner.

An obvious requirement is to dispose of an efficient optimisation procedure. If the number of features is not big, then the simplex algorithm (called also creeping amoeba) can be applied. The algorithm is authored by Nelder and Mead—compare Chap. 10 in [398]. Application of a genetic algorithm to optimise the indicator (3.72) was presented in [228].

More details about the equivalence of solutions obtained in both cases are presented in paper [240].

3.3.5 Modifications of the FCM Algorithm

The base FCM algorithm assumes that the distance between an object and a prototype is an Euclidean one. But this is not the best way of measuring the similarity of the multidimensional objects to distinguished prototypes (see Sect. 2.2). Below we present several different modifications of the target function having the general form

$$J_\alpha(U, M) = \sum_{j=1}^{k} \sum_{i=1}^{m} u_{ij}^\alpha d^\gamma(\mathbf{x}_i, \boldsymbol{\mu}_j) \tag{3.73}$$

In Sect. 3.3.5.1 in place of $d(\mathbf{x}_i, \boldsymbol{\mu}_j)$ we consider Minkowski distance. The cluster allocation applied in FCM is sufficient if the clusters are Voronoi-shaped. Therefore, in Sect. 3.3.5.2 in place of $d(\mathbf{x}_i, \boldsymbol{\mu}_j)$ we assume Mahalanobis distance. It enables the discovery of clusters with varying shapes and densities. Another modification, adopted for such a situation was proposed by Gath and Geva in [197]. If the shape of the clusters is known in advance (e.g. line segments, circles, ellipsoids), the algorithms presented in Sects. 3.3.5.3 and 3.3.5.4 may prove useful. In Sect. 3.3.5.5 we present a generalisation of the spherical algorithm from Sect. 3.1.5.3. We will present also two kernel variants of the base FCM algorithm (Sect. 3.3.5.6).

---

[29]This problem is discussed in depth in Sect. 3.3.5.1.

Constraints imposed on matrix U (see Eq. (3.53) from page 101) cause that the algorithm is sensitive to the presence of non-typical data (*outliers*). Wu and Yang [508] attempted to overcome this shortcoming by suggesting the extension of the set of k groups with an additional cluster containing all non-typical data. Another idea is to give up the requirement that the sum of membership degrees to all groups be equal 1. This leads to so-called possibilistic clustering algorithm that we discuss in Sect. 3.3.5.7.

Finally, in Sect. 3.3.5.8 an algorithm is presented in which instead of distance, the matrix of dissimilarities R is used, in which entries quantify the dissimilarity between pairs of objects.

An exhaustive review of all FMC algorithm modifications is beyond the scope of this book. An interested reader is kindly advised to consult, e.g., the books [70, 73, 140, 255]. A systematic review of other modifications of the FCM algorithms, including the ones for observations stemming from a mixture of various probability distributions, is presented in [520].

The basic version of FCM requires storing the allocation matrix U, which increases its memory complexity and, in consequence, also time complexity. But with knowledge of the prototype vectors, elements of this matrix can be restored quite efficiently, if needed. Such a solution was presented by Kolen and Hutcheson in the paper [298]. Hore, Hall and Goldgof present in [256] the so-called weighted FCM algorithm, developed for very big data sets. Another big data analysis variant of the algorithm was proposed in paper [113].

Let us note at the end the interesting attempts to merge self-organising networks of Kohonen[30] with FCM algorithm. First attempts were presented in [267], and their refinement is the FKCN algorithm (*Fuzzy Kohonen Clustering Network*), described in papers [381, 468].

3.3.5.1   FCM Algorithm with Minkowski Metric

As announced in Sect. 2.2, let us start with the most natural modification of the algorithm, consisting in replacement of the Euclidean distance d_2 with its generalisation, that is, with Minkowski distance d_p, $p > 0$, defined by the formula (2.4), in particular—the Manhattan distance d_1. This modification was deemed to make the algorithm resistant against outliers. So, in the general case, the function (3.73) can have the form (see also [241])

---

[30]T. Kohonen. *Self-Organization and Associative Memory*, Springer-Verlag, Berlin, Heidelberg, 1898.

$$J_{\alpha,p}(U, M) = \sum_{i=1}^{m} \sum_{j=1}^{k} u_{ij}^{\alpha} \|\mathbf{x}_i - \boldsymbol{\mu}_j\|_p^p$$

$$= \sum_{i=1}^{m} \sum_{j=1}^{k} \sum_{l=1}^{n} u_{ij}^{\alpha} |x_{il} - \mu_{jl}|^p \qquad (3.74)$$

We will seek here a solution by alternating computation of the elements of the U and M matrices just like in the case of the base FCM algorithm. Elements of the U matrix are determined according to Eq. (3.59) except that the Euclidean distance used there is substituted with Minkowski distance $\|\mathbf{x}_i - \boldsymbol{\mu}_j\|$.

Determining components of prototypes is, however, more expensive than in the classical case of $p = 2$, because one has to find kn minima of one-dimensional functions of the form

$$f_{jl}(\mu_{jl}) = \sum_{i=1}^{m} u_{ij}^{\alpha} |x_{il} - \mu_{jl}|^p, \quad j = 1, \ldots, k, l = 1, \ldots, n \qquad (3.75)$$

A function defined in this way is not convex for $0 < p < 1$; its diagram has "spikes" at points $\mu_{jl} = x_{il}$, compare Fig. 3.12a. Determining the value μ_{jl}^* minimising the function f consists, therefore, in computing its values for all arguments $x_{il}, i = 1, \ldots, m$ and choosing the proper one among them, that is

$$\mu_{jl}^* = \underset{1 \le i \le m}{\arg\min} \ f_{jl}(\mu_{jl}) \qquad (3.76)$$

We obtain a piece-wise linear function in case of $p = 1$, compare Fig. 3.12b, but the solution is found by the same procedure. Finally, in case of $p > 1$ we have to do with a convex function with a single minimum, Fig. 3.12c.

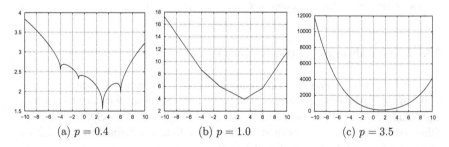

(a) $p = 0.4$ (b) $p = 1.0$ (c) $p = 3.5$

Fig. 3.12 Diagrams of the function (3.75) for three values of the parameter p. Function is of the form $f(x) = 0.6^{\alpha}|-4 - x|^p + 0.5^{\alpha}|-1 - x|^p + 0.8^{\alpha}|3 - x|^p + 0.7^m|6 - x|^p$, where x means μ_{jl}, and $\alpha = 2.5$

Jajuga[31] presented an effective method of minimising function (3.74) for $p = 1$. By changing the summation order and by introducing weight

$$w_{ijl} = \frac{u_{ij}^{\alpha}}{|x_{il} - \mu_{jl}|}$$ (3.77)

the target function (3.74) can be represented in an equivalent form

$$J_{\alpha,1}(W, M) = \sum_{j=1}^{k} \sum_{l=1}^{n} \sum_{i=1}^{m} w_{ijl}(x_{il} - \mu_{jl})^2$$ (3.78)

resembling the original target function (3.54) if we replace u_{ij}^{α} with w_{ijl}. Cluster membership degrees can be determined in the following (already known) way

$$u_{ij} = \left(\sum_{l=1}^{k} \frac{d_{ij}^{1/(1-\alpha)}}{d_{il}^{1/(1-\alpha)}} \right)^{-1} , d_{ij} = \sum_{l=1}^{n} |x_{il} - \mu_{jl}|$$ (3.79)

But determining prototypes requires solving kn equations of the form

$$\min \sum_{i=1}^{m} w_{ijl}(x_{il} - \mu_{jl})^2, \quad j = 1, \ldots, k, l = 1, \ldots, n$$

i.e. knowing the actual values of u_{ij}, the weights are computed according to Eq. (3.77), and then prototype coordinates are calculated

$$\mu_{jl} = \frac{\sum_{i=1}^{m} w_{ijl} x_{il}}{\sum_{i=1}^{m} w_{ij}}$$ (3.80)

in analogy to Eq. (3.58). A review of other methods of fuzzy clustering for various values of the parameter p is contained in the paper [241]).

3.3.5.2   Gustafson-Kessel (GK) Algorithm

Both FCM algorithm and its crisp counterpart prefer Voronoi ("spherical") clusters of balanced cardinality. If we have to deal with ellipsoid shaped clusters, then the algorithm proposed by Gustafson and Kessel [218] would be helpful. This algorithm exploits adaptively modified Mahalanobis distance, defined in Sect. 2.2.1.2. The target function (3.73) has now the form

---

[31] K. Jajuga, L_1-norm based fuzzy clustering. *Fuzzy Sets and System* **39**, 43–50, 1991. See also P.J.F. Groenen, U. Kaymak and J. van Rosmalen: Fuzzy clustering with Minkowski distance functions, Chap. 3 in [140].

$$J_{\alpha, A_1, \ldots, A_k}(U, M) = \sum_{j=1}^{k} \sum_{i=1}^{m} u_{ij}^{\alpha} \|\mathbf{x}_i - \boldsymbol{\mu}_j\|_{A_j}^2 \qquad (3.81)$$

where $A_j \in \mathbb{R}^{n \times n}$ is a positive definite matrix defining the distances between the elements of the j-th cluster:

$$\|\mathbf{x}_i - \boldsymbol{\mu}_j\|_{A_j}^2 = (\mathbf{x}_i - \boldsymbol{\mu}_j)^{\mathrm{T}} A_j (\mathbf{x}_i - \boldsymbol{\mu}_j)$$

Matrix A_j is determined from the covariance matrix Σ_j, computed for objects of j-th cluster

$$A_j^{-1} = |\rho_j \Sigma_j|^{1/(r+1)} \Sigma_j^{-1} \qquad (3.82)$$

where r is a user-defined parameter,[32] ρ_j is the volume of the j-th cluster (usually $\rho = 1$), and Σ_j is the covariance matrix of the j-th cluster:

$$\Sigma_j = \frac{\sum_{i=1}^{m} u_{ij}^{\alpha} (\mathbf{x}_i - \boldsymbol{\mu}_j)(\mathbf{x}_i - \boldsymbol{\mu}_j)^{\mathrm{T}}}{\sum_{i=1}^{m} u_{ij}^{\alpha}} \qquad (3.83)$$

and $|\Sigma_j|$ is the determinant of this matrix.

The algorithm consists of the following steps

1. Compute the prototypes of clusters $\boldsymbol{\mu}_j$ by applying the formula (3.58).
2. For each cluster compute the covariance matrix Σ_j from Eq. (3.83).
3. Calculate distances $d_{A_j}^2(\mathbf{x}_i, \boldsymbol{\mu}_j) = (\mathbf{x}_i - \boldsymbol{\mu}_j)^{\mathrm{T}} |\Sigma_j|^{1/n} \Sigma_j^{-1} (\mathbf{x}_i - \boldsymbol{\mu}_j)$
4. Update the assignment matrix U by assuming

$$u_{ij} = \left[\sum_{l=1}^{k} \left(\frac{d_{A_j}(\mathbf{x}_i, \boldsymbol{\mu}_j)}{d_{A_l}(\mathbf{x}_i, \boldsymbol{\mu}_l)} \right)^{\frac{2}{\alpha-1}} \right]^{-1}$$

This algorithm has been widely applied, in particular in digital image analysis[33] and in engineering tasks, [36]. However, if the coordinates of some of the observations constituting one group are strongly correlated then the covariance matrix suffers from singularities, which results in incorrectness of cluster assignments. A method immunising the algorithm against such situations was presented in [37].

Example 3.3.2 Let us consider a synthetic data set consisting of three differently shaped clusters, presented in Fig. 3.13. Samples representing the individual groups stem from two-dimensional normal distributions with parameters presented in Table 3.1.

The table to the right presents the quality indicators for partitions generated by the FCM and GK (Gustafson-Kessel) algorithms. Both the partition coefficient and the

---

[32]Its role is similar to that of the fuzziness exponent α. Usually we assume $r = n - 1$.

[33]R.N. Dave. Boundary detection through fuzzy clustering. In *IEEE International Conf. on Fuzzy Systems*, pp. 127–134, San Diego, USA, 1992.

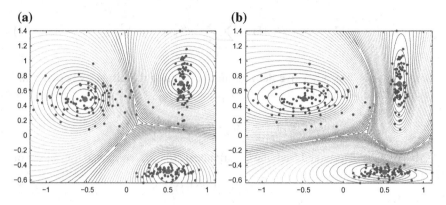

Fig. 3.13 Comparison of partitions generated by the **a** FCM algorithm and **b** GK algorithm for a synthetic data set consisting of three clusters with different shapes

Table 3.1 Table to the left: Parameters of Gaussian distributions used to generate samples belonging to the three groups. Table to the right: Quality indicators of the partitions generated by the algorithms FCM and GK (Gustafson-Kessel)

group	μ_X	μ_Y	σ_X	σ_Y
1	-0.5	0.5	0.35	0.15
2	0.7	0.7	0.05	0.25
3	0.5	-0.5	0.20	0.05

	$F_3(U)$	$H_3(U)$
FCM	0.8602	0.2865
GK	0.9133	0.1923

partition entropy are better for the partition obtained from GK algorithm. Figure 3.13 shows silhouette diagrams of the membership functions generated by both algorithms.   □

3.3.5.3 FCV Algorithm: Fuzzy c-varietes

The FCM algorithm identifies spherical clusters by computing the centre and radius of each of them. The resultant spheres constitute *manifolds* describing data subsets. A natural research direction was to extend this idea to manifolds of other shapes. An extension to hyper-ellipsoids was presented in the preceding section. Bezdek introduced in [72] linear manifolds of dimension $r \in \{0, \ldots, s\}$, spanned over vectors $\{\mathbf{b}_1, \ldots, \mathbf{b}_r\}$ and passing through a given point $\mathbf{a} \in \mathbb{R}^s$. The subsequent equation describes the general form of such a manifold

$$\mathcal{V}^r(\mathbf{a}; \mathbf{b}_1, \ldots, \mathbf{b}_r) = \left\{ \mathbf{y} \in \mathbb{R}^s : \mathbf{y} = \mathbf{a} + \sum_{l=1}^{r} \xi_l \mathbf{b}_l, \xi_l \in \mathbb{R} \right\} \qquad (3.84)$$

If $r = 0$ then $\mathcal{V}^0(\mathbf{a})$ is a point in s-dimensional Euclidean space; if $r = 1$ then $\mathcal{V}^1(\mathbf{a}; \mathbf{b})$ is a straight line in this space, and if $r = 2$ then $\mathcal{V}^2(\mathbf{a}; \mathbf{b}_1, \mathbf{b}_2)$ is a plane passing through the point $\mathbf{a} \in \mathbb{R}^s$.

Like in Sect. 3.3.5.2, one defines the distance $d_A(\mathbf{x}, \mathcal{V}^r)$ of the point $\mathbf{x} \in \mathbb{R}^s$ to the manifold $\mathcal{V}^r$:

$$d_A(\mathbf{x}, \mathcal{V}^r) = \min_{\mathbf{y} \in \mathbb{R}^s} d_A(\mathbf{x}, \mathbf{y}) = \left(\|\mathbf{x} - \mathbf{a}\|_A^2 - \sum_{l=1}^{r} \left((\mathbf{x} - \mathbf{a})^\mathrm{T} A \mathbf{b}_l \right)^2 \right)^{1/2} \qquad (3.85)$$

If we have now k manifolds $\mathcal{V}_1^r, \ldots, \mathcal{V}_k^r$ described by vectors $\mathfrak{a} = (\mathbf{a}_1, \ldots, \mathbf{a}_k)$, $\mathfrak{b}_j = (\mathbf{b}_{1j}, \ldots, \mathbf{b}_{kj})$, $j = 1, \ldots, r$, i.e.

$$\mathcal{V}_j^r = \left\{ \mathbf{y} \in \mathbb{R}^s : \mathbf{a}_j + \sum_{l=1}^{r} \xi_l \mathbf{b}_{jl}, \ \xi_l \in \mathbb{R} \right\}, \ j = 1, \ldots, k$$

then the generalised target function (3.73) has the form

$$J_\alpha(U, \mathfrak{a}; \mathfrak{b}_1, \ldots, \mathfrak{b}_r) = \sum_{j=1}^{k} \sum_{i=1}^{m} u_{ij}^m D_{ij}^2 \qquad (3.86)$$

where $D_{ij} = d_A(\mathbf{x}_i, \mathcal{V}_j^r)$.

Once again the minimisation of this indicator is achieved via a variant of the alternating algorithm from Sect. 3.3.2. Vectors $\mathbf{a}_j$ and $\mathbf{b}_{jl}$, $l = 1, \ldots, r$, spanning the j-th manifold are determined from the equation[34]

$$\mathbf{a}_j = \frac{\sum_{i=1}^{m} u_{ij}^\alpha \mathbf{x}_i}{\sum_{i=1}^{m} u_{ij}^\alpha}, \quad \mathbf{b}_{jl} = A^{-1/2} s_{jl} \qquad (3.87)$$

where s_{jl} is the l-th eigenvector of the scatter matrix

$$S_j = A^{1/2} \left(\sum_{i=1}^{m} u_{ij}^\alpha (\mathbf{x}_i - \mathbf{a}_j)(\mathbf{x}_i - \mathbf{a}_j)^\mathrm{T} \right) A^{1/2}$$

corresponding to the l-th maximal eigenvalue[35] of this matrix. Membership degree of i-th object in j-th manifold is determined using an analogue of the Eq. (3.61), i.e.

---

[34] See J.C. Bezdek, C. Coray, R. Gunderson and J. Watson. Detection and characterization of cluster substructure: I. Linear structure: Fuzzy c-lines. *SIAM J. Appl. Math.* **40**(2), 339–357, 1981; Part II Fuzzy c-varieties and convex combinations thereof, *SIAM J. Appl. Math.* **40**(2), 358–372, 1981.

[35] Note that S_j is a symmetric matrix of dimension $n \times n$. Hence it has n (not necessarily distinct) real-valued eigenvalues. See Sect. B.3.

$$u_{ij} = \frac{D_{ij}^{\frac{2}{1-\alpha}}}{\sum_{l=1}^{k} D_{il}^{\frac{2}{1-\alpha}}} \tag{3.88}$$

The algorithm presented here is very sensitive to initialisation, particularly for bigger values of the parameter k. This topic has been investigated more deeply in e.g. [307, 515].

The reader may pay attention to the fact that the FCV algorithm can be viewed as a precursor of approaches like k-planes algorithm presented in [85], projective clustering [158], or manifold learning [417]. It is also worth mentioning that a simple clustering algorithm for straight-line-segment shaped clusters was presented in [178].

3.3.5.4   FCS Algorithm: Fuzzy c-shells

Still another generalisation of the basic FCM algorithm was proposed by Dave [135, 137]. His algorithm was devoted to detection of spherical and elliptical edges in digital images, as an alternative approach to Circle Hough Transform for circle detection. Though Hough transform, [205], is a useful edge detection technique, it has the drawback of high memory and time complexity.

The FCS algorithm seeks a fuzzy partition into k circular groups. Each group is determined by the circle centre $\mathfrak{a} = \{\mathbf{a}_1, \ldots, \mathbf{a}_k\}$ and radius $\mathbf{r} = \{r_1, \ldots, r_k\}$. The algorithm strives to minimize the indicator

$$J_\alpha(U, \mathfrak{a}, \mathbf{r}) = \sum_{j=1}^{k} \sum_{i=1}^{m} u_{ik}^\alpha \left(\|\mathbf{x}_i - \mathbf{a}_j\|_{A_j} - r_j \right)^2 \tag{3.89}$$

We assume here that $\mathbf{a}_j \in \mathbb{R}^s, r_j > 0, j = 1, \ldots, k$.

Under these circumstances, the FCS algorithm consists of the following steps:

(i) Fix the algorithm parameters α, k and the precision ϵ.
(ii) Initialise the partition matrix $U^{(t)}$
(iii) Determine circle centres $\mathbf{a}_j$ and radiuses r_j by solving the equation system

$$\sum_{i=1}^{m} u_{ij}^\alpha \left(1 - \frac{r_j}{\|\mathbf{x}_i - \mathbf{a}_j\|_{A_j}} \right) (\mathbf{x}_i - \mathbf{a}_j) = 0 \tag{3.90}$$

and

$$\sum_{i=1}^{m} u_{ij}^\alpha \left(\|\mathbf{x}_i - \mathbf{a}_j\|_{A_j} - r_j \right) = 0 \tag{3.91}$$

(iv) Determine the new membership degrees u_{ij} by applying Eq. (3.88) with $D_{ij} = \left(\|\mathbf{x}_i - \mathbf{a}_j\|_{A_j} - r_j \right)$.
(v) if $\|U_{old} - U_{new}\| < \epsilon$ then STOP, else return to step (iii).

If $A_j = \mathbb{I}$, $j = 1, \ldots, k$, then the above algorithm detects clusters, in which objects are located on or in close vicinity to circles, representing individual groups. By assigning each group a unique matrix A_j one obtains a generalised (adaptive) FCS algorithm. In such a case, when new values $\mathbf{a}_j$, r_j are determined, one needs to update the matrix A_j using e.g. the Eq. (3.82).

A careful convergence analysis performed for the FCS algorithm in [71] suggests that a single iteration of the Newton method should be used to determine subsequent approximations of the vector $\mathbf{a}_j$ and the radius r_j. On the other hand, Dave finds in [137] that already three iterations of the Newton method are sufficient to reach precise results.

One can approach manifolds of other shapes in a way analogous to the one presented above.[36]

3.3.5.5   SFCM: Spherical FCM Algorithm

If both $\mathbf{x}_i$ and μ_j are unit vectors, we can recall the observation (3.12) and define the target function as follows

$$J_s(U, M) = \sum_{i=1}^{m} \sum_{j=1}^{k} u_{ij}^{\alpha} \left(1 - \sum_{l=1}^{n} x_{il} \mu_{jl} \right) \tag{3.92}$$

We seek such a pair (U^*, M^*), for which the above function attains the minimum. This time, we require not only that $U \in \mathcal{U}_{fk}$ but also that $\|\mu_j\| = 1$. The latter condition means that

$$\sum_{l=1}^{n} \mu_{jl}^2 = 1, \ j = 1, \ldots, k \tag{3.93}$$

Here, the Lagrange function has the form

$$L(J_s, \lambda, \Lambda) = \sum_{i=1}^{m} \sum_{j=1}^{k} u_{ij}^{\alpha} \left(1 - \sum_{l=1}^{n} x_{il} \right) - \sum_{i=1}^{m} \lambda_i \left(\sum_{j=1}^{k} u_{ij} - 1 \right) - \sum_{j=1}^{k} \Lambda_j \left(\sum_{l=1}^{n} \mu_{jl}^2 - 1 \right) \tag{3.94}$$

By applying a procedure analogous to the one from Appendix A we can state that the entries of the matrix U are of the form

---

[36]See e.g. F. Klawonn, R. Kruse, and H. Timm. Fuzzy shell cluster analysis. *Courses and Lectures—International Centre for Mechanical Sciences*, Springer 1997, pp. 105–120; http://public. fh-wolfenbuettel.de/~klawonn/Papers/klawonnetaludine97.pdf.

$$u_{ij} = \frac{D^{\frac{1}{1-\alpha}}(\mathbf{x}_i, \boldsymbol{\mu}_j)}{\sum_{l=1}^{k} D^{\frac{1}{1-\alpha}}(\mathbf{x}_i, \boldsymbol{\mu}_l)} = \frac{\left(1 - \sum_{t=1}^{n} x_{it}\mu_{jt}\right)^{\frac{1}{1-\alpha}}}{\sum_{l=1}^{k} \left(1 - \sum_{t=1}^{n} x_{it}\mu_{lt}\right)^{\frac{1}{1-\alpha}}} \tag{3.95}$$

The above formula holds only if $D(\mathbf{x}_i, \boldsymbol{\mu}_j) = (1 - \mathbf{x}_i^{\mathrm{T}}\boldsymbol{\mu}_j) \neq 0$. Otherwise, if the constraint does not hold, one applies a variant of the Eq. (3.59).

By solving the equation system

$$\begin{cases} \dfrac{\partial}{\partial \mu_{jl}} L(J_s, \lambda, \Lambda) = 0 \\[4mm] \dfrac{\partial}{\partial \Lambda_j,} L(J_\alpha, \lambda, \Lambda) = 0 \end{cases} \quad j = 1, \ldots, k, \quad l = 1, \ldots, n \tag{3.96}$$

one determines the prototype components, see [353], as

$$\mu_{jl} = \sum_{i=1}^{m} u_{ij}^{\alpha} x_{il} \sqrt{\sum_{r=1}^{n} \left(\sum_{t=1}^{m} u_{jt}^{\alpha} x_{tr}\right)^{-2}} \tag{3.97}$$

3.3.5.6 Kernel-Based Variants of the FCM Algorithm

Like classical k-means algorithm, also FCM algorithm can be kernelised. Two variants of the kernel-based fuzzy c-means algorithm are distinguished. The KFCM-X variant constructs the class prototypes in the original feature space. The other variant, called KFCM-$\mathcal{F}$, assumes that the prototypes are defined in the kernel feature space, in analogy to the KHCM algorithm from Sect. 3.1.5.5.

KFCM-X Algorithm

The algorithm seeks the minimum of the target function defined as

$$J_\alpha^K(U, M) = \sum_{i=1}^{m} \sum_{j=1}^{k} u_{ij}^{\alpha} \|\Phi(\mathbf{x}_i) - \Phi(\boldsymbol{\mu}_j)\|^2 \tag{3.98}$$

We know already that

$$\|\Phi(\mathbf{x}_i) - \Phi(\boldsymbol{\mu}_j)\|^2 = \mathsf{K}(\mathbf{x}_i, \mathbf{x}_i) - 2\mathsf{K}(\mathbf{x}_i, \boldsymbol{\mu}_j) + \mathsf{K}(\boldsymbol{\mu}_j, \boldsymbol{\mu}_j) \tag{3.99}$$

where $\mathsf{K}(\mathbf{x}, \mathbf{y}) = \Phi^{\mathrm{T}}(\mathbf{x})\Phi(\mathbf{y})$ is a kernel function. The most frequently applied function is in practice the Gaussian kernel, i.e. $\mathsf{K}(\mathbf{x}, \mathbf{y}) = \exp\left(\|\mathbf{x} - \mathbf{y}\|^2/\sigma^2\right)$, where $\sigma > 0$ is a parameter. In such a case $\mathsf{K}(\mathbf{x}, \mathbf{x}) = 1$, hence[37]

---

[37]In fact the formulas below remain valid for any kernel function matching the condition $K(\mathbf{x}, \mathbf{x}) = 1$.

$$\|\Phi(\mathbf{x}_i) - \Phi(\boldsymbol{\mu}_j)\|^2 = 2(1 - K(\mathbf{x}_i, \boldsymbol{\mu}_j))$$

which means that the quality index (3.98) is of the form

$$J_\alpha^K(U, M) = 2 \sum_{i=1}^{m} \sum_{j=1}^{k} u_{ij}^\alpha (1 - K(\mathbf{x}_i, \boldsymbol{\mu}_j)) \tag{3.100}$$

The resulting algorithm resembles the original FCM algorithm where one alternates between updating the membership degrees u_{ij} and the prototype coordinates according to the equations below,[38] see e.g. [214, 532]:

$$u_{ij} = \frac{\left(1 - K(\mathbf{x}_i, \boldsymbol{\mu}_j)\right)^{\frac{1}{1-\alpha}}}{\sum_{l=1}^{k} \left(1 - K(\mathbf{x}_i, \boldsymbol{\mu}_l)\right)^{\frac{1}{1-\alpha}}} \tag{3.101}$$

$$\boldsymbol{\mu}_j = \frac{\sum_{i=1}^{m} u_{ij}^\alpha K(\mathbf{x}_i, \boldsymbol{\mu}_j) \mathbf{x}_i}{\sum_{i=1}^{m} u_{ij}^\alpha K(\mathbf{x}_i, \boldsymbol{\mu}_j)} \tag{3.102}$$

If K is a Gaussian kernel and $\sigma \to \infty$, then $K(\mathbf{x}_i, \boldsymbol{\mu}_j) \to 1 - \|\mathbf{x}_i - \boldsymbol{\mu}_j\|^2/\sigma^2$ and the algorithm resembles the performance of the classical FCM algorithm.

The procedure was applied in [532] to cluster incomplete data. Under such scenario the algorithm consists of the following steps

(a) Initialise prototypes $\boldsymbol{\mu}_j$, $j = 1, \ldots, k$.
(b) Substitute $x_{ib} = 0$, if b-th component of i-th observation is not known.
(c) As long as prototypes are not stable, execute:

(c1) update membership degrees according to Eq. (3.101),
(c2) update prototypes according to Eq. (3.102),
(c3) reconstruct values of unknown features by applying a variant of Eq. (3.68):

$$x_{ib} = \frac{\sum_{j=1}^{k} u_{ij}^\alpha K(\mathbf{x}_i, \boldsymbol{\mu}_j) \mu_{jb}}{\sum_{j=1}^{k} u_{ij}^\alpha K(\mathbf{x}_i, \boldsymbol{\mu}_j)}$$

KFCM-$\mathcal{F}$ Algorithm

In this case the target function is of the form

$$J_\alpha^{\mathcal{F}}(U) = \sum_{i=1}^{m} \sum_{j=1}^{k} u_{ij}^\alpha \|\Phi(\mathbf{x}_i) - \boldsymbol{\mu}_j^\Phi\|^2 \tag{3.103}$$

---

[38]If $K(\mathbf{x}_i, \boldsymbol{\mu}_l) = 1$ for some index l then, upon determining u_{il}, we proceed analogously as when applying Eq. (3.59).

Following the guidelines of Sect. 3.1.5.5 and [214] and the references cited therein, the membership degrees u_{ij} are determined from the equation

$$u_{ij} = \frac{\|\Phi(\mathbf{x}_i) - \boldsymbol{\mu}_j^\Phi\|^{\frac{2}{1-\alpha}}}{\sum_{l=1}^k \|\Phi(\mathbf{x}_i) - \boldsymbol{\mu}_l^\Phi\|^{\frac{2}{1-\alpha}}} \tag{3.104}$$

Because of

$$\|\Phi(\mathbf{x}_i) - \boldsymbol{\mu}_j^\Phi\|^2 = \Phi(\mathbf{x}_i)^{\mathrm{T}}\Phi(\mathbf{x}_i) - 2\Phi(\mathbf{x}_i)\boldsymbol{\mu}_j^\Phi + (\boldsymbol{\mu}_j^\Phi)^{\mathrm{T}}\boldsymbol{\mu}_j$$

and

$$\boldsymbol{\mu}_j^\Phi = \frac{\sum_{i=1}^m u_{ij}^\alpha \Phi(\mathbf{x}_i)}{\sum_{i=1}^m u_{ij}^\alpha} \tag{3.105}$$

we obtain

$$\|\Phi(\mathbf{x}_i) - \boldsymbol{\mu}_j^\Phi\|^2 = \mathsf{K}(\mathbf{x}_i, \mathbf{x}_i) - 2\frac{\sum_{l=1}^m u_{lj}^\alpha \mathsf{K}(\mathbf{x}_i, \mathbf{x}_l)}{\sum_{l=1}^m u_{lj}^\alpha} + \frac{\sum_{l=1}^m \sum_{t=1}^m u_{lj}^\alpha u_{tj}^\alpha \mathsf{K}(\mathbf{x}_i, \mathbf{x}_t)}{\left(\sum_{l=1}^m u_{lj}^\alpha\right)^2} \tag{3.106}$$

Like in the case of the kernel-based variant of the k-means algorithm, there exists also the possibility to construct the matrix K with elements $k_{ij} = \mathsf{K}(\mathbf{x}_i, \mathbf{x}_j)$ in advance. Therefore the above-mentioned equation can be simplified to

$$\|\Phi(\mathbf{x}_i) - \boldsymbol{\mu}_j\|^2 = k_{ii} - 2\frac{\sum_{l=1}^m u_{lj}^\alpha k_{il}}{\sum_{l=1}^m u_{lj}^\alpha} + \frac{\sum_{l=1}^m \sum_{t=1}^m u_{lj}^\alpha u_{tj}^\alpha k_{lt}}{\left(\sum_{l=1}^m u_{lj}^\alpha\right)^2} \tag{3.107}$$

or in vector form (consult Eq. (3.28) from Sect. 3.1.5.5):

$$\|\Phi(\mathbf{x}_i) - \boldsymbol{\mu}_j\|^2 = k_{ii} - 2(\tilde{\mathbf{u}}_j^{\mathrm{T}} K)_i + \tilde{\mathbf{u}}_j^{\mathrm{T}} K \tilde{\mathbf{u}}_j \tag{3.108}$$

where $\mathbf{u}_j$ is the j-th column of matrix U, and $\tilde{\mathbf{u}}_j = \mathbf{u}_j^\alpha / \|\mathbf{u}_j^\alpha\|_1$.

So, the kernel-based analogue of the FCM algorithm, similar to the Algorithm 3.6 from p. 90 runs as follows: squares of distances of objects to prototypes, computed from the above formula, are substituted into the Eq. (3.104) and the computations are repeated until stability is reached of all values of u_{ij}.

Zhou and Gan proposed in [541] an interesting idea of determining group prototypes. $\boldsymbol{\mu}_j^\Phi$ in (kernel) feature space has the counterimage $\mathbf{m}_j$ in the original feature space in which the expression $\|\Phi(\mathbf{m}_j) - \boldsymbol{\mu}_j^\Phi\|$ arrives at its minimum. By substituting (3.105) into this equation and setting to zero the gradient (with respect to $\mathbf{m}_j$) of an equation modified in this way one gets the formula

$$\mathbf{m}_j \sum_{i=1}^{m} u_{ij}^{\alpha} \mathsf{K}(\mathbf{x}_i, \mathbf{m}_j) = \sum_{i=1}^{m} u_{ij}^{\alpha} \mathsf{K}(\mathbf{x}_i, \mathbf{m}_j) \mathbf{x}_i \qquad (3.109)$$

If K is a Gaussian kernel, we get the expression

$$\mathbf{m}_j = \frac{\sum_{i=1}^{m} u_{ij}^{\alpha} \mathsf{K}(\mathbf{x}_i, \mathbf{m}_j) \mathbf{x}_i}{\sum_{i=1}^{m} u_{ij}^{\alpha} \mathsf{K}(\mathbf{x}_i, \mathbf{m}_j)} \qquad (3.110)$$

After initiating the vector $\mathbf{m}_j \in \mathbb{R}^n$, the computations are repeated following the above equation until the solution is stable.

The paper [246] proposes modifications of the kernel-based fuzzy algorithms, meant to adapt them to processing big data sets.

Application of the algorithms described above should be cautious. An exhaustive study [214] compares the algorithms FCM, GK (Gustafson and Kessel) with KFCM-X and KFCM-$\mathfrak{F}$ for a large set of test datasets. The authors of that paper draw quite an important conclusion that kernel-based methods generally only slightly outperform FCM methods when measured in terms of clustering classification rates or reconstruction error, but the difference is statistically not significant. They (in particular KFCM-$\mathfrak{F}$) have only advantage over FCM or GK in case of specific non-spherical clusters like ring clusters, but only if the number of clusters is known in advance.

This remark is especially important in the context of much greater computational complexity of kernel-based methods.

3.3.5.7   PCM: Possibilistic Clustering Algorithm

Much attention has been paid to methods strengthening the resistance of FCM algorithm to noise and outliers. One of such approaches was presented in [136], where the set of k groups was extended by one additional cluster that is intended to contain all non-typical data. This approach redefines the quality indicator as follows:

$$J_{\alpha}(U, V) = \sum_{i=1}^{m} \sum_{j=1}^{k} u_{ij}^{\alpha} d^2(\mathbf{x}_i, v_j) + \sum_{i=1}^{m} u_{i,k+1}^{\alpha} \delta^2 \qquad (3.111)$$

where δ represents the distance of an object from the "noisy" cluster. Note that this distance is identical for all objects.

Wu and Yang, on the other hand, introduced in [508] the target function of the form

$$J_{\alpha}(U, M; \beta) = \sum_{j=1}^{k} \sum_{i=1}^{m} u_{ij}^{\alpha} \left(1 - \exp(-\beta \|\mathbf{x} - \mathbf{y}\|^2)\right) \qquad (3.112)$$

where

$$\beta = \frac{m}{\sum_{i=1}^{m} \|\mathbf{x}_i - \bar{\mathbf{x}}\|^2}, \quad \bar{\mathbf{x}} = \frac{1}{m} \sum_{i=1}^{m} \mathbf{x}_i$$

The authors claim that not only the resulting algorithm is resistant to the presence of noise and outliers but also it tolerates groups with diverse cardinalities.

Another interesting proposal is to resign from the requirement that the total membership degree of an objects to all k clusters be equal 1, $\sum_{j=1}^{k} u_{ij} = 1$, as this requirement forces each outlier to be assigned to one or more clusters.

We say that the matrix U represents a possibilistic partition of an object set if the following three conditions are met:

(a) $0 \leq u_{ij} \leq 1, \ i = 1, \ldots, m, \ j = 1, \ldots, c$

(b) $\sum_{i=1}^{m} u_{ij} > 0, \ j = 1, \ldots, k$

(c) $\sum_{j=1}^{c} u_{ij} > 0, \ i = 1, \ldots, m$

In order to obtain such a non-trivial partition, Krishnapuram and Keller, [306] introduced a target function of the form:

$$J_P(U, M) = \sum_{j=1}^{k} \sum_{i=1}^{m} u_{ij}^{\alpha} \|\mathbf{x}_i - \boldsymbol{\mu}_j\|^2 + \sum_{j=1}^{k} \eta_j \sum_{i=1}^{m} (1 - u_{ij})^{\alpha} \tag{3.113}$$

where $\eta_j > 0, \ j = 1, \ldots, k$. The role of the second summand is to enforce a large value of the element u_{ij}. Elements u_{ij} are now treated as degrees of typicality and each column of the matrix U represents the so called distribution of possibilities over the set of objects.

Parameter η_j controls the "typicality" of objects: it indicates the distance from the prototype below which it is permissible to assign high membership values u_{ij}. The value u_{ij} is obtained from the formula

$$u_{ij} = \left(1 + \frac{\|\mathbf{x}_i - \boldsymbol{\mu}_j\|^2}{\eta_j}\right)^{\frac{1}{\alpha-1}} \tag{3.114}$$

The membership degree of the object $\mathbf{x}_i$ in the class j in the standard FCM depends on the distance from each prototype. But this is different in the PCM algorithm. Now, u_{ij} depends on the distance from the prototype $\boldsymbol{\mu}_j$ alone, with this distance being additionally modified by the parameter η_j. If $\eta_j = \|\mathbf{x}_i - \boldsymbol{\mu}_j\|^2$ then $u_{ij} = 1/2$. Hence η_j points at the distance from the prototype $\boldsymbol{\mu}_j$, for which u_{ij} reaches the value $1/2$. When $u_{ij} > 1/2$, then $\mathbf{x}_i$ is treated as a typical representative of the cluster C_j. Krishnapuram and Keller propose in [306] to compute η_j from the equation

$$\eta_j = \frac{\sum_{i=1}^{m} u_{ij}^\alpha \|\mathbf{x}_i - \boldsymbol{\mu}_j\|^2}{\sum_{i=1}^{m} u_{ij}^\alpha} \tag{3.115}$$

This value can be updated in each iteration or fixed during initialisation. The first variant may be the source of instability of the algorithm, therefore the second variant is suggested in [306] with u_{ij} being substituted with values returned by the FCM algorithm. After initialising the matrix U and determining the value η_j, the prototype coordinates are computed as prescribed by Eq. (3.58) and new values of assignments are computed from Eq. (3.114). These last two steps are repeated till the values of the matrix U are stable.

The PCM algorithm has two important drawbacks: it is sensitive to initialisation and the function (3.113) reaches minimum when the coordinates of all prototypes are identical,[39] which is a consequence of the fact that the degree u_{ij} depends only on the distance of the i-th object from the j-th cluster. Authors cited in the footnote proposed to enrich the function (3.113) with a summand preventing collapsing of prototypes:

$$J_{RP}(U, M) = \sum_{j=1}^{k} \sum_{i=1}^{m} u_{ij}^\alpha \|\mathbf{x}_i - \boldsymbol{\mu}_j\|^2 + \sum_{j=1}^{k} \eta_j \sum_{i=1}^{m} (1 - u_{ij})^\alpha$$

$$+ \sum_{i=1}^{k} \gamma_i \sum_{j=1, j \neq i}^{k} \|\boldsymbol{\mu}_i - \boldsymbol{\mu}_j\|^{-2} \tag{3.116}$$

In this case the coordinates of the prototypes are updated as follows:

$$\boldsymbol{\mu}_j = \frac{\sum_{i=1}^{m} u_{ij}^\alpha \mathbf{x}_i - \gamma_j \sum_{l=1, l \neq j}^{k} \boldsymbol{\mu}_l r_{jl}}{\sum_{i=1}^{m} u_{ij}^\alpha - \gamma_j \sum_{l=1, l \neq j}^{k} r_{jl}} \tag{3.117}$$

where $r_{jl} = \|\boldsymbol{\mu}_j - \boldsymbol{\mu}_l\|^{-4}$ means the repulsion force between respective prototypes. Values u_{ij} are determined according to Eq. (3.114). If the repulsion force exceeds the attraction force, i.e. $\sum_{i=1}^{m} u_{ij} < \gamma_j \sum_{l \neq j} r_{il}$ of one (or more) prototype $\boldsymbol{\mu}_j$, then such a prototype should be initialised anew.

Pal et al. proposed in [382] to combine the FCM algorithm with the PCM algorithm in such a way that at the same time the membership degrees u_{ij} and the typicality degrees t_{ij} are computed and the target function is formalised as follows:

$$J_{RP}(U, M) = \sum_{j=1}^{k} \sum_{i=1}^{m} (a u_{ij}^\alpha + b t_{ij}^\beta) \|\mathbf{x}_i - \boldsymbol{\mu}_j\|^2 + \sum_{j=1}^{k} \eta_j \sum_{i=1}^{m} (1 - t_{ij})^\beta \tag{3.118}$$

---

[39]This problem has been analysed in the paper by H. Timm, C. Borgelt, C. Doring, and R. Kruse, "An extension to possibilistic fuzzy cluster analysis", *Fuzzy Sets Syst.*, **147**(1), 2004, pp. 3–16.

where the parameter $a > 0$ characterises the importance of the membership degrees, and the parameter $b > 0$ stresses the importance of typicality degrees. If $b = 0$ then the algorithm behaves like FCM. It is required that the exponents respect the constraint $\alpha, \beta > 0$. It is recommended that they be equal $\alpha = \beta = 2.0$.

More information on possibilistic clustering algorithms can be found in [382] and in the bibliography cited therein.[40] Kernel based variants of such algorithms are discussed in [177].

3.3.5.8 Relational Variant of the FCM Algorithm

Let us consider now the task of clustering objects in a situation where their description in terms of a feature vector is not available and, instead, we can access only their dissimilarity matrix R. Its elements r_{ij} are interpreted as dissimilarity degrees of the i-th object from the j-th object. We assume that r is a symmetric matrix with non-negative elements, and additionally $r_{ii} = 0$ for all $i = 1, \ldots, m$.

This problem was investigated, among others, by Roubens [410] and Windham [503]. The algorithms they propose suffer, however, from a strong dependence on initialisation, which influences the stability of the algorithms and the quality of the generated clusters which is usually poor.

Hathaway, Davenport and Bezdek [244] developed a more satisfactory algorithm, but at the expense of additional constraints on the matrix R. They require that the dissimilarity matrix be Euclidean, meaning that in some n'-dimensional space there exists such a set of points $\{\mathbf{x}_1, \ldots, \mathbf{x}_m\}$ that $r_{ij} = \|\mathbf{x}_i - \mathbf{x}_j\|^2$. Let also U be an (arbitrarily initialised) matrix of dimension $m \times k$, representing the group membership of objects, i.e. a matrix fulfilling the conditions (3.53). Under these assumptions the prototype ("centre") of the j-th cluster is determined from the equation

$$\boldsymbol{\mu}_j = \frac{(u_{1j}^{\alpha}, \ldots, u_{mj}^{\alpha})^{\mathrm{T}}}{\sum_{i=1}^{m} u_{ij}^{\alpha}}, \quad j = 1, \ldots, k \qquad (3.119)$$

and the distance[41] d_{ij} between the i-th object and the j-th prototype is calculated from the equation

$$d_{ij} = (R\boldsymbol{\mu}_j)_i - \frac{1}{2}\boldsymbol{\mu}_j^{\mathrm{T}} R \boldsymbol{\mu}_j, \quad i = 1, \ldots, m, \quad j = 1, \ldots, k \qquad (3.120)$$

where the symbol $(\mathbf{v})_i$ denotes the i-th element of the vector $\mathbf{v}$.

New values u_{ij} of the membership of i-th object in the j-th group are obtained from Eq. (3.59), i.e.

---

[40] We recommend also the paper X. Wu, B. Wu, J. Sun and H. Fu, "Unsupervised possibilistic fuzzy clustering", *J. of Information & Computational Science*, **7**(5), 2010, pp. 1075–1080.

[41] Formally it is the squared distance.

$$u_{ij} = \begin{cases} \left[\sum_{l=1}^{k} \left(\frac{d_{ij}}{d_{il}} \right)^{\frac{1}{\alpha-1}} \right]^{-1} & \text{if } Z_i = \emptyset \\ 1 & \text{if } j \in Z_i \neq \emptyset \\ 0 & \text{if } j \notin Z_i \neq \emptyset \end{cases} \tag{3.121}$$

where $Z_i = \{j : 1 \leq j \leq k, d_{ij} = 0\}$, as before.

If $r_{ij} = \|\mathbf{x}_i - \mathbf{x}_j\|^2$ then the partition returned by the above algorithm is identical with the partition returned by the classical FCM algorithm. If, however, R is not a Euclidean dissimilarity relation then the values d_{ij}^2, obtained from Eq. (3.120) may be negative. Non-Euclidean relation R can be transformed to a Euclidean one using a simple mapping, [242]

$$r_{ij}^{\beta} = \begin{cases} r_{ij} + \beta \text{ if } i \neq j \\ 0 \quad \text{ if } i = j \end{cases}$$

where $\beta \geq \beta_0$, and β_0 is some positive number. In Sect. 3 of the paper [242] an adaptive procedure was presented, enabling to estimate the appropriate value of the parameter β.

Another, simpler solution of this problem was proposed in [126]. The respective authors suggested that information about relations between the objects allows to describe the i-th object with the vector $\mathbf{x}_i = (r_{i,1}, \ldots, r_{i,m})^{\mathrm{T}}$ expressing the dissimilarity of this object with respect to the other ones. Also, the prototype $\boldsymbol{\mu}_j$ of the j-th class can be represented in the same way. Hence, one can define the gap between the dissimilarity of the i-th object with respect to the other objects and the profile of the j-th class, that is

$$\delta^2(\mathbf{x}_i, \boldsymbol{\mu}_j) = \sum_{l=1}^{m} (r_{il} - \mu_{jl})^2 \tag{3.122}$$

and then try to minimise the indicator

$$J_{\alpha}(U, C) = \sum_{i=1}^{m} \sum_{j=1}^{k} u_{ij}^{\alpha} \delta^2(\mathbf{x}_i, \boldsymbol{\mu}_j) \tag{3.123}$$

where, as previously, elements u_{ij} of the assignment matrix U fulfil the conditions $u_{ij} \in [0, 1]$, $\sum_{j=1}^{k} u_{ij} = 1$, and $\sum_{i=1}^{m} u_{ij} > 0$.

Equation (3.123) has the same shape as Eq. (3.54), which means that Eqs. (3.58) and (3.59) can be applied to determine the prototypes and the assignment (allocation) matrix. Under the assumed representation of prototypes, the first of these equations will have the form

$$\mu_{j,l} = \frac{\sum_{s=1}^{m} u_{lj}^{\alpha} \cdot r_{ls}}{\sum_{s=1}^{m} u_{sj}}, \, j = 1, \ldots, k, l = 1, \ldots, m \tag{3.124}$$

Iterative repetitions of computations, prescribed by Eqs. (3.124) and (3.59) constitute the ARCA algorithm (*Any Relation Clustering Algorithm*). Like FCM, it requires initialisation of the matrix U. Authors of the paper [126] assumed, following Windham in [503], that $u_{i,1} = 0.5(1 + 1/k)$ for the first m/k objects, $u_{i,2} = 0.5(1 + 1/k)$ for the next m/k objects, ..., $u_{i,k} = 0.5(1 + 1/k)$ for the remaining m/k objects, and membership of all objects in other classes is equal $u_{ij} = 0.5/k$.

More information on the relational variant of the FCM algorithm can be found in Chap. 3 of the book [72]. Its applications in data mining, in particular in personalisation of search results, are discussed, among others, in [305, 365].

3.4 Affinity Propagation

We now deal with another example of a relational algorithm returning a set of prototypes.[42] Input data consist of similarities between pairs of objects, s_{ij}. Authors of [191] assume that the number s_{ij}, $i \neq j$ indicates how well the point j is suitable to represent the point i. In the simplest case $s_{ij} = \|\mathbf{x}_i - \mathbf{x}_j\|^2$, though one can apply other similarity measures (see examples in the paper [191]). The numbers s_{ii} describe so called preferences, the bigger is s_{ii}, the greater is the chance that the object i will become a prototype. s_{ii} can be equal e.g. to the median of the set of values s_{ij}, $i \neq j$ over all j (resulting in a larger number of clusters) or the minimal value from this set (whereby a smaller number of clusters is induced). Neither symmetry nor metricity of the similarities is required (i.e. the condition $s_{ik} < s_{ij} + s_{jk}$ is not imposed).

The essence of the algorithm is to maximise the function

$$E(\mathbf{c}) = -\sum_{i=1}^{m} s_{i,c_i} \tag{3.125}$$

where $\mathbf{c} = (c_1, \ldots, c_m)$ is a set of labels. Label c_i points at the representative of the object i. A task formulated in this way is $\mathcal{NP}$-hard, therefore it is formulated as maximisation of the *pure* similarity of the form

$$E(\mathbf{c}) = -\sum_{i=1}^{m} s_{i,c_i} + \sum_{j=1}^{m} \delta_j(\mathbf{c}) \tag{3.126}$$

---

[42]Its authors call those prototypes *exemplars* (of data groups) [191].

where

$$\delta_j(\mathbf{c}) = \begin{cases} -\infty \text{ if } c_j \neq j \text{ and } \exists i : c_i = j \\ 0 \text{ otherwise} \end{cases} \qquad (3.127)$$

is the penalty imposed in case object i chooses object $j = c_i$ as its prototype and object j does not consider itself as its own prototype, that is $c_j \neq j$. It is an essential condition that a prototype has to fulfil: not only several objects should choose j as their representative, but also j must choose itself as its own prototype, i.e. the following condition must hold: $c_j = j$.

The problem (3.126) is solved in a process of message passing, hence its name *affinity propagation*. Message r_{ij} sent by the object i to the object j reflects *responsibility* of being the prototype for object i. *Availability*, a_{ij}, is a message sent by the object j to the object i informing about the readiness to take over the task of being the prototype for the object i.

If we disregard technicalities, related to the message exchange,[43] then the algorithm has the form shown as pseudo-code 3.10. Most important operations therein are the updates of responsibility and availability, described by Eqs. (3.128) and (3.130). In the first iteration, the new values of responsibility are equal

$$r'_{ij} = \begin{cases} s_{ij} - s_{i,j_1(i)} \text{ if } j \neq j_1(i) \\ s_{ij} - s_{i,j_2(i)} \text{ otherwise} \end{cases} , \ i, j = 1, \ldots, m$$

where $j_1(i)$ is the id of the object most similar to object i, while $j_2(i)$ is the id of the object second most similar to object i.

A "smoothing" operation, described by Eqs. (3.129) and (3.131), was introduced in order to avoid numerical oscillations of the values of both messages.

The algorithm terminates when either a predefined number of **while** loop iterations was performed or when the object assignments to the prototypes do not change over t consecutive iterations (in [191] $t = 10$ was assumed). Values c_i, defined by the Eq. (3.132), indicate the prototypes (if $c_i = i$) or membership of the object i in the class represented by the appropriate prototype (if $c_i \neq i$).

Please note that the number of classes does not need to be specified in advance, contrary to the algorithms discussed previously. The affinity propagation algorithm determines it automatically.

---

[43]Interested reader is recommended to get acquainted with the Ph.D. Thesis [165].

Algorithm 3.10 Affinity propagation, [191]

Input: Similarity matrix $S = [s_{ij}]_{m \times m}$.
Output: Set of prototypes together with the objects represented by them.
1: $a_{ij} = 0$ for $i, j = 1, \ldots, m$
2: **while** (**not** termination condition) **do**
3:    update responsibilities

$$r'_{ij} = s_{ij} - \max_{v \neq j} (a_{iv} + s_{iv}), \ i, j = 1, \ldots, m \tag{3.128}$$

4:    determine the final values of responsibilities

$$r_{ij} = \lambda r_{ij} + (1 - \lambda) r'_{ij}, \ i, j = 1, \ldots, m \tag{3.129}$$

5:    update availabilities

$$a'_{ij} = \begin{cases} \displaystyle\sum_{u \neq j} \max (0, r_{u,j}) \ \text{if } i = j \\ \displaystyle\min \left[0, r_{jj} + \sum_{u \notin \{i,j\}} \max(0, r_{uj})\right] \text{otherwise} \end{cases}, \ i, j = 1, \ldots, m \tag{3.130}$$

6:    determine the final availability values

$$a_{ij} = \lambda a_{ij} + (1 - \lambda) a'_{ij}, \ i, j = 1, \ldots, m \tag{3.131}$$

7: **end while**
8: determine indexes

$$c_i = \arg\max_{1 \leq j \leq m} (r_{ij} + a_{ij}), \ i = 1, \ldots, m \tag{3.132}$$

3.5 Higher Dimensional Cluster "Centres" for k-means

k-means is an algorithm that splits the data into k clusters in such a way that for k points in space, called centres, the sum of squared distances of data elements to the closest centre is minimised.

One way to generalise this concept is to allow for different types of centres [469], not only points, but also lines, planes, in general any flats.[44] This makes particularly sense if the objects to be clustered do not concentrate around specific measurable properties, but rather can be grouped on the grounds of various laws they are following. For example different elastic materials would respond differently to stress depending on stress they are exposed to and other features. So the data points may arrange around q-flats related to different materials they are made of. So for the simplest dependencies it may make sense to look at data points arranged along a line, or a plane or a higher q-flat.

---

[44]In geometry, a flat is a subset of n-dimensional Euclidean space that is congruent to an Euclidean sub-space of lower dimension. A point is a 0-flat, a line is a 1-flat, a plane is a 2-flat, because they are congruent with 0,1,2 dimensional Euclidean subspace resp.

To be uniquely defined, a q-flat would require $q + 1$ points in the n-dimensional space. Instead, another representation is used, consisting of a matrix $W \in \mathbb{R}^{n \times n-q}$ of rank $n - q$ and a vector $\gamma \in R^{n-q}$. A flat (W, γ) is a set of points $\mu = \{\mathbf{x}; W^T \mathbf{x} = \gamma\}$. This means that vectors $\mathbf{x} - \gamma$ for $\mathbf{x} \in \mu$ are orthogonal to γ. The distance between a data point $\mathbf{x}$ and a q-flat μ is defined as $d(\mathbf{x}, \mu) = \min_{\mathbf{y} \in \mu} \|\mathbf{y} - \mathbf{x}\|$ and was demonstrated to be equal $d(\mathbf{x}, \mu) = \|W (W^T W)^{-1} (W^T \mathbf{x} - \gamma)\|$. The same q-flat μ may be defined via different pairs (W, γ), but one type of such pairs is of particular interest (which can always be found), that is such (W_o, γ_o) that W_o is orthonormal, that is $W_o^T W_o = I$. Under such representation we get a simplification for the distance $d(\mathbf{x}, \mu) = \|W_o^T \mathbf{x} - \gamma\|$.

Now consider a set of points $X = (\mathbf{x}_1 \cdots \mathbf{x}_m)^{\mathrm{T}}$ and ask the question about a q-flat $\mu = \mu(W, \gamma)$, such that $\sum_{i=1}^{m} d(x_i, \gamma)^2$ is minimised, while we are interested only in orthonormal W matrices.

Given orthonormality assumption

$$
\begin{aligned}
\sum_{i=1}^{m} d(x_i, \gamma)^2 &= \sum_{i=1}^{m} \|W^T \mathbf{x}_i - \gamma^2\| \\
&= \sum_{i=1}^{m} \sum_{c=1}^{n-q} (w_c^T \mathbf{x}_i - \gamma_c)^2 \\
&= \sum_{c=1}^{n-q} \sum_{i=1}^{m} (\mathbf{x}_i^T w_c - \gamma_c)^2 \\
&= \sum_{c=1}^{n-q} \|X w_c - e \gamma_c\|^2
\end{aligned}
$$

(3.133)

w_c is the c-th column of W.

Generalised Rayleigh-Ritz theorem allows to conclude that minimisation of $\sum_{c=1}^{n-q} \|X w_c - e \gamma_c\|^2$ is obtained via finding $n - q$ least eigenvalues of (the symmetric matrix) $X^T (I - e^T e/m) X$ and forming W as consisting of columns being eigenvectors corresponding to these eigenvalues. W is orthonormal. Then $\gamma = e^T X W / m$.

Under these circumstances we are ready to formulate the so-called k-q-flats algorithm that minimises the quality function:

$$
J(U, M) = \sum_{i=1}^{m} \sum_{j=1}^{k} u_{ij} d(\mathbf{x}_i, \mu_j)^2
$$

(3.134)

Though the form of Eq. (3.134) is similar to (3.1), there occur the following differences. Now M is a set of representations of k flats in n-dimensional space. To perform the algorithm in a meaningful way, the number of data points m should exceed $k \cdot (q + 1)$.

The computational algorithm starts, like in k-means, with an initialisation of k q-flats, and then the steps of cluster assignment update and cluster centre update are alternated.

In the assignment update step, u_{ij} is set to 1, if $j = \arg\min_j d(\mathbf{x}_i, \boldsymbol{\mu}_j)$, and set to 0 otherwise.

In the centre update step, we take all the data points assigned to a given cluster and proceed as above outlined to find the q-flat that minimises the sum of squares of distances of the points of the cluster to this q-flat.

Both in the initialisation and in the update step one has to keep in mind that one needs at least $q + 1$ data points to find a unique q-flat.[45] This means that clusters with fewer points have to be abandoned and some random other q flat initiated.

This kind of clustering would be of particular interest if the goal of clustering is to partition the data into such groups that within a group the property of mutual (linear) predictability of object features is increased. If it is in fact the case then one can guess that the data points lie in a lower-dimensional subspace of the feature space and one hopes using q flats that this relationship is linear.

3.6 Clustering in Subspaces via k-means

It happens frequently, that the dataset contains attributes that are not relevant for the problem under investigation. In such a case clustering may be distorted in a number of ways, as the non-relevant attributes would act like noise.

To annihilate such effects, a number of approaches have been tried out, including so-called *Tandem Clustering* (described and discussed e.g. by [232]) that consists of: (1) Principal Component Analysis to filter out appropriate subspaces, in which the actual clustering can be performed, (2) applying conventional k-means clustering on the first few eigenvectors, corresponding to highest absolute eigenvalues. Chang [232] criticised this approach demonstrating that sometimes eigenvectors related to low eigenvalues may contain the vital information for clustering.

De Soete and Carroll [141] proposed an interesting extension of k-means algorithm that can handle this issue in a more appropriate way, a method called *reduced k-means*, (or *RKM* for short). Its stability properties are discussed by Terada [457]. The algorithm seeks to minimise the objective function

$$J(U, M, A) = \sum_{i=1}^{m} \sum_{j=1}^{k} u_{ij} \|\mathbf{x}_i - A\boldsymbol{\mu}_j\|^2 = \|X - UMA^T\|_F^2 \qquad (3.135)$$

where A is a matrix of dimensions $n \times q$ ($q < \min(n, k-1)$), responsible for search of clusters in lower dimensional space, it must be column-wise orthonormal, and

---

[45]For degenerate sets of points the number may be even larger.

M is a matrix containing, as rows, the k cluster means $\mu_j \in \mathbb{R}^q$ in the lower, q dimensional space. $\|.\|_F$ denotes the Frobenius norm.

The algorithm consists of steps:

1. Initialise A, M, U.
2. Compute the singular value decomposition $Q \Sigma P^T$ of the matrix $(UF)^T X$. where Q is a $q \times q$ orthonormal matrix, Σ is a $q \times q$ diagonal matrix, and P is a $n \times q$ column-wise orthonormal matrix.
3. Compute next approximation of A as $A := PQ^T$.
4. Compute a new U matrix so that each $\mathbf{x}_i$ belongs to the transformed centre $A\mu_j$ of the cluster j.
5. Compute a new M matrix as $M = (UTU)^{-1}U^T X A$.
6. Compute a new value of $J(U, M, A)$ according to Eq. (3.135). If this value is decreased, replace old U, M, A with the new ones and go to step 2. Otherwise stop.

As the matrix U is constrained to be a binary one, the algorithm may stick in local minima. Hence multiple starts of the algorithm are recommended.

Note that this algorithm not only finds a partition U around appropriate cluster centres U, but also identifies the appropriate subspace (via A) where the clusters are best identified.

A drawback is that both k and q must be provided by the user. However, [457] proposes a criterion for best choice of q.[46]

Coates and Ng [124] exploit the above idea in an interesting way in order to learn extremely simplified method for representing data points, with application to image processing. One may say that they look for a clustering of data points in such a way, that each data point belongs to some degree to only one cluster. So their optimisation objective is to minimise:

$$J(M, S) = \sum_{i=1}^{m} \|\mathbf{x}_i - M^T \mathbf{s}_i\|^2 \qquad (3.136)$$

where $S = (\mathbf{s}_1, \mathbf{s}_2, \ldots, \mathbf{s}_m)^T$ is a matrix of m so-called code vectors $\mathbf{s}_i$, each of k rows, but with only one non-zero entry, each encoding the corresponding data vector $\mathbf{x}_i$, and M, called "dictionary", is a matrix containing the k cluster centres $\mu_j \in \mathbb{R}^n$ as rows. As the cluster centres are normalised to unit length (lie on a unit sphere), we can talk about spherical k-means here.

The computational algorithm, after initialisation, consists of alternating computation of M and S, till some stop condition.

---

[46]Still another approach, called *subspace clustering*, was proposed by Timmerman et al. [460]. In this approach, the coordinates of cluster centres are no more required to lie in a single q-flat, but have rather two components, one "outer" (in a q_b-flat) and one inner (q_w-flat), that are orthogonal. So not only subspaces are sought, where clustering can be performed better, but for each cluster a different subspace may be detected. Regrettably, this method is equivalent to $q_w + q_b$-flat based reduced k-means clustering in terms of found clusters, but additionally ambiguities are introduced when identifying both flats.

Given M we compute for each i the vector $\mathbf{s}_i$ as follows: Compute the vector $\mathbf{s}' = M\mathbf{x}_i$. Let l be the index of the maximal absolute value of of a component of $\mathbf{s}'$, $l = \arg\max_l \|s'_l\|$. Let $\mathbf{s}''$ be a vector such that $s''_l = s'_l$ and equal zero otherwise. Then $\mathbf{s}_i = \mathbf{s}''$.

Given S we compute M as follows: $M' = X^T S$. Then normalise M' row-wise and you obtain M.

In this case, instead of representing cluster centres in a low dimensional subspace, the data points are represented in a one-dimensional subspace, separate for each cluster.

Witten et al. [504] approached this problem in a still another way. They propose to assign weights to features in their *sparse clustering k-means* algorithm. Weights are subject to squared sum constraint and they alternate application of traditional k-means with weight optimisation for fixed cluster assignment.

They maximize the between-cluster-sum-of-squares:

$$ J(U, M, \mathbf{w}) = \sum_{i=1}^{m}\sum_{j=1}^{k}\sum_{l=1}^{n} w_l(x_{il} - \mu_l)^2 - \sum_{i=1}^{m}\sum_{j=1}^{k}\sum_{l=1}^{n} u_{ij}w_l(x_{il} - \mu_{j_l})^2 \quad (3.137) $$

where μ_l is the overall mean for feature l, $\mathbf{w}$ is a vector of non-negative weights of individual features of objects $\mathbf{x}_i$, where $\sum_{l=1}^{n} w_l^2 \leq 1$, and additionally the constraint $\sum_{l=1}^{n} w_l < s$ is imposed, where s is a user defined parameter. Both constraints are preventing elimination of too many features Initially, the weights $\mathbf{w}$ are set to $\frac{1}{\sqrt{n}}$. Then in a loop first k-means algorithm is performed while keeping constant $\mathbf{w}$. Upon reaching an optimum, the cluster assignment is fixed and then $\mathbf{w}$ is optimised via a quadratic optimisation procedure under the mentioned constraints (upon fixing clusters, this is a convex optimisation problem).

As a result we obtain both a partition of the data set and a selection (weighting) of individual features.

Hastie et al. [238, Sect. 8.5.4] point at the problem that Witten's approach performs optimisation in a space that is bi-convex (cluster centres, weights), but is not jointly convex, and hence guarantees for obtaining a global solution are hard to achieve. Hence they propose a modification based on prototypes. It may be deemed as a kind of k-means with each element belonging to a separate cluster and one attempts to collapse upon one another the cluster centres (so that the number of clusters drops below m). They minimise

$$ J(M) = 0.5 \sum_{i=1}^{m} \|\mathbf{x}_i - \mu_i\|^2 + \lambda \sum_{i<j} w_{ij}\|\mu_i - \mu_j\|_q \quad (3.138) $$

q=1 or 2, $\lambda > 0$ is a user-defined parameter. Weights w_{ij} may be equal 1 or fixed as a function of distances between observations $\mathbf{x}_i$, $\mathbf{x}_j$, for example $w_{ij} = e^{-\|\mathbf{x}_i, \mathbf{x}_j\|^2}$. See the book by Hastie et al. [238], Sects. 8.4 and 8.5 for details.

Regrettably, the sparse clustering k-means algorithm is quite sensitive to outliers.[47] Therefore Kondo et al. [299] propose a modification, called *robust sparse k-means algorithm*. The modification consists in changing the procedure of computing cluster centres twice: once in the traditional manner and then after rejecting the elements most distant from their cluster centres either in the original space or in the space with weighted coordinates.

3.7 Clustering of Subsets—k-Bregman Bubble Clustering

If we look at the data matrix X, then formally the previous section was dealing with dropping away some columns from this matrix (not usable features). But we may quite well proceed the other way around, that is we can dismiss rows (data elements that constitute the general background, that cannot be associated with any real cluster). The k-Bregman Bubble Clustering algorithm,[48] proposed by Gupta et al. in [217], proceeds in exactly this way. In their approach they assume that only s data elements out of all m constitute real k clusters and hence they minimise the cost function

$$J(U, M) = \sum_{i=1}^{m} \sum_{j=1}^{k} b_i u_{ij} \|x_i - \mu_j\|^2 \qquad (3.139)$$

where b_i is an indicator equal 1 if data element i is assigned to any cluster and 0 otherwise, subject to the constraint that $\sum_{i=1}^{m} b_i = s$.

The computing algorithm follows the k-means algorithm, with one difference, however. After cluster assignment of the data points, they are sorted by increasing distance from the closest cluster centre. Only the first s of them participate in computation of new cluster centres and are regarded as cluster members, the other $m - s$ ones are considered as "background noise".

The algorithm tends to identify densest fragments of the data space. The name "Bubble Clustering" indicates that the clusters are bubble-shaped. If distant, they take the form of balls, when close to one another, the balls coagulate with clusters being separated by hyper-plane.

Note that this algorithm, with $k = 1$, called also *BBOCC*, is suited for tasks called "one class clustering" in which one is not interested so much in clustering all data but rather of a small segment of interest, around a designated point (we seek the cluster containing one particular point).

---

[47]This sensitivity was dealt already for the classical k-means by Cuesta-Albertos et al. [128] in that at each step, after cluster assignment was performed, $\alpha 100\%$ of elements most distant to their cluster centres were removed and the cluster centres were recomputed.

[48]The algorithm is of broader applicability than described here. It extends the Bregman Hard Clustering algorithm, which itself is an extension of k-means algorithm by replacing the Euclidean distance in the cost function with the more general Bregman Divergence, discussed in Sect. 2.2.1.3.

It suffers from similar problems as k-means itself: sticking in local minima. Hence proper initialisation is needed. One approach to handle the issue is to apply "pressure", that initially allows for big bubbles, and then the pressure is increased to result in smaller bubbles. This is implemented as allowing for dynamic variation of the number of elements permitted to form clusters in each step. In iteration j of the algorithm $s_j = s + (m - s)\gamma^j$, $j = 1, \ldots$, where γ is a learning parameter. So in iteration 1 all $s + (m - s) = m$ data elements are allowed to participate in the computation of cluster centres and then their amount is exponentially diminished till it differs from s by less than 1. In this way one prohibits that a poor initial seeding excludes some dense regions.

3.8 Projective Clustering with k-means

In Sect. 3.5 we looked at a generalisation of k-means consisting in selecting a q-flat as a cluster centre instead of a point. Each flat there had to have the same dimension q. In Sect. 3.6 we considered a different restriction in that all cluster centres were placed in a single q-flat. In this section, we investigate a generalisation of the problem of Sect. 3.5 in that each cluster j has its own q_j-flat with q_j differing from cluster to cluster.

Projective clustering assumes that we can project elements x_{ij} of the original data matrix X into a subspace that is lower dimensional so that the clustering within this lower dimensional space will provide with better clusters. We face here the additional problem of which subspace to choose. Furthermore, it may turn out that one may beneficially use a different subspace for each of the clusters.

Let us present the approach suggested by Agarwal et al. [7]. They proposed the so-called *k-means projective clustering* algorithm that minimises the quality function:

$$J(U, M(U)) = \sum_{i=1}^{m} \sum_{j=1}^{k} u_{ij} d(\mathbf{x}_i, \boldsymbol{\mu}_j)^2 \tag{3.140}$$

which is clearly similar to Eq. (3.134), but now we have a significant difference in the representation of M, which cannot be represented as a matrix any more but is rather a function of cluster assignment. For each cluster j the optimal q_j dimensionality of its central flat needs to be determined.

The algorithm to compute the clustering does not differ from k q-flats algorithm in an important way, only the q_j-flats are computed based on local optimal values of q_j, that is the "dimensionality" of the jth cluster. [7] proposes determination of the q_j in two different ways: density-based and range-of-change-based.

The algorithm uses a user-defined parameter α, called the improvement. Define the function $flatdist(X, q)$ as being the sum of squares of distances of the data points X to the closest q-flat. Define $q_{(\alpha)}$ as the smallest q such that $flatdist(X, q) \leq \alpha flatdist(X, 1)$.

The *density based dimension function* is defined as follows: Let us draw in the two-dimensional plane the points $(q_{(\alpha)}, flatdist(X, q_{(\alpha)}))$ and $(n, 0)$. where n is the dimensionality of points in the set X. Among all q such that $q_{(\alpha)} \leq q \leq n$ we choose the one for which the point $(q, flatdist(X, q))$ is most distant from the line passing through the two aforementioned points.

The *range-of-change based dimensionality function* is defined as follows: Select among the qs such that $q_{(\alpha)} \leq q \leq n$ the one for which the quantity $flatdist(X, q - 1) + flatdist(X, q + 1) - 2flatdist(X, q)$ takes the minimal value.

As the second measure is quite sensitive to local changes, the authors suggest to use some kinds of mixtures of both. The mixtures engage introducing additional user-defined parameters. For example, if the difference between the two dimension functions is too large, then prefer the density based function, otherwise use the range-of-change based function.

3.9 Random Projection

The problem of choice of the subspace has been surpassed by several authors by so-called random projection, applicable in particularly highly dimensional spaces (tens of thousands of dimensions) and correspondingly large data sets (of at least hundreds of points).

The starting point here is the Johnson-Lindenstrauss Lemma.[49]

Roughly speaking it states that there exists a linear mapping from a higher dimensional space into a sufficiently high dimensional subspace that will preserve the distances between points, as needed e.g. by k-means algorithm.

A number of proofs of this theorem have been proposed which in fact do not prove the theorem as such but rather create a probabilistic version of it, like e.g. [133]

So it is proven that the probability of reconstructing the length of a random vector from a projection onto a subspace within a reasonable error boundaries is high.

One then inverts the thinking and states that the probability of reconstructing the length of a given vector from a projection onto a (uniformly selected) random subspace within a reasonable error boundaries is high.

But uniform sampling of high dimensional subspaces is a hard task. So instead n' vectors with random coordinates are sampled from the original n-dimensional space and one uses them as a coordinate system in the n'-dimensional subspace which is a much simpler process. One hopes that the sampled vectors will be orthogonal (and hence the coordinate system will be orthogonal) which in case of vectors with thousands of coordinates is reasonable. Then the mapping we seek is the projection multiplied by a suitable factor.

---

[49] W.B. Johnson and J. Lindenstrauss. Extensions of Lipschitz mappings into a Hilbert space. In Conference in modern analysis and probability (New Haven, Conn., 1982), volume 26 of Contemp. Math., pages 189–206. Amer. Math. Soc., Providence, RI, 1984.

It is claimed that this mapping is distance-preserving not only for a single vector, but also for large sets of points.

Let us present the process in a more detailed manner.

Let us consider a vector $\mathbf{x} = (x_1, x_2, \ldots, x_n)$ of n independent random variables drawn from the normal distribution $\mathcal{N}(0, 1)$ with mean 0 and variance 1. Let $\mathbf{x}' = (x_1, x_2, \ldots, x_{n'})$, where $n' < n$, be its projection onto the first n' coordinates.

Dasgupta and Gupta [133] in their Lemma 2.2 demonstrated that for a positive β

- if $\beta < 1$ then

$$Pr(\|\mathbf{x}'\|^2 \le \beta \frac{n'}{n} \|\mathbf{x}\}|^2) \le \beta^{n'/2} \left(1 + \frac{n'(1-\beta)}{n-n'}\right)^{(n-n')/2}$$

- if $\beta > 1$ then

$$Pr(\|\mathbf{x}'\|^2 \ge \beta \frac{n'}{n} \|\mathbf{x}\}|^2) \le \beta^{n'/2} \left(1 + \frac{n'(1-\beta)}{n-n'}\right)^{(n-n')/2}$$

Now imagine we want to keep the error of squared length of $\mathbf{x}$ bounded within a range of $\pm\delta$ relative error upon projection, where $\delta \in (0, 1)$. Then we get the probability

$$Pr\left((1-\delta)\|\mathbf{x}\|^2 \le \frac{n}{n'}\|\mathbf{x}'\|^2 \le (1+\delta)\|\mathbf{x}\|^2\right)$$

$$\ge 1 - (1-\delta)^{\frac{n'}{2}}\left(1 + \frac{n'\delta}{n-n'}\right)^{(n-n')/2} - (1+\delta)^{n'/2}\left(1 - \frac{n'\delta}{n-n'}\right)^{(n-n')/2}$$

This implies

$$Pr\left((1-\delta)\|\mathbf{x}\|^2 \le \frac{n}{n'}\|\mathbf{x}'\|^2 \le (1+\delta)\|\mathbf{x}\|^2\right)$$

$$\ge 1 - 2\max\left((1-\delta)^{n'/2}\left(1 + \frac{n'\delta}{n-n'}\right)^{(n-n')/2}, (1+\delta)^{n'/2}\left(1 - \frac{n'\delta}{n-n'}\right)^{(n-n')/2}\right)$$

$$= 1 - 2\max_{\delta^* \in \{-\delta, +\delta\}}\left((1-\delta^*)^{n'/2}\left(1 + \frac{\delta^* n'}{n-n'}\right)^{(n-n')/2}\right)$$

The same holds if we scale the vector $\mathbf{x}$.

Now if we have a sample consisting of m points in space, without however a guarantee that coordinates are independent between the vectors then we want that the probability that squared distances between all of them lie within the relative range $\pm\delta$ is higher than

$$1 - \epsilon \le 1 - \binom{m}{2} \left(1 - Pr\left(1 - \delta\right)\|\mathbf{x}\|^2 \le \frac{n}{n'}\|\mathbf{x}'\|^2 \le (1 + \delta)\|\mathbf{x}\|^2\right)\right)$$

for some error term $\epsilon \in (0, 1)$.

To achieve this, the following must hold:

$$\epsilon \ge 2\binom{m}{2} \max\left((1 - \delta^*)^{n'/2}\left(1 + \frac{\delta^* n'}{n - n'}\right)^{(n-n')/2}\right)$$

Taking logarithm

$$\ln \epsilon \ge \ln(m(m-1)) + \max\left(\frac{n'}{2}\ln(1 - \delta^*) + \frac{(n - n')}{2}\ln\left(1 + \frac{\delta^* n'}{n - n'}\right)\right)$$

$$\ln \epsilon - \ln(m(m-1)) \ge \max\left(\frac{n'}{2}\ln(1 - \delta^*) + \frac{(n - n')}{2}\ln\left(1 + \frac{\delta^* n'}{n - n'}\right)\right)$$

We know that $\ln(1 + x) = x - x^2/2 + x^3/3\ldots$ hence both for positive and negative x in the range (-1,1) $\ln(1 + x) < x$

$$\ln \epsilon - \ln(m(m-1)) \ge \max\left(\frac{n'}{2}\ln(1 - \delta^*) + \frac{(n - n')}{2}\frac{\delta^* n'}{n - n'}\right)$$

$$\ln \epsilon - \ln(m(m-1)) \ge \max\left(\frac{n'}{2}\ln(1 - \delta^*) + \frac{1}{2}(\delta^*)n'\right)$$

We know that $\ln(1 - x) = -x - x^2/2 - x^3/3\ldots$ so $\ln(1 - x) + x = -x^2/2 - x^3/3\ldots < 0$

$$\max\left(2\frac{\ln \epsilon - \ln(m(m-1))}{\ln(1 - \delta^*) + \delta^*}\right) \le n'$$

So finally, realising that $-\ln(1 - \delta) - \delta \ge -\ln(1 + \delta) + \delta$, we get

$$n' \ge 2\frac{-\ln \epsilon + 2\ln(m)}{-\ln(1 + \delta) + \delta}$$

Note that this expression does not depend on n that is the number of dimensions in the projection is chosen independently of the number of dimensions.

Note that if we would set ϵ (close) to 1, and expand by Taylor method the ln function in denominator to up to three terms then we get the value of n' from the paper [133]:

$$n' \ge 4\frac{\ln m}{\delta^2 - \delta^3}$$

Note, however, that setting ϵ to a value close to 1 does not make sense as we want to make rare the event that the data does not fit the interval we are imposing.

3.10 Subsampling

An orthogonal problem to dimensionality reduction is the sample reduction. If m, the number of data points, is huge, we may pose the question whether a randomly selected, relatively small subset of size $m^* \ll m$ would be sufficient for performing the clustering operation.

One answer was proposed by Bejarano et al. [58], as already mentioned when discussing acceleration approaches to k-means. They suggest to use a sample of the size

$$m^* = k \left[\frac{1}{m} + \left(\frac{w}{2z_{1-\alpha}\sigma} \right)^2 \right]^{-1} \tag{3.141}$$

where w is the user defined width of confidence interval for each of the cluster means, and $z_{1-\alpha}$ is the percentage point of the standard normal distribution $\mathcal{N}(0, 1)$ for probability $1 - \alpha$ for a user-defined confidence level α, and σ is the standard deviation of the population.

The above estimation can be justified as follows. Assume that the data points in a cluster j stem from a normal distribution around cluster centre with ("normally distributed") distances from this centre having a standard deviation of σ_j. If in the original data set there were m_j data points belonging to this cluster, then the variance of the estimated cluster centre would amount to σ_j^2/m_j. If we would use a subsample of size m_j^*, this variance would amount to σ_j^2/m_j^*, which will be larger as $m_j^* < m_j$. Hence the variance increase due to subsampling amounts to $\sigma_j^2/m_j^* - \sigma_j^2/m_j = \sigma_j^2(\frac{1}{m_j^*} - \frac{1}{m_j})$. As the width of confidence interval for significance level α amounts to $2z_{1-\alpha}$ times standard deviation, we have the requirement:

$$w > 2z_{1-\alpha}\sigma_j \sqrt{\frac{1}{m_j^*} - \frac{1}{m_j}}$$

$$\frac{w}{2z_{1-\alpha}\sigma_j} > \sqrt{\frac{1}{m_j^*} - \frac{1}{m_j}}$$

$$\left(\frac{w}{2z_{1-\alpha}\sigma_j} \right)^2 > \frac{1}{m_j^*} - \frac{1}{m_j}$$

$$\frac{1}{m_j} + \left(\frac{w}{2z_{1-\alpha}\sigma_j} \right)^2 > \frac{1}{m_j^*}$$

Hence

$$m_j^* > \left[\frac{1}{m_j} + \left(\frac{w}{2z_{1-\alpha}\sigma_j} \right)^2 \right]^{-1} \tag{3.142}$$

The expression on the right hand grows with m_j and σ_j and clearly $m_j \leq m, \sigma_j \leq \sigma$, therefore to match the above condition, it is sufficient that:

$$m_j^* > \left[\frac{1}{m} + \left(\frac{w}{2z_{1-\alpha}\sigma} \right)^2 \right]^{-1}$$

As we have k clusters, then it is sufficient to have m^* data points amounting to

$$m^* \geq k \left[\frac{1}{m} + \left(\frac{w}{2z_{1-\alpha}\sigma} \right)^2 \right]^{-1}$$

Bejarano et al. [58] used this property in their *Sampler* clustering algorithm which is based on k-means. It samples randomly m^* data elements from the large collection (e.g. database) and subsequently performs the clustering on this sample only. Additionally, however, they imposed the constraint about keeping proportions between clusters. Let $p_1, \ldots, p_k$ being the proportions of clusters cardinalities in the full data set. Let w_p denote the confidence interval width acceptable for proportion error (the true proportion p in the sample should be in the range $\hat{p}_j \pm w_p/2$ for each j, where $\hat{p}_j$ is the proportion of cluster j in the subsample). For proportions, we know that the true portion lies within the interval $\hat{p}_j \pm z_{1-\alpha/2}\sqrt{\frac{1}{m^*}\hat{p}_j(1 - \hat{p}_j)}$. Hence they require

$$w_p/2 \geq z_{1-\alpha/2}\sqrt{\frac{1}{m^*}\hat{p}_j(1 - \hat{p}_j)}$$

Hence

$$m^* \geq \left(\frac{2z_{1-\alpha/2}}{w_p} \right)^2 \hat{p}_j(1 - \hat{p}_j)$$

for each j. This implies under worst case scenario $\hat{p}_j = 0.5$ that

$$m^* \geq \left(\frac{z_{1-\alpha/2}}{w_p} \right)^2 \tag{3.143}$$

So in the end, two conditions for m^* have to be matched. These properties are used for an optimization of k-means for large datasets. The *Sampler* algorithm starts with obtaining, from the database, a sample of size m^* being the maximum of those implied by Eqs. (3.141) and (3.143). This sample is used by the algorithm for later steps and no other sampling is performed. From this sample the first m^* according to

Eq. (3.143) is taken and the first step (initial clustering) of k-means is performed. In later steps for each cluster σ_j is estimated from data and then according to (3.142) the m_j^* is computed and in the next step of the algorithm the first $\sum_{j=1}^{k} m_j^*$ data elements from the sample are used in order to (1) assign elements to the cluster centres and to (2) compute new cluster centres. Then again σ_j and m_j^* are computed and the procedure is repeated till convergence.

3.11 Clustering Evolving Over Time

In a number of applications data to be clustered is not available simultaneously, but rather data points become available over time, for example when we handle data streams, or when the clustered objects change position in space over time etc. So we can consider a number of distinct tasks, related to a particular application domain:

- produce incrementally clusters, as the data is inflowing, frequently with the limitation that the storage is insufficient to keep all the data or it is impractical to store all the data
- generate partitions of data that occurred in a sequence of (non-overlapping) time frames
- generate clusters incrementally, but data points from older time points are to be taken into account with lower weights
- detect new emerging clusters (novelty detection) and vanishing ones
- detecting of clusters moving in data space (topical drift detection)
- the above under two different scenarios: (a) each object is assumed to occur only once in the data stream, (b) objects may re-occur in the data stream cancelling out their previous occurrence.

Especially in the latter case the task of clustering is not limited just to partitioning the data, but also tracing of corresponding clusters from frame to frame is of interest. Furthermore, it may be less advantageous to the user to obtain best clusterings for individual windows, but rather less optimal but more stable clusters between the time frames may be preferred.

3.11.1 Evolutionary Clustering

Chakrabarti et al. [101] introduces the task of evolutionary clustering as a clustering under the scenario of periodically arriving *new* data portion that needs to be incorporating into an existent clustering, which has to be modified if the new data exhibits a shift in cluster structure. It is necessary under such circumstances to balance the consistency of clustering over time and the loss of precision of the clustering of the current data. Note that we are interested in clusterings obtained at each step and want to observe how they evolve over time.

Formally, at any point in time t, $\mathcal{C}^{(t)}$ be the clustering of all data seen by the algorithm up to time point t. If $\mathcal{C}^{(t)}$ is the clustering at the point of time t, then the snapshot quality $sq(\mathcal{C}^{(t)}, X^{(t)})$ measures how well $\mathcal{C}^{(t)}$ reflects the cluster structure of data $X^{(t)}$ that arrived at time point t (neglecting the earlier data), while the history cost $hc(\mathcal{C}^{(t-1)}, \mathcal{C}^{(t)})$ measures the distance between $\mathcal{C}^{(t-1)}$ and $\mathcal{C}^{(t)}$ and is computed for all data seen by the algorithm up to the time point $t-1$, that is $X^{(1)} \cup \cdots \cup X^{(t-1)}$. They develop algorithms maximising

$$sq(\mathcal{C}^{(t)}, X^{(t)}) - \gamma hc(\mathcal{C}^{(t-1)}, \mathcal{C}^{(t)}) \tag{3.144}$$

where γ is a user-defined parameter balancing historic consistency and current accuracy.

Chakrabarti et al. illustrate their approach among others by spherical k-means algorithm, but it may be applicable quite well to the classical k-means algorithm. Let us recall the cost function of k-means clustering that can be formulated as follows (a rewrite version of (3.1)):

$$J(M, X) = \sum_{i=1}^{m} \min_{j \in \{1, \ldots, k\}} \|\mathbf{x}_i - \boldsymbol{\mu}_j\|^2 \tag{3.145}$$

In their framework we would set $sq(\mathcal{C}^{(t)}, X^{(t)}) = -J(M^{(t)}, X^{(t)})$ and $hc(\mathcal{C}^{(t-1)}, \mathcal{C}^{(t)}) = J(M^{(t)}, X^{(1)} \cup \cdots \cup X^{(t-1)}) - J(M^{(t-1)}, X^{(1)} \cup \cdots \cup X^{(t-1)})$. As at time point t the clustering $\mathcal{C}^{(t-1)}$ is already fixed, maximising $sq(\mathcal{C}^{(t)}, X^{(t)}) - \gamma hc(\mathcal{C}^{(t-1)}, \mathcal{C}^{(t)})$ is equivalent to minimising $J(M^{(t)}, X^{(t)}) + \gamma J(M^{(t)}, X^{(1)} \cup \cdots \cup X^{(t-1)})$ that is performing the classical k-means for which the historical data is down-weighted by the factor γ.

However, as one wants to keep as little as possible of historical data, one may replace the data up to the point $t-1$ with the cluster centres $M^{(t-1)}$ weighted by cardinalities of these clusters and perform the clustering on data redefined this way.

3.11.2 Streaming Clustering

Ailon et al. [11] consider the problem of an efficient approximation of the k-means clustering objective when the data arrive in portions, as in the preceding subsection. But this time we are interested only in the final result and not in the in-between clusterings. So the historical consistency plays no role. Rather we are interested in guarantees on the final clustering while we investigate a single data snapshot $X^{(t)}$ at a time.

They introduce the concept of (α, β)-approximation to k-means problem (their Definition 1.2) as an algorithm B such that for a data set X it outputs a clustering $\mathcal{C}$ containing αk clusters with centres described by the matrix M such that

$J(M, X) \leq \beta J_{opt}(k, X)$ where $J_{opt}(k, X)$ is the optimal value for X being clustered into k clusters, $\alpha > 1, \beta > 1$.

Obviously, the $k$means++ algorithm from Sect. 3.1.3.1 is an example of a $(1, 8$ $(\ln k + 2))$ approximation. They modify this algorithm to obtain the so-called k-means# algorithm, described by the pseudo-code 3.11.

Algorithm 3.11 k-means# algorithm, [11]

Input: Data set X and number of groups k.
Output: Final number of groups k', gravity centres of classes $\{\mu_1, \ldots, \mu_{k'}\}$ along with the assignment of objects to classes $U = [u_{ij}]_{m \times k'}$.
1: Pick randomly $3 \ln k$ objects $\mathbf{x} \in X$ and make them to be the set C.
2: **for** $j = 2$ to k **do**
3:    For each object $\mathbf{x}$ compute the distance to the nearest centre from the set C, that is $u(\mathbf{x}) = \min_{\mu \in C} \|\mathbf{x} - \mu\|^2$.
4:    Create a set C_j consisting of $3 \ln k$ objects $\mathbf{x} \in X$ $\mathbf{x}$ picked randomly according to the distribution (3.5).
5:    Add C_j to the set of centres, $C \leftarrow C \cup C_j$.
6: **end for**
7: Run the $|C|$-means algorithm. Set $k' = |C|$.

Ailon et al. [11] prove that their algorithm, with probability of at least 1/4, is a $(3 \ln k, 64)$-approximation to the k-means problem.

Based on $k$means++ and $k$means# they develop a Streaming divide-and-conquer clustering (pseudo-code 3.12).

Algorithm 3.12 Streaming divide-and-conquer clustering, version after [11]

Input: (a) Stream of data sets $X^{(1)}, \ldots, X^{(T)}$.
   (b) Number of desired clusters, $k \in \mathbb{N}$
   (c) A, being an (α, β)-approximation algorithm to the k-means objective. (d) A', being an (α', β')-approximation algorithm to the k-means objective.
Output: Final number of groups k', set M of gravity centres of classes $\{\mu_1, \ldots, \mu_{k'}\}$.
1: create an empty set R of pairs (object, weight).
2: **for** $t = 1$ to T **do**
3:    Run the algorithm A for the set $X^{(t)}$, obtaining the set $R^{(t)}$ of αk cluster centres with weights being the cardinality of the respective cluster.
4:    Add $R^{(t)}$ to the set of centres, $R \leftarrow R \cup R^{(t)}$.
5: **end for**
6: Run the A' algorithm on R to obtain $\alpha' k$ cluster centres M.

This algorithm uses algorithms A and A' being (α, β) and (α', β') approximations to k-means objective and is itself an $(\alpha', 2\beta + 4\beta'(\beta + 1))$-approximation of k-means objective (see their Theorem 3.1). Hereby the algorithm A is defined as follows: "Run k-means# on the data $3 \ln m$ times independently, and pick the clustering with the smallest cost. where m is the number of data items in $X^{(1)}, \ldots, X^{(T)}$ taken together". Apparently with probability $(1 - (3/4)^{3\ln m})$ it is a $(3\ln k, 64)$ algorithm. The algorithm A' is defined as "Run the k-means++ algorithm on the data".

In all, the algorithm in over 50% of runs yields an $(1, O(\ln k))$—approximation to k-means objective.

3.11.3 Incremental Clustering

When speaking about incremental clustering, we assume that the intrinsic clustering does not change over time and the issue is whether or not we can detect properly the clustering using limited resources for storage of intermediate results.

Reference [5] sheds some interesting light on the problem induced by algorithms that are sequential in nature, that is when data is presented one object at a time. They consider a sequential k-means that during its execution stores only k cluster centres and the cardinalities of each of the cluster and assigns a new element to the closest cluster updating the respective statistics. If one defines a clustering in which in a cluster any two elements are closer to each other than to any other point, as "nice clustering", then this algorithm is unable to detect such a clustering even if it exists in the data. What is more, it cannot even detect a "perfect clustering", that is one in which the distance of two elements in a cluster is always lower than the distance between any two elements in different clusters.

Another sequential algorithm, nearest neighbour clustering algorithm, can surprisingly discover the perfect clustering, but the nice one remains undetected. This algorithm maintains k data points, and upon seeing a new data object it stores it in its memory, seeks the pair of closest points (in the set of the previous objects plus the new one) and randomly throws out one of the two objects of the pair.

The "nice-clustering" can be discovered by incremental algorithms only if 2^{k-1} cluster centres are maintained. For high k this requirement may be prohibitive even if a clear data structure exists in the data.

3.12 Co-clustering

The problem of co-clustering has been called in the past *direct clustering, bi-dimensional clustering, double clustering, coupled clustering* or *bimodal clustering*.

With data described by a data matrix X, one usually clusters the rows representing objects/elements. But from the formal point of view one could transpose such a matrix and with same clustering algorithms one may obtain the clustering of features describing objects. While in general it may be pointless, there exist still special application areas where clustering of at least some of the object features may be beneficial. So for example, instead of clustering text documents, one may cluster the words of these documents to reveal some fixed phrases, or specific subareas of a language etc. One may have a number of alternative ways of measuring properties of objects, like different radiological methods of determining age of rocks in geology, or different methods of determining red cells counts in blood samples in medicine, different corrosion speed measurement methods in chemistry etc. One may be interested which of these methods return matching results and which diverge.

But one may be particularly interested in clustering the objects and the features at the same time. Commonly, such a clustering of both features and objects at the same time is called co-clustering. So when co-clustering of Web document one may discover that documents are grouped by languages in which they are written even if one does not know these languages. In case of rock age measurement methods one can detect both methods yielding similar results and types of rocks for which this is the case. Other interesting applications of co-clustering concern detection of groups of customers buying certain products, even though their overall buying patterns are very different. In micro array analysis, one may be interested in gene co- expression under certain experimental conditions that are a priori unknown. In social network analysis we may want to detect social groups engaged in some kinds of social activities. In general, with co-clustering a deeper insight into the data is obtained than by separate clustering of objects and features. Subsequently we will explain how co-clustering works.

Whereas k-means algorithm seeks to minimize the cost function of the form (3.1) minimising distances between rows $\mathbf{x}_i^T$ of the data matrix X and the vector cluster centres μ_j, co-clustering assumes that there exists a meaningful notion of (scalar) distance between individual elements x_{ij} of this data matrix so that one seeks to minimize

$$J(I, J, M) = \sum_{i \in I} \sum_{j \in J} \sum_{i \in i} \sum_{j \in j} u_{ij} \|x_{ij} - \mu_{ij}\|^2 \qquad (3.146)$$

where μ_{ij} denotes the (scalar) centre of the cluster described by the row and column index sets $\mathbf{i}, \mathbf{j}$. The difference here is that in the classical clustering in each cluster we had a vector of distinct "central" values for each separate feature. Now, as each feature is meant to measure the same quantity (say blood pressure), one scalar common central values is used for all features. This makes perfectly sense as we expect that for the same say patient the measurement result should be essentially the same.

Furthermore, it is assumed that we seek to divide the objects into k_o clusters and the features into k_f clusters. So we can speak about (k_o, k_f)-means co-clustering.

We can speak in this formulation about a special case of problems discussed in the previous section—we want to obtain clusters with centres in a lower dimensional space.

One can proceed as in iterative k-means algorithm by alternating k_o-means on objects and k_f-means on features. As proposed e.g. by Banerjee et al. [48] one starts with an initial (k_o, k_f)-co-clustering. In step 1 one computes the co-cluster means μ_{ij}, In step 2 one updates the row clusters (objects). In step 3 one updates the column clusters (features). In step 4 one checks for convergence and eventually returns to step 1.

As indicated by [48], any form of Bregman divergence may be applied instead of Euclidean norm as used in classical case.

3.13 Tensor Clustering

An n-th order tensor would be a generalisation of the concept of a matrix. A matrix can be thought of as a generalisation of a vector, being a vector of vectors. A vector would be viewed as a 1-st order tensor. A matrix is a second order tensor. A third order tensor would be a vector of matrices. An nth order tensor would be a vector of $(n-1)$-st order tensors.

The simplest set of data is one where each data element is described by a single number. Such a data set would be represented by a single vector, that is 1-st order tensor.

If each data elements is described by a feature vector, then the data set may be represented by a matrix, that is a 2-nd order tensor.

If a single element has a richer (but somehow regular) structure, one can look at higher order tensors as appropriate representations of the data set. For example a patient can be subject of regular examinations over a period of time (one dimension) and each examination may consist of multiple tests (a second dimension). So the data of a single patient may be described by a matrix and so in consequence the representation of the data set on the patients would be a 3rd order tensor. If the patient data are actually 2-d images, then the problem may be described as 4th order tensor and so on. In general, an N-th order tensor may be thought of as a generalised matrix $T \in \mathbb{R}^{n_1 \times n_2 \times \cdots \times n_N}$, elements of which are indexed with multiple indexes, e.g. $T_{j_1, j_2, \dots, j_N}$. The tensors may be combined by various operators, foremost by the outer product operator $\circ$. If $T^{(1)}$ is a N_1 order tensor and $T^{(2)}$ is an N_2 order tensor, then $T = T^{(1)} \circ T^{(2)}$ is a $N_1 + N_2$ order tensor with components:
$T_{j_1, \dots, j_{N_1}, k_1, \dots, k_{N_2}} = T^{(1)}_{j_1, \dots, j_{N_1}} \cdot T^{(2)}_{k_1, \dots, k_{N_2}}$. If two tensors $T^{(1)}, T^{(2)} \in \mathbb{R}^{n_1 \times n_2 \times \cdots \times n_N}$, then they can be combined by the inner product operation $T^{(1)} \cdot T^{(2)}$ such that $T^{(1)} \cdot T^{(2)} = \sum_{j_1=1}^{n_1} \sum_{j_2=1}^{n_2} \cdots \sum_{j_N=1}^{n_N} T^{(1)}_{j_1, \dots, j_N} T^{(2)}_{j_1, \dots, j_N}$. The F norm of a tensor T is defined as $\|T\|_F = \sqrt{T \cdot T}$.

Also a matrix multiplication called $mode - p$ operation is defined as a multiplication of a tensor $T \in \mathbb{R}^{n_1 \times \cdots \times n_p \times \cdots \times n_N}$ by the (ordinary) matrix $M \in \mathbb{R}^{n_p \times k_p}$ yielding a tensor $T' = T \otimes_p M \in \mathbb{R}^{n_1 \times \cdots \times k_p \times \cdots \times n_N}$ where $T'_{j_1, \dots, j_{p-1}, j_p, j_{p+1}, \dots, j_N} = \sum_{i=1}^{n_p} T_{j_1 \dots j_{p-1}, i, j_{p+1}, \dots, j_N} M_{i, j_p}$.

Still another important operation on tensors is their *unfolding*, that is transformation to an ordinary matrix. For a tensor $T \in \mathbb{R}^{n_1 \times \cdots \times n_p \times \cdots \times n_N}$ the matrix $A \in \mathbb{R}^{n_k \times (n_{k+1} n_{k+2} \dots n_N n_1 \times \dots n_{k-1}')}$ is its (k)th unfolding (denoted $A = T_{(k)}$) if the element $T_{j_1, j_2, \dots, j_{k-1}, j_k, j_{k+1}, \dots, j_N}$ of the tensor T is equal to the element $A_{j_k, c}$ of A, where
$c = (j_{k+1} 1)n_{k+1} \dots n_N n_1 \dots n_{k-1} + (j_{k+2} - 1)n_{k+3} \dots n_N n_1 \dots n_{k-1} \dots + (j_{N-1} - 1)n_N n_1 \dots n_{k-1} + (j_N - 1)n_1 \dots n_{k-1} + (j_1 - 1)n_2 \dots n_{k-1} \dots + (j_{k-2} - 1)n_{k-1} + j_{k-1}$.

Tensor clustering may be deemed as a generalisation of co-clustering to multiple dimensions—we co-cluster the data simultaneously over several dimensions (e.g. patients, time series, tests, images).

In its simplest form one can view N-th order tensor clustering as a task when we split a tensor along one of its dimensions (for example patients) into a set of m $(N-1)$st order tensors, define a measure of distance between these $(N-1)$st order tensors and perform a clustering into k clusters, e.g. using k-means algorithm. A more complex task one could split the data along p dimensions into $(N-p)$th order tensors, define some optimization function and perform "co-clustering" along multiple dimensions at the same time, following guidelines from the previous section.

Two big challenges must be faced during such a process. First, the appropriate definition of distance, which is domain specific (distances between medical images and between particle trajectory images in particle accelerators may be strongly different). The other issue is the tractability of tensor processing. Usually some kind of decomposition of the tensor into low order tensors is performed so that the whole tensor (and cluster centres too in centric methods) is approximated from projections of the tensor onto lower order tensors, and the distances are then defined based on these decompositions.

Two major brands of tensor decompositions are worth mentioning here: the *ParaFac* and *HOSVD*.

The ParaFac decomposition of a tensor (called also CANDECOMP or Canonical polyadic decomposition (CPD)) is its approximation via so-called *Kruskal form tensor*. This may be deemed as a kind of approximation via margins.

An N-th order tensor $T \in \mathbb{R}^{n_1 \times n_2 \times \cdots \times n_N}$ is in *Kruskal form* if there exist coefficients $\lambda_1, \ldots, \lambda_R$ and vectors $\mathbf{a}^{(1,1)} \in \mathbb{R}^{n_1}, \mathbf{a}^{(2,1)} \in \mathbb{R}^{n_1}, \ldots, \mathbf{a}^{(R,1)} \in \mathbb{R}^{n_1}, \mathbf{a}^{(1,2)} \in \mathbb{R}^{n_2}, \mathbf{a}^{(2,2)} \in \mathbb{R}^{n_2}, \ldots, \mathbf{a}^{(R,2)} \in \mathbb{R}^{n_2}, \ldots, \mathbf{a}^{(1,N)} \in \mathbb{R}^{n_N}, \mathbf{a}^{(2,N)} \in \mathbb{R}^{n_N}, \ldots \mathbf{a}^{(R,N)} \in \mathbb{R}^{n_N}$ such that

$$T = \sum_{j=1}^{R} \lambda_j \mathbf{a}^{(j,1)} \circ \mathbf{a}^{(j,2)} \circ \cdots \circ \mathbf{a}^{(j,N)}$$

R is called hereby the rank of the Kruskal form tensor.

For a tensor T_2 its ParaFac decomposition is a tensor T_1 in Kruskal form for which the distance $\|T_1 - T_2\|_F^2$ is minimised.

We speak about a *Tucker decomposition*, if an Nth order tensor T is approximated by a core tensor B and a series of ordinary matrices $M^{(1)}, \ldots, M^{(N)}$ via the formula $B \otimes_1 M^{(1)} \otimes_2 \ldots, \otimes_N M^{(N)}$.

It is known that the following *HODSV decomposition* (Higher Order Singular Value) Decomposition) of each tensor $T \in \mathbb{R}^{n_1 \times \cdots \times n_p \times \cdots \times n_N}$ is possible (see e.g. Theorem 2 in [318]):

$$T = S \otimes_1 U^{(1)} \otimes_2 \ldots, \otimes_N U^{(N)}$$

where

- $U^{(j)} \in \mathbb{R}^{n_j \times n_j}$ is a unitary matrix (that is with rows and columns orthogonal)

- $S \in \mathbb{R}^{n_1 \times \cdots \times n_p \times \cdots \times n_N}$ is a tensor such that if its subtensors $S_{j_k=\alpha}$ (that is tensors obtained from S by fixing kth index to the value α have the properties:

  - orthogonality: any two subtensors $S_{j_k=\alpha}$, $S_{j_k=\beta}$ are orthogonal for any k, α, β where $\alpha \neq \beta$: that is $S_{j_k=\alpha} \cdot S_{j_k=\beta} = 0$
  - ordering: for any k $\| \| S_{j_k=1} \| \|_F \geq \| \| S_{j_k=2} \| \|_F \geq \cdots \geq \| \| S_{j_k=n_k} \| \|_F$

The Frobenius norm $\| S_{j_k=i} \|_F$, denoted also as $\lambda_i^{(k)}$ is called ith k-mode singular value of tensor T, and the ith column of $U^{(k)}$ is called ith k-mode singular vector.

One can simplify the representation of a tensor, via its approximation by neglecting low value singular values and the corresponding singular vectors.

3.14 Manifold Clustering

Manifold clustering may be considered as a generalisation of projective clustering in that we seek the clusters in manifolds instead of subspaces.

A subspace is globally linear.

A manifold is a topological space that is *locally* Euclidean (that is around every point, there is a neighbourhood that is topologically the same as the open unit ball in $\mathbb{R}^k$ for some k). A topological space is a set X together with a collection of open subsets T that satisfies the four conditions: 1. The empty set $\emptyset$ is in T. 2. X is in T. 3. The intersection of a finite number of sets in T is also in T. 4. The union of an arbitrary number of sets in T is also in T.

The first question is of course if we can discover the manifolds to which the dataset belongs. the second is to separate different manifolds forming distinct clusters.

The first question can be answered by application of the k-means or k q-flats algorithms as presented e.g. in [95]. With sufficient high number of of gravity centres piecewise linear approximations can be found. The question is of course how dense the centres should be chosen. On the one hand increase of the number of centres decreases the discretisation errors, but on the other the estimation error of respective centres increases. A balance needs to be found.

On the other hand [456] proposes the so-called Isomap algorithm extracting a manifold from a high-dimensional space. It consists of three steps: construction of a neighbourhood graph, computing of shortest distances and construing a q-dimensional embedding. The first step, construction of a neighbourhood graph, requires user specific parameter: either $\epsilon > 0$ specifying a ball in which to seek neighbours or p—the number of nearest neighbours which will become neighbours in the graph. So for the set of points X in a high (n-dimensional space) one creates a graph Γ with vertices from X and weighted edges. An edge e is added between nodes x_i, x_j if either $\|x_i, x_j\| < \epsilon$ or x_j belongs to the set of p nearest neighbours of x_i. The edge e gets the weight $d_\Gamma(x_i, x_j) = \|x_i, x_j\|$. In the next step shortest distances between elements of the set X are computed in the weighted graph Γ. In this way a distance $d_G(x_i, x_j)$ is defined even if these elements x_i, x_j are not neighbours in Γ. The result is a matrix D such that $D_{ij} = d_\Gamma(x_i, x_j)$. To represent data points in the coordinate system of the manifold, compute the eigenvalues and eigenvectors of the matrix K derived from D as follows. Let D_{sq} be the matrix of squared elements of D (the squared distance matrix).

A matrix K is constructed as $K = 0.5(I - s\mathbb{1}^T)D_{sq}(I - \mathbb{1}s^T)$, where $s^T\mathbb{1} = 1$.

Upon sorting eigenvalues in decreasing order, pth coordinate of element i in this space would be $\sqrt{\lambda_p}v_{p_i}$, where v_p is the eigenvector associated with the eigenvalue λ_p. Dimensionality reduction is obtained via taking the most significant eigenvalues and eigenvectors that reconstruct D with "sufficient" precision.

After this transformation one could apply the k-means algorithm in the reduced manifold space (as the distances in this space would be identical as in D).

However, [471] suggests not to go to the spectral coordinate system, but rather to apply k-medoids method on the distance matrix D when determining centres of clusters, and to apply random walk in order to determine cluster membership. He uses hereby the concept of a tiring random walker. A tiring random walker walks continuously through the graph, with a fixed transition probability matrix P, but with each step it looses its energy at a fixed rate α so that it is with each step less probable that it will move. So at step 1 the probability that it moves to i given it is at node j amounts to $(\alpha P)_{i,j}$. The probability, that it does so in exactly t steps t equals $(\alpha P)_{i,j}^t$. Therefore the probability, that starting at node j it reaches node i at any point of time amounts to $\left(\sum_{t=0}^{\infty}(\alpha P)^t\right)_{i,j} = (1 - \alpha P)_{i,j}^{-1}$.

The matrix $\tilde{P} = (1 - \alpha P)^{-1}$ is used to decode on cluster membership. Each non-centroid is assigned to the cluster for which the probability of reaching it from this centroid is the biggest.

The only issue is how to define P for the graph G. One may follow PageRank or have a measure inversely proportional to distances in the matrix D.

Criminisi et al. [127, Chaps. 5, 6] demonstrate that random forests, a well-known method of supervised learning, can be applied for manifold learning. They introduce the concept of so-called density forests. A *density forest* consists of a set of randomly trained clustering trees. A *clustering tree* resembles a classical decision tree except that the tree leaves contain simple prediction models such as Gaussians. When building the decision tree, no target variable is available. Therefore assumptions must be made as to the nature of what is predicted by such a tree. Prediction of all the variables is intended. They assume for this purpose that the data at each node stem from a multi-variate Gaussian distributions at the nodes. This assumption is unrealistic, as it implies that the original data consists of a mixture of Gaussian distributions, but it works in practice. It implies that the differential (continuous) entropy of such a node (with n-variate Gaussian) may be expressed as

$$H(X_j) = \frac{1}{2}\ln((2\pi e)^n \|Cov(X_j)\|)$$

where $Cov(X_j)$ is the covariance matrix for the data set X_j associated with a given node j. Hence at each node j of the tree a split is chosen that maximises the information gain

$$
\begin{aligned}
I_j &= H(X_j) - \sum_i \frac{\|X_j^i\|}{\|X_j\|} H(X_j^i) \\
&= \frac{1}{2} \ln((2\pi e)^n \|Cov(X_j)\|) - \sum_i \frac{\|X_j^i\|}{\|X_j\|} \frac{1}{2} \ln((2\pi e)^n \|Cov(X_j^i)\|)
\end{aligned}
\tag{3.147}
$$

where X_j^i is the dataset associated with the ith subnode of the node j.

This split criterion tends to separate high density Gaussian distributions constituting a mix from which the data is drawn.

The leaf nodes are assigned a multivariate Gaussian distribution, summarizing the data set that reached the given leaf.

The leaves of a single clustering tree constitute a partition of the data set (into Gaussian components). Their population may be viewed as an approximation of the probability density of the sampling space.

A density forest, constituting of a set of parallely trained clustering trees may be viewed as an approximation of this probability density by taking average of individual tree densities.

To facilitate manifold learning, they transform the density forest into a similarity graph, to be further clustered using the spectral analysis methods (see Chap. 5). The transformation relies on accessing density forest based on similarity between elements of X. The forest similarity is the average similarity within each clustering tree. The similarity within a clustering tree is zero if elements belong to different leaves of the tree. If they belong to the same leaf, the similarity is set to 1 in the simplest case, and more elaborate approaches may take into account the leaf density, leaf correlations etc. to modify the distance between elements.

3.15 Semisupervised Clustering

While much research effort goes to elaboration of both clustering and classification algorithms (called unsupervised and supervised learning respectively), there exists big area for investigations, mixing both of them, where part of the elements of the sample are pre-labelled and the other part is not. Depending on the perspective one can speak about partially supervised classification or semi-supervised clustering, as the classification algorithms may benefit from unlabelled elements via better exploration of the geometry of the sample space (via exploitation of the similarity function) and clustering may benefit not only for application of external quality criteria but also by better understanding the aesthetic rules underlying the similarity function. The labelling hereby may be direct assignment to some predefined classes or have the form of hints of the form: these two elements belong to the same cluster, these two cannot belong to the same cluster. Whatever labelling method is used, the information contained therein and in the similarity function should not contradict one another.

The basic idea behind exploitation of both labelled and unlabelled examples may be generally described as follows: On the one hand there exits an underlying probability distribution $P(y|\mathbf{x})$ of assignment of a label $y \in Y$ to the exemplar $\mathbf{x} \in \mathbf{X}$. On the other hand the data lie on a low dimensional manifold in a high dimensional space so that it makes sense to consider distances different from the ones induced by the high dimensional space. One makes the following assumption: Given a similarity function over $\mathbf{X}$ one hopes that the aforementioned conditional probability changes continuously with the similarity.

Under these circumstances a labelling function is sought that is penalized both for deviations from labels on the labelled part of the examples and from continuity for the whole set of examples.

One may distinguish several fundamental approaches to semisupervised clustering:

- similarity-adapting methods, where the similarity function is adapted to fit the label information (Sect. 3.15.1)
- search-adapting methods where the clustering algorithm itself is modified to respect the label information (Sect. 3.15.2)
- target variable controlled methods (Sect. 3.15.3)
- weakened classification methods, where the unlabelled data is used to modify class boundaries (Sect. 3.15.4)
- information spreading algorithms, where the labels are spread with similarity based "speed" over the set of object (Sect. 3.15.5).

3.15.1   Similarity-Adapting Methods

As most clustering algorithms rely on some measure of similarity, dissimilarity or distance, the simplest way to take into account the pre-classified elements is to modify the similarity matrix (or more generally the similarity function).

Klein et al. [292] developed an algorithm called Constrained Complete-Link (CCL), that takes into account various kinds of information on common cluster membership via specific modification of similarity matrix, which can be later used by a clustering algorithm like k-means and in this way take advantage of user knowledge about relations between clustered objects. The Klein's method consists in modifying the similarity matrix to reflect imposed constraints (e.g. increasing the dissimilarity between elements for which we know that need to lie in distinct clusters) and to propagate these constraints wherever applicable also to other elements of similarity matrix (if e.g. points x_i and x_j are close to one another, then a point close to x_i must also be close to x_j; on the other hand if points x_i and x_j are far apart, then a point close to x_i must also be far away from x_j). Klein's algorithm takes into account two types of labelling information

- must-link (two data points must belong to the same cluster)
- cannot-link (two data items must not belong to the same cluster).

The must-link constraint is imposed via setting the dissimilarity between the elements to zero. If the dissimilarity has to be a (metric) distance, corrective actions are applied. From the original distance matrix with modified entries the shortest path matrix is constructed which is metric and is relatively close to the original distance matrix, so it can replace it. The cannot-link constraint is not so easy to impose because even identifying if all such constraints may be respected can be an NP-complete task. Therefore Klein et al. propose a heuristic approach. After imposing must-link constraints and replacing the distance with shortest path distance, they propose to model the cannot-link constraint simply by replacing the distance between the data points with a maximum distance plus one. Subsequently they suggest to use an algorithm like complete link for clustering so that the metricity is partially restored.

Xing et al. [513] consider constraints of the type "data point x_i must be close to data point x_j". They suggest to look for a (semi-positive definite) matrix A such that the Mahalanobis distance

$$d_A(x_i, x_j) = \|x_i - x_j\|_A = \sqrt{(x_i - x_j)^T A(x_i - x_j)}$$

would reflect this requirement. In particular, if S is the set of (all) pairs of points which should be "close" to one another, then we need to minimize

$$\sum_{(x_i, x_j) \in S} d_A(x_i, x_j)^2$$

subject to the constraint

$$\sum_{(x_i, x_j) \in D} d_A(x_i, x_j)^2 \geq 1$$

where D is a set of pairs of objects known to be dissimilar. This constraint is necessary to avoid the trivial solution $A = 0$. The problem is solved by gradient descent method.

3.15.2 Search-Adapting Methods

The search adapting methods mean modification of the clustering algorithm itself. Examples of such approaches are the constrained k-means algorithm, seeded k-means algorithm, k-mean with preferential data selection, k-means with penalizing via Gini-Index etc. Though many authors illustrate the approaches using k-means modifications, the underlying principles apply generally to other algorithms.

The *seeded k-means algorithm*, proposed by Basu et al. [55] modifies the basic k-means algorithm in the initialisation step. Instead of some kind of randomised initialisation, initial clusters are formed from labelled data only and these initial clusters form the basis for initial cluster centroid determination. Further steps of the algorithm are the same as in k-means.

The *constrained k-means algorithm* [55] takes into account labels of the labelled examples also during iterations, that is they constitute a fixed part of the cluster even if some of them may become closer to the centre of some other cluster.

Demiriz et al. [143] propose to amend the k-means target function with a penalty factor based in Gini index, representing a mismatch of class membership of labelled examples.

$$ J(U, M, P) = \alpha \left(\sum_{i=1}^{m} \sum_{j=1}^{k} u_{ij} \| \mathbf{x}_i - \boldsymbol{\mu}_j \|^2 \right) + \beta \sum_{j=1}^{k} \left(\sum_{i=1}^{m} u_{ij} - \frac{(\sum_{c=1}^{rc} \sum_{i=1}^{m} p_{ic})^2}{\sum_{i=1}^{m} u_{ij}} \right) $$
(3.148)

where P is a matrix with entries $p_{i,c} = 1$ stating that element i belongs to the intrinsic class c. $p_{i,c} = 0$ otherwise. $\sum_{c=1}^{rc} p_{ic} \leq 1$. If $\sum_{c=1}^{rc} p_{ic} = 1$ then the element is labelled, otherwise it is unlabelled. α and β are some non negative constants, specific for a given problem, which determine which part of data (labelled or unlabelled) shall play a stronger role.

Instead of classical k-means algorithm, a genetic algorithm is applied, with target function equal to $J(U, M, P)$. For each chromosome, one step of the classical k-means algorithm is performed with the exception of the cluster assignment step where the genetic operations of mutation and crossover replace the classic assignment to the closest cluster centre for non-labelled elements. Also clusters with too few labelled elements are discarded and new random clusters are formed instead.

Wagstaff and Cardie [487] take into account slightly weaker information about cluster membership: *must-link constraints* that specify that two instances have to be in the same cluster, and *cannot-link constraints* that specify that two instances cannot be in the same cluster. They use the CONWEB algorithm to illustrate their proposal, which is incremental in nature. They propose also a modification for the k-means algorithm in this spirit (called *COP-KMEANS*). A respective extension to k-means is quite straight forward. Prior to running a weighted version of k-means, all elements bound by "must link" constraints are pulled together and will be represented during the rest of the process by the respective mean vector, weighted by the number of element pulled together. Then during the iterations, the cluster assignment is performed as follows: we take the unassigned elements and seek the one that is closest to any cluster centre and assign it to the closest cluster. Then we consider iteratively the remaining ones taking the one that is closest to any cluster centre not containing an element that it is in the "cannot link" relation. This algorithm will work if no element is engaged in more than $k - 1$ "cannot link" relations. If the constraints cannot be satisfied. The algorithm fails.

Basu et al. [56] proposed another algorithm *PCKmeans* taking into account these weaker types of constraints, that is pairwise "must-link" and "cannot-link" relations. Their algorithm exhibits a kind of behaviour similar to either seeded or constrained k-means, depending on parameter setting. The algorithm starts with creating a consistent set of "must-link" relations, that is adding the appropriate transitive closures. Then the "cannot-link" relations are extended appropriately (if x cannot link y and y must link z, then x cannot link z). Groups of elements are formed that are connected by "must-link" relation. If their number equals to k, the centres of these groups will become the initial cluster centres. If there are more of them, then only the groups of k highest cardinalities will become cluster centres. If there are fewer than k such groups, other cluster centres will be formed of elements having "cannot-link" relation to these groups. Remaining clusters will be initialised at random. Then iteratively cluster assignment and computation of cluster centres is performed. Cluster centres are computed as in traditional k-means. Cluster assignment takes into account the squared distance to the cluster centre and the number of violations of the "must-link" and "cannot-link" constraints with respect to elements already assigned to clusters, with user-defined weights. As this step is dependent on the order of element consideration, the authors use a random sequence of elements at each step.

3.15.3 Target Variable Driven Methods

The methods of this type may be considered as a variation on both previously mentioned approaches.

In methods of this type we can take into account not only labelling of some elements of the data set, but more generally data on any variable considered as an "outcome" variable (or rather a noisy version of the true outcome of interest). The "outcome" variable is the one that a clustering should help to predict. It is not taken into account when clustering, but preferentially the clusters should say something about the value of this variable for objects classified into individual clusters.

Bair and Tibshirani [42] proposed subspacing methods, where features are selected to reflect the intended clustering prior to application of the proper algorithm. They suggest the following *targeted feature selection* procedure: For each feature compute its correlation with the outcome variable and use for clustering only the variables that are significantly correlated and the correlation level is above some threshold.

Thereafter, either unsupervised clustering or one of the semi-supervised clustering methods can be applied.

Still another approach relies on the so-called *information bottleneck approach* [461]. The general idea is that our data shall predict some variable of interest, say Y. So we can compute mutual information between the data X and the variable of interest, $I(X; Y) = \sum_x \sum_y p(x, y) \log_2 \frac{p(x,y)}{p(x)p(y)}$. The clustering will now be understood as a data compression technique performed in such a way that cluster representations $\hat{X}$ can be used instead of the original data—the mutual information

$I(\hat{X}, Y) \leq I(X; Y)$ between this target variable and the representations will be necessarily decreased but we are interested in keeping it close (up to a user-defined parameter) to the mutual information of the target variable and the original data. That is one seeks to minimize the function

$$\mathcal{L}[p(\hat{x}|x)] = I(\hat{X}, X) - \beta I(\hat{X}, Y) \tag{3.149}$$

where β is a parameter weighing the extent to which information on Y shall not be lost while performing the compression. Such an understanding can drive both the number of clusters and other aspects of cluster representation. An algorithm for the minimization is proposed by Tioshby et al. in [461]. Its relation to k-means is explored in [445].

3.15.4 Weakened Classification Methods

A number of classification algorithms, like SVM, have been adapted to work under incomplete data labelling (semi-supervised classification) so that they can serve also as semi-supervised clustering algorithms.

A classifier $h : \mathbf{X} \to \mathbf{L}$ assigns each element of a domain $\mathbf{X}$ a label from a set $\mathbf{L}$. One defines a loss function $\updownarrow : \mathbf{L} \times \mathbf{L} \to \mathbb{R}$ defining the punishment for mismatch between the real label of an instance and the one defined by the classifier. A classification algorithm seeks to minimise the value of $\sum_{(\mathbf{x}_i, l_i) \in \mathbf{X}_t} \updownarrow(l_i, h(\mathbf{x}_i))$ where $\mathbf{X}_t$ is the labelled training set containing examples $\mathbf{x}_i$ and their intrinsic labels l_i. Under semisupervised settings, an additional loss function $\updownarrow_\sqcap : \mathbf{L} \to \mathbb{R}$ punishing the classifier for assigning a label to an unlabelled instance. So one seeks not to minimise $\sum_{(\mathbf{x}_i, l_i) \in \mathbf{X}_t} \updownarrow(l_i, h(\mathbf{x}_i)) + \sum_{(\mathbf{x}_j) \in \mathbf{X}_u} \updownarrow_\sqcap(h(\mathbf{x}_j))$ where $\mathbf{X}_u$ is the unlabelled training set containing examples $\mathbf{x}_j$ without labels.

In algorithms based on *manifold assumption* like label propagation or Laplacian SVMs the unlabelled loss function can take the form $\updownarrow_\sqcap(h(\mathbf{x}_j)) = \sum_{\mathbf{x}_i \in \mathbf{X}_t, \mathbf{x}_i \neq \mathbf{x}_j}$ $s(\mathbf{x}_i, \mathbf{x}_j) \|h(\mathbf{x}_i) - h(\mathbf{x}_j)\|^2$.

Zhung [544] proposes to train a classifier *classif* using the labelled data whereby this classifier returns some *confindence* level for each unseen example. Then cleverly chosen subsets of these originally unlabelled data labelled by the classifier are used to extend the training set of the classifier in order to improve classifier accuracy. The data selection process proceeds as follows: First the originally unlabelled examples are clustered e.g. using k-means. Within each cluster the examples are ordered by the *confindence* and split using the "quantile" method into the same number z of subsets and the subsets are ranked from 1 to z. Then subsets of the same rank r from all the clusters are joined into r-th bin, so that we get z "bins". Then iteratively one picks each bin and checks its suitability to be added to the labelled set for training of a new classifier. The process is stopped, if the classification accuracy cannot be improved (is worsened).

Tanha et al. [454] apply a strategy in the same spirit, but to specifically decision trees. They train a base classifier on the labelled data. Then they classify the unlabelled data with this classifier and pick those that were correctly classified with highest confidence and add them to the labelled set and repeat the process till a stopping criteria. The major problem they need to cope with is the issue of how to estimate the confidence in a label assigned by the classifier. Decision trees have the characteristic that decisions are made at the leaf nodes where there are frequently quite few examples so that an estimate of the confidence (the ratio of the main class of a leaf to the total number of examples in the leaf) is unreliable. Various proposals for such estimates are made, like Laplacian correction, Naive Bayes classifier etc.

Criminisi et al. [127, Chap. 7] deal with extension of random forest algorithms to semi-supervised learning. They introduce the concept of a *transductive tree* that generalises the concept of the decision tree and the clustering tree. When growing the transductive tree, the mutual information

$$I_j = I_j^U + \alpha I_j^S$$

is optimised (maximised) where I_j^U is the clustering mutual information as defined by Eq. 3.147 and is computed for all the data elements, and I_j^S is the classical mutual information compuyted only for labelled examples. α is a used-provided parameter.

Another extension of a classifier for semi-supervised learning is so-called Transductive Support Vector Machine Support (TSVM) method proposed by Vapnik [478]. The Support Vector Machine (SVM) classifier seeks to separate two classes of objects in space by a hyperplane in such a way that there exists a large margin between them. So one seeks to minimize the inner product of weight vector $\mathbf{w}^T \mathbf{w}$ subject to constraints $(\mathbf{w}^T \mathbf{x}_i + b)l_i \geq 1$ for each data vector $\mathbf{x}_i$ with label l_i, where the labels can be either 1 or -1 (two-class classification problem). Transductive approach, with unlabelled data, consists in imposing additional constraints $(\mathbf{w}^T \mathbf{x}_u + b)l_u \geq 1$ where $\mathbf{x}_u$ are unlabelled examples and $l_u \in \{-1, 1\}$ is an additional variable created for each unlabelled example. This approach will work only if the labelled examples are linearly separable. If they are not, so-called slack variables $\eta \geq 0$ are introduced and one minimises the sum $\mathbf{w}^T \mathbf{w} + c_L \sum_i \eta_i + c_U \sum_u \eta_u$ subject to $\mathbf{w}^T \mathbf{x}_i + b)y_i \geq 1 - \eta_i$ for each labelled data vector $\mathbf{x}_i$ with label y_i, $\mathbf{w}^T \mathbf{x}_u + b)y_u \geq 1 - \eta_u$ and $y_u \in \{-1, 1\}$ for each unlabelled example $\mathbf{x}_u$. c_L, c_U are user-defined constants. The unlabelled loss function is here of the form $\updownarrow_\sqcap = \max(0, 1 - \|h(\curvearrowleft\lrcorner)\|)$.

3.15.5 Information Spreading Algorithms

In this category of algorithms the labels are spread with similarity based "spread" over the set of object. As described in previous chapter, there exist many ways to transform distances to similarity measures, and any can be applied.

One of the approaches, proposed by Kong and Ding [300] relies on the computation of so-called Personalized PageRank for random walks on a neighbourhood graph.

As a first step we need to obtain measures of similarity between the objects to be clustered. One can build a graph with nodes being the clustered objects and edges connecting objects with similarity above some threshold, eventually with edge weights reflecting the degree of similarity.

In this graph we have $k+1$ types of nodes: that is the set of nodes can be split into disjoint parts $V = V_1 \cup V_2 \cup \ldots \cup V_k \cup V_u$ where $V_1, \ldots, V_k$ contain objects labelled as $1, \ldots, k$ respectively, while V_u contains all the unlabelled objects.

Now k random walkers will move from node v to node $u, u, v \in V$ with probability equal to the weight of the link (v, u) divided by the sum of weights of all outgoing links of node v times $1 - b$. With probability b the random walker i jumps to any node in the set V_i. Under these circumstances the probability distribution of the walker i being at any node reaches some stationary distribution π_i. Probability of walker i being at node v will amount to $\pi_i(v)$.

As a result of this process any unlabelled object $v \in V_u$ will be assigned probabilities $\pi_1(v), \ldots, \pi_k(v)$ and it will be assigned a label equal $\arg\max_i \pi_i(v)$.

To formalise it, let S be a similarity matrix. and D be its diagonal matrix $D = diag(S1)$. Define the (column-stochastic) transition probability matrix as $P = SD^{-1}$. Hence for each walker

$$\pi_i = (1 - b)P\pi_i + b\mathbf{h_i}$$

where the preferential jump vector $h_i(v) = 1/\|V_i\|$ for $v \in V_i$ and is equal 0 otherwise.

So for the population of the random walkers we have:

$$\Pi = (1 - b)P\Pi + bH$$

where Π consists of stationary distribution column vectors π_i and H of the preferential jump vectors $\mathbf{h_i}$.

Hence

$$\Pi = bH(\mathbf{I} - (1 - b)P)^{-1}$$

Another approach, proposed by Zhu et al. [542] assumes that a smooth function f can be defined over the graph nodes $f : V \to \mathbb{R}$ assigning values to nodes in such a way that for labelled nodes $f(v) = l_v$ where l_v is the label of the node v and that the labels of similar nodes are similar that is the sum

$$E(f) = \frac{1}{2} \sum_{i,j \in V} s_{ij}(f(i) - f(j))^2$$

is minimised.

The function f minimising $E(f)$ is harmonic, that is for unlabelled examples $Lf = 0$, where L is the Laplacian $L = D - S$, where $D = diag(S1)$.

As f is harmonic, the value of f at each unlabelled point is the average of f at neighbouring point, that is:

$$f(j) = \frac{1}{d_j - s_{ii}} \sum_{i \neq j} s_{ij} f(i)$$

for j being unlabelled point. This implies that $f = (D - maindiag(S))^{-1}(S - maindiag(S))f$. For simplicity, denote $D - maindiag(S)$ with F, $S - maindiag(S)$ with W. Let $W_{ll}, W_{ul}, W_{lu}, W_{uu}$ denote submatrices of W related to pairs (labelled, labelled), (labelled, unlabelled), (unlabelled, labelled), (unlabelled, unlabelled). Similar notation should apply to F, and f_l, f_u should mean vectors of function values for labelled and unlabelled examples. Now one can see that

$$f_u = (F_{uu} - W_{uu})^{-1} W_{ul} f_l$$

3.15.6 Further Considerations

The ideas of semi-supervised clustering may also be combined with soft clustering, as done e.g. by [81, 82, 388]. other possibilities of generalisation are may concern the relationship between a class and a cluster. So far we assumed that clusters are to be generated that related somehow to classes indicated in labelled data. Reference [81] considers the case when several clusters constitute one class. Still another assumption made so far is that the distance between objects is independent of the actual clustering. Reference [82] drops this assumption and develops an algorithm adapting distance measure to the actual content of clusters.

In order to incorporate semi-supervised learning into the FCM algorithm, [388] proposes to extend optimisation goal given by Eq. (3.54) to the form

$$J_\alpha(U, M) = \sum_{i=1}^{m} \sum_{j=1}^{k} u_{ij}^{\alpha} \|\mathbf{x}_i - \boldsymbol{\mu}_j\|^2 + \beta \sum_{i=1}^{m} \sum_{j=1}^{k} \|u_{ij} - f_{ij} b_i\|^{\alpha} \|\mathbf{x}_i - \boldsymbol{\mu}_j\|^2 \quad (3.150)$$

where b_i tells whether the sample element i is labelled ($b_i = 1$) or not ($b_i = 0$), f_{ij} is meaningful for labelled sample elements and reflects the degree of membership of the labelled element i to the class/cluster j. The parameter β governs the impact of the supervised part of the sample. The algorithm reduces to FCM, if β is zero. For $\beta > 0$ the second sum punishes for diversion of cluster membership degrees u_{ij} from the prescribed class membership f_{ij}.

The above approach is valid if the number of classes is identical with the number of clusters that are formed, so that distinct clusters are labelled with distinct labels.

As [81] points out, however, it may be advantageous to form more clusters than classes (k>c) so that the clusters may have more natural form and classes a more valid business meaning. In this case, besides determining the matrices U and M, we need to find a (hard) partition π of clusters among the classes. So π_l will indicate the set of clusters j that belong to class l. Furthermore, one needs to distribute the known class l membership f_{il} of labelled data element I onto membership $\tilde{u}_{ij}$ in clusters $j \in \pi_l$ in such a way that $f_{il} = \sum_{j \in \pi_l} \tilde{u}_{ij}$. This distribution will be also subject to optimisation.

In this case one needs to minimise

$$J_\alpha(U, M, \pi) = \sum_{i=1}^m \sum_{j=1}^k u_{ij}^\alpha \|\mathbf{x}_i - \boldsymbol{\mu}_j\|^2 + \beta \sum_{i=1}^m \sum_{j=1}^k \|u_{ij} - \tilde{u}_{ij}\|^\alpha \|\mathbf{x}_i - \boldsymbol{\mu}_j\|^2 \quad (3.151)$$

The actual approach of [81] is more subtle, due to computational burden and cluster balancing, but the above equation reflects the general idea behind their considerations.

Reference [82] extends the above by pointing at possibilities of adapting the distance measure within a cluster to the actual content of the cluster. Let us consider the fuzzy covariance matrix F_j of the cluster j:

$$F_j = \frac{\sum_{i=1}^m u_{ij}^\alpha (\mathbf{x}_i - \boldsymbol{\mu}_j)(\mathbf{x}_i - \boldsymbol{\mu}_j)^T}{\sum_{i=1}^m u_{ij}^\alpha} \quad (3.152)$$

Then, for FCM, a norm-inducing matrix for the cluster may be expressed as

$$A_j = \rho_j (det(F_j))^{1/n} F_j^{-1} \quad (3.153)$$

where ρ_j is a constraint on the "volume" of A_j (volume of A_j amounts to $|det(A_j)|$) and n is the dimensionality of the elements in the clustered set. Then an adaptive distance measure for cluster j can be defined as

$$\|\mathbf{x}_i - \boldsymbol{\mu}_j\|_{A_j}^2 = (\mathbf{x}_i - \boldsymbol{\mu}_j)^T A_j (\mathbf{x}_i - \boldsymbol{\mu}_j) \quad (3.154)$$

which is a kind of modified Mahalanobis distance (by a factor).

Please note that the further the elements of a cluster j lie, the bigger the values of entries in F_j (higher covariance). This implies higher values of $|det(F_j)|$ (a higher volume) and smaller volumes for A_j.

$$J_\alpha(U, M, \pi) = \sum_{i=1}^m \sum_{j=1}^k u_{ij}^\alpha \|\mathbf{x}_i - \boldsymbol{\mu}_j\|_{A_j}^2 + \beta \sum_{i=1}^m \sum_{j=1}^k \|u_{ij} - \tilde{u}_{ij}\|^\alpha \|\mathbf{x}_i - \boldsymbol{\mu}_j\|_{A_j}^2 \quad (3.155)$$

would be unjustly minimised by broadening the clusters. Therefore [82] insists on either punishing or restricting the growth of individual clusters ($def(A_j) \geq \rho$).

3.15.7   Evolutionary Clustering

The term *evolutionary clustering* expresses the the concerns of clustering data that *evolves* over time. The evolution may mean hereby either occurrence of new data, or data points being shifted in some space or the similarities being changed for some other reason. The concept was proposed by Chakrabarti et al. [101].

The major concern here is that the data structure should not change radically between successive points in time. Hence we have here a notion of another type of semi-supervised clustering—the clustering of tomorrow should "resemble" the clustering of yesterday. So the new clustering has to be consistent with the former one, the process of re-clustering should be noise tolerant, the clusters relocated in some space shall move therein smoothly and a kind of correspondence between former and later clusters should be possible to establish.

Their application of this concept to k-means algorithm consists essentially in modification of cluster centre computation in each iteration of the algorithm. While at the time point $t = 0$ the traditional k-means is applied to the dataset, at any later time point t, the k-means is seeded with the cluster centres of the previous timepoint $t - 1$ and during iterations, after cluster membership assignment the new cluster centres as computed as follows: For each cluster j a "suggested" cluster centre μ_j^s is computed as cluster centre in the conventional k-means. Then each cluster is assigned the closest cluster centre μ_j^p from the previous timepoint $t - 1$ (each cluster a distinct one). μ_j is computed as a weighted average of the former two, while the weights are proportional to timepoint t and $t - 1$ corresponding cluster cardinalities and a user specified "change parameter" cp expressing the degree to which user is ready to accept μ_j^s.

Chapter 4
Cluster Quality Versus Choice of Parameters

Abstract This chapter is devoted to actions to be performed in order to get maximum insights into the data by application of clustering algorithms. For data preprocessing stage, methods for choosing the appropriate set of features and algorithms for selection of the proper number of clusters are presented. For post-processing of cluster analysis algorithm results, criteria evaluating the quality of the obtained clusters, both for the output of a single clustering algorithm and in case of applying multiple ones. Multiple internal and external quality measures are suggested.

In previous chapters we presented various algorithms designed for data clustering. In order to use them efficiently though, we have to solve some basic issues namely:

(a) preparing the data,
(b) establishing the proper number of clusters,
(c) estimating the quality of the applied algorithm,
(d) comparing the results obtained by using alternative algorithms.

In this chapter these issues will be considered.

4.1 Preparing the Data

When preparing the data for the analysis it is essential to decide on the set of features used to characterise the data and to apply their transformations, if necessary. Sufficient description of these issues exceeds the book's scope. However, we would like to draw the reader's attention to the importance of this stage of data analysis.

The possibility of using specified algorithms and the quality of results obtained highly depend on the set of features used for describing the real objects. Usually, we would like to operate on an efficient representation, which is a small set of features well distinguishing objects coming from different groups. The majority of the measures of similarity discussed in the Sect. 2.2 lose their discriminant abilities as the dimension of the vectors describing the analysed objects increases. On the other hand, the so called predictive analytics[1] sets the requirement of using the biggest possible set of features in order to make proper predictions. A careful choice of features influences not only the algorithm quality and performance, but also allows for better understanding of the data generating process. A range of various methods applied to choose features is the area of interest for statisticians[2] and for machine learning researchers. A wide range of remarks on this topic can be found in monographs [221, 331], in the paper [220] and in the 18-th chapter of the monograph [237]. A classical reference in this area is [164].

There is a distinction in the literature between feature selection and feature extraction. The former consists in selecting a small subset from a feature set and the latter—in establishing a small subset of *new* (artificial) features. In case of k-means algorithm there are two proven methods of feature extraction: one using random projection[3] and another consisting in SVD. A brief overview of important feature selection methods is presented in the work [262]. In the same a simple feature weighing method there was suggested consisting in the generalisation of the objective function, applied in k-means algorithm, of the form

$$J_\beta(U, M, \mathbf{w}) = \sum_{i=1}^{m} \sum_{j=1}^{k} \sum_{l=1}^{n} u_{ij} w_l^\beta (x_{il} - \mu_{jl})^2 \tag{4.1}$$

where $\mathbf{w}$ is a vector whose components represent the weights of particular features, with

$$\sum_{l=1}^{n} w_l = 1, \ 0 \le w_j \le 1$$

and β being parameter. When $\beta = 0$, the above function is identical with to function (3.1). The authors suggest assuming $\beta < 0$ or $\beta > 1$.

An attempt to combine feature selection with feature extraction is presented in the work [83].

Having a set of measurements, we often apply normalisation, that is, we transform the data so that the values of the j-th feature belong either to the set [0, 1], or to the set [−1, 1]. In turn, standardisation consists in transforming

---

[1]Cf. e.g. E. Siegel, *Predictive Analytics: The power to predict who will click, buy, lie, or die*. Wiley 2013.

[2]We recommend the procedure VARCLUST described abundantly in Chap. 104 of *SAS/STAT® 13.1 User's Guide*. Cary, NC: SAS Institute Inc.

[3]Cf. A. Blum, "Random projection, margins, kernels, and feature-selection". In: C. Saunders, M. Grobelnik, S. Gunn, and J. Shawe-Taylor, eds. *Subspace, Latent Structure and Feature Selection*. LNCS 3940. Springer Berlin Heidelberg, 2006, 52–68.

$$x_j \leftarrow \frac{x_j - \bar{x}_j}{\sigma_j}, \ j = 1, \dots, n$$

where $\bar{x}_j$ is the mean value of j-th feature and σ_j represents its standard deviation.

4.2 Setting the Number of Clusters

It has been assumed up till now that the number of clusters k is known, which, in fact, is not always the case. The most common method of establishing the proper value of k is using certain partition quality measure, $m(k)$ and establishing such a value of k parameter, which optimises the measure. Nonetheless, it must be emphasized that in the majority of cases the values of quality indicators depend on concrete data. The fact that for particular data certain indicator allows for establishing the proper number of clusters does not mean that in case of different data it will also indicate the proper value of k. For this reason, various methods are applied, which can be divided into following groups, [515]

(a) Data visualisation. In this approach multidimensional data projection on bi- or tridimensional space is applied. Typical representatives of this direction are principal component analysis (PCA) and multidimensional scaling.[4]
(b) Optimisation of some criterion function characterising the properties of mixtures of the probability distributions. E.g. the algorithm EM, discussed in Sect. 3.2, optimises the parameters θ of the mixture for a given value of k. The value of k for which the quality index attains the optimum value is supposed to be the probable number of clusters. The typical indicators i this domain are

   – Akaike information criterion

$$AIC(k) = \frac{1}{m}\left(-2(m-1) - n_k - \frac{k}{2}l(\theta)\right) + 3n_p$$

   where n_k denotes the number of cluster parameters, n_p—total number of para- meters being optimised and $l(\theta)$—likelihood function logarithm.
   – Bayesian criterion

$$BIC(k) = l(\theta) - \frac{n_p}{2}\ln(n)$$

   – Variance based criteria, for example maximising Fisher-like quotient of between-cluster variance and within cluster variance
   – Comparison of within-cluster cohesion and between-cluster separation, for example the point-biserial correlation (Pearson's correlation between the vec- tor of distances between pairs of elements and the vector of indicators assign-

---

[4]Cf. e.g. I. Borg, and Patrick JF Groenen. *Modern multidimensional scaling: Theory and applica- tions.* Springer, 2005.

ing each pair 1, if they belong to the same cluster, and 0 otherwise), or silhouette widths $s(i) = \frac{b(i)-a(i)}{\max(a(i),b(i))}$, with $a(i)$ being the average distance between i and all other entities of its cluster, and $b(i)$ is the minimum of the average distances between i and all the entities in each other cluster (If the silhouette width value is close to -1, it means that the entity is misclassified. If all the silhouette width values are close to 1, it means that the set is well clustered).

- Consensus based methods. They assume that under the "right" number of clusters the partitions generated by randomised clustering algorithms should vary less than under wrong number of clusters.
- Resampling methods are in their spirit similar to consensus based ones, except that the variation in partitioning results not from non-determinism of the algorithm but rather from the random process of sample formation (subsampling, bootstrapping, random division into training and testing subsets).
- Hierarchical methods. Hierarchical clustering (divisive or agglomerative algorithms) are applied till a stopping criterion (too small improvement or worsening) occurs. The resulting number of clusters is assumed to be right one and used as a parameter for partitional methods. An interesting example is here the "intelligent k-means" (ik-means) algorithm,[5] designed as special initialisation for k-means clustering, that takes the sample centre and the most remote point as starting cluster centres for 2-means run, in which only the second one is updated till stabilisation. Thereafter the second cluster is removed from the sample and the process is repeated without changing the position of the very first cluster centre. The clusters are then post-processed by refuting the smallest ones (of cardinality 1). The number of clusters is known to be overestimated in this process. Still another approach (purely divisive) is used in the x-means" algorithm[6]

(c) Heuristic methods.

The spectral methods, which are discussed in the next chapter, also provide tools allowing to decide on the number of clusters.

Generally, it must be said that, according to many researchers, determining the proper number of clusters is the basic issue concerning the credibility of cluster analysis results. Excessive number of clusters leads to results which are non-intuitive and difficult to interpret, while too small value of k results in information loss and wrong decisions.

---

[5]M. Ming-Tso Chiang and B. Mirkin: Intelligent Choice of the Number of Clusters in K-Means Clustering: An Experimental Study with Different Cluster Spreads. Journal of Classification 27 (2009).

[6]D. Pelleg and A. Moore: X-means: Extending k-means with efficient estimation of the number of clusters. Proc. 17th International Conf. on Machine Learning, 2000.

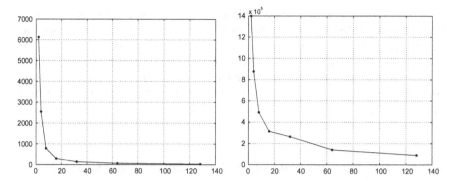

Fig. 4.1 Relation between an average cluster radius (left figure) and the total distance of objects from the prototypes depending on the number of clusters (right figure)

4.2.1 Simple Heuristics

One of the simplest rules says that[7]

$$k \approx \sqrt{m/2} \qquad (4.2)$$

Another, so called *Elbow Method*,[8] consists in examining the fraction of the explained variance in the function of the number of clusters. According to this heuristics we take for k such a number of clusters, that adding yet another cluster increases the fraction only slightly. The decision is based on the diagram: on the coordinate axis one puts subsequent values of k, and on the ordinate axis—the percentage of the explained variance corresponding to them. The point where the curvature flexes is the candidate k value. The percentage of the explained variance is conceived as the ratio of the respective within-group variance to the total of the whole data set. However, one should remember that setting the inflection point is not always unambiguous. Another variant of this method consists in examining the variability of an average cluster radius depending on the number of classes k. The radius r_j of the cluster C_j is defined as the maximum distance of a point from C_j to the prototype μ_j. The mean value of the radii of all clusters is indeed the length of an average radius $\bar{r}$. A typical diagram of dependence of $\bar{r}$ on the number of clusters k is presented in Fig. 4.1. Usually, this kind of diagram is plotted for multiples of the value of k, e.g. as indicated in figures below, for $k = 2, 4, 8, \dots$. If a given value changes slightly in the interval $\{k, 2k\}$, then as the number of clusters we select the value of k^* from this interval [402, Sect. 7.3.3].

---

[7]Cf. K. Mardia et al. *Multivariate Analysis*. Academic Press 1979, p. 365.

[8]The basic work of reference, usually cited while this method is discussed, is the following: R.L. Thorndike "Who Belong in the Family?". *Psychometrika* 18 (4) 1953. Some modification of this method was applied in the paper C. Goutte, P. Toft, E. Rostrup, F.A. Nielsen, L.K. Hansen. "On clustering fMRI time series". *NeuroImage* 9 (3): 298–310 (March 1999).

A more formal approach to this problem, together with a survey of other methods can be found in [459]. The authors propose there the so called gap statistics, measuring the change of the within-group variance $W(k, m)$, determined after partition of m-elements set into k clusters (using a chosen clustering algorithm) in relation to the expected variance obtained from an m dimensional sample coming from an exemplary distribution. It is assumed that the value of k, for which the range is maximal, is the probable estimation of the number of clusters.

4.2.2   Methods Consisting in the Use of Information Criteria

For determining the value of k, one applies Bayesian information criterion (BIC), already mentioned, Akaike information criterion (AIC), or, finally, the so called deviance information criterion (DIC), which is a generalisation of both previous criteria.

Reference [449] proposes yet another method. For a given partition of the set X into k clusters (knowing their dimensions) one determines an averaged Mahalanobis distance (cf. Sect. 2.2)

$$\widehat{d}(k) = \frac{1}{n} \min_{j=1,\dots,k} d_\Sigma(X, \mu_j) \qquad (4.3)$$

where $d_\Sigma(X, \mu_j)$ denotes Mahalanobis distance between the elements of the set X and the j-th centre obtained e.g. by using the k-means algorithm.

Subsequently, we determine the range

$$J_k = \widehat{d}^{-\alpha}(k) - \widehat{d}^{-\alpha}(k-1) \qquad (4.4)$$

where $\widehat{d}^{-\alpha}(0) = 0$ and $\alpha = n/2$. The value

$$k^* = \arg \max_j J_j \qquad (4.5)$$

is assumed to be the proper number of clusters.

In particular, when X is a set without any internal structure, then $k^* = 1$.

In practice, the value of $\widehat{d}(k)$ is approximated by the minimum sum of error squares (3.1), computed by running the k-means algorithm.

4.2.3   Clustergrams

By clustergrams we understand the diagrams in the parallel coordinate system, where a vector was assigned to each observation, [420]. The components of a vector correspond to the cluster membership of the observation for a given number of clusters. The diagrams thus obtained allow for observing the changes in the assignment of objects into clusters in line with the increase of the number of clusters. The author of the idea claims that clustergrams are useful not only in the partitioning methods of cluster analysis but also in hierarchical grouping, when the number of observations increases. At the address http://www.r-statistics.com/2010/06/clustergram-visualization-and-diagnostics-for-cluster-analysis-r-code/ the code (in the R language) of a programme generating clustergrams is available.

Internal indexes such as the Bayesian information criterion (BIC) and the sum-of-squares encounter the difficulties regarding finding the inflection point of the respective curves and so, detection methods through BIC in partition-based clustering are proposed in the mentioned study. A new sum-of-squares based index is also proposed, where the minimal value is considered to correspond to the optimal number of clusters. External indexes, on the other hand, need a reference clustering or ground-truth information on data and therefore cannot be used in cluster validity. Consequently, an external index was extended into an internal index in order to determine the number of clusters by introducing a re-sampling method.

4.2.4 Minimal Spanning Trees

Several methods of determining the number of clusters refer to spanning a minimum weight tree. One of them, proposed by Galluccioa et al. [196] is based on the assumption that clusters one is looking for are dense regions in the sample space, surrounded by low frequency areas. Consequently, when constructing the minimum weight spanning tree (MST) using so-called Prim algorithm one will observe "low frequency" areas in the so-called Prim-trajectory. Assume we proceed as follows when constructing MST: Pick a node to be the initialisation of a partial tree and consider the remaining nodes as non-tree nodes. Now in each step $t = 1, \ldots, m-1$, where m is the total number of nodes, look for the pair of nodes, one from the tree and one from non-tree that has the minimum weight. Define $g(t)$ as this minimum weight found in step t. Remove the respective node from the non-tree set and add it to the tree set. If one draws $g(t)$, one will observe "valleys" with low g values, separated by picks of distances. The number of such valleys is the number of clusters, after some filtering process. The centre of the element set constituting a valley is the cluster mean candidate (seed for subsequent k-means iterations). The filtering consists in statistical testing whether or not a sequence of low distance values can be considered high density areas that is if we can say that the sequence is unlikely to come from a uniform distribution together with the surrounding peaks of g. The method works fine if the assumptions are met, but it may fail wherever single linkage clustering methods have difficulties. It differs from single linkage by making statistical tests whether or not a tree may be "broken" due to not conforming to the hypothesis that some Poisson process with uniform distribution generated the points.

4.3 Partition Quality Indexes

It is expected that a clustering algorithm would assign objects which are similar one to another to a common group and those that differ from one another—to different groups. We would like the partition returned by the algorithm, at least for a definite matrix of observations X, to be *optimal* in the sense of the adopted quality index. If the actual assignment of objects into groups is known, the so called *external* quality indexes are constructed. They quantify conformity of the partition returned by an algorithm with the actual assignment of objects into groups. In the opposite case, one attempts to characterise the immanent features of clusters proposed by the algorithm. The basic criteria, applied in constructing *internal* quality indexes belonging to this category, are [225]: *compactness* (groups should be well clustered in the feature space, which means that they should have, e.g., a possibly small diameter) and *separability* (clusters should be easily discernible from each other).

While external indexes allow for selecting the algorithm attaining the highest conformity of the partition with the given partition, internal indexes are used additionally for estimating the number of clusters. A survey of different criteria, to be taken into account during the construction of the external and internal indexes is presented in [231]. Although the work concerns postgenomic research, the notes presented there have general character. In [356], Milligan and Cooper presented a comparison of 30 internal indexes, which are used in hierarchical grouping, while Wu et al. outlined in [507] 16 external indexes, which are used for describing k-means algorithm properties. A survey of various quality indexes can be found in [104, 155, 167, 224, 225, 273, 274, 511]. We recommend also the 3rd chapter of the thesis [535]. Conform to the claim of Kannan, Vempala and Vetta [279], all these indexes, although – because of their simplicity—highly attractive, may lead to wrong conclusions. That is the case mainly when the clusters concerned are not Voronoi clusters (cf. p. 45). This is why the authors quoted, proposed a certain graph-theoretic index of partition quality.

At this point we limit ourselves to measures evaluating the quality of the partition returned by combinatorial algorithms. Milligan and Cooper, already mentioned, compared in [356] 30 different indicators. For this purpose, they constructed testing sets consisting of 2–5 clusters containing 50 elements each, described using 4–8 features. In their opinion, the best properties are exhibited by the indicator proposed by Caliński and Harabasz, having the form

$$q_{CH}(k) = \frac{tr(B)}{k-1} \frac{m-k}{tr(W)} \tag{4.6}$$

where W and B are matrices of inter- and within-group correlations, defined by the equations (2.29) and (2.30). The optimal number of clusters corresponds to the value of k for which q_{CH} attains the maximum value.

We present below some other often used measures.[9]

It is assumed that the most natural cluster compactness measure is the quantisation error, defined as

$$q(k) = \frac{1}{k} \sum_{j=1}^{k} \frac{1}{n_j} \sum_{\mathbf{x} \in C_j} d(\mathbf{x}, \boldsymbol{\mu}_j) \tag{4.7}$$

where $\boldsymbol{\mu}_j$ denotes the cluster centre C_j, n_j denotes its cardinality, while $d(\mathbf{x}, \boldsymbol{\mu}_j)$ denotes the distance of the point $\mathbf{x}$ from the center $\boldsymbol{\mu}_j$.

Then, a simple and very popular measure of cluster separation is, [284], the coefficient s_{KR} introduced by Kaufmann and Rousseeuw (called by its authors the *silhouette coefficient*). Let us, namely, denote by $a(\mathbf{x})$ the average distance between some object $\mathbf{x}$ from the class C_j and the rest of objects from this class, and by $b(\mathbf{x})$—the distance between this point $\mathbf{x}$ and elements of the closest cluster different from C_j:

$$a(\mathbf{x}) = \frac{1}{n_j - 1} \sum_{\mathbf{y} \in C_j} d(\mathbf{x}, \mathbf{y}) \qquad b(\mathbf{x}) = \min_{l=1,\ldots,k, l \neq j} \frac{1}{n_l} \sum_{\mathbf{y} \in C_l} d(\mathbf{x}, \mathbf{y})$$

Then

$$s(\mathbf{x}) = [b(\mathbf{x}) - a(\mathbf{x})] / \max[a(\mathbf{x}), b(\mathbf{x})] \tag{4.8}$$

indicates whether the given partition is good (the value of $s(\mathbf{x})$ is close to 1) or bad (negative values of $s(\mathbf{x})$). This indicator is not suitable for large data set analysis; its computational complexity is $O(m^2)$.

The Dunn index, $D(k)$, mentioned in p. 100, is used to measure the compactness and the separability of clusters and is defined as

$$D(k) = \min_{j=1,\ldots,k} \left\{ \min_{l=j+1,\ldots,k} \frac{d(C_j, C_l)}{\max_{v=1,\ldots,k} diam(C_v)} \right\} \tag{4.9}$$

It is, in effect, a variant of the Eq. (3.51), where $d(C_j, C_l)$ represents the distance between two clusters, calculated according to the nearest neighbour scheme (in p. 29), and $diam(C_v)$ denotes the diameter of the cluster C_v. Its disadvantages are: high computational complexity and sensitivity to outliers. The "optimal" value of k is the one which maximises the index value.

Another popular measure is the Davies-Bouldin index, $DB(k)$, defined as follows:

---

[9]The interested reader should visit the website http://cran.r-project.org/web/packages/clusterCrit/ which continues to be maintained by Bernard Desgraupes. Description of 42 quality indexes can be found there, as well as clusterCrit package in R, providing their implementation.

$$DB(k) = \frac{1}{k} \sum_{j=1}^{k} \max_{l=1,\dots,k,l\neq j} \frac{\rho(C_j) + \rho(C_l)}{d(C_j, C_l)} \qquad (4.10)$$

where $\rho(C_j)$ denotes the average distance between points from the cluster C_j and the gravity centre of the cluster, that is

$$\rho(C_j) = \frac{1}{|C_j|} \sum_{\mathbf{x} \in C_j} \|\mathbf{x} - \boldsymbol{\mu}_j\|$$

while $d(C_j, C_l)$ denotes the distance between the gravity centres of clusters C_j and C_l. The value of k, for which $DB(k)$ takes on the minimum value, indicates the optimal number of clusters.

Another frequently used quality index, introduced by Xie and Beni [511] has the following form

$$\chi = \frac{\sum_{i=1}^{m} \sum_{j=1}^{k} u_{ij}^{\alpha} d^2(\mathbf{x}_i, \boldsymbol{\mu}_j)}{m \cdot \min_{j\neq l} d^2(\boldsymbol{\mu}_j, \boldsymbol{\mu}_l)} = \frac{J_{\alpha}(U, M)}{m \cdot \min_{j\neq l} d^2(\boldsymbol{\mu}_j, \boldsymbol{\mu}_l)} \qquad (4.11)$$

Here, aspects of both compactness and separability are taken into account. Yet another measure was proposed by Fukuyama and Sugeno [193]:

$$w_{FS} = \sum_{i=1}^{m} \sum_{j=1}^{k} (u_{ij}^{\alpha})[d^2(\mathbf{x}_i, \boldsymbol{\mu}_j) - d^2(\boldsymbol{\mu}_j, \overline{\boldsymbol{\mu}})] \qquad (4.12)$$

where $\overline{\boldsymbol{\mu}}$ is the gravity centre of the data set, i.e. $\overline{\boldsymbol{\mu}}_l = (1/m) \sum_{i=1}^{m} \mathbf{x}_{il}, l = 1, \dots, n$. The first minimum that the index attains indicates the number of clusters.

Finally, the indicators authored by Gath and Geva [197] are worth mentioning. They suggested two criteria of comparing and selecting the optimal partitioning, assuming that a good partition is characterised by clear separation of clusters, and clustering results contain as many as possible of points centred around the prototype, which means that their volume is minimal. These indicators apply to fuzzy partitions. The first of them, "fuzzy volume", has the form

$$F_{HV} = \sum_{j=1}^{c} [det(F_j)]^{1/2}$$

where

$$F_j = \frac{\sum_{i=1}^{N} u_{ij}(\mathbf{x}_i - v_j)(\mathbf{x}_i - v_j)^T}{\sum_{i=1}^{N} u_{ij}}$$

in the situation when the defuzzification parameter is equal to 2. The second indicator, describing the mean density of partition, has the form

$$D_{PA} = \frac{1}{k} \sum_{j=1}^{k} \frac{S_j}{[det(F_j)]^{1/2}}$$

where $S_j = \sum_{i=1}^{N} u_{ij}$. Finally, the density of partition is defined as: $P_D = \frac{S}{F_{HV}}$, where $S = \sum_{j=1}^{k} \sum_{i=1}^{N} u_{ij}$.

The Γ index of Baker-Huber

$$\Gamma = \frac{s^+ - s^-}{s^+ + s^-} \qquad (4.13)$$

tries to capture the frequency with which the natural intuition of elements within a cluster being closer to ones outside is fulfilled. s^+ is the number of quadruples (p, q, u, v) of elements such that p, q belong to the same cluster and u, v belong to different clusters and condition $d(p, q) < d(u, v)$—such a quadruple is called "concordance". s^- is the number of quadruples (p, q, u, v) of elements such that p, q belong to the same cluster and u, v belong to different clusters and condition $d(p, q) > d(u, v)$—such a quadruple is called "disconcordance".

Recall the cophenetic correlation coefficient, described on p. 30, being a quality measure for hierarchical clustering. An analogous correlation coefficient has been proposed for partition methods. Consider a table the rows of which corresponding to each pair of different elements of the set X ($\frac{m(m-1)}{2}$ rows in all.). Let the first column be a measure of distance, dissimilarity or similarity between the pair of elements of the row, and the second column shall contain 1, if both elements lie in the same cluster, and 0 otherwise. The Pearson correlation coefficient between these columns can be considered as a quality measure of the clustering. Values close to 1 or -1 would tell that the clustering reflects the (dis)similarity well, values close to 0 would say that the clustering performed poorly.

4.4 Comparing Partitions

Estimating the quality of partition is just one side of the exploratory procedure. As we dispose of many algorithms, we would like to know [486]:

- Is a given algorithm sensitive to outliers (noise)?
- Is the result of algorithm operation dependent on the order in which it processes the data?
- Are the solutions returned by alternative algorithms similar, and if yes, then to what extent?
- Does, for a given task, an optimal solution exist, and if yes, then how close to it are the solutions returned by different algorithms?

The majority of methods of comparing partitions use the contingency table, called also the cross tabulation, the frequency distribution table, or the two way contingency table (in case of comparing two partitions). Let $\mathcal{P}^i = \{C_1^i, \ldots, C_k^i\}$, $i = 1, 2, \ldots$ denote the partition of the set $\mathfrak{X}$ into k classes returned by i-th algorithm. In particular, it can be assumed that we compare two partitions and each of them divides the set $\mathfrak{X}$ into different number of groups. Without loss of generality, we assume, however, that we are interested in comparing the results of the operation of two algorithms, so $i \in \{1, 2\}$, while each of them divides the set of objects $\mathfrak{X}$ into an established number of classes k. Let

$\mathfrak{X}_{11}$ – set of pairs of objects assigned in both partitions to the same cluster
$\mathfrak{X}_{00}$ – set of pairs of objects assigned in both partitions to different clusters
$\mathfrak{X}_{01}$ – set of pairs of objects assigned in partition $\mathcal{P}^1$ to different clusters
        and in partition $\mathcal{P}^2$ to the same cluster
$\mathfrak{X}_{10}$ – set of pairs of objects assigned in partition $\mathcal{P}^1$ to the same cluster
        and in partition $\mathcal{P}^2$ to different clusters

$$(4.14)$$

Let, further on, $n_{\alpha\beta} = |\mathfrak{X}_{\alpha\beta}|$. Of course

$$n_{00} + n_{11} + n_{10} + n_{01} = m(m - 1)/2$$

and

$$m_j^i = |C_j^i|, \qquad m_{ij} = |C_i^1 \cap C_j^2| \qquad (4.15)$$

The measure

$$proj(\mathcal{P}^1, \mathcal{P}^2) = \sum_{C^1 \in \mathcal{P}^1} \max_{C^2 \in \mathcal{P}^2} |C^1 \cap C^2| \qquad (4.16)$$

is called the value of projection of the partition $\mathcal{P}^1$ on $\mathcal{P}^2$ [475]. Note that $proj(\mathcal{P}^1, \mathcal{P}^2) \neq proj(\mathcal{P}^2, \mathcal{P}^1)$. The distance $d(\mathcal{P}^1, \mathcal{P}^2)$ between partitions is now defined as

$$d(\mathcal{P}^1, \mathcal{P}^2) = m - proj(\mathcal{P}^1, \mathcal{P}^2) - proj(\mathcal{P}^2, \mathcal{P}^1) \qquad (4.17)$$

More remarks on this topic are presented in van Dongen's work [475].

Classical reference here is the work [265]. Referring to [349, 486], we present a short survey of approaches applied in comparing partitionings. The reader interested in a more thorough discussion should refer to these works and to the bibliographical references[10] quoted there.

---

[10] Another work of reference is G. Saporta, G. Youness, Comparing two partitions: Some proposals and experiments. *Compstat 2002*, pp. 243-248. It is also worth becoming familiar with the website http://darwin.phyloviz.net/ComparingPartitions/.

4.4.1 Simple Methods of Comparing Partitions

The oldest and the simplest method consists in using the chi-square statistics

$$\chi(\mathcal{P}^1, \mathcal{P}^2) = \sum_{i=1}^{k} \sum_{j=1}^{k} \frac{(m_{ij} - m_i^1 m_j^2)^2}{m_i^1 m_j^2} \tag{4.18}$$

Another method lies in the use of the Rand coefficient

$$\mathcal{R}(\mathcal{P}^1, \mathcal{P}^2) = \frac{2(n_{00} + n_{11})}{m(m-1)} \tag{4.19}$$

which varies from 0 (completely different partitions) to 1 (identical partitions). It is, thus, an equivalent of the recall, i.e. the measure applied in evaluating binary classifiers. The $\mathcal{R}$ coefficient depends on the number of elements and on the number of classes. In particular, if both partitions are independent, then with the increase of k the value of $\mathcal{R}$ tends to 1, which disqualifies this coefficient as a measure of similarity. This is why the so called Adjusted Rand coefficient is introduced, though even that modification is not perfect, [349].

The subsequent indicator is the modified Fowlkes-Mallows coefficient

$$\mathcal{FM}(\mathcal{P}^1, \mathcal{P}^2) = \frac{n_{11}}{\sqrt{(n_{11} + n_{10})(n_{11} + n_{01})}} \tag{4.20}$$

Fowlkes and Mallows introduced their measure in order to compare the results of hierarchical grouping, but its modification, presented above, is used for comparing partitions. As observed in [486], in the context of information retrieval this indicator corresponds to the geometric mean of precision and recall. The measure $\mathcal{FM}(\mathcal{P}^1, \mathcal{P}^2)$ represents a divergence of the "empty" model where both partitions are independent. However, a disadvantage of $\mathcal{FM}$ is that for small k it takes on very high values.

The Mirkin metric

$$\mathcal{M}(\mathcal{P}^1, \mathcal{P}^2) = \sum_{i=1}^{k} (n_i^1)^2 + \sum_{j=1}^{k} (n_j^2)^2 - 2 \sum_{i=1}^{k} \sum_{j=1}^{k} m_{ij}^2 \tag{4.21}$$

denotes (the non-normalised) Hamming distance between vectors representing the partitions being compared. The particular elements of vectors correspond to pairs of objects being compared (i, j) and if both objects belong to the same cluster, then we assign 1 to the proper position, and in the opposite case we assign 0, [475]. Furthermore, the following relation exists

$$\mathcal{M}(\mathcal{P}^1, \mathcal{P}^2) = m(m-1)[1 - \mathcal{R}(\mathcal{P}^1, \mathcal{P}^2)] \tag{4.22}$$

The Jackard coefficient

$$J(\mathcal{P}^1, \mathcal{P}^2) = \frac{n_{11}}{n_{01} + n_{10} + n_{11}} \qquad (4.23)$$

measures the share of pairs belonging to the same cluster in both partitions with respect to the number of pairs in the same cluster in any partition.

4.4.2   Methods Measuring Common Parts of Partitions

One of the simplest and the most commonly used measures from this category is the purity (*purity*)

$$\mathfrak{P}(\mathcal{P}^1, \mathcal{P}^2) = \frac{1}{m} \sum_{i=1}^{k_1} \max_{1 \le j \le k_2} m_{ij} \qquad (4.24)$$

where k_1 (resp. k_2) denotes the number of groups in the partition $\mathcal{P}^1$ (resp. $\mathcal{P}^2$), and $m_{ij} = |C_i^1 \cap C_j^2|$. We introduced this measure in Sect. 3.3.3.

A subsequent measure, commonly used in information retrieval, is the so called F-measure, that is, the weighted harmonic mean of the precision and recall [347, p. 156]. Suppose we treat the clusters of the partition $\mathcal{P}^1$ as predefined classes of documents and clusters of the partition $\mathcal{P}^2$ as the results of queries. Then the F-measure of the dependance between cluster C_j^2 and another cluster C_i^1 indicates how well C_j^2 describes the class C_i^1

$$F(C_i^1, C_j^2) = \frac{2 re_{ij} pe_{ij}}{re_{ij} + pr_{ij}} = \frac{2 m_i^1 m_j^2}{m_i^1 + m_j^2} \qquad (4.25)$$

where $re_{ij} = m_{ij}/m_i^1$ denotes the recall (*recall*) and $pr_{ij} = m_{ij}/m_j^2$ the precision (*precision*). The complete F-measure is defined as weighted sum of

$$F(\mathcal{P}^1, \mathcal{P}^2) = \frac{1}{m} \sum_{i=1}^{k} m_i^1 \max_{j=1,\ldots,k} F(C_i^1, C_j^2) \qquad (4.26)$$

This is an asymmetric measure, so it is appropriate e.g. for comparing how well the obtained partition $\mathcal{P}^2$ approximates the actual partition $\mathcal{P}^1$. The only problem is that usually the actual partition is not known. Meilă [349] mentions a modification, introduced by Larsen

$$F_L(\mathcal{P}^1, \mathcal{P}^2) = \frac{1}{k} \sum_{i=1}^{k} \max_{j=1,\ldots,k} F(C_i^1, C_j^2) \qquad (4.27)$$

Yet, the authors of [486] note that Larsen did not apply such modification.

Another asymmetric measure

$$MH(\mathcal{P}^1, \mathcal{P}^2) = \frac{1}{m} \sum_{j=1}^{k} \max_{i=1,\ldots,k} m_{ij} \tag{4.28}$$

was proposed by Meilă and Heckerman in [350]. Again, $\mathcal{P}^1$ is an exemplary partition. In order to adapt this measure for comparing arbitrary partitions the authors introduced a symmetric measure having the form

$$MH_m(\mathcal{P}^1, \mathcal{P}^2) = \frac{1}{m} \sum_{j=1}^{k} m_{ij*} \tag{4.29}$$

where m_{ij*} is determined as follows. Let M_1 denote the original contingency table with entries m_{ij} as defined above. In a matrix M_j let m_{ij*} be a cell with coordinates $(i(j^*), j^*)$ holding a maximum entry of M_j, that means these coordinates mean that the cluster C_{j*}^2 matches the best the cluster $C_{i(j^*)}^1$. M_{j+1} is obtained them by "deleting" the corresponding line $(i(j^*))$ and column (j^*) from the contingency table M_j (e.g. by setting them to -1). While in previous definitions the partitions could have different number of clusters, in this one they should have the identical numbers of clusters.

Then, Wallace [490] postulated introducing an index, representing the probability that a pair of objects taken from a certain cluster $C_i^1 \in \mathcal{P}^1$, in the partition $\mathcal{P}^2$ also belongs to the same cluster

$$WI(\mathcal{P}^1, \mathcal{P}^2) = \frac{M}{\sum_{i=1}^{k} m_i^1 (m_i^1 - 1)/2} \tag{4.30}$$

Here $m_i^1 = |C_i^1|$ and M denotes the number of all the pairs of objects which belong to the same cluster in both partitions. If both partitions are identical, then $WI(\mathcal{P}^1, \mathcal{P}^2) = 1$, and if they are completely different, i.e. $C_i^1 \cap C_j^2 = \emptyset$ for all (i, j), then $WI(\mathcal{P}^1, \mathcal{P}^2) = 0$.

4.4.3 Methods Using Mutual Information

Let m_j denote the cardinality of the j-th cluster $C_j \in \mathcal{P}$ and let $p_j = m_j/m$ denote the probability that a randomly selected element $\mathbf{x} \in X$ belongs to that cluster. The entropy of the partition $\mathcal{P}$ is calculated as

$$H(\mathcal{P}) = -\sum_{j=1}^{k} p_j \log p_j \tag{4.31}$$

In case of two partitions, mutual information between the partitions equals

$$I(\mathcal{P}^1, \mathcal{P}^2) = \sum_{i=1}^{k} \sum_{j=1}^{k} p_{ij} \log \frac{p_{ij}}{p_i^1 p_j^2} \tag{4.32}$$

where $p_{ij} = m_{ij}/m$ and m_{ij} denotes the number of objects belonging to the class $C_i \in \mathcal{P}^1$ and to the class $C_j \in \mathcal{P}^2$. Thus, I can be calculated as

$$I(\mathcal{P}^1, \mathcal{P}^2) = H(\mathcal{P}^1) - H(\mathcal{P}^1 | \mathcal{P}^2) \tag{4.33}$$

where the conditional entropy is defined as

$$H(\mathcal{P}^1 | \mathcal{P}^2) = - \sum_{i=1}^{k} \sum_{j=1}^{k} p_{ij} \log \frac{p_{ij}}{p_j^2} \tag{4.34}$$

Even though the mutual information is a metric in the space of all partitions, it is not limited, which gives rise to problems of comparing different partitions. However, the following estimation is true

$$0 \le I(\mathcal{P}^1, \mathcal{P}^2) \le \min \left[H(\mathcal{P}^1), H(\mathcal{P}^2) \right] \tag{4.35}$$

Strehl and Ghosh proposed in [447] the normalised mutual information

$$NMI_{SG}(\mathcal{P}^1, \mathcal{P}^2) = \frac{I(\mathcal{P}^1, \mathcal{P}^2)}{\sqrt{H(\mathcal{P}^1) H(\mathcal{P}^2)}}$$

$$= \frac{\displaystyle\sum_{i=1}^{|\mathcal{P}^1|} \sum_{j=1}^{|\mathcal{P}^2|} m_{ij} \log \left(\frac{m m_{ij}}{m_i} m_j \right)}{\sqrt{\left(\displaystyle\sum_i m_i \log \frac{m_i}{m} \right) \left(\displaystyle\sum_j m_j \log \frac{m_j}{m} \right)}} \tag{4.36}$$

The thus defined measure satisfies the conditions $NMI_{SG}(\mathcal{P}^1, \mathcal{P}^2) = 1$ if $\mathcal{P}^1 = \mathcal{P}^2$ and $NMI_{SG}(\mathcal{P}^1, \mathcal{P}^2) = 0$ if $m_{ij} = m_i^1 m_j^2$ or if $m_{ij} = 0$ for all the pairs (i, j).

Then, Fred and Jain [188] normalised the mutual information via the transformation

$$NMI_{FJ}(\mathcal{P}^1, \mathcal{P}^2) = \frac{2I(\mathcal{P}^1, \mathcal{P}^2)}{H(\mathcal{P}^1) + H(\mathcal{P}^2)} \tag{4.37}$$

Like before, $0 \le NMI_{FJ}(\mathcal{P}^1, \mathcal{P}^2) \le 1$. Furthermore, $NMI_{FJ}(\mathcal{P}^1, \mathcal{P}^2) = 0 \Leftrightarrow NMI_{SG}(\mathcal{P}^1, \mathcal{P}^2) = 0$ and $NMI_{FJ}(\mathcal{P}^1, \mathcal{P}^2) = 1 \Leftrightarrow NMI_{SG}(\mathcal{P}^1, \mathcal{P}^2) = 1$.

On the other hand, in the work [314], the following normalisation is used

$$NMI_{LFK}(\mathcal{P}^1, \mathcal{P}^2) = \frac{H(\mathcal{P}^1) + H(\mathcal{P}^2) - H(\mathcal{P}^1, \mathcal{P}^2)}{H(\mathcal{P}^1) + H(\mathcal{P}^2)} \tag{4.38}$$

where $H(\mathcal{P}^1, \mathcal{P}^2) = -\sum_{ij} p_{ij} \log p_{ij}$ denotes the joint entropy.

A measure, which was the subject of detailed analysis, is the disturbance of information, introduced by Meilă [349]

$$\begin{aligned} VI(\mathcal{P}^1, \mathcal{P}^2) &= H(\mathcal{P}^1) + H(\mathcal{P}^2) - I(\mathcal{P}^1, \mathcal{P}^2) \\ &= H(\mathcal{P}^1|\mathcal{P}^2) + H(\mathcal{P}^2|\mathcal{P}^1) \end{aligned} \tag{4.39}$$

It is a metric in the space of all partitions of the $\mathfrak{X}$ set, and if we consider only the partitions into $k \leq \sqrt{m}$ classes, then

$$\frac{2}{m} \leq VI(\mathcal{P}^1, \mathcal{P}^2) \leq 2 \log k \tag{4.40}$$

When the number of objects m is a multiplication of k^2, then $VI(\mathcal{P}^1, \mathcal{P}^2) = 2 \log k$, while in general, if $k > \log k$ then $VI(\mathcal{P}^1, \mathcal{P}^2) \leq \log m$. The value of $VI(\mathcal{P}^1, \mathcal{P}^2)$ can be calculated in time $O(m+k^2)$, which includes: setting the value of contingency table (in time $O(m)$) and setting of the proper values of VI in time $O(k^2)$.

In the work [282] the following normalisation of the index VI was proposed:

$$VI_{KLN}(\mathcal{P}^1, \mathcal{P}^2) = \frac{1}{2}\left[\frac{H(\mathcal{P}^1|\mathcal{P}^2)}{H(\mathcal{P}^1)} + \frac{H(\mathcal{P}^2|\mathcal{P}^1)}{H(\mathcal{P}^2)}\right] \tag{4.41}$$

It represents the averaged defect of information if we reason about $\mathcal{P}^1$ knowing $\mathcal{P}^2$ and vice versa.

4.5 Cover Quality Measures

Studies concerning this topic are very rare, although we face the problem of cover while generating fuzzy partitions and while analysing the ensembles in empirical graphs (cf. e.g. [182]). Below we present the preliminary results obtained by Lancichinetti, Fortunato and Kertész [314].

Let us remind that by cover $\mathcal{C}$ of the set $\mathfrak{X}$ with k subsets we understand a family of sets $\{C_1, \ldots, C_k\}$ such that

$$\begin{aligned} &(a)\ C_i \neq \emptyset, i = 1, \ldots, k \\ &(b)\ \bigcup_{i=1}^{k} C_i = \mathfrak{X} \\ &(c)\ C_i \cap C_j \neq \emptyset \end{aligned} \tag{4.42}$$

The (c) condition means that the sets forming the cover are not disjoint.

Assignment of objects into classes is now represented by the table $T^1 = [t_{ij}^1]_{m \times k}$, where $t_{ij}^1 = 1$ if i-th object belongs to j-th class, and $t_{ij}^1 = 0$ in the opposite case. Let

$\mathbf{t}_j^1$ denote j-th column of the matrix T^1 representing the cover $\mathcal{C}^1$. It can be treated as a realisation of dichotomous random variable X_j having probability distribution

$$P(X_j = 1) = m_j^1/m, \qquad P(X_j = 0) = 1 - m_j^1/m \qquad (4.43)$$

The random variable Y_j, representing the assignment of objects to the j-th class in the cover $\mathcal{C}^2$ is defined analogously.

Let us determine, after [314], four probability distributions

$$
\begin{aligned}
P(X_i = 1, Y_j = 1) &= \tfrac{1}{m}|C_i^1 \cap C_j^2| \\
P(X_i = 1, Y_j = 0) &= \tfrac{1}{m}(|C_i^1| - |C_i^1 \cap C_j^2|) \\
P(X_i = 0, Y_j = 1) &= \tfrac{1}{m}(|C_j^2| - |C_i^1 \cap C_j^2|) \\
P(X_i = 0, Y_j = 0) &= \tfrac{1}{m}(m - |C_i^1| - |C_j^2| + |C_i^1 \cap C_j^2|)
\end{aligned}
\qquad (4.44)
$$

Let further

$$H(X_i|Y_j) = H(X_i, Y_j) - H(Y_j) \qquad (4.45)$$

denote the quantity of information, crucial to conclude about X_i, knowing the distribution of the variable Y_j. In particular, if $C_1^1 = C_{j*}^2$ then $H(X_i|Y_{j*}) = 0$ and one can say that Y_{j*} is the best candidate for concluding about the distribution X_i. Thus, it is assumed that

$$H(X_i|\mathcal{C}^2) = H(X_i|\{Y_1, \ldots, Y_k\}) = \min_{j=1,\ldots,k} H(X_i|Y_j) \qquad (4.46)$$

This measure can be normalised

$$H_{norm}(X_i|\mathcal{C}^2) = \frac{H(X_i|\mathcal{C}^2)}{H(X_i)} \qquad (4.47)$$

and we get the final formula

$$H_{norm}(\mathcal{C}^1|\mathcal{C}^2) = \frac{1}{k} \sum_{i=1}^{k} \frac{H(X_i|\mathcal{C}^2)}{H(X_i)} \qquad (4.48)$$

For comparing the covers of $\mathcal{C}^1$ and $\mathcal{C}^2$, the arithmetic mean of conditional entropies is applied

$$N(\mathcal{C}^1|\mathcal{C}^2) = 1 - \frac{1}{2}[H_{norm}(\mathcal{C}^1|\mathcal{C}^2) + H_{norm}(\mathcal{C}^2|\mathcal{C}^1)] \qquad (4.49)$$

Further information concerning the application and the implementation of the above formula can be found in the appendix to the work [314].

Chapter 5
Spectral Clustering

Abstract This chapter discusses clustering methods based on similarities between pairs of objects. Such a knowledge does not imply that the entire objects are embedded in a metric space. Instead, the local knowledge supports a graphical representation displaying relationships among the objects from a given data set. The problem of data clustering transforms then into the problem of graph partitioning, and this partitioning is acquired by analysing eigenvectors of the graph Laplacian, a basic tool used in spectral graph theory. We explain how various forms of graph Laplacian are used in various graph partitioning criteria, and how these translate into particular algorithms. There is a strong and fascinating relationship between graph Laplacian and random walk on a graph. Particularly, it allows to formulate a number of other clustering criteria, and to formulate another data clustering algorithms. We briefly review these problems. It should be noted that the eigenvectors deliver so-called spectral representation of data items. Unfortunately, this representation is fixed for a given data set, and adding or deleting some items destroys it. Thus we discuss recently invented methods of out of sample spectral clustering allowing to overcome this disadvantage. Although spectral methods are successful in extracting non-convex groups in data, the process of forming graph Laplacian is memory consuming and computing its eigenvectors is time consuming. Thus we discuss various local methods in which only relevant part of the graph are considered. Moreover, we mention a number of methods allowing fast and approximate computation of the eigenvectors.

5.1 Introduction

There exists a number of reasons why clustering algorithms based on embedded data representation cannot be used for analysis:

- the data may not be available in the embedded form (either because it cannot be collected in such a way or for data protection reasons),

© Springer International Publishing AG 2018
S. T. Wierzchoń and M. A. Kłopotek, *Modern Algorithms of Cluster Analysis*,
Studies in Big Data 34, https://doi.org/10.1007/978-3-319-69308-8_5

- the data is represented in a natural way as relationships between objects rather than featured objects (for example in social networks),
- the data does not occupy the whole embedding space, but are rather present on a low-dimensional manifold, which may be additionally folded,
- the data is not linearly separable, as expected by many such methods, or does not form compact ball-shaped clusters.

For these reasons, a relational representation, or in particular graph representation may be preferred. Due to the hardness of purely graph-based clustering methods, spectral clustering methods, solving relaxed graph clustering tasks, were found to an attractive choice because they are easy to implement, and they are reasonably fast especially for sparse data sets up to several thousands.

Spectral clustering treats the data clustering as a graph partitioning problem without making any assumption on the form of the data clusters.

In these methods a proximity (or distance) matrix S, which is an m by m matrix containing the pairwise (dis-)similarities plays an important role. The (dis-)similarities can be computed by means of a mathematical formula (e.g. a transformation of the distance between any pair of points; see also [245] for a more general case), or even they can be defined subjectively. As noted in Sect. 2.1, a similar approach was proposed for the first time, in 1909 by Polish anthropologist, ethnographer, demographer and statistician—Jan Czekanowski. The similarity matrix can be converted to so-called similarity graph being another representation of Czekanowski's diagram. Two nodes representing entities i and j are joined by an edge if the corresponding entry in the Czekanowski's diagram is marked by a non-white symbol (i.e. they are "sufficiently" similar to each another), and the weight of this edge reflects the shade of gray used to paint the entry in the diagram. What is more important, using graphs we can discover clusters in which connectivity (and not compactness) plays dominant role, see Fig. 5.1.

Using graph-theoretic formalism we gain new possibilities of defining and extracting clusters. Unfortunately, expressed in purely graph-theoretic language these prob-

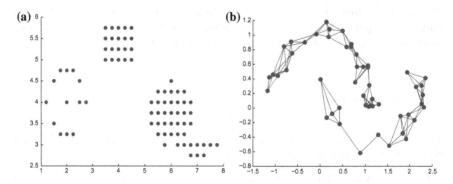

Fig. 5.1 Compact and well separated clusters (**a**) vs. clusters characterized by a similarity relationship (**b**)

lems usually appear to be $\mathcal{NP}$-hard problems. Spectral methods offer a way to solve in polynomial time relaxed versions of these problems. Although there is no a single "canonical" spectral clustering algorithm we can distinguish four main steps involved in each of its instantiation:

(a) For a given data set X containing m items determine a similarity matrix S. Its elements describe either a strength of relationship between pairs of objects (usually $s_{ij} \in [0, 1]$), or they only indicate whether such a relationship exists. In this last case the similarity matrix becomes the adjacency matrix A with elements $a_{ij} \in \{0, 1\}$. Thus, the matrix S can be treated as a generalized adjacency matrix. Usually we assume that S is symmetric; hence it describes an undirected and weighted graph. Then, the matrix S is transformed to another matrix $\mathbb{L}$; it usually corresponds to the normalized or unnormalized Laplacian (see Sect. 5.2.2 for the definitions) of the graph associated to S.
(b) Compute eigendecomposition of the matrix $\mathbb{L}$, i.e. $\mathbb{L} = W \Lambda W^{-1}$, where W is the square $m \times m$ matrix whose i-th column is the eigenvector $\mathbf{w}_i$ of $\mathbb{L}$ and Λ is the diagonal matrix whose diagonal elements are the corresponding eigenvalues of $\mathbb{L}$.
(c) Determine the so-called spectral mapping $x_i \mapsto (w_{i,1}, \ldots, w_{i,k'}), i = 1, \ldots, m$, i.e. the k'-dimensional representation of the d-dimensional data items. Typically, k' equals to the number of groups k, but see e.g. Sect. 3.3 in [275] for a deeper discussion.
(d) Use the low dimensional representation to partition the data, via an algorithm for embedded data representation.

In summary, we can say that such a procedure partitions the data according to a low-dimensional representation that is obtained from computing eigenvectors of the graph Laplacian matrix implied by these data. Spielman and Teng have shown in [439] that the spectral clustering procedure works well on bounded-degree planar graphs and finite element meshes—the classes of graphs to which they are usually applied.

Remark 5.1.1 The methods discussed in this chapter play an important role not only in clustering, but also in various related problems, like semi-supervised learning [103], dimensionality reduction (LLE, Laplacian Eigenmaps, Isomap,...), and more generally, in manifold learning [446]. As noted by Belkin and Niyogi [61], the central modelling assumption in all of these methods is that the data resides on or near a low-dimensional submanifold embedded in a higher-dimensional space. Such an assumption seems natural for a data-generating source with relatively few degrees of freedom.

In almost all practical situations, one does not have access to the underlying manifold but instead approximates it from a point cloud. Hence, the common approximation strategy in these methods it to construct a similarity (or adjacency-based) graph associated to a point cloud forming the set X. Most manifold learning algorithms then proceed by exploiting the structure of this graph. It should be noted that the entire graph is an empirical object, constructed from available data. Therefore any graph-theoretic technique is only applicable, when it can be related to the underlying process generating the data [61].   □

Remark 5.1.2 Kernel methods, introduced in Sect. 2.5.7, form an intriguing bridge between partitional and relational approaches. Here an ability to identify the non-linear structures, is gained by mapping the data points to a higher-dimensional feature space $\mathfrak{F}$, and next, by applying the *distance kernel trick*. Accordingly, computing Euclidean distance in the feature space reduces to elementary operations on the elements of a (kernel) matrix $K \in \mathbb{R}^{n \times n}$, whose elements represent similarity degrees between pairs of objects. This leads to the kernel variants of the k-means algorithm as discussed in Chap. 3; see also e.g. [176]. Dhillon, Guan and Kulis proposed in [150] a framework unifying standard k-means with its spectral and kernel variants. Other relationship between spectral and kernel methods is discussed later in Sect. 5.5.   □

In this chapter we give technical comments to the procedure of spectral clustering and its hottest variants. In Sect. 5.2.1 we give some hints how to choose the weights W, and how to create similarity graphs. In Sect. 5.3 we explain why do we need graph Laplacian and how its variants relate to different graph cutting criteria. With this machinery it is possible to relax the $\mathcal{NP}$-complete discrete problem to much easier to solve continuous problem. However, the solution of the relaxed problem and that of the discrete $\mathcal{NP}$-hard problem can differ substantially. In Sect. 5.3.5 a hottest variant of spectral clustering leading to a tight relaxation that has a solution that closely matches the solution of the original discrete problem is briefly presented. A spectral clustering procedure allows to find clusters in a given data set X but it is not fitted to the situations when new observations are coming. Section 5.3.6 is devoted to this problem. As the graph construction and followed eigen-decomposition are not cheap,[1] in Sect. 5.5 some methods designed for coping with large data sets are mentioned. Other way of reducing time complexity in case of huge data sets is to use local methods—these are described in Sect. 5.4.

5.2 Basic Notions

Let $\Gamma = (V, E)$ denotes an undirected and connected graph. Here $V = \{v_1, \ldots, v_m\}$ is the set of nodes, and E specifies the set of m edges $e_{ij} = \{v_i, v_j\}$. Such a graph, if all its edges are equally weighted and unit valued, can be described by the adjacency matrix, i.e. a symmetric matrix A with the elements

$$a_{ij} = \begin{cases} 1 & \text{if } \{v_i, v_j\} \in E \\ 0 & \text{otherwise} \end{cases}$$

---

[1] I.e. time complexity of the eigen-decomposition is $O(m^3)$ and the matrix S has in general $O(m^2)$ entries.

In this book we assume that Γ is a simple graph, i.e. $a_{ii} = 0$ for all i's, or equivalently, $\text{diag}(A) = \mathbf{0}$. In this case the number of edges of the graph can be determined as

$$\mathfrak{m} = \frac{1}{2} \sum_{i=1}^{m} \sum_{j=1}^{m} a_{ij} = \sum_{i=1}^{m} \sum_{i<j\leq m} a_{ij}$$

If Γ is a weighted graph, then its *generalized* adjacency matrix is denoted by the S letter. Its elements have the form

$$s_{ij} = \begin{cases} \mathsf{W}(v_i, v_j) & \text{if } \{v_i, v_j\} \in E \\ 0 & \text{otherwise} \end{cases}$$

where $\mathsf{W}(v_i, v_j)$ is a function assigning weights to the edges. Typically, if the data items corresponding to the nodes v_i and v_j are described by n-dimensional real-valued feature vectors, $\mathbf{x}_i$ and $\mathbf{x}_j$, then $\mathsf{W}(v_i, v_j)$ is a function of the distance $\|\mathbf{x}_i - \mathbf{x}_j\|$ between these objects, see Sect. 5.2.1. Without loss of generality we assume that $s_{ij} \in [0, 1]$. Obviously, A is an instance of the matrix S. However, in some situations such a distinction is important.

The sum of S's i-th row, denoted d_i, is said to be (generalized) degree of the node v_i, and the diagonal matrix D whose (i, i)-th element is the degree d_i is said to be the degree matrix, i.e.:

$$\mathsf{d}_i = \sum_{j=1}^{m} s_{ij}, \quad D = \text{diag}(S\mathbf{e}) = \text{diag}(\mathsf{d}_1, \ldots, \mathsf{d}_m) \tag{5.1}$$

For any subset $B \subseteq V$ we define its volume, $\text{vol } B$, as the sum of the degrees of nodes belonging to this set,

$$\text{vol } B = \sum_{v_i \in B} \mathsf{d}_i \tag{5.2}$$

Particularly, if $B = V$ and S is the adjacency matrix A, then $\text{vol } V = 2\mathfrak{m}$.

5.2.1 Similarity Graphs

Let X be a set of m data items. Relationships between the pairs of data items are represented by a weighted undirected graph $\Gamma = (V, E, S)$, where $V = \{1, \ldots, n\}$ is the set of nodes representing the data items, E is the set of edges and S is a generalized adjacency matrix. The entries s_{ij} of matrix S represent degrees of similarity between the nodes (items) i and j. Two nodes are joined by an edge $\{i, j\}$ if they are sufficiently similar. It is important to note, that transforming the set X to the graph Γ corresponds to converting *local* information (i.e. pairwise similarities) into *global* information expressed by the whole graph, and more particularly, by its generalized adjacency matrix.

Different methods of constructing such graphs are described in Sect. 2.2 of [482] (see also [275] and Chap. 3 of [446] for a review of other approaches). Among them let us mention *mutual K-nearest neighbour graph*, or simply K-NN graph, in which two nodes i and j are joined by an edge if both i is among the K-nearest neighbours of j and j is among the K-nearest neighbours of the node i. In this case the elements of similarity matrix S take the form

$$S(i, j) = \begin{cases} W(i, j) \text{ if } i \in N_K(j) \text{ and } j \in N_K(i) \\ 0 \qquad \text{otherwise} \end{cases} \tag{5.3}$$

where W is a weighting function (discussed below) and $N_K(i)$ stands for the K-neighbourhood of node i; it consists of K nodes most similar[2] to the node i. Such a construction favours links joining the nodes from cluster of similar densities, but does not connect nodes from clusters of different densities [482]. Thus the resulting graph may be disconnected. To overcome this problem we can e.g. augment the graph with a minimum spanning tree spanned on the set of nodes.[3] Another simple scenario is to join two nodes by an edge if they are sufficiently similar, i.e. if $W(u, v) \geq \tau$, where τ is a user defined parameter. The resulting graph is referred to as unweighted τ-NN graph. A variant of this graph is the r-neighbourhood graph, in which every point is connected to all other points within a distance of r. The impact of graph construction on final graph clustering is studied in [344]. Focusing on the K-NN graph and r-neighbourhood graph these authors observed that even if we fix graph cutting criterion, both graphs lead to systematically different results. This finding can be easily illustrated both with toy and real-world data. These authors conclude that graph clustering criteria cannot be studied independently of the kind of graph they are applied to.

The weights $W(i, j)$ are typically computed by using Gaussian kernel with the parameter σ, i.e.

$$W(i, j) = \exp\left(-\frac{\|x_i - x_j\|^2}{2\sigma^2}\right) \tag{5.4}$$

A justification for such a formula—based on the relationship between random walking and solution of the diffusion equation—was provided by Belkin and Niyogi

---

[2]The problem of finding K nearest neighbours in high dimensional spaces is far from trivial. Fortunately, there are several efficient approximate techniques allowing to solve this problem. See e.g. P. Indyk, Dimensionality reduction techniques for proximity problems. *Proceedings of the 11-th Annual ACM-SIAM Symposium on Discrete Algorithms*, SODA'00, pp. 371–378. Society for Industrial and Applied Mathematics, Philadelphia, PA, USA 2000. Another promising solution is FLANN library described in [363].

[3]See e.g. J.A. Albano and D.W. Messinger, Euclidean commute time distance embedding and its application to spectral anomaly detection. *Proceedings of SPIE 8390, Algorithms and Technologies for Multispectral, Hyperspectral, and Ultraspectral Imagery XVIII*, 83902G, (May 8, 2012); https://doi.org/10.1117/12.918411.

in [60, Sect. 3.3]. Of course, the Eq. (5.4) is applicable only if the data items are represented as points in some n-dimensional Euclidean space. For a more general approach see e.g. [335].

It is obvious that with a fixed parameter σ, the similarity between two items is only a function of their Euclidean distance. So, one can expect that such a procedure is inefficient in proper reflecting the distribution of complex data. A number of alternative approaches to construct efficient similarity matrix is reviewed in [275, Sect. 3.1]. A simple extension of (5.4) to the form

$$W_1(i, j) = \exp\left(-\frac{\|x_i - x_j\|^2}{2\sigma_i\sigma_j}\right) \tag{5.5}$$

was proposed in [528]. Here σ_i stands for the distance between i-th data point and its K-th nearest neighbour, where the suggested value of K is $\lfloor \ln m \rfloor$. Another variation of (5.4) was proposed in [542], namely

$$W_2(i, j) = \exp\left(-\sum_{\ell=1}^{n}\frac{(x_{\ell i} - x_{\ell j})^2}{2\sigma_\ell^2}\right) \tag{5.6}$$

where $\sigma_1, \ldots, \sigma_n$ are length scale hyperparameters for each dimension.

The same authors consider in [543] another weighting function

$$W_3(i, j) = \frac{1}{2}\left(\tanh\left(\alpha_1(\|x_i - x_j\| - \alpha_2)\right) + 1\right) \tag{5.7}$$

where the hyperparameters α_1 and α_2 control slope and cutoff value, respectively. The function W_3 mimics τ-NN rule: when $\|x_i - x_j\| \gg \alpha_2$, then $W_3(i, j) \approx 0$, and when $\|x_i - x_j\| \ll \alpha_2$, then $W_3(i, j) \approx 1$. Unlike τ-NN the function W_3 is continuous with respect to α_1 and α_2 what makes it amenable to learning with gradient methods.

Instead of considering pairwise relationships (5.4) or (5.7), Wang and Zhang proposed in [493] locally linear reconstruction-based weighting scheme. It is inspired by the locally linear embedding of Roweis and Saul [411].

Even if the use of single value of σ is sufficient, we still have problems with choosing its proper value. Some solutions to this problem are discussed in [502, Sect. 5.2.8.1], while methods of amplifying the block-diagonal structure of similarity matrix are reviewed in [502, Sect. 5.2.8.2].

5.2.2 Graph Laplacian

Graph Laplacian plays important role is spectral clustering, and its eigenvectors are used to extract appropriate partition of a similarity graph representing a given data set. In this section various forms of the graph Laplacian are discussed, and some fundamental properties of its eigenvectors are presented in next section. Many other properties of graph Laplacian are discussed, among others, in [76, 115, 358, 359]. Particularly, see [438] for quick review of various properties and applications of the graph Laplacian.

The three variants of graph Laplacian are of interest[4]:

$$L = D - S \tag{5.8a}$$

$$\mathcal{L} = D^{-1/2} L D^{-1/2} \tag{5.8b}$$

$$\Delta = D^{-1} L \tag{5.8c}$$

They are termed, respectively: combinatorial (or un-normalized) Laplacian, L, normalized Laplacian, $\mathcal{L}$, and discrete Laplacian, Δ, called also random-walk Laplacian [336]. Note, that in the field of electrical engineering, the combinatorial Laplacian is known as the admittance matrix or Kirchhoff matrix.

All these variants of graph Laplacian are related to each other, that is

$$L = D^{1/2} \mathcal{L} D^{1/2} = D \Delta \tag{5.9a}$$

$$\mathcal{L} = D^{-1/2} L D^{-1/2} = D^{1/2} \Delta D^{-1/2} \tag{5.9b}$$

$$\Delta = D^{-1} L = D^{-1/2} \mathcal{L} D^{1/2} \tag{5.9c}$$

Interestingly, both the normalized and the discrete Laplacian are, in fact, the complements of appropriately normalized similarity matrix, that is

$$\mathcal{L} = \mathbb{I} - D^{-1/2} S D^{-1/2} = \mathbb{I} - \mathcal{S} \tag{5.10a}$$

$$\Delta = \mathbb{I} - D^{-1} S = \mathbb{I} - P \tag{5.10b}$$

A reader should note that the normalization $P = D^{-1} S$ reweighs the edges of graph Γ so that the degree of each node is equal to 1. Hence the relationship (5.10b) is equiform with (5.8a) if the similarity S is replaced by the matrix P. Obviously, the matrix $\mathcal{S} = D^{-1/2} S D^{-1/2}$ is symmetric, while the matrix P is nonsymmetric, but row-stochastic, i.e. $\sum_{j=1}^{m} p_{ij} = 1$ for $i = 1, \ldots, m$. In effect, P can be viewed as a transition probability matrix describing random walk on the graph Γ, what justifies the name "random-walk Laplacian" used for Δ. Interestingly, the Laplacians $\mathcal{L}$ and Δ have related eigenvectors (see Theorem 5.2.3 below) what justifies a statement that the normalized Laplacian is closely related to random walks in the graphs, see [115] for details.

It is easy to verify that the elements of these matrices, denoted respectively as l_{ij}, ℓ_{ij} and δ_{ij}, are of the form[5]

---

[4]In fact all these matrices can be viewed as the instances of a generalized Laplacian called also discrete Schrödinger operator: it is a matrix M with the elements $m_{uv} < 0$ if $\{u, v\}$ is an edge of Γ and $m_{uv} = 0$ if the nodes u and $v \neq u$ are not adjacent. There are no constraints on the diagonal entries of M [76, 139].

[5]Recall that Γ is a simple graph.

Fig. 5.2 A simple graph

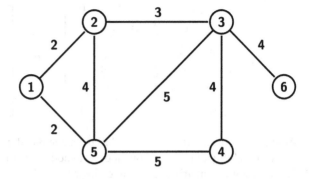

$$l_{ij} = \begin{cases} d_i & \text{if } j = i \\ -s_{ij} & \text{if } \{i, j\} \in E \\ 0 & \text{otherwise} \end{cases} \tag{5.11a}$$

$$\ell_{ij} = \begin{cases} 1 & \text{if } j = i \\ \dfrac{-s_{ij}}{\sqrt{d_i d_j}} & \text{if } \{i, j\} \in E \\ 0 & \text{otherwise} \end{cases} \tag{5.11b}$$

$$\delta_{ij} = \begin{cases} 1 & \text{if } j = i \\ -\dfrac{s_{ij}}{d_i} & \text{if } \{i, j\} \in E \\ 0 & \text{otherwise} \end{cases} \tag{5.11c}$$

Example 5.2.1 Consider a small graph depicted on Fig. 5.2. Below the matrices S, D, $D^{-1/2}$, L, $\mathcal{L}$ and Δ are shown:

$$S = \begin{bmatrix} 0 & 2 & 0 & 0 & 2 & 0 \\ 2 & 0 & 3 & 0 & 4 & 0 \\ 0 & 3 & 0 & 4 & 5 & 4 \\ 0 & 0 & 4 & 0 & 5 & 0 \\ 2 & 4 & 5 & 5 & 0 & 0 \\ 0 & 0 & 4 & 0 & 0 & 0 \end{bmatrix}, \quad D = \begin{bmatrix} 4 & 0 & 0 & 0 & 0 & 0 \\ 0 & 9 & 0 & 0 & 0 & 0 \\ 0 & 0 & 16 & 0 & 0 & 0 \\ 0 & 0 & 0 & 9 & 0 & 0 \\ 0 & 0 & 0 & 0 & 16 & 0 \\ 0 & 0 & 0 & 0 & 0 & 4 \end{bmatrix}, \quad D^{-1/2} = \begin{bmatrix} \frac{1}{2} & 0 & 0 & 0 & 0 & 0 \\ 0 & \frac{1}{3} & 0 & 0 & 0 & 0 \\ 0 & 0 & \frac{1}{4} & 0 & 0 & 0 \\ 0 & 0 & 0 & \frac{1}{3} & 0 & 0 \\ 0 & 0 & 0 & 0 & \frac{1}{4} & 0 \\ 0 & 0 & 0 & 0 & 0 & \frac{1}{2} \end{bmatrix}$$

$$L = \begin{bmatrix} 4 & -2 & 0 & 0 & -2 & 0 \\ -2 & 9 & -3 & 0 & -4 & 0 \\ 0 & -3 & 16 & -4 & -5 & -4 \\ 0 & 0 & -4 & 9 & -5 & 0 \\ -2 & -4 & -5 & -5 & 16 & 0 \\ 0 & 0 & -4 & 0 & 0 & 4 \end{bmatrix}, \quad \mathcal{L} = \begin{bmatrix} 1 & -\frac{2}{6} & 0 & 0 & -\frac{2}{8} & 0 \\ -\frac{2}{6} & 1 & -\frac{3}{12} & 0 & -\frac{4}{12} & 0 \\ 0 & -\frac{3}{12} & 1 & -\frac{4}{12} & -\frac{5}{16} & -\frac{4}{8} \\ 0 & 0 & -\frac{4}{12} & 1 & -\frac{5}{12} & 0 \\ -\frac{2}{8} & -\frac{4}{12} & -\frac{5}{16} & -\frac{5}{12} & 1 & 0 \\ 0 & 0 & -\frac{4}{8} & 0 & 0 & 1 \end{bmatrix}$$

$$\Delta = \begin{bmatrix} 1 & -\frac{1}{2} & 0 & 0 & -\frac{1}{2} & 0 \\ -\frac{2}{9} & 1 & -\frac{3}{9} & 0 & -\frac{4}{9} & 0 \\ 0 & -\frac{3}{16} & 1 & -\frac{4}{16} & -\frac{5}{16} & -\frac{4}{16} \\ 0 & 0 & -\frac{4}{9} & 1 & -\frac{5}{9} & 0 \\ -\frac{2}{16} & -\frac{4}{16} & -\frac{5}{16} & -\frac{5}{16} & 1 & 0 \\ 0 & 0 & -1 & 0 & 0 & 1 \end{bmatrix}, \quad P = \begin{bmatrix} 0 & \frac{1}{2} & 0 & 0 & \frac{1}{2} & 0 \\ \frac{2}{9} & 0 & \frac{3}{9} & 0 & \frac{4}{9} & 0 \\ 0 & \frac{3}{16} & 0 & \frac{4}{16} & \frac{5}{16} & \frac{4}{16} \\ 0 & 0 & \frac{4}{9} & 0 & \frac{5}{9} & 0 \\ \frac{2}{16} & \frac{4}{16} & \frac{5}{16} & \frac{5}{16} & 0 & 0 \\ 0 & 0 & 1 & 0 & 0 & 0 \end{bmatrix}$$

Additionally, the matrix $P = \mathbb{I} - \Delta$ has been computed. Its entries p_{ij} can be viewed as the probability of moving from node v_i to node v_j in random walk on the graph Γ. Properties of this matrix are discussed later.   □

Some elementary and important properties of the combinatorial Laplacian are presented in the next lemma.

Lemma 5.2.1 *The matrix L has the following properties:*

(a) It is a symmetric, positive semi-definite and diagonally dominant matrix.
(b) For any vector $\mathbf{f} \in \mathbb{R}^m$ there holds

$$(L\mathbf{f})_i = \sum_{j=1}^{m} s_{ij}(f_i - f_j) \tag{5.12a}$$

$$\langle \mathbf{f}, L\mathbf{f} \rangle = \mathbf{f}^T L \mathbf{f} = \frac{1}{2} \sum_{i=1}^{m} \sum_{j=1}^{m} s_{ij}(f_i - f_j)^2 \tag{5.12b}$$

Here the symbol $\langle \mathbf{a}, \mathbf{b} \rangle = \mathbf{a}^T \mathbf{b}$ denotes the scalar product of the vectors $\mathbf{a}$ and $\mathbf{b}$.   □

These properties were noted by many authors, see e.g. [482], hence we give a sketch of the proof only. Note first, that the symmetry of L follows from the symmetry of the matrix S, or equivalently, from the Eq. (5.11a). Also, from this equation it follows that

$$|l_{ii}| = \sum_{j \neq i} |l_{ij}|, \quad i = 1, \ldots, m \tag{5.13}$$

what implies that L is diagonally dominant.[6] It is known that the diagonally dominant matrices are positive semi-definite, see e.g. [354].

The Eq. (5.12a) implies that the combinatorial Laplacian can be viewed as an operator on the space of functions $f: V \rightarrow \mathbb{R}$, or equivalently, on the space of m-dimensional vectors $\mathbf{f} = (f_1, \ldots, f_m)^T$. The i-th entry of the vector $L\mathbf{f}$ is equal to

---

[6]Formally, a square and real matrix A is said to be diagonally dominant if $|a_{ii}| \geq \sum_{j \neq i} |a_{ij}|$, and strictly diagonally dominant if $|a_{ii}| > \sum_{j \neq i} |a_{ij}|$, see e.g. [354].

$$(L\mathbf{f})_i = \mathsf{d}_i f_i - \sum_{j \neq i} s_{ij} f_j = \sum_{j=1}^{m} s_{ij}(f_i - f_j)$$

In fact, the Eq. (5.12a) can be viewed as an alternative definition of the combinatorial Laplacian, see e.g. [115].

Multiplying the vector $\mathbf{f}^{\mathrm{T}}$ by the vector $L\mathbf{f}$ we obtain Eq. (5.12b). Since $\mathbf{f}^{\mathrm{T}}L\mathbf{f} \geq 0$ for any real-valued vector $\mathbf{f}$, we get another justification of positive semi-definiteness of the matrix L. The quadratic form $\mathbf{f}^{\mathrm{T}}L\mathbf{f}$ can be viewed as a measure of "smoothness" of the vector $\mathbf{f}$ (or equivalently, the function $f: V \to \mathbb{R}$): $\mathbf{f}^{\mathrm{T}}L\mathbf{f} = 0$ only if $\mathbf{f}$ is a constant vector. In general, $\mathbf{f}^{\mathrm{T}}L\mathbf{f}$ is small when the entries of the vector $\mathbf{f}$ are similar at neighbouring vertices connected by an edge with a large weight. See Remark 5.2.4 for further interpretation of this quadratic form.

Remark 5.2.1 The discrete mapping $f: V \to \mathbb{R}$ is termed also a graph signal, and $f(i)$ is the value[7] of the signal on node i, see e.g. [432]. Another interpretation of f is so-called fitness landscape; it plays an important role in molecular evolution and in the analysis of combinatorial optimization algorithms, see e.g. [442]. In the sequel the mapping f is represented as an m-dimensional vector $\mathbf{f}$ with indices corresponding to the node indices in the graph. However, the reader should note that the vector $\mathbf{f} = (f_1, \ldots, f_m)^{\mathrm{T}}$ is not just a list, as the elements f_i are related by various dependencies (e.g. similarity, physical proximity) encoded in the graph [415, 432]. □

The Eq. (5.12) is a special case of a more general discrete version of Green formula noted for the first time in [442, Prop. 1(v)], and then reproduced in [76]—see part (b) of Lemma 5.2.2 below. To introduce this formula, define the incidence matrix ∇, first. It is a matrix of size $\mathsf{m} \times m$ with the elements[8]

$$\nabla_{e_{ij}, v_k} = \begin{cases} +1 \text{ if } k = i \\ -1 \text{ if } k = j \\ 0 \quad \text{otherwise} \end{cases}, \; e_{i,j} \in E, v_k \in V \qquad (5.14)$$

The incidence matrix can be interpreted as a combinatorial gradient operator,[9] and its transpose ∇^{T}, as a combinatorial divergence operator, see [211, 212] or [76] for more formal discussion of this topic.

Lemma 5.2.2 *Let L be the combinatorial Laplacian of an undirected and unweighted graph Γ, and let ∇ be the incidence matrix. Then:*

(a) $L = \nabla^T \nabla$,
(b) $\langle \mathbf{f}, L\mathbf{g} \rangle = \langle \mathbf{g}, L\mathbf{f} \rangle = \langle \nabla \mathbf{f}, \nabla \mathbf{g} \rangle$ for any real vectors $\mathbf{f}, \mathbf{g} \in \mathbb{R}^m$.

---

[7]In general this value may be a complex number, like in [415].
[8]You can imagine that Γ is a graph with (arbitrarily) oriented edges, for instance the edge $\{i, j\}$ is converted into $i \to j$ if $i < j$, see e.g. [354].
[9]Hence we used the symbol ∇ to denote this matrix.

□

For proof of (a) see e.g. [74, Prop. 4.8]. Interestingly, this representation of L does not depend on the orientation of graph edges. Part (b) of this lemma, being a discrete version of Green's formula, can be obtained by applying the identity (a), i.e.

$$\langle \mathbf{f}, Lg \rangle = \mathbf{f}^{\mathrm{T}} \nabla^{\mathrm{T}} \nabla g = \langle \nabla \mathbf{f}, \nabla g \rangle$$

The property (a) exemplifies the close relationship between the incidence matrix and graph Laplacian and their continuous counterparts.

Now we are ready to discuss an analogy between the un-normalized Laplacian L and the Laplace-Beltrami operator $\nabla^2_{\mathcal{M}}$ defined on a Riemannian manifold $\mathcal{M}$ [115]. If $\mathcal{M} = \mathbb{R}^n$, then $\nabla^2_{\mathcal{M}}$ reduces to the Laplace operator, ∇^2. It is a second order differential operator in the n-dimensional Euclidean space, defined as the divergence of the gradient of a twice-differentiable real-valued function f, $\nabla^2 f = -\sum_{j=1}^{n} \frac{\partial^2 f}{\partial x_i^2}$. In general case, $\nabla^2_{\mathcal{M}} = \mathrm{div}(\mathrm{grad} f)$.

The representation mentioned in Lemma 5.2.2(a) of the Laplacian matrix of an un-weighted graph can be interpreted as the composition of the combinatorial divergence and the combinatorial gradient. In case of a weighted graph we define the diagonal constitutive matrix C of size $m \times m$, with the weights of each edge along the diagonal. The matrix C may be interpreted as representing a metric [211]. In this case

$$L = \nabla^{\mathrm{T}} C \nabla \qquad (5.15)$$

i.e. the matrix L can be viewed indeed as a discrete counterpart of Laplace-Beltrami operator. Obviously, if Γ is an unweighted graph, then $C = \mathbb{I}$.

Example 5.2.2 Figure 5.3 presents a directed variant of the weighted graph from Fig. 5.2. Let (i, j) denotes an ordered pair $i \rightarrow j$. The incidence matrix ∇ and the constitutive matrix C are as follows

Fig. 5.3 Graph from Fig. 5.2 with (arbitrarily) oriented edges

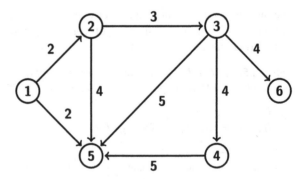

$$\nabla = \begin{array}{c} \\ (v_1,v_2) \\ (v_2,v_3) \\ (v_3,v_6) \\ (v_3,v_5) \\ (v_3,v_4) \\ (v_4,v_5) \\ (v_2,v_5) \\ (v_1,v_5) \end{array} \begin{array}{cccccc} v_1 & v_2 & v_3 & v_4 & v_5 & v_6 \\ \begin{bmatrix} 1 & -1 & 0 & 0 & 0 & 0 \\ 0 & 1 & -1 & 0 & 0 & 0 \\ 0 & 0 & 1 & 0 & 0 & -1 \\ 0 & 0 & 1 & 0 & -1 & 0 \\ 0 & 0 & 1 & -1 & 0 & 0 \\ 0 & 0 & 0 & 1 & -1 & 0 \\ 0 & 1 & 0 & 0 & -1 & 0 \\ 1 & 0 & 0 & 0 & -1 & 0 \end{bmatrix} \end{array}, \quad C = \begin{bmatrix} 2 & 0 & 0 & 0 & 0 & 0 & 0 & 0 \\ 0 & 3 & 0 & 0 & 0 & 0 & 0 & 0 \\ 0 & 0 & 4 & 0 & 0 & 0 & 0 & 0 \\ 0 & 0 & 0 & 5 & 0 & 0 & 0 & 0 \\ 0 & 0 & 0 & 0 & 4 & 0 & 0 & 0 \\ 0 & 0 & 0 & 0 & 0 & 5 & 0 & 0 \\ 0 & 0 & 0 & 0 & 0 & 0 & 4 & 0 \\ 0 & 0 & 0 & 0 & 0 & 0 & 0 & 2 \end{bmatrix}$$

It is easy to observe that if $\mathbf{f} \in \mathbf{R}^m$ represents a real-valued function defined on the set of nodes, then the value of $\nabla \mathbf{f}$ at any edge $e = (i, j)$ is simply the difference of the values of $\mathbf{f}$ at the two endpoints of the edge e, i.e. $(\nabla \mathbf{f})_{(i,j)} = f_i - f_j$. This illustrates that the matrix ∇ can be viewed as a counterpart of a gradient (difference) operator. The mapping $\mathbf{f} \to \nabla \mathbf{f}$ is referred to as a co-boundary mapping.

Similarly, if $g \in \mathbf{R}^m$ represents a real valued function over the edges, then for any node v we have

$$(\nabla^T g)_v = \sum_{\substack{e=(a,b)\in E \\ a=v}} g_e - \sum_{\substack{e=(a,b)\in E \\ b=v}} g_e$$

For instance, $(\nabla^T g)_{v_2} = g_2 + g_7 - g_1$ as the edges $e_2 = (v_2, v_3)$, $e_7 = (v_2, v_5)$ leave from v_2 and the edge $e_1 = (v_1, v_2)$ enters the node v_2. If we treat g as describing some kind of a flow on the edges, then $(\nabla^T g)_v$ is the net outbound flow on the vertex v, i.e. ∇^T is just a divergence operator. More details on such a discrete calculus can be found in [76] or [212]. □

Remark 5.2.2 If the Laplacian has the representation (5.15), then the co-boundary mapping has the form $\mathbf{f} \to C^{1/2}\nabla \mathbf{f}$. The value

$$(C^{1/2}\nabla \mathbf{f})_{(i,j)} = \sqrt{s_{ij}}(f_i - f_j)$$

is termed also to be the edge derivative of the vector $\mathbf{f}$ along the edge (i, j) at the node v_i [539]. It is denoted as $\partial \mathbf{f}/\partial e|_i$. Then the local variation of $\mathbf{f}$ at each vertex v is defined to be

$$\|\nabla_v f\| = \sqrt{\sum_{e \vdash v} \left(\frac{\partial f}{\partial e}\Big|_u\right)^2}$$

where $e \vdash u$ denotes the set of edges containing the node u. Then the smoothness of the vector $\mathbf{f}$ is the sum of the local variations at each vertex, i.e.:

$$V(\mathbf{f}) = \frac{1}{2}\sum_{i=1}^{m} \|\nabla_u f\|^2 = \langle \mathbf{f}, L\mathbf{f}\rangle \tag{5.16}$$

□

A normalized version of the incidence matrix, $\widetilde{\nabla}$ is a matrix of size $\mathsf{m} \times m$ with the elements, see [115]

$$\widetilde{\nabla}_{e_{ij}, v_k} = \begin{cases} +1/\sqrt{\mathsf{d}_i} & \text{if } k = i \\ -1/\sqrt{\mathsf{d}_j} & \text{if } k = j, \ e_{i,j} \in E, v_k \in V \\ 0 & \text{otherwise} \end{cases} \tag{5.17}$$

Lemma 5.2.3 *Let $\widetilde{\nabla}$ stands for the normalized incidence matrix characterizing an undirected and weighted graph Γ, and let C be the constitutive matrix. The matrix $\mathcal{L}$ has the following properties:*

(a) It is symmetric and positive semi-definite matrix.
(b) For any vector $\mathbf{f} \in \mathbb{R}^m$ there holds

$$(\mathcal{L}\mathbf{f})_i = \frac{1}{\sqrt{\mathsf{d}_i}} \sum_{j=1}^{m} s_{ij} \left(\frac{f_i}{\sqrt{\mathsf{d}_i}} - \frac{f_j}{\sqrt{\mathsf{d}_j}} \right) = f_i - \sum_{j=1}^{m} \frac{s_{ij}}{\sqrt{\mathsf{d}_i \mathsf{d}_j}} f_j \tag{5.18a}$$

$$\langle \mathbf{f}, \mathcal{L}\mathbf{f} \rangle = \mathbf{f}^T \mathcal{L} \mathbf{f} = \frac{1}{2} \sum_{i=1}^{m} \sum_{j=1}^{m} s_{ij} \left(\frac{f_i}{\sqrt{\mathsf{d}_i}} - \frac{f_j}{\sqrt{\mathsf{d}_j}} \right)^2 \tag{5.18b}$$

(c) $\mathcal{L} = \widetilde{\nabla}^T C \widetilde{\nabla}$
(d) $\langle \mathbf{f}, \mathcal{L}g \rangle = \langle g, \mathcal{L}\mathbf{f} \rangle = \langle \widetilde{\nabla}\mathbf{f}, \widetilde{\nabla}g \rangle$ for any vectors $\mathbf{f}, g \in \mathbb{R}^m$.   □

Again, the representation (c) suggest that $\mathcal{L}$ can be treated as a discrete analogue of the Laplace-Beltrami operator. We can ask which of the Laplacian matrices mimics the continuous operator? This problem was analyzed by Belkin and Niyogi [61], Hein et al. [248] and Giné and Koltchinskii [199], among others. An extensive review of various results concerning this topic is given [248]. All these authors assume that a set $X = \{\mathbf{x}_1, \ldots, \mathbf{x}_m\}$ of points is chosen randomly from a given manifold $\mathcal{M} \subset \mathbb{R}^n$. The similarity relationships among these points are expressed in the form of a similarity graph, constructed as described in Sect. 5.2.1. Typically, as $\mathsf{W}(i, j)$ the Gaussian kernel

$$\mathsf{W}(i, j) = \exp \left(\frac{\|x_i - x_j\|^2}{4t} \right) \tag{5.19}$$

is taken. Let L_m^t denotes the combinatorial Laplacian computed for the similarity graph. Belkin and Niyogi noted in [61] that if the points $\mathbf{x}_1, \ldots, \mathbf{x}_m$ were chosen with uniform distribution from a compact manifold $\mathcal{M} \subset \mathbb{R}^n$ of dimension $k < n$, then there exits a sequence of real numbers $t_m \to 0$, and a positive constant C such that

$$\lim_{n \to \infty} \sup_{\substack{\mathbf{x} \in \mathcal{M} \\ f \in \mathcal{F}}} |C \frac{(4\pi t_m)^{1+k/2}}{m} L_m^{t_m} f(\mathbf{x}) - \nabla_{\mathcal{M}}^2 f(\mathbf{x})| = 0 \tag{5.20}$$

where $\mathcal{F}$ denotes the set of functions $f \in \mathcal{C}^\infty(\mathcal{M})$ such that $\nabla^2 f$ is a Lipschitz function. Hein et al. have shown in [248] that for a uniform measure on the manifold all graph Laplacians have the same limit up to constants. However in the case of a non-uniform measure on the manifold only the discrete Laplacian converges to the weighted Laplace-Beltrami operator.

5.2.3 Eigenvalues and Eigenvectors of Graph Laplacian

Recall that if A is a square matrix then the vector $\mathbf{x}$ satisfying the equation $A\mathbf{x} = \lambda\mathbf{x}$ is said to be the eigenvector, and $\lambda \in \mathbb{C}$ is referred to as the eigenvalue (see e.g. [354]). The pair $(\lambda, \mathbf{x})$ is said to be the eigenpair of the matrix A, and the problem of solving the equation $A\mathbf{x} = \lambda\mathbf{x}$ is said to be the eigenproblem. If additionally A is a symmetric and real-valued matrix,[10] then all its eigenvalues are real numbers and all its eigenvectors are real-valued.[11] Such a matrix has m not necessarily unique eigenpairs. The eigenvectors corresponding to distinct eigenvalues are orthogonal.[12] If we stack all the eigenvectors $\mathbf{u}_1, \ldots, \mathbf{u}_m$ in columns of a matrix U and form the diagonal matrix with the eigenvalues, $\Lambda = \text{diag}(\lambda_1, \ldots, \lambda_m)$, then the matrix A can be represented as

$$A = U\Lambda U^\mathrm{T} = \sum_{i=1}^{m} \lambda_i \mathbf{u}_i \mathbf{u}_i^\mathrm{T}$$

The right hand side of this equation is said to be eigen-decomposition of the matrix A, or diagonalization of A. Other properties of eigenvalues and eigenvectors are discussed e.g. in [354].

In the sequel we will assume that the eigen-decompositions of the combinatorial and normalized Laplacian are of the form

$$L = \Psi\Lambda\Psi^\mathrm{T} = \sum_{i=1}^{m} \lambda_i \psi_i \psi_i^\mathrm{T} \tag{5.21a}$$

$$\mathcal{L} = \Phi\Lambda'\Phi^\mathrm{T} = \sum_{i=1}^{m} \lambda'_i \phi_i \phi_i^\mathrm{T} \tag{5.21b}$$

---

[10]More generally, a square matrix A with complex entries is said to be hermitian, if each its entry a_{ij} is equal to the complex conjugate of the element a_{ji}. In such a case the eigenvalues of A are real numbers. If all the elements of A are real numbers, "hermitian" means simply symmetric matrix.

[11]Note that the eigenproblem, if it has a solution $(\lambda, \mathbf{x})$, and as $\mathbf{x}$ is always assumed to be non-zero, then it has infinitely many solutions of the form $(\lambda, r\mathbf{x})$, where r is any non-zero real number. Therefore, unless we specify otherwise, we choose among them one eigenvector of unit length, $\mathbf{x}^\mathrm{T}\mathbf{x} = 1$.

[12]Even if there are degenerate eigenvalues, it is always possible to find an orthogonal basis of $\mathbb{R}^m$ consisting of m eigenvectors of A.

where Λ is a diagonal matrix with the eigenvalues of the matrix L on the diagonal, and the columns of the matrix Ψ are the eigenvectors $\psi_1, \ldots, \psi_m$ corresponding to these eigenvalues. We assume that the eigenvalues are ordered in increasing order[13]:

$$0 = \lambda_1 \leq \cdots \leq \lambda_m$$

Also the eigenvalues of symmetric Laplacian are indexed in ascending order:

$$0 = \lambda'_1 \leq \cdots \leq \lambda'_m \leq 2$$

and the eigenvectors $\phi_1, \ldots, \phi_m$ corresponding to these eigenvalues are columns of the matrix Φ.

Theorem 5.2.1 *Let $\Gamma = (V, E)$ be an undirected graph, and L be the combinatorial Laplacian of this graph. Then:*

(a) *L has m (not necessarily different) real-valued eigenvectors $0 = \lambda_1 \leq \lambda_2 \leq \cdots \leq \lambda_m \leq \max_{1 \leq j \leq m} \sqrt{2\mathsf{d}_j(\mathsf{d}_j + \overline{\mathsf{d}}_j)}$, where $\overline{\mathsf{d}}_j = \left(\sum_{\ell=1}^m s_{j\ell} \mathsf{d}_\ell \right)/\mathsf{d}_j$. This and other upper bounds on λ_m are reviewed in [435].*
(b) *The minimal eigenvalue of the matrix L is $\lambda_1 = 0$, and its corresponding (normalized) eigenvector is $\psi_1 = \mathbf{e}/\sqrt{m}$.*
(c) *The components of the non-constant eigenvectors of the combinatorial Laplacian satisfy the following constraints:*

  (i) $\sum_{i=1}^m \psi_{ij} = 0, \quad j = 2, \ldots, m$
  (ii) $-1 < \psi_{ij} < 1, \quad i = 1, \ldots, m, j = 2, \ldots, m$ □

The proof of this theorem can be found e.g. in [482]. Thus only some remarks to this theorem are given.

From the Eq. (5.11a) it follows that if ψ_1 is a constant vector, i.e. $\psi_1 = c\mathbf{e}$, where $c \neq 0$, then the Equation $L\psi_1 = \lambda_1 \psi_1$ is satisfied if (and only if) $\lambda_1 = 0$. Particularly, if $c = 1/\sqrt{m}$ then ψ_1 has length 1. The condition (c-i) follows from the orthogonality of the eigenvectors; combining this with the fact that $\psi_j^T \psi_j = 1$ we obtain the last inequality.

Theorem 5.2.2 *The matrix $\mathcal{L}$ has the following properties:*

(a) *$\mathcal{L}$ has m (not necessarily different) real eigenvalues $0 = \lambda'_1 \leq \lambda'_2 \leq \cdots \leq \lambda'_m \leq 2$, wherein $\lambda'_m = 2$ if Γ is a bipartite graph.*
(b) *$\lambda'_1 = 0$ is the minimal eigenvalue of $\mathcal{L}$ and its corresponding eigenvector is $\phi_1 = \sqrt{\mathsf{d}/\mathrm{vol}\,\Gamma}$.*   □

Properties of normalized Laplacian are studied in Chap. 1 of the monograph [115]. We only explain property (b). Since $\mathcal{L}$ is positive semi-definite real matrix, cf. (5.12b), all its eigenvalues are real and non-negative. If we set $\phi_1 = D^{1/2}\mathbf{e}$, then we obtain

---

[13] The nodes can be always rearranged so that the matrix S has an appropriate form for this ordering of eigenvalues to hold.

$$\mathcal{L}\phi_1 = (D^{-1/2}LD^{-1/2})(D^{1/2}\mathbf{e}) = D^{-1/2}(L\mathbf{e}) = \mathbf{0}$$

Thus $\lambda'_1 = 0$, and the corresponding eigenvector equals $\phi_1 = D^{1/2}\mathbf{e} = \sqrt{\mathbf{d}}$, and the normalized eigenvector is $\phi_1 = \sqrt{\mathbf{d}/\mathrm{vol}\,\Gamma}$.

A reader should note, that properties of the discrete Laplacian Δ are, in a sense, dual to the properties of the normalized Laplacian. For the completeness let us mention them in the following form:

Theorem 5.2.3 *The discrete Laplacian Δ has the following properties:*

(a) Δ is an nonsymmetric and positive semi-definite matrix.
(b) Δ has m (not necessarily different) real eigenvalues and eigenvectors with $0 = \lambda_1 \leq \cdots \leq \lambda'_m$.
(c) If (λ', η) is an eigenpair of the matrix Δ, then $(\lambda', D^{1/2}\eta)$ is the eigenpair of the normalized Laplacian $\mathcal{L}$.
(d) The eigenpair (λ', η) of Δ solves the generalized eigenproblem $L\eta = \lambda' D\eta$.
(e) 0 is an eigenvalue of Δ with the constant vector $\mathbf{e}$ as eigenvector.

These properties[14] follow from the relationship (5.9) between Δ and $\mathcal{L}$, L. For instance, if (λ', φ) is an eigenpair of the matrix Δ, then—by the correspondence (5.9c)—the equality $\Delta\varphi = \lambda'\varphi$ can be transformed to the form

$$(D^{-1/2}\mathcal{L}D^{1/2})\varphi = \lambda'\varphi \Leftrightarrow \mathcal{L}(D^{1/2}\varphi) = \lambda'(D^{1/2}\varphi) \qquad (5.22)$$

i.e. $(\lambda', D^{1/2}\varphi)$ is an eigenpair of the normalized Laplacian $\mathcal{L}$. Thus, both matrices have identical eigenvalues.

Similarly, the relationship $\Delta = D^{-1}L$ implies statement (d) above. From this statement it follows that the eigenvectors of Δ are identical with the eigenvectors of the generalized eigen system $L\eta = \lambda'D\eta$. It should be noted that computing Δ involves multiplication by D^{-1} and this matrix could contain very small values.[15] Thus solving the generalized eigenproblem is numerically more stable method to compute the eigenvectors.

In practice, instead of the discrete Laplacian, its complement

$$P = \mathbb{I} - \Delta = D^{-1/2}(\mathbb{I} - \mathcal{L})D^{1/2} = D^{-1}S \qquad (5.23)$$

is used. P can be viewed as a transition matrix of a Markov chain describing random walk on the graph Γ. A reader can verify that any eigenpair $(\bar{\lambda}, \eta)$ of P solves the generalized eigenproblem $S\eta = \lambda D\eta$ and simultaneously solves the eigenproblem $\Delta\eta = (1 - \bar{\lambda})\eta$. Some other formal properties of the matrix P are discussed in the next section.

---

[14]Note that we refrained from normalising eigenvectors of asymmeric matrices. It is due to the issue of left-handed and right-handed eigenvectors.

[15]For instance an outlier would have a small degree as it is not close to another point.

Remark 5.2.3 The Eqs. (5.12a) and (5.18a) can be rewritten, respectively, in the form

$$\psi_i(\mathsf{d}_i - \lambda) = \sum_{j \in N(i)} s_{ij} \psi_j \tag{5.24a}$$

$$\phi_i((1 - \lambda')\sqrt{\mathsf{d}_i}) = \sum_{j \in N(i)} \frac{s_{ij}}{\sqrt{\mathsf{d}_j}} \phi_j \tag{5.24b}$$

where $N(i)$ is the set of nodes adjacent to the node i. Similarly, if (λ', ϕ) is an eigenpair of the normalized Laplacian, then the eigenvector $\eta = D^{-1/2}\phi$ of the discrete Laplacian satisfies the condition

$$\eta_i(1 - \lambda')\mathsf{d}_i = \sum_{j \in N(i)} s_{ij} \eta_j \tag{5.25}$$

This shows that the value of each eigenvector at the node i is equal to the weighted average of the values of this function in neighbouring nodes. In other words, the eigenvectors exhibit a harmonic property [115].   □

5.2.4   Variational Characterization of Eigenvalues

The eigenvalues of a real symmetric and square matrix[16] M can be characterized as the solution of a series of optimization problems. Define the Rayleigh quotient as, see e.g. [354]

$$R(M; \mathbf{w}) = (\mathbf{w}^{\mathsf{T}} M \mathbf{w})/(\mathbf{w}^{\mathsf{T}} \mathbf{w}) \tag{5.26}$$

If $\mathbf{w}^{\mathsf{T}}\mathbf{w} = \|\mathbf{w}\| = 1$ the Rayleigh quotient has the simpler form $R(M; \mathbf{w}) = \mathbf{w}^{\mathsf{T}} M \mathbf{w}$. If $\mathbf{w}$ is an eigenvector of M, then the corresponding eigenvalue λ is obtained as

$$R(M, \mathbf{w}) = \frac{\mathbf{w}^{\mathsf{T}} M \mathbf{w}}{\mathbf{w}^{\mathsf{T}} \mathbf{w}} = \frac{\lambda \mathbf{w}^{\mathsf{T}} \mathbf{w}}{\mathbf{w}^{\mathsf{T}} \mathbf{w}} = \lambda$$

Remark 5.2.4 It should be noted, that the eigenvalues can be thought of as frequencies—see Fig. 5.4, where four eigenvectors corresponding to four different eigenvalues of the path graph consisting of 22 nodes are shown. These eigenvectors resemble basic modes of a string vibrating with different frequencies. Since $R(L, \phi_j) = \lambda_j$, we say that the eigenvalues indicate the variation in the eigenvectors: a high eigenvalue implies higher variation in the corresponding eigenvector [432]. To illustrate this variability let us compute the number $\mathcal{Z}(\phi_\ell)$ of edges connecting nodes that have different sign in the eigenvector ϕ_ℓ, i.e.

---

[16]More generally this applies to any hermitian matrix.

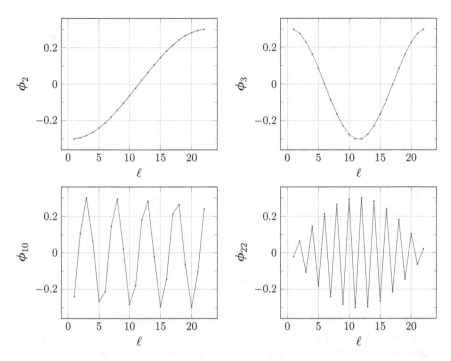

Fig. 5.4 Eigenvectors of the combinatorial Laplacian of the path graph consisting of 22 nodes

$$\mathcal{Z}(\phi_\ell) = |\{i, j\} \in E : \phi_{i,\ell}\phi_{j,\ell} < 0| \qquad (5.27)$$

$\mathcal{Z}(\phi_\ell)$ is referred to as the number of zero crossings of the graph Γ [432]. Figure 5.5 illustrates the number of zero crossings for subsequent eigenvectors of the combinatorial and normalized Laplacian characterising jazz musician network.[17] It is apparent that the eigenvectors associated with larger eigenvalues cross zero more often, what confirms the interpretation of the Laplacian eigenvalues as frequencies.   □

If M, N are two square real and symmetric matrices, we define the generalized Rayleigh quotient

$$R(M, N; \mathbf{w}) = \frac{\mathbf{w}^{\mathrm{T}} M \mathbf{w}}{\mathbf{w}^{\mathrm{T}} N \mathbf{w}} \qquad (5.28)$$

Note that if $N = \mathbb{I}$, then $R(M, \mathbb{I}; \mathbf{w}) = R(M; \mathbf{w})$. Furthermore, if we substitute for M the combinatorial Laplacian L, and for N—the degree matrix D, then

---

[17]This dataset can be obtained e.g. from the repository maintained by Alex Arenas, http://deim.urv. cat/~alexandre.arenas/data/welcome.htm.

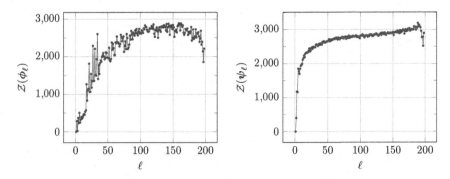

Fig. 5.5 The number of zero crossings of: **a** the combinatorial and **b** normalized Laplacian eigenvectors for the jazz musician network

$$R(L, D; \mathbf{w}) = \frac{\mathbf{w}^{\mathrm{T}} L \mathbf{w}}{\mathbf{w}^{\mathrm{T}} D \mathbf{w}} = \frac{\mathbf{w}^{\mathrm{T}} D^{1/2} \left(D^{-1/2} L D^{-1/2} \right) D^{1/2} \mathbf{w}}{(\mathbf{w}^{\mathrm{T}} D^{1/2})(D^{1/2} \mathbf{w})}$$
$$= \frac{\mathbf{y}^{\mathrm{T}} \mathcal{L} \mathbf{y}}{\mathbf{y}^{\mathrm{T}} \mathbf{y}} = R(\mathcal{L}; \mathbf{y}) \tag{5.29}$$

where $\mathbf{y} = D^{1/2} \mathbf{w}$. This means that the generalized Rayleigh quotient can be reduced to the "normal" Rayleigh quotient. This is in agreement with already noted fact that if an eigenpair $(\lambda, \mathbf{w})$ is a solution to the generalized eigenproblem $L\mathbf{w} = \lambda D\mathbf{w}$, then the pair $(\lambda, D^{1/2}\mathbf{w})$ is the solution to the eigenproblem $\mathcal{L}\mathbf{y} = \lambda \mathbf{y}$, where $\mathbf{y} = D^{1/2}\mathbf{w}$. A reader should note that if $(\lambda, D^{1/2}\mathbf{w})$ is an eigenpair solving the generalized eigenproblem, then $(1 - \lambda, D^{1/2}\mathbf{w})$ is an eigenpair of the discrete Laplacian, see the Eq. (5.22).

The Rayleigh quotient plays an important role in the variational characterization of the eigenvalues, known as the mini-max Courant-Fischer theorem

Theorem 5.2.4 *Let M be a real symmetric matrix of size $m \times m$ with the eigenvalues $\lambda_1 \leq \lambda_2 \leq \cdots \leq \lambda_m$. Let $\mathfrak{S}_k$ denotes the space spanned over the eigenvectors $\mathbf{v}_1, \ldots, \mathbf{v}_k$, with the convention that $\mathfrak{S}_0 = \{0\}$. Then*

$$\lambda_k = \min_{\substack{\mathbf{u} \neq 0 \\ \mathbf{u} \perp \mathfrak{S}_{k-1}}} \frac{\mathbf{u}^T M \mathbf{u}}{\mathbf{u}^T \mathbf{u}} = \min_{\substack{\|\mathbf{u}\|=1 \\ \mathbf{u} \perp \mathfrak{S}_{k-1}}} \mathbf{u}^T M \mathbf{u} \tag{5.30}$$

where the symbol $\mathbf{u} \perp \mathfrak{S}_k$ means that the vector $\mathbf{u}$ is orthogonal to each vector $\mathbf{v} \in \mathfrak{S}_k$. □

From this theorem it follows that

$$\lambda_1 = \min_{\mathbf{u} \neq 0} \frac{\mathbf{u}^T M \mathbf{u}}{\mathbf{u}^T \mathbf{u}} \tag{5.31a}$$

$$\lambda_2 = \min_{\substack{\mathbf{u} \neq 0 \\ \mathbf{u} \perp \mathbf{v}_1}} \frac{\mathbf{u}^T M \mathbf{u}}{\mathbf{u}^T \mathbf{u}} \tag{5.31b}$$

$$\lambda_m = \max_{\mathbf{u} \neq 0} \frac{\mathbf{u}^T M \mathbf{u}}{\mathbf{u}^T \mathbf{u}} \tag{5.31c}$$

Particularly, if M is equal to the combinatorial Laplacian L, then

$$\lambda_2 = \min_{\substack{\mathbf{u} \neq 0 \\ \mathbf{u} \perp \mathbf{v}_1}} \frac{\sum_{i=1}^{m} \sum_{j \in N(i)} s_{ij} (u_i - u_j)^2}{\sum_{i=1}^{m} u_i^2} \tag{5.32}$$

and if $M = \mathcal{L}$, then

$$\lambda_2' = \min_{\substack{\mathbf{u} \neq 0 \\ \mathbf{u} \perp v_1}} \frac{\sum_{i=1}^{m} \sum_{j \in N(i)} s_{ij} (u_i - u_j)^2}{\sum_{i=1}^{m} u_i^2 \mathsf{d}_i} \tag{5.33}$$

The next, important for us, theorem is Ky Fan's trace min/max principle.[18] It binds eigenvectors of a symmetric matrix to the trace minimization/maximization problem of the matrix.

Theorem 5.2.5 *Let $M \in \mathbb{R}^{m \times m}$ be a symmetric matrix with the eigenvalues $0 \leq \lambda_1 \leq \cdots \leq \lambda_m \in \mathbb{R}$ and the corresponding eigenvectors $\mathbf{v}_1, \ldots, \mathbf{v}_m$. Let $U \in \mathbb{R}^{m \times k}$, $1 \leq k \leq m$, be a unitary matrix,[19] and let $\mathbb{I}_k$ denotes the identity matrix of size $k \times k$. Then the solution to the problem*

$$Y = \arg\min_{U^T U = \mathbb{I}_k} tr(U^T M U) \tag{5.34}$$

is the matrix $Y = (\mathbf{v}_1, \ldots, \mathbf{v}_k) Q$, where $Q \in C^{k \times k}$ is a unitary matrix. Similarly, the solution to the problem

$$Z = \arg\max_{U^T U = \mathbb{I}_k} tr(U^T M U) \tag{5.35}$$

is the matrix $Z = (\mathbf{v}_{m-k+1}, \ldots, \mathbf{v}_m) Q$, where $Q \in C^{k \times k}$ is a unitary matrix. ☐

---

[18] See e.g. R. Bhatia, *Matrix Analysis*, Springer, 1997.
[19] I.e. such a matrix that $U^T = U^{-1}$, or equivalently, to $U^T U = \mathbb{I}$. If U is a hermitian matrix this condition extends to $U^* = U^{-1}$, where U^* stands for the conjugate transpose of U.

It is easy to observe, that substituting for U the matrix Y we obtain

$$
\begin{aligned}
\operatorname{tr}(Y^{\mathrm{T}} M Y) &= \operatorname{tr}\left(Q^{\mathrm{T}}(\mathbf{v}_1, \ldots, \mathbf{v}_k)^{\mathrm{T}} M(\mathbf{v}_1, \ldots, \mathbf{v}_k) Q\right) = \operatorname{tr}\left(Q^{\mathrm{T}} \operatorname{diag}(\lambda_1, \ldots, \lambda_k) Q\right) \\
&= \operatorname{tr}\left(\operatorname{diag}(\lambda_1, \ldots, \lambda_k)\right) = \sum_{i=1}^{k} \lambda_i
\end{aligned}
\tag{5.36}
$$

Similarly, substituting for U the matrix Z we obtain

$$
\operatorname{tr}(Z^{\mathrm{T}} M Z) = \sum_{j=m-k+1}^{m} \lambda_j
\tag{5.37}
$$

A reader should note that when $k = 1$ then Y reduces to the vector $\mathbf{v}_1$ and the matrix Z—to the vector $\mathbf{v}_m$, i.e. Fan's theorem reduces to the corollaries (5.31a) and (5.31c) of Courant-Fischer theorem.

Let us close this section with some remarks on Eckart-Young-Mirsky theorem,[20] [204]. Recall first that any rectangular matrix $X \in \mathbb{R}^{m \times n}$ can be represented as the product of three matrices $U \Sigma V^{\mathrm{T}}$, where $U \in \mathbb{R}^{m \times m}$ and $V \in \mathbb{R}^{n \times n}$ are orthonormal matrices, whose columns are referred to as, respectively, left and right singular vectors, and $\Sigma \in \mathbb{R}^{m \times n}$ is a pseudo-diagonal matrix; its diagonal elements $\sigma_1 \geq \cdots \geq \sigma_r > 0$, where $r = \min(m, n)$, are referred to as singular values. Eckart and Young considered the problem of approximation of the matrix X by the matrix $\widehat{X}_k$ of rank $k < \min(m, n)$ and such that

$$
\|X - \widehat{X}_k\|_F = \min_{\substack{X' \in \mathbb{R}^{m \times n} \\ rank(X')=k}} \|X - X'\|_F
\tag{5.38}
$$

The solution has the form $\widehat{X}_k = U \widehat{\Sigma}_k V^{\mathrm{T}}$, where $\widehat{\Sigma}_k = \operatorname{diag}(\sigma_1, \ldots, \sigma_k, 0, \ldots, 0)$. From the computational point of view, the matrix $\widehat{X}_k$ can be expressed in the form

$$
\widehat{X}_k = U_k \Sigma_k V_k^{\mathrm{T}}
\tag{5.39}
$$

where $\Sigma_k = \operatorname{diag}(\sigma_1, \ldots, \sigma_k)$ is the diagonal matrix of size $k \times k$, and $U_k \in \mathbb{R}^{m \times k}$ and $V_k \in \mathbb{R}^{n \times k}$ are matrices whose columns are the corresponding singular vectors. The Eq. (5.39) is referred to as the truncated SVD. From the Eq. (5.38) it follows that $\widehat{X}_k$, which is constructed from the k-largest singular triplets of X, is the closest rank-k matrix to X. The quality of such an approximation is

$$
\|X - \widehat{X}_k\|_F^2 = \sum_{i=k+1}^{r} \sigma_i^2
\tag{5.40}
$$

---

[20] See G. Eckart and G. Young, The approximation of one matrix by another of lower rank, *Psychometrika* 1: 211–218 (1936) and L. Mirsky, Symmetric gauge functions and unitarily invariant norms, *Quart. J. Math. Oxford* 1156–1159 (1966).

and

$$\|X - \widehat{X}_k\|_2 = \sigma_{k+1} \tag{5.41}$$

Mirsky proved that this solution is valid for any unitarily invariant norm.[21]

As [342] points out, there are several important reasons for popularity of SVD and its broad use in education. Though there exist many other, frequently more efficient matrix decomposition methods, they are usually much more complex, if stability is required, and their runtime beats that of SVD only by some constant factor. Additionally, expressing results of those decompositions via the SVD or best rank-k approximation of the SVD makes them more comprehensive.[22]

5.2.5 Random Walk on Graphs

A random walk is a stochastic process characterizing a path that consists of a sequence of random steps on some mathematical space. In this section we describe its two variants: classical random walk on graphs, which in fact is a stationary Markov chain [14, 161, 336], and so-called Ruelle-Bowens, or maximum entropy, random walk [142]. The first variant occurs naturally when studying spectral clustering based on the discrete Laplacian [352], while the second has found an intriguing application in link mining [325].

5.2.5.1 Classical Random Walk

The entries of the matrix $P = D^{-1}S$ introduced in Eq. (5.23) can be viewed as the probabilities of moving, in one step, from a node (or state) v_i to the node v_j, i.e.

$$p_{ij} = \begin{cases} s_{ij}/\mathsf{d}_i & \text{if } \{i, j\} \in E \\ 0 & \text{otherwise} \end{cases} \tag{5.42}$$

Hence P is a row stochastic matrix, i.e. $\sum_j p_{ij} = 1$. The matrix P can be interpreted as a transition probability of a Markov chain over states V, see e.g. [336] or [14].

Recall—see Theorem 5.2.3(c)—that since $P = D^{-1/2}(\mathbb{I} - \mathcal{L})D^{1/2}$, if $\overline{\lambda}$ is an eigenvalue of the matrix P, and $\mathbf{u}$ is the corresponding eigenvector, then $\lambda' = 1 - \overline{\lambda}$

---

[21] A norm $\| \cdot \|$ is called unitarily invariant if $\|UAV\| = \|A\|$ for any orthogonal pair of matrices U and V.

[22] These terms are quoted in: G.H. Golub, M.W. Mahoney, P. Drineas, and L.-H. Lim. "Bridging the gap between numerical linear algebra, theoretical computer science, and data applications". *SIAM News*, 39(8), October 2006.

is an eigenvalue of the matrix $\mathcal{L}$, and $\phi = D^{1/2}\mathbf{u}$—the corresponding eigenvector of the matrix $\mathcal{L}$. Thus, the eigenvalues of the matrix P characterizing random walk on a connected graph satisfy the condition

$$1 = \overline{\lambda}_1 > \overline{\lambda}_2 \geq \cdots \geq \overline{\lambda}_m \geq -1 \tag{5.43}$$

and $\overline{\lambda}_m = -1$ only if Γ is a bipartite graph. $\overline{\lambda}_1 = 1$ is the dominating eigenvalue of P.

Let $\mathbf{p}$ be an m-dimensional stochastic vector, i.e. $\sum_j p_j = 1$, and $p_j \geq 0$ for $j = 1, \ldots, m$. The vector $\mathbf{p}^{\mathrm{T}} P$ describes states distribution after one step of a random walk, i.e. the i-th element of this vector

$$(\mathbf{p}^{\mathrm{T}} P)_i = \sum_{j=1}^{m} \frac{s_{ij}}{\mathsf{d}_j} p_j$$

describes the probability of being in state i. In general, the states distribution after $t > 0$ steps is described by the vector $(\mathbf{p}^t)^{\mathrm{T}} = \mathbf{p}_0^{\mathrm{T}} P^t$, where $\mathbf{p}_0$ denotes an initial states distribution. When Γ is a connected, and not bipartite, graph, then the stochastic vector

$$\boldsymbol{\pi} = \frac{D\mathbf{e}}{\mathrm{vol}\,\Gamma} = \frac{\mathsf{d}}{\mathrm{vol}\,\Gamma} = \frac{\mathsf{d}}{\sum_{j=1}^{m} \mathsf{d}_j} \tag{5.44}$$

is called the stationary distribution. Intuitively, the probability of being in node i is proportional to the degree of this node. If the initial distribution is $\boldsymbol{\pi}$ then

$$\boldsymbol{\pi}^{\mathrm{T}} P = \boldsymbol{\pi}^{\mathrm{T}} \tag{5.45}$$

and, $\boldsymbol{\pi}^{\mathrm{T}} P^t = \boldsymbol{\pi}$ for any $t \geq 0$.

A characteristic feature of random walk on graphs is so-called reversibility described by the equation

$$\pi_i p_{ij} = \pi_j p_{ji} \tag{5.46}$$

termed also the detailed (or local) balance condition. An immediate consequence of this property is that the distribution $\boldsymbol{\pi}$ is invariant, i.e.

$$\sum_{i=1}^{m} \pi_i p_{ij} = \pi_j \sum_{i=1}^{m} p_{ji} = \pi_j \tag{5.47}$$

The stationary distribution $\boldsymbol{\pi}$ can be characterized in several ways. First of all, due to the Eq. (5.45), $\boldsymbol{\pi}$ can be viewed as the left dominating eigenvector of the matrix P, or equivalently, as a right eigenvector of the matrix P^{T}; both these eigenvectors correspond to the maximal eigenvalue $\lambda(P) = 1$. Second, from the above properties it follows that if Γ is a connected and not bipartite graph, then

$$\lim_{t \to \infty} \mathbf{p}^t = \boldsymbol{\pi} \qquad (5.48)$$

A Markov chain possessing stationary distribution is said to be ergodic [14]. The time at which the convergence (5.48) takes place is called the mixing time. More precisely, if $d(t)$ stands for a distance between the two distributions, then the mixing time is defined as, see e.g. [323]

$$t_{mixing}(\epsilon) = \min\{t : d(t) \le \epsilon\} \qquad (5.49)$$

where ϵ is a user defined constant. Of course this time depends on how the distance $d(t)$ is defined. Below we give some examples.

Example 5.2.3 As $(\mathbf{p}')^{\mathrm{T}} = \mathbf{p}P^t$, where $\mathbf{p}$ is any initial states distribution, and—in case of ergodic Markov chain—the matrix P^t tends to the matrix Q with elements $q_{ij} = \pi_j$, let us estimate the ratio with which the elements p'_{ij} tend to π'_js. From the Eq. (5.10), see also Proposition 5.2.3(c), it follows that the asymmetric matrix P can be expressed in terms of the normalized similarity matrix $\mathcal{S} = \mathbb{I} - \mathcal{L} = D^{-1/2}SD^{-1/2}$, as $P = D^{-1/2}\mathcal{S}D^{1/2}$. The matrix $\mathcal{S}$ is symmetric and has the spectral decomposition

$$\mathcal{S} = \sum_{j=1}^{m} \overline{\lambda}_j \boldsymbol{\phi}_j \boldsymbol{\phi}_j^{\mathrm{T}}$$

Thus

$$P^t = D^{-1/2}\mathcal{S}^t D^{1/2} = \sum_{j=1}^{m} \overline{\lambda}_j^t D^{-1/2}\boldsymbol{\phi}_j \boldsymbol{\phi}_j^{\mathrm{T}} D^{1/2} = Q + \sum_{j=2}^{m} \overline{\lambda}_j^t D^{-1/2}\boldsymbol{\phi}_j \boldsymbol{\phi}_j^{\mathrm{T}} D^{1/2}$$

where $Q = D^{-1/2}\boldsymbol{\phi}_1 \boldsymbol{\phi}_1^{\mathrm{T}} D^{1/2}$ is the already defined matrix with the entries $q_{ij} = \pi_j$.

From the last equation we conclude, that the elements of the matrix P^t are of the form

$$p'_{ij} = \pi_j + \sum_{\ell=2}^{m} \overline{\lambda}_\ell^t \phi_{\ell i} \phi_{\ell j} \sqrt{\frac{\mathsf{d}_j}{\mathsf{d}_i}} \qquad (5.50)$$

From Theorem 5.2.2(a) it follows that if Γ is not bipartite graph, then $|\overline{\lambda}_j| = |1 - \lambda'_j| < 1$, $j = 2, \dots, m$, hence

$$\lim_{t \to \infty} p'_{ij} = \pi_j$$

Assume now, that the initial distribution has the form χ_i, i.e. the random walk starts from the node v_i. Denote $\mathbf{p}'_i = (P^t)^{\mathrm{T}} \chi_i$. Then from Eq. (5.50) it follows that

$$|p'_{ij} - \pi_j| \le \sqrt{\frac{\mathsf{d}_j}{\mathsf{d}_i}} \widehat{\lambda}^t, \quad j = 1, \dots, m \qquad (5.51)$$

where $\widehat{\lambda} = \max(|\overline{\lambda}_2|, |\overline{\lambda}_m|) = \max(1 - \lambda_2', \lambda_m' - 1)$. The number

$$\gamma = 1 - \overline{\lambda}_2 = \lambda_2' \tag{5.52}$$

is said to be spectral gap, while the number

$$\gamma^* = 1 - \max_{i=2,\dots,m} |\overline{\lambda}_i| \tag{5.53}$$

is said to be absolute spectral gap. In case of lazy random walk (introduced below) $\gamma^* = \gamma$. Define now

$$d_1(t) = \frac{1}{2} \max_{1 \le i \le m} \sum_{j=1}^{m} |p_{ij}^t - \pi_j|$$

Then the mixing time is, see [323, Theorem 12.3]

$$t_1(\epsilon) = \log\left(\frac{1}{\epsilon \pi_{min}}\right) \frac{1}{\gamma^*}$$

where $\pi_{min} = \min_j \pi_j$. On the other hand, if we define

$$d_2(t) = \|\mathbf{p}_0^{\mathsf{T}} P^t - \boldsymbol{\pi}\|_2$$

then the mixing time is equal, see [115]

$$t_2(\epsilon) = \frac{1}{\lambda_2'} \log \frac{\sqrt{\mathsf{d}_{max}}}{\epsilon \sqrt{\mathsf{d}_{min}}}$$

Here $\mathsf{d}_{min}, \mathsf{d}_{max}$ denote the minimal and maximal node degree. Other approaches to mixing time are discussed in [115, 323].   □

If Γ is a connected and bipartite graph, then the corresponding Markov chain can be transformed to ergodic Markov chain by adding to each node self-loop with the weight d_v. The new generalized adjacency matrix takes the form

$$\widetilde{S} = D + S \tag{5.54}$$

and the corresponding transition matrix is

$$\widetilde{P} = \frac{1}{2}(\mathbb{I} + P) \tag{5.55}$$

This matrix defines lazy random walk on the graph. It is easy to verify that the vector $\widetilde{\pi}$ is as described by the Eq. (5.44), and that $\pi = \widetilde{\pi}$ is the stationary distribution of $\widetilde{P}$, i.e. $\widetilde{P}^t \to Q$.

Taking into account the definition (5.55) we state that

Conclusion 5.2.1 *If $(\overline{\lambda}, \mathbf{u})$ is an eigenpair of the matrix $P = D^{-1/2}(\mathbb{I} - \mathcal{L})D^{1/2}$, then $((1+\overline{\lambda})/2, \mathbf{u})$ is an eigenpair of the matrix $\widetilde{P} = (\mathbb{I}+P)/2$. Thus, the eigenvalues of the matrix $\widetilde{P}$ are real numbers from the interval $[0, 1]$, i.e.*

$$1 = \lambda_1(\widetilde{P}) > \lambda_2(\widetilde{P}) \geq \cdots \geq \lambda_m(\widetilde{P}) \geq 0$$

and $\lambda_m(\widetilde{P}) = 0$ if Γ is a bipartite graph. □

Note also that $\lambda_i(\widetilde{P}) = \lambda_i'/2$, where λ_i' is the i-th (among increasingly ordered) eigenvalue of the normalized Laplacian $\mathcal{L}$.

In general we can define α-lazy random walk introducing the matrix

$$\widetilde{P}_\alpha = \alpha\mathbb{I} + (1 - \alpha)P, \quad 0 \leq \alpha < 1 \tag{5.56}$$

Clearly, if $(\overline{\lambda}, \mathbf{u})$ is an eigenpair of P, then

$$\widetilde{P}_\alpha \mathbf{u} = \alpha\mathbf{u} + (1 - \alpha)\overline{\lambda}\mathbf{u} = \left(\alpha + (1 - \alpha)\overline{\lambda}\right)\mathbf{u}$$

that is $\mathbf{u}$ jest also the eigenvector of the matrix $\widetilde{P}_\alpha$, and the corresponding eigenvalue is $\lambda(\widetilde{P}_\alpha) = \alpha + (1 - \alpha)\overline{\lambda}$. If $\alpha = 1/2$, the transition matrix will be denoted by $\widetilde{P}$.

5.2.5.2 Ruelle-Bowens Random Walk

Let Γ be an unweighted and undirected[23] graph described by a symmetric adjacency matrix A. Let τ_{ij}^t denotes a path of length $t \geq 1$ from node v_i o node v_j. If this path goes through the nodes $v_i, v_{i_1}, \ldots, v_{i_{t-1}}, v_{i_t} = v_j$, then the probability $p(\tau_{ij}^t)$ of the path is equal to $p(\tau_{ij}^t) = p_{ii_1}p_{i_1i_2} \cdot \cdots \cdot p_{i_{t-1}j}$, where $p_{i_\ell i_{\ell+1}}$ is an element of the transition matrix $P = D^{-1}A$.

To make the long time behaviour of the walk as unpredictable as possible we re-weight the elements of the transition matrix such that all the possible paths of the walker are (almost) identical. Namely, an edge $\{i, j\}$ obtains the weight $\mathbf{W_u}(i, j) = \gamma u_i u_j$, where u_i's are elements of the normalized eigenvector $\mathbf{u}$ corresponding to the maximal eigenvalue λ_{max} of the adjacency matrix A and γ is a real number (parameter); in the sequel we will set $\gamma = 1/\lambda_{max}$. Let $S_\mathbf{u}$ stands for the resulting similarity matrix, and $D_\mathbf{u} = \text{diag}(u_1, \ldots, u_m)$. The transition probability takes now the form [325]

---

[23]The case of directed graph is considered in [142] and in [112].

$$P_{\mathbf{u}} = \left(\operatorname{diag}(S_{\mathbf{u}}\mathbf{e})\right)^{-1} S_{\mathbf{u}} = \frac{D_{\mathbf{u}}^{-1} A D_{\mathbf{u}}}{\lambda_{max}}$$

The random walk with the transition probability $P_{\mathbf{u}}$ is referred to as the Ruelle-Bowens random walk, or as the maximum entropy random walk as it maximizes so-called entropy rate of a walk.

It can be verified that the stationary distribution of this walk is $\pi^* = (u_1^2, \ldots, u_m^2)^{\mathrm{T}}$. Moreover, for an undirected graph it is possible to define the maximal entropy Laplacians as follows

$$L_{\mathbf{u}} = D_{\mathbf{u}}^2 - \frac{1}{\lambda_{max}} D_{\mathbf{u}} A D_{\mathbf{u}} \tag{5.57a}$$

$$\mathcal{L}_{\mathbf{u}} = D_{\mathbf{u}}^{-1} L_{\mathbf{u}} D_{\mathbf{u}}^{-1} = \mathbb{I} - \frac{1}{\lambda_{max}} A \tag{5.57b}$$

$$\Delta_{\mathbf{u}} = D_{\mathbf{u}}^{-2} L_{\mathbf{u}} = \mathbb{I} - \frac{1}{\lambda_{max}} D_{\mathbf{u}}^{-1} A D_{\mathbf{u}} \tag{5.57c}$$

Other properties of the Ruelle-Bowen random walk can be found in [142, 325], and in [112]. Particularly, in [142] so-called Entropy Rank was proposed as well as its generalized version termed Free Energy Rank. This last measure can can be used in disconnected networks. The authors of [142] note that the Entropy Rank takes into account not only the paths leading to a node, but also the paths emanating from this node. In case of undirected networks the Entropy Rank provides the ranking identical to the ranking induced by the left dominant eigenvector of A. The authors of [112] consider the problem of transportation over a strongly connected, and directed graph. They show that using this apparatus we obtain a transportation plan that tends to lessen congestion and appears to be robust with respect to links/nodes failure. Lastly, Li et al. define in [325] new kernels and similarity measures desirable for the link prediction problem.

Let us close this section with a numerical example.

Example 5.2.4 Consider the graph from Fig. 5.2. Assuming that the weight of each edge is equal to one, we obtain

$$P_{\mathbf{u}} = \begin{bmatrix} 0.00000 & 0.45681 & 0.00000 & 0.00000 & 0.54319 & 0.00000 \\ 0.24093 & 0.00000 & 0.36459 & 0.00000 & 0.39448 & 0.00000 \\ 0.00000 & 0.30186 & 0.00000 & 0.22914 & 0.35894 & 0.11006 \\ 0.00000 & 0.00000 & 0.48031 & 0.00000 & 0.51969 & 0.00000 \\ 0.20261 & 0.27899 & 0.30662 & 0.21178 & 0.00000 & 0.00000 \\ 0.00000 & 0.00000 & 1.00000 & 0.00000 & 0.00000 & 0.00000 \end{bmatrix}$$

and

$$L_\mathrm{u} = \begin{bmatrix} 0.10856 & -0.04959 & -0.00000 & -0.00000 & -0.05897 & -0.00000 \\ -0.04959 & 0.20583 & -0.07505 & -0.00000 & -0.08120 & -0.00000 \\ -0.00000 & -0.07505 & 0.24861 & -0.05697 & -0.08924 & -0.02736 \\ -0.00000 & -0.00000 & -0.05697 & 0.11860 & -0.06163 & -0.00000 \\ -0.05897 & -0.08120 & -0.08924 & -0.06163 & 0.29104 & -0.00000 \\ -0.00000 & -0.00000 & -0.02736 & -0.00000 & -0.00000 & 0.02736 \end{bmatrix}$$

There are, for instance, three paths of length 3 from the node 1 to the node 3: 1-2-5-3, 1-5-2-3, and 1-5-4-3. Probability of each path is equal to 0.055253. In fact there holds the following equality [112]

$$\pi_i^* p(\tau_{ij}^t) = \frac{\pi_i^* \pi_j^*}{\lambda_{max}^t}$$

The left hand side of this equation is termed the Ruelle-Bowen path measure $\mathfrak{m}_{RB}(\tau i j^t)$. □

5.3 Spectral Partitioning

One of important problems in graph theory is so called isoperimetric problem defined as follows, see e.g. Chap. 2 of [115]: remove as little of the graph as possible to separate out a subset of vertices of some desired "size". In its simplest case, the size of a subset of vertices, is defined as the number of vertices belonging to this subset, but other choices are also possible. A typical variant of the isoperimetric problem is to remove as few edges as possible to disconnect the graph into two parts of almost equal size. Such problems are usually called separator problems and are particularly useful in a number of areas including network design, parallel architectures for computers, or computer vision, see e.g. [159, 223, 427].

If Γ is a similarity graph representing relationships among the objects constituting a given data set X, then solving the isoperimetric problem is often identified with the graph cutting problem, and in consequence—with the problem of partitioning of this data set. A useful tool in solving such stated problem offers so-called spectral graph theory presented e.g. in [115]. Here an important role plays the (un)-normalized graph Laplacian defined in Sect. 5.2.2; its eigenvectors are used to extract appropriate partition. Different forms of the Laplacian, introduced in Sect. 5.2.2, correspond to different graph cutting criteria. This is the theme of this section. We start by showing that classical graph bi-partitioning problem can be formulated as searching for a vector $\chi \in \{-1, 1\}^m$ that minimizes the quadratic form $\chi^T L \chi$, where L is the combinatorial Laplacian. Next we define the fundamental concepts used in spectral clustering, that is: Fiedler value and Fiedler vector, and we show how different graph cutting criteria lead to different optimization problems that exploit different variants of graph Laplacian. Lastly, various instances of spectral clustering algorithm are described.

5.3.1 Graph Bi-Partitioning

Suppose first that we are interested in finding a partition of the nodes of Γ into two disjoint sets A and $B = V \setminus A$. The quality of this partition is expressed as total weight of the edges connecting the nodes belonging to the parts A and B. In graph theoretic language, it is called the cut:

$$\text{cut}(A, B) = \sum_{i \in A, j \in B} s_{ij} \tag{5.58}$$

Hence, by the "optimal partition" we understand a partition with minimal cut value. Searching for such a solution is referred to as the *MINcut* problem.

To give a manageable procedure allowing to compute the value of cut for a given matrix S, describe the partition (A, B) by the vector χ whose j-th entry χ_j equals $+1$ if the node j belongs to A and $\chi_j = -1$ if j belongs to B. Let, as in previous section, d_j stands for the degree of node j, i.e. the sum of S's j-th row, and D be the degree matrix, i.e. the diagonal matrix whose (j, j)-th element is the degree d_j. One can verify that

$$
\begin{aligned}
\chi^{\mathsf{T}} D \chi = \sum_{j=1}^{n} \mathsf{d}_j \chi_j^2 &= \sum_{\substack{i \in A \\ j \in V}} s_{ij} + \sum_{\substack{i \in B \\ j \in V}} s_{ij} \\
&= \left(\sum_{\substack{i \in A \\ j \in A}} s_{ij} + \sum_{\substack{i \in A \\ j \in B}} s_{ij} \right) + \left(\sum_{\substack{i \in B \\ j \in B}} s_{ij} + \sum_{\substack{i \in B \\ j \in A}} s_{ij} \right) \\
&= assoc(A, A) + assoc(B, B) + 2\text{cut}(A, B)
\end{aligned}
\tag{5.59}
$$

where $assoc(A, A)$ and $assoc(B, B)$ are total weights of edges connecting nodes within A and B, respectively.[24]

---

[24] $assoc(A, B) = \sum_{\substack{i \in A \\ j \in B}} s_{ij}$ measures the strength of association between sets A and B. Note that in an undirected graph $assoc(A, A)$ counts edge weights between nodes i, j twice: once as s_{ij} and once as s_{ji}. If we have two distinct sets A, B, then $assoc(A, B) = \text{cut}(A, B)$.

Similarly

$$\chi^T S \chi = \sum_{i=1}^{n} \sum_{j=1}^{n} s_{ij} \chi_i \chi_j$$

$$= \sum_{\substack{i \in A \\ j \in A}} s_{ij} + \sum_{\substack{i \in B \\ j \in B}} s_{ij} - 2 \sum_{\substack{i \in A \\ j \in B}} s_{ij} \qquad (5.60)$$

$$= assoc(A, A) + assoc(B, B) - 2\mathsf{cut}(A, B)$$

Thus

$$\chi^T(D - S)\chi = 4\mathsf{cut}(A, B) \qquad (5.61)$$

i.e. the cut value can be computed by multiplying the matrix $(D - S)$ by the vector representing cluster membership. From previous section we know that $D - S = L$ is the combinatorial Laplacian of the graph Γ, hence the above equation can be rewritten as

$$\mathsf{cut}(A, B) = \frac{1}{4}\chi^T L \chi \qquad (5.62)$$

Let us imagine for a moment that χ is an (un-normalized) eigenvector of L, and λ is the corresponding eigenvalue. In such a case $\chi^T L \chi = \lambda \chi^T \chi = \lambda m$, that is the cut is proportional to the eigenvalue. This explains informally why we are interested in second lowest eigenvalue of un-normalized Laplacian. Note that the first lowest eigenvalue 0 corresponds to a constant eigenvector that is a "clustering" in which A contains all elements and B is empty, which we discard in advance as uninteresting. Also, a second lowest eigenvalue of zero would indicate that the cutting cost is zero, that means that there are two disconnected components. Though the situation in which there exists an un-normalized eigenvector with entries equal either 1 or -1 are very rare in practice, the spectral clustering based on analysis of eigenvalues provides still with useful approximations, or good starting points for a fine-tuned local search. This intuitive picture is formalized in two sub-sections below.

5.3.1.1 Fiedler Value and Fiedler Vector

Let $\lambda_1 \le \lambda_2 \le \cdots \le \lambda_m$ be the eigenvalues of the combinatorial Laplacian. Fiedler proved that the smallest second eigenvalue has two important properties:

(a) If Γ is a connected graph then the second smallest eigenvalue λ_2 of the Laplacian that describes this graph is positive [174]. If the graph consists of k disjoint subgraphs, then the multiplicity of λ_1 is k, i.e. $0 = \lambda_1 = \cdots = \lambda_k$ and $\lambda_{k+1} > 0$.
(b) If $\Gamma' = (V, E')$ is a subgraph of $\Gamma = (V, E)$ obtained by deleting some edges from E, then $\lambda_2(\Gamma') \le \lambda_2(\Gamma)$ [175].

(a)   **(b)**

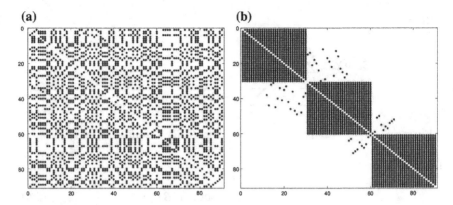

Fig. 5.6 An application of the Fiedler vector to visualization of the structure of a dataset: **a** an exemplary similarity matrix, **b** the same matrix after its rows and columns have been reordered according to the order obtained by sorting the entries of the Fiedler vector

This second smallest eigenvalue λ_2 is often referred to as the algebraic connectivity or simply as *Fiedler value*. It informs not only about the graph connectivity, but also about the intensity of the connections between the nodes of this graph. Thus low values of λ_2 suggest existence of well separated subgraphs (clusters) in a given graph. The eigenvector ψ_2, called *Fiedler vector*, or characteristic valuation [175], has both positive and negative entries, see Theorem 5.2.1(c). There are two important applications of the Fiedler vector:

(i) Its entries can be treated as the degrees of membership in appropriate group, i.e. the nodes of the graph can be partitioned into two groups: 'negative' and 'nonnegative'. More precisely, Fiedler proved in [175] that when Γ is a connected and undirected graph, and the set A (resp. $B = V \backslash A$) contains the nodes, for which $\psi_{j2} \geq 0$ (resp. $\psi_{j2} < 0$), then the subgraph (B, E_B), spanned over the nodes belonging to the set B, is connected. Similarly, the subgraph spanned over the nodes from the set A is connected if $\psi_{j2} > 0$ for all the nodes from the set A.

(ii) The difference between the elements of the Fiedler vector represents a "distance" between appropriate nodes of the graph. This property can be used in graph drawing, and, what is more important, in graphical analysis of the structure of the dataset, represented by a similarity graph [53]. It suffices to sort the entries of the Fiedler vector and to use the resulting ordering to rearrange rows and columns of the similarity (i.e. generalized adjacency) matrix. Thus, the Fiedler vector seems to be quite useful tool for rearranging rows and columns of similarity matrix, as already suggested by Czekanowski. The resulting ordering is often referred to as *Fiedler ordering*—see Fig. 5.6. It produces often better orderings than more traditional combinatorial methods, albeit at a somewhat increased cost.

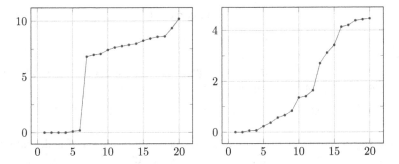

Fig. 5.7 First 20 eigenvalues of the Laplacian characterizing: linearly separable clusters (left panel), and moon-shaped clusters that are not linearly separable (right panel). In the first case Gaussian kernel has been used with $\sigma = 0.63$, and in the second case the K-NN grap has been constructed with $K = 10$ and $\sigma = 1$

The fact that in connected graph $\lambda_1 = 0$ implies the existence of so-called *spectral gap*. Namely, if there are k well defined clusters in the dataset, then the similarity matrix will have, after appropriate rearrangement of rows and columns, the (almost) block diagonal structure, i.e. $s_{ij} > 0$ if the objects i and j belong to the same cluster, and $s_{ij} \approx 0$ otherwise. The Laplacian L, corresponding to this matrix, will also be block diagonal. It is known that the spectrum (i.e. the set of eigenvalues) of a block diagonal matrix is the union of the spectra of its blocks, [354]. The minimal eigenvalue of each block in the Laplacian is $\lambda_1^{(j)} \approx 0$, $j = 1, \ldots, k$, hence the first k minimal eigenvalues of the Laplacian will be close to zero, and the next, $(k + 1)$-th eigenvalue will be $\lambda_{k+1} = \min_{1 \leq j \leq k} \lambda_2^{(j)}$. When L has a well-defined block-diagonal structure, λ_{k+1} must be definitely greater than λ_k. This phenomenon is illustrated in Fig. 5.7. On the left panel 20 first (i.e. smallest) eigenvalues of the Laplacian of the set consisting of six well separated clusters are depicted. The gap between the 6-th and 7-th eigenvalues pretty well reflects the true structure of the dataset. On the right panel the eigenvalues of the Laplacian corresponding to the set consisting of two moon-shaped clusters. These clusters are not linearly separable, but also in this case we state that $0 = \lambda_1 < \lambda_2 \approx 0$, and $\lambda_3 = 0.060065$. Hence $\lambda_3 - \lambda_2 = 0.060065$ what again suggest proper number of clusters present in the data.

5.3.1.2 Graph Partitioning by Using Fiedler Vector

Property (i) from previous subsection is used in bisecting a given graph. That is, denoting $m(\psi_2)$ the median of the Fiedler vector, we assign the node j to A if $\psi_{j2} \geq m(\psi)$, and to B otherwise. This procedure can be generalized by defining thresholded partition described by the characteristic vector

$$\chi(\tau) = \text{sign}(\psi_2 - \tau) \tag{5.63}$$

Then we search for the value τ^* providing clustering with smallest cut value, i.e.

$$\tau^* = \underset{-1 < \tau < 1}{\arg\min}\ \mathsf{cut}(C_\tau, \overline{C}_\tau) \tag{5.64}$$

where $C_\tau = \{j \in V : \chi_j(\tau) \geq 0\}$.

To be illustrative, consider the small graph from Example 5.2.1. A reader can check that the Fiedler vector equals

$$\psi_2 = (-0.6378,\ -0.1940,\ 0.1577,\ 0.0357,\ -0.0843,\ 0.7227)^\mathsf{T}$$

and the vector $\chi(0) = \mathrm{sign}(\psi_2)$ induces the partition of the set of nodes into two sets: $C_1 = \{1, 2, 5\}$, $C_2 = \{3, 4, 6\}$ with the cost $\mathsf{cut}(C_1, C_2) = \frac{1}{4}\chi^\mathsf{T} L \chi = 13$. However the Fiedler value is rather high, $\lambda_2 = 3.1272$, what suggest that such a partition is rather artificial.

Interesting, and even more efficient rounding procedure is proposed in [462].

Having two subgraphs, we can further split them until desired number k of subgraphs will be obtained. Suppose we divided entire graph Γ into $k' < k$ subgraphs (clusters) $\Gamma_1, \ldots, \Gamma_{k'}$. To choose a candidate subgraph for splitting we compute the Fiedler values $\lambda_2(\Gamma_i)$, $i = 1, \ldots, k'$ and—in agreement with the property (i) from previous subsection—we choose the component with smallest Fiedler value.

Other properties of Fiedler value and Fiedler vector are reported in [286].

5.3.2 k-way Partitioning

Assume now that we cut the graph Γ into $k > 2$ separate subgraphs spanned over the subsets $C_1, \ldots, C_k$. In this case the *MINcut* criterion takes the form

$$\mathsf{cut}(C_1, \ldots, C_k) = \sum_{j=1}^{k} \mathsf{cut}(C_j, \overline{C}_j) \tag{5.65}$$

where $\overline{C}_j$ stands for the complement of C_j in V. Unfortunately, *MINcut* often results in a skewed cut, i.e. a very small subgraph is cut away. To avoid this disadvantage, a number of different criteria have been introduced. Below we present some of them together with the corresponding Laplacian-based representation. A reader interested in more detailed discussion is referred to e.g. [275, 482], or [502]. Some other graph cutting criteria are discussed in Sect. 5.3.5.

5.3.2.1 Balanced Graph Cutting Criteria

The criteria discussed in this section are of general form

$$J(C_1, \ldots, C_k) = \sum_{j=1}^{k} \frac{\mathsf{cut}(C_j, \overline{C}_j)}{f(C_j)} \tag{5.66}$$

where $f(C_j)$ is a measure on subsets of vertices. The most popular variants of this formula are:

$$Rcut(C_1, \ldots, C_k) = \sum_{j=1}^{k} \frac{\text{cut}(C_j, \overline{C}_j)}{|C_j|} \tag{5.67a}$$

$$Ncut(C_1, \ldots, C_k) = \sum_{j=1}^{k} \frac{\text{cut}(C_j, \overline{C}_j)}{\text{vol } C_j} \tag{5.67b}$$

$$Mcut(C_1, \ldots, C_k) = \sum_{j=1}^{k} \frac{cut(C_j, \overline{C}_j)}{assoc(C_j, C_j)} \tag{5.67c}$$

where vol C_j, the volume of C_j, is the sum of the degrees of the nodes in C_j.

The criterion (5.67a), introduced in [223], is known as the *Ratio cut* while (5.67b), introduced in [427] defines so-called *Normalized cut*. Both of these criteria force that the clusters are "balanced" in terms of the number of nodes or edge weights, respectively. Shi and Malik noted in [427] that *Ncut* takes into account both inter- and intra-cluster connections. The *Max-min cut* criterion (5.67c) was proposed in [156] to avoid the clusters containing only a few nodes. Unfortunately, the algorithm used to optimize *Mcut* is rather complex with slow running speed. Thus in the sequel we will consider only *Rcut* and *Ncut*.

5.3.2.2 Spectral Relaxations of Graph Cutting Problem

Following [102], see also [482, Sect. 5.2], let us repeat the reasoning leading to the Eq. (5.61). To express (5.67a) define k indicator vectors $\mathbf{h}_1, \ldots, \mathbf{h}_k$ with the elements

$$h_{ij} = \begin{cases} 1/\sqrt{|C_j|} & \text{if } i \in C_j \\ 0 & \text{otherwise} \end{cases} \tag{5.68}$$

and form the matrix $H = (\mathbf{h}_1, \ldots, \mathbf{h}_k)$ by stacking these vectors in columns of H. Let $\mathcal{C}^k$ be the set of all possible partitions of the set V into k subsets $C_1, \ldots, C_k$.

As $\|\mathbf{h}_i\| = \mathbf{h}_i^T \mathbf{h}_i = 1$ and $\mathbf{h}_i^T \mathbf{h}_j = 0$ if $i \neq j$, we state that any matrix with columns defined as above satisfies the condition $H^T H = \mathbb{I}$. Furthermore,

$$\mathbf{h}_i^T L \mathbf{h}_i = (H^T L H)_{ii} = \frac{1}{2} \sum_{p=1}^{m} \sum_{q=1}^{m} s_{pq} (h_{pi} - h_{qi})^2 = \frac{\text{cut}(C_i, \overline{C}_i)}{|C_i|}$$

Thus

$$Rcut(C_1, \ldots, C_k) = \sum_{i=1}^{k} \mathbf{h}_i^T L \mathbf{h}_i = \sum_{i=1}^{k} (H^T L H)_{ii} = \text{tr}\,(H^T L H) \tag{5.69}$$

In other words the problem of minimizing *Rcut* can be rewritten as

$$\min_{(C_1,\ldots,C_k)\in\mathcal{C}^k}\ \mathrm{tr}\,(H^{\mathrm{T}}LH),\ \text{s.t. } H \text{ as in Eq. 5.68} \tag{5.70}$$

This is a discrete optimization problem as the entries of the solution vectors $\mathbf{h}_i$ are only allowed to take two particular values. Obviously, it is $\mathcal{NP}$-hard. But it can be easily solved if we assume that the elements of the matrix H are arbitrary real numbers. The relaxed version of *Rcut* minimization takes the form

$$\min_{\substack{H\in\mathbb{R}^{n\times k}\\ H^{\mathrm{T}}H=\mathbb{I}}}\ \mathrm{tr}\,(H^{\mathrm{T}}LH) \tag{5.71}$$

According to Fan's Theorem 5.2.5 the optimal value of this objective is equal to $\sum_{i=1}^{k}\lambda_i$ if

$$H^* = \operatorname*{arg\,min}_{H:\ H^{\mathrm{T}}H=\mathbb{I}}\ \mathrm{tr}\,(H^{\mathrm{T}}LH) = (\psi_1,\ldots,\psi_k)R \tag{5.72}$$

where ψ_i is the eigenvector corresponding to i-th smallest eigenvalue λ_i of the Laplacian L, and $R\in\mathbb{C}^{k\times k}$ is an arbitrary unitary matrix.

Similarly, if the elements of the characteristic vector are defined as

$$h_{ij} = \begin{cases} 1/\sqrt{\mathrm{vol}\,C_j} & \text{if } i\in C_j \\ 0 & \text{otherwise} \end{cases} \tag{5.73}$$

then one can verify that $\mathbf{h}_j^{\mathrm{T}}L\mathbf{h}_j = \mathrm{cut}(C_j,\overline{C}_j)/(\mathrm{vol}\,C_j)$ and the matrix H satisfies $H^{\mathrm{T}}DH = \mathbb{I}$, where D is the degree matrix of Γ. Thus the relaxed variant of $Ncut$ minimization problem takes the form

$$\min_{\substack{H\in\mathbb{R}^{n\times k}\\ H^{\mathrm{T}}DH=\mathbb{I}}}\ \mathrm{tr}\,(H^{\mathrm{T}}LH)$$

Introducing the matrix $Q = D^{1/2}H$ we transform this problem to the problem

$$\min_{\substack{Q\in\mathbb{R}^{n\times k}\\ Q^{\mathrm{T}}Q=\mathbb{I}}}\ \mathrm{tr}\,\left[Q^{\mathrm{T}}\left(D^{-1/2}LD^{-1/2}\right)Q\right] = \min_{\substack{Q\in\mathbb{R}^{n\times k}\\ Q^{\mathrm{T}}Q=\mathbb{I}}}\ \mathrm{tr}\,\left(Q^{\mathrm{T}}\mathcal{L}Q\right) \tag{5.74}$$

which is almost identical with (5.71) with only one exception: instead of graph Laplacian, now we use the normalized Laplacian. Again, using Fan's Theorem 5.2.5 we obtain the solution of the form

$$Q^* = (\phi_1,\ldots,\phi_k)R \tag{5.75}$$

where ϕ_i is the eigenvector corresponding to i-th smallest eigenvalue λ_i' of $\mathcal{L}$. The minimum value of the trace of $Q^{\mathrm{T}}\mathcal{L}Q$ is equal to $\sum_{i=1}^{k}\lambda_i'$.

Note that since $\mathcal{L}$ can be expressed as the complement of normalized similarity matrix $\mathcal{S}$, i.e.

$$\mathcal{L} = \mathbb{I} - D^{-1/2} S D^{-1/2} = \mathbb{I} - \mathcal{S}$$

we can rewrite (5.74) as

$$\min_{\substack{Q \in \mathbb{R}^{n \times k} \\ Q^T Q = \mathbb{I}}} \operatorname{tr} \left(Q^T \mathcal{L} Q \right) = k - \min_{\substack{Q \in \mathbb{R}^{n \times k} \\ Q^T Q = \mathbb{I}}} \operatorname{tr} \left(Q^T \mathcal{S} Q \right) \tag{5.76}$$

i.e. minimization of the trace of $Q^T \mathcal{L} Q$ is equivalent to maximization of the trace of $Q^T \mathcal{S} Q$. Denote $\overline{\Lambda}_k = \operatorname{diag}(1 - \lambda_1', \ldots, 1 - \lambda_k')$. From the Eckart-Young-Mirsky theorem (discussed in Sect. 5.2.4) we conclude that

$$\| \mathcal{S} - Q^* \overline{\Lambda}_k (Q^*)^T \|_F = \min_{\substack{M \in \mathbb{R}^{m \times m} \\ rank(M) = k}} \| \mathcal{S} - M \|_F \tag{5.77}$$

i.e. the matrix $Q^* \overline{\Lambda}_k (Q^*)^T$ is the best approximation of rank k of the matrix $\mathcal{S}$.

5.3.2.3 From Eigenvectors to Partitions

A real valued solutions of a relaxed cutting problem can be converted to a hard partition into k subsets in a number of ways.

Kannan, Vempala and Vetta suggested in [279] a simplest solution: the nodes representing observations are assigned to clusters according to the maximal entry in the corresponding row of the eigenvector matrix $M = S U_k$, where U_k contains k right singular eigenvectors of the similarity matrix S. As noted in [528], this works well when there are no repeated eigenvalues.

In a more elaborated approach, described by Shi and Malik in [427], and called repeated bisection, the nodes of a similarity graph are partitioned into two clusters first. To do so, the eigenvector corresponding to the second smallest eigenvalue in the generalized eigenproblem $L\phi = \lambda D\phi$ is used as described in Sect. 5.3.1.2. Then one of these clusters is selected and is further bisected, and so on. This process continues $k-1$ times, leading to k clusters. Pseudocode 5.1 summarizes this algorithm. Likewise, Kanan, Vempala and Vetta analyzed in [279] a similar procedure: the only modification is that in step 3 they they used Cheeger's conductance (defined in the next section) instead of *Ncut*. However, this procedure had rather poor behavior in the experiments reported in [352].

Although simple, repeated bisection algorithm has two main disadvantages: (a) final partition depends on the decisions made in early stages of the process, and (b) focusing on single eigenvector we lose some useful partitioning information contained in next few small eigenvectors. Thus the most popular approach, called reclustering, relies upon using the approximate solutions H^* or Q^* defined by the

---

Algorithm 5.1 Repeated bisection spectral clustering [427]

Input: Data $X = (\mathbf{x}_1, \ldots, \mathbf{x}_m)^{\mathrm{T}}$, number od groups k.
Output: Partition $\mathcal{C} = \{C_1, \ldots, C_k\}$.
1: Transform a given data set X to the similarity graph characterized by a similarity matrix S, and
   compute the combinatorial Laplacian $L = D - S$, where $D = \mathrm{diag}(S\mathbf{e})$.
2: Solve the generalized eigenproblem $L\phi = \lambda D\phi$ for eigenvectors with the smallest eigenvalues.
3: Use the eigenvector ϕ_2 with the second smallest eigenvalue to bipartition the graph by finding
   the splitting point τ such that $Ncut(C_\tau, \overline{C}_\tau)$ is minimized. Here $C_\tau = \{j \in V : \phi_{j2} \geq \tau\}$.
4: Recursively repartition the subgraphs if necessary.

Eqs. (5.72) and (5.75). Typically the matrix R, suggested by Fan's theorem, is simply assumed to be the unit matrix. In this case

$$H^* = \Psi_k = (\psi_1, \ldots, \psi_k)$$

and

$$Q^* = \Phi_k = (\phi_1, \ldots, \phi_k)$$

The rows of these matrices can be viewed as the k-dimensional "spectral" coordinates of the nodes of the similarity graph. The mapping $V \rightarrow H^*$, or $V \rightarrow Q^*$ (or equivalently $X \rightarrow H^*$, or $X \rightarrow Q^*$) is refereed to as the spectral (or Laplacian) embedding—see [60] for details. The whole clustering procedure, called partitional spectral clustering and described by pseudocode 5.2, can be summarized as a clustering of the data embedded into the space spanned by the eigenvectors of a Laplacian.

Algorithm 5.2 A k-way partitional spectral clustering

Input: Data $X = (\mathbf{x}_1, \ldots, \mathbf{x}_m)^{\mathrm{T}}$, number of groups k.
Output: Partition $\mathcal{C} = \{C_1, \ldots, C_k\}$.
1: Transform a given data set X to the similarity graph characterized by a similarity matrix S.
2: Compute the (combinatorial, normalized or discrete) Laplacian, $\mathfrak{L}$.
3: Find k eigenvectors $\mathbf{w}_1, \ldots, \mathbf{w}_k$ corresponding to the smallest eigenvalues of the Laplacian $\mathfrak{L}$
   and stack these eigenvectors in columns of the matrix Y.
4: Treating each row of Y as a k-dimensional point, partition the data into k clusters $C_1, \ldots, C_k$.

The only problem is: "How to extract final clusters from the information contained in the matrix Y constructed in step 3 of the algorithm?". An answer to this question can be obtained by using one of (at least) four recipes:

(1) Use a clustering algorithm of the data represented by these new coordinates. Typically the k-means algorithm is used to recluster these data, although other choices are possible, see e.g. [480] for details.
(2) The pivoted QR decomposition as suggested by Zha et al. in [529] can be applied to extract cluster indices. Algorithm 5.3 presents the MATLAB code implementing this idea. In line 1 the QR decomposition of the matrix Y^{T} is performed. Here

$Q \in \mathbb{R}^{k \times k}$ is an orthogonal matrix, R is a matrix of size $k \times m$ and $P \in \mathbb{R}^{m \times m}$ denotes a permutation matrix, such that $Y^T P = QR = Q[R_{11}, R_{12}]$, where R_{11} is a an $k \times k$ upper triangular matrix. In line 2 an auxiliary matrix $U \in \mathbb{R}^{k \times m}$ is constructed. Then, in line 3 of the algorithm, the row vector C of length m is determined by the row index of the largest element in absolute value of the corresponding column of U. Damle et al. describe in [130] a fastened version of this algorithm stemmed from the computational quantum chemistry.

(3) The cluster indices can be obtained by applying a rotation matrix R, mentioned in Fan's theorem, that transforms the eigenvector matrix to a cluster indicator matrix (i.e. the matrix with only one nonzero entry per row). More precisely, if Y is the matrix containing k eigenvectors of a Laplacian as its columns and $\mathfrak{I} = \{\mathcal{I} \in \{0, 1\}^{m \times k} : \mathcal{I}e_k = e_m\}$ is the set of indicator matrices, the task is to find an indicator matrix $\mathcal{I}^*$ and a rotation matrix R^* solving the problem

$$\min_{\substack{R^T R = \mathbb{I} \\ \mathcal{I} \in \mathfrak{I}}} J(R, \mathcal{I}) = \min_{\substack{R^T R = \mathbb{I} \\ \mathcal{I} \in \mathfrak{I}}} \|YR - \mathcal{I}\|_F \tag{5.78}$$

To solve this problem, Yu and Shi proposed in [524] an alternating iterated algorithm making use of the SVD. However, it was noted in [528] that this iterative method can easily get stuck in local minima. That is why an incremental gradient descent algorithm was proposed in this last paper. It finds near-optimal rotation matrix as well as it can be used to identify the optimal number of clusters. However, finding this rotation matrix is rather time consuming. Huang et al. described in [260] a simpler method competitive (both with respect to time complexity and precision of final results) to the k-means algorithm.

(4) Alzate and Suykens described in [17] an application of the Error Correcting Output Codes method, introduced in [154], to solve this problem. We discuss it in Sect. 5.3.6.2.

Algorithm 5.3 Pivoted QR decomposition used to extract clusters from the matrix Y [529]

Input: Matrix $Y \in \mathbb{R}^{m \times k}$ computed in step 3 of the Algorithm 5.2.
Output: Cluster assignment C.
```
1: [Q,R,P] = qr(Y^T);
2: U=inv(R(1:k,1:k))*R*P^T;
3: [tmp,C]=max(abs(U),[],1);
```

In the sequel we discuss two important specializations of the Algorithm 5.2 (for other possible variants see e.g. [482]). The first one, proposed by Meilă and Shi in [352] emerged from studying relationships between spectral clustering and random walks. Thus it is described in Sect. 5.3.4. The second efficient implementation of the partitional spectral clustering algorithm is that one proposed by Ng, Jordan and Weiss, and hereafter called the NJW algorithm, in [374]. It is characterized by pseudocode 5.4. The NJW algorithm differs from the baseline algorithm in several points. First of all, its authors used the fact that the eigenvectors corresponding to k smallest eigenvalues of the normalized Laplacian are identical to the eigenvectors corresponding to k largest eigenvalues of the normalized similarity matrix $S = \mathbb{I} - \mathcal{L}$. This is not a technical trick only, as in many cases (e.g. when using the deflation method) it is easier to compute eigenvectors corresponding to the largest eigenvalues than the eigenvectors corresponding to the smallest eigenvalues. Another difference is the additional step 4 where the matrix Φ_k is transformed to another matrix Y. A rationale for this is given below. Lastly, in step 5 the data represented by Y are clustered by using k-means algorithm with the Orthogonal Initialization described in point (d) of Sect. 3.1.3.

Algorithm 5.4 Spectral clustering proposed by Ng et al. in [374]

Input: Data $X = (\mathbf{x}_1, \ldots, \mathbf{x}_m)^T$, number od groups k.
Output: Partition $C = \{C_1, \ldots, C_k\}$.
1: Transform a given data set X to the similarity graph characterized by a similarity matrix S.
2: Compute the complement of the normalized Laplacian, $S = D^{-1/2} S D^{-1/2}$.
3: Find the k largest eigenvectors $\phi_1, \ldots, \phi_k$ of S and form the matrix $\Phi_k = (\phi_1, \ldots, \phi_k) \in \mathbb{R}^{m \times k}$.
4: Transform the matrix Φ_k to the matrix Y by normalizing each row of Φ_k to have unit length.
5: Treating each row of Y as a k-dimensional point, partition the data into k clusters $C_1, \ldots, C_k$.

To justify the algorithm, its authors considered the ideal case first, when the similarity matrix $\hat{S}$ (after appropriate reordering of its rows and columns) is block diagonal, i.e. $\hat{s}_{ij} = 0$ if $\mathbf{x}_i$ and $\mathbf{x}_j$ belong to distinct clusters (and $\hat{s}_{ij} > 0$ otherwise). Assuming for simplicity that there are $k = 3$ clusters, the matrices $\hat{S}$ and $\hat{\mathcal{L}}$ take the form

$$\hat{S} = \begin{bmatrix} \hat{S}^{(1)} & 0 & 0 \\ 0 & \hat{S}^{(2)} & 0 \\ 0 & 0 & \hat{S}^{(3)} \end{bmatrix} \quad \Rightarrow \quad \hat{\mathcal{L}} = \begin{bmatrix} \hat{\mathcal{L}}^{(1)} & 0 & 0 \\ 0 & \hat{\mathcal{L}}^{(2)} & 0 \\ 0 & 0 & \hat{\mathcal{L}}^{(3)} \end{bmatrix}$$

Stacking $k = 3$ dominant eigenvectors of the matrix $\hat{S} = \mathbb{I} - \hat{\mathcal{L}}$ (they are identical with the eigenvectors corresponding to the minimal eigenvalues of $\hat{\mathcal{L}}$) we obtain the matrix

$$\hat{\Phi}_k = \begin{bmatrix} \hat{\phi}^{(1)} & 0 & 0 \\ 0 & \hat{\phi}^{(2)} & 0 \\ 0 & 0 & \hat{\phi}^{(3)} \end{bmatrix}$$

where $\hat{\phi}^{(j)} = \sqrt{\hat{D}^{(j)}} \mathbf{e}_{C_\ell}$, $j = 1, \ldots, k$, is the dominating eigenvector of j-th block, $\hat{D}^{(j)} = \text{diag}(\hat{S}\mathbf{e})$ is the degree matrix of j-th block, and $\mathbf{e}_{C_j}$ is the unit vector of size $|C_j|$. In other words, in the ideal case these eigenvectors are piecewise constant on the connected components of the similarity graph. Finally, when each row of $\hat{\Phi}_k$ is renormalized to have unit length, the points $\hat{\mathbf{y}}_i$ (i.e. rows of the renormalized matrix $\hat{Y}$) have the form $(0, \ldots, 0, 1, 0, \ldots, 0)$, where the position of the 1 indicates the connected component this point belongs to. Then, the authors of [374] formulated sufficient conditions under which the k leading eigenvectors of a "real" similarity matrix $S = \hat{S} + E$, where E is a perturbation matrix, reflect the underlying structure characterized by the matrix $\hat{S}$. In particular, they shown that this will be true if the eigengap $\delta = \lambda_k - \lambda_{k+1}$, i.e. the difference between k-th and $(k+1)$-th dominating eigenvalue of $\hat{S}$, is sufficiently large. Interestingly, empirical studies show that the method is successful even in cases where the matrix S is far from block diagonal.

Remark 5.3.1 The vector $\phi_1 = \sqrt{d/\text{vol}\,\Gamma}$ computed in Step 3 of the Algorithm 5.4 is rather trivial. Hence some authors suggest that the matrix Φ_k should have the eigenvectors $\phi_2, \ldots, \phi_{k+1}$ as its columns. $\Box$

Remark 5.3.2 The transformation $V \to H^*$ (or $X \to Q^*$) can equally well be regarded as an embedding of the graph Γ into k dimensional Euclidean space. Such a mapping was introduced by Hall [227] in order to solve the problem of placing m connected points (or nodes) in k-dimensional Euclidean space. In such a context the entries s_{ij} of similarity matrix S specify "connection" between nodes v_i and v_j, and, when $k = 1$, the problem relies upon finding locations x_i for the m points which minimize the weighted sum of squared distances between the points

$$z = \frac{1}{2} \sum_{i=1}^{m} \sum_{j=1}^{m} s_{ij}(x_i - x_j)^2$$

Obviously z can be represented as the quadratic form $\mathbf{x}^T L \mathbf{x}$, where $\mathbf{x} = (x_1, \ldots, x_m)^T$ and $L = \text{diag}(S\mathbf{e}) - S$. To avoid trivial solution $\mathbf{x} = \mathbf{0}$, the constraint $\mathbf{x}^T\mathbf{x} = 1$ is introduced. The author shows that in this one-dimensional case the nontrivial solution is the Fiedler vector of the Laplacian L, and if $k > 1$ then the solution has the form $H = (\psi_2, \ldots, \psi_{k+1})$. Particularly, when $k = 2$ or $k = 3$ this idea can be used to draw a graph—see Fig. 5.8. Other interesting examples illustrating this approach can be find in [437]. A similar approach was used by Pisanski and Shawe-Taylor in [393] for drawing fullerene molecules. $\Box$

Let us close this section with few observations. It was observed empirically that the spectral images (i.e. rows of the matrix Φ_k) of points belonging to the same cluster are distributed around straight lines, and the more compact the cluster, the better alignment of the spectral images—see Fig. 5.9a, c. If the similarity matrix is generated by using the Gussian kernel, this observation can be used to choose a proper value of the hyperparameter σ, see Sect. 4 in [374].

Usually spectral clustering is used to partition data that are connected but not necessarily compact nor located within convex regions. If the data exhibit clear structure, see e.g. Fig. 5.10a, c, then the similarity matrix S—and consequently Laplacian matrix L—is approximately block-diagonal, with each block defining a cluster, see Fig. 5.10b, d. Thus the case of "clear structure" seem to be trivial one. If the structure becomes more "fuzzy", the block diagonal structure becomes less aparent and

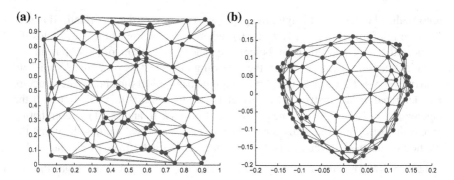

Fig. 5.8 Exemplary graph embedding: **a** original graph and **b** its embedding into spectral the space spanned by the eigenvectors (ψ_2, ψ_3)

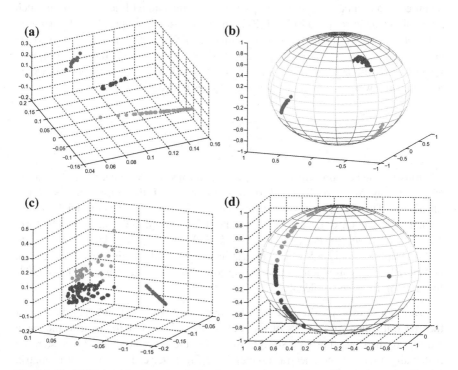

Fig. 5.9 The NJW algorithm in action. Left column: data in spectral coordinates determined by the eigenvectors Φ_k. Right column: the same data in the coordinates described by the matrix Y. Panel (**a**) corresponds to the data consisting of three linearly separable groups, and panel (**c**) represents iris data set

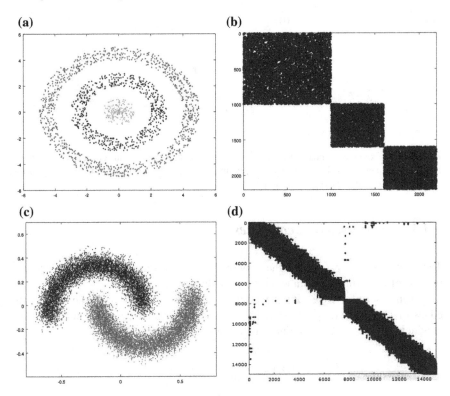

Fig. 5.10 Examples of data with clear structure, (**a**) and (**c**), and similarity matrices representing these data, (**b**) and (**d**)

more effort is needed to fit the parameters providing near block diagonal similarity matrix.[25] In this context it is worth mentioning that von Luxburg [482] attributed the success of spectral relaxation not to the good solutions obtained but rather to the elegant reformulation of the clustering problem in terms of standard linear algebra. This reformulation allows for application of known and simple linear solvers.

It should be noted also that even in simple problems spectral clustering can be advantageous over other classical algorithms. Figure 5.11a shows an example where three groups of different cardinalities are linearly separable. Unfortunately, standard k-means algorithm cannot classify these data properly, see Fig. 5.11b, while spectral clustering does it pretty well, see Fig. 5.11c.

---

[25]Particularly, the matrix from Fig. 5.10d is rather block-band matrix. Its block diagonal structure can be recovered by using the method described in [178].

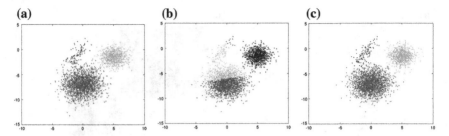

Fig. 5.11 An example of unbalanced clusters: **a** Original data with three clusters of size 2000 (red), 1000 (green), and 100 (blue). **b** Clustering obtained by the k-means algorithm, and **c** clustering returned by spectral clustering algorithm

5.3.3 Isoperimetric Inequalities

To to get a deeper insight into the previously presented results let us introduce the notions of isoperimetric number and the conductance.

Given a set of vertices $Z \subset V$, denote ∂Z the boundary of Z, i.e. the set of edges of Γ with exactly one vertex in Z. The isoperimetric number $i(Z)$ and the conductance $c(Z)$ of Z are defined as follows

$$i(Z) = \frac{|\partial Z|}{\min(|Z|, |\bar{Z}|)} \tag{5.79a}$$

$$c(Z) = \frac{\text{cut}(Z, \bar{Z})}{\min(\text{vol } Z, \text{vol } \bar{Z})} \tag{5.79b}$$

The isoperimetric number of a graph and the conductance of a graph are defined to be the minima of these quantities over subsets of vertices, i.e.

$$i_\Gamma = \min_{Z \subset V} i(Z) \tag{5.80a}$$

$$c_\Gamma = \min_{Z \subset V} c(Z) \tag{5.80b}$$

Cheeger's inequality establishes a tight connection between conductance and the Fiedler value of the normalized Laplacian [115]:

$$\frac{c_\Gamma^2}{2} \leq \lambda_2' \leq 2c_\Gamma \tag{5.81}$$

We already know, see e.g. Example 5.2.3, that the time required for a random walk on a graph to mix is inversely proportional to λ_2'. Hence, sets of vertices of small conductance are obvious obstacles to rapid mixing.

Let $P_{AB} = Pr(A \to B|A)$ denote the probability of moving in one step from a set $A \subset V$ to a $B \subset V$ if the current state is in A. Since the random walk starts from a set A, and the random walk is done according to the stationary distribution, the initial probability distribution $\mathbf{p}^A$ has the form[26]

$$p_i^A = \begin{cases} \dfrac{d_i}{\sum_{j \in A} d_j} & \text{if } i \in A \\ 0 & \text{otherwise} \end{cases} \tag{5.82}$$

and the probability P_{AB} equals to

$$P_{AB} = \sum_{i \in A} \sum_{j \in B} p_i^A p_{ij} = \frac{\sum_{i \in A} \sum_{j \in B} s_{ij}}{\text{vol } A} = \frac{\text{cut}(A, B)}{\text{vol } A} \tag{5.83}$$

As a consequence, we obtain the following lemma:

Lemma 5.3.1 *Let Γ be be an undirected and connected graph. Let $\{Z, \bar{Z}\}$ be a partition of the set of nodes, and let $P_{Z\bar{Z}}$ denotes the probability of leaving the set of states Z (and moving to the set $\bar{Z}$) defined in Eq. (5.83). Then:*

(i) $NCut(Z, \bar{Z}) = P_{Z\bar{Z}} + P_{\bar{Z}Z}$
(ii) $P_{\bar{Z}Z} = \mathfrak{c}(Z)$ if Γ is a connected graph, and vol $Z \le \frac{1}{2} vol\, V$.   □

This lemma offers an interesting probabilistic interpretation of the quantities used to measure the quality of a partition. Part (i) of this Lemma has been proven in [352]. It states that if $\{Z, \bar{Z}\}$ is a partition with low normalized cut, then the two probabilities $P_{Z\bar{Z}}$ and $P_{\bar{Z}Z}$ are also small. In other words, if the walker goes to a state belonging to cluster Z or $\bar{Z}$, then the probability of leaving this cluster is small. To be more illustrative, imagine that the undirected graph is a city map: the links of the graph represents streets and the nodes of this graph correspond to the streets intersections. If the walker went to the quarter covered by many local streets and only a few of them lead to other districts,[27] he will spend a lot of time there before he gets out of there.

The second part of the Lemma, used e.g. in [440], provides a probabilistic interpretation for the conductance of a "small" set, i.e. a set of nodes with small volume.

5.3.4 Clustering Using Random Walk

Random walk and its various characteristics imply planty of new definitions of a cluster. For the first time this approach was suggested by Hagen and Kahn who proposed in [222] a self-tuning clustering method. Unfortunately, their algorithm is rather time and space consuming.

---

[26] See Eq. (5.44).

[27] An excellent example of this situation is the old town (*medina*) of Fez in Morocco. There is almost 9000 streets!.

Van Dongen proposed in his Ph.D. dissertation the MCL (Markov Cluster Algorithm)—a scalable, fast and easy to implement algorithm which simulates random walk on a graph. The underlying idea is based on the fact that if a random walker is placed in a node located in a dense area of a sparse graph then she/he is able to leave that area after visiting the majority of nodes from this area (see Lemma 5.3.1 from previous page). Obviously, the random walking idea as such, does not allow for extracting the clusters as after having walked sufficiently long, the walker will visit all the nodes of a (connected) graph. Thus, in the MCL algorithm, after few steps of random walking, the most important paths are identified, and later on the walker considers only these paths. Such a procedure converges very fast (typically after 10–20 repetitions) to the idempotent and sparse matrix P_I. Only few of its rows contain non-zero elements which are used to reconstruct clusters in the graph. Because of its simplicity and efficiency the MCL algorithm has found many applications in bioinformatics, social network analysis, linguistics, and so on—see [474] for further information, source codes, and publications.

Meilă and Shi proposed in [352] an algorithm called *Modified NCut*, or MNcut for brevity, see pseudocode 5.5, with one major modification of the baseline Algorithm 5.2: in step 2 they construct stochastic matrix $P = D^{-1}S$ and thus in step 3 they compute k eigenvectors corresponding to the largest eigenvalues of P. The matrix P describes the random walk on graph (briefly characterized in Sect. 5.2.5). Although MNcut is related to the NJW algorithm, taking such a point of view allows to observe some subtlety of a solution to the clustering problem. Mahoney [343] has already observed that eigenvector methods and random walks on graphs optimize nearly the same objective. But the advantage of random walks lies in the fact that they can be performed mostly locally, while eigenvectors need to be computed globally, for the entire graph.

Algorithm 5.5 Spectral clustering algorithm MNCut [352]

Input: Data $X = (\mathbf{x}_1, \ldots, \mathbf{x}_m)^\mathsf{T}$, number od groups k.
Output: Partition $\mathcal{C} = \{C_1, \ldots, C_k\}$.
1: Transform a given data set X to the similarity graph characterized by a similarity matrix S.
2: Compute the complement of the discrete Laplacian, $P = \mathbb{I} - \Delta D^{-1}S$.
3: Find the k largest eigenvectors $\phi_1, \ldots, \phi_k$ of P and form the matrix $Z = (\phi_2, \ldots, \phi_k) \in \mathbb{R}^{m \times k-1}$.
4: Treating each row of Z as a $k-1$-dimensional point, partition the data into k clusters $C_1, \ldots, C_k$.

In particular, Meilă has shown[28] that if $\mathcal{C} = \{C_1, \ldots, C\}$ is a partition of the set of nodes V, then

$$NCut(\{C_1, \ldots, C_k\}) = \sum_{j=1}^{k} \left(1 - Pr(C_j \to C_j | C_j)\right)$$

---

[28] See M. Meilă: The multicut lemma. Technical Report 417, Department of Statistics, University of Washington, 2001.

where $Pr(C_j \rightarrow C_j | C_j) = P_{C_j,C_j}$, defined in Eq. (5.83), denotes the probability that a random walker starting at C_j will stay, after one step, at this subset. Hence, due to the above equation the algorithm minimizing $NCut$ may be interpreted as minimization of the probability that a random walk jumps to another cluster. Moreover she formulated the conditions under which the algorithm returns the optimal value of $NCut$ and she proved that this value cannot be smaller than $k - \sum_{j=1}^{k} \bar{\lambda}_j$, where $\bar{\lambda}_j$, $j = 1, \ldots, k$ are k dominant eigenvalues of the matrix P.

If the matrix P factorizes as $P = \Upsilon \bar{\Lambda} \Upsilon^{\mathrm{T}}$, where $\bar{\Lambda} = \mathrm{diag}(\bar{\lambda}_1, \ldots, \bar{\lambda}_m)$, and columns of Υ are the corresponding eigenvectors of P, then $P^t = \Upsilon \bar{\Lambda}^t \Upsilon^{\mathrm{T}}$ for any t. Azran and Grahramani noted in [35] that if t is odd then the ordering of the eigenvalues of P is left unchanged and the same eigenvectors are picked to cluster the data, i.e. the matrix P used in step 2 of the MNCut algorithm can be replaced by P^t for any odd t. This equivalence between P and P^t reveals two important key ideas: (i) spectral clustering implicitly searches for "good" paths between points, and (ii) the eigenvalues can be used to indicate the scale and quality of partitioning, by separating between the eigenvalues that survive t steps and those that don't. This last property can be used to find a value of t which yields a transition matrix P^t that efficiently reveals k clusters.

Another tool to analyze the structure of graphs is the hitting time and the commute time. For two vertices i and j, the hitting time h_{ij} is the expected time it takes a random walk to travel from i to j. The commute time c_{ij} is defined as the average number of steps a random walker, starting in node $i \neq j$, will take before entering a given node j for the first time, and go back to the node i, i.e. $c_{ij} = h_{ij} + h_{ji}$.

It is known, see e.g. [161], that the vector of hitting times can be computed as

$$\mathbf{h}_j = (\mathbb{I} - Q)^{-1} \mathbf{e} \tag{5.84}$$

where Q is the matrix obtained by deleting j-th row and j-th column from the transition probability matrix P. The node j is known to be absorbing state in Markov chain nomenclature. Thus the entries of $\mathbf{h}_j$ express the expected number of steps before absorbing random walk starting at the state $i \neq j$ by the state j. Alternatively these times can be computed by using the Moore-Penrose pseudoinverse $L^\dagger$ of the graph Laplacian[29] defined as

$$L^\dagger = \Psi \Lambda^\dagger \Psi^{\mathrm{T}} \tag{5.85}$$

where the columns of Ψ are the eigenvectors corresponding to the eigenvalues $\lambda_1, \ldots, \lambda_m$ of the graph Laplacian L, see the Eq. (5.21a), and $\Lambda^\dagger = \mathrm{diag}(\lambda_1^\dagger, \ldots, \lambda_m^\dagger)$, where $\lambda_i^\dagger = 1/\lambda_i$ if $\lambda_i > 0$, and $\lambda_i^\dagger = 0$ otherwise. If Γ is a connected graph then the pseudoinverse takes the form

$$L^\dagger = \sum_{j=2}^{m} \frac{1}{\lambda_j} \psi_j \psi_j^{\mathrm{T}}$$

---

[29]Since the minimal eigenvalue of L, $\lambda_1 = 0$, this matrix is singular. Fouss et al. describe in [184] an efficient procedure for computing the pseudoinverse of L in case of large data sets.

as only $\lambda_1 = 0$. The expected number of steps needed by a random walker for reaching state j for the first time, when starting from a state i is, see e.g. [184]

$$h_{ji} = \sum_{k=1}^{m} (l_{ik}^\dagger - l_{ij}^\dagger - l_{jk}^\dagger + l_{jj}^\dagger) \mathsf{d}_k \qquad (5.86)$$

The notion of hitting time is exploited in clustering in various algorithms. For instance, using the analogy between random walks and electric networks, Wu and Huberman proposed in [506] a bisection clustering algorithm. Grady took in [211] another perspective. Assume namely that A is the set of absorbing states and each state $i \in A$ has assigned group label. Then we can construct semi-supervised learning algorithm by assigning to each node $j \in V \backslash A$ the label of the node in A which absorbs j with highest probability.[30]

Similarly the hitting time can be computed by using pseudoinverse $L^\dagger$ as follows

$$c_{ij} = \text{vol}_\Gamma (l_{ii}^\dagger + l_{jj}^\dagger - 2l_{ij}^\dagger) = \text{vol}_\Gamma (\mathbf{e}_i - \mathbf{e}_j)^{\mathsf{T}} L^\dagger (\mathbf{e}_i - \mathbf{e}_j) \qquad (5.87)$$

where $\mathbf{e}_i$ denotes the i-th column of the unit matrix, and $\text{vol}_\Gamma = \text{vol} V = \sum_{j=1}^{m} \mathsf{d}_j$ is the volume of graph Γ.

It is known that the commute time is proportional to so-called resistance distance[31] which is a metric, i.e. (a) $c_{ii} = 0$, (b) $c_{ij} = c_{ji}$, and (c) $c_{ij} \leq c_{ik} + c_{kj}$. Furthermore, using the representation (5.85) we obtain

$$\begin{aligned}
c_{ij} &= \text{vol}_\Gamma (\mathbf{e}_i - \mathbf{e}_j)^{\mathsf{T}} \Psi \Lambda^\dagger \Psi^{\mathsf{T}} (\mathbf{e}_i - \mathbf{e}_j) \\
&= \sqrt{\text{vol}_\Gamma} (\mathbf{e}_i - \mathbf{e}_j)^{\mathsf{T}} \Psi (\Lambda^\dagger)^{1/2} (\Lambda^\dagger)^{1/2} \Psi^{\mathsf{T}} (\mathbf{e}_i - \mathbf{e}_j) \\
&= \left[\sqrt{\text{vol}_\Gamma} (\Lambda^\dagger)^{1/2} \Psi^{\mathsf{T}} (\mathbf{e}_i - \mathbf{e}_j) \right]^{\mathsf{T}} \left[\sqrt{\text{vol}_\Gamma} (\Lambda^\dagger)^{1/2} \Psi^{\mathsf{T}} (\mathbf{e}_i - \mathbf{e}_j) \right] \\
&= (\boldsymbol{\xi}_i - \boldsymbol{\xi}_j)^{\mathsf{T}} (\boldsymbol{\xi}_i - \boldsymbol{\xi}_j) = \|\boldsymbol{\xi}_i - \boldsymbol{\xi}_j\|^2
\end{aligned}$$

where $\boldsymbol{\xi}_i = \sqrt{\text{vol}_\Gamma} (\Lambda^\dagger)^{1/2} \Psi^{\mathsf{T}} \mathbf{e}_i$, what shows that the square root of the commute time is also Euclidean distance. This distance is called the Euclidean Commute-Time Distance (ECTD) in [184]. These authors state also that the ECTD is nothing else than a Mahalanobis distance with a weighting matrix $L^\dagger$.

The commute-time distance between two nodes has few nice properties, namely it decreases when the number of paths connecting the two nodes increases and when the length of paths decreases. This is in contrast to the the shortest path distance which does not capture the fact that strongly connected nodes are closer than weakly connected nodes. Thus, having the matrix $C = [c_{ij}]_{m \times m}$ we can use e.g. the k-medoids

---

[30]These probabilities can be computed as $(\mathbb{I} - Q)^{-1} R$, where R is the matrix of size $|V \backslash A| \times |A|$ obtained by deleting from P the rows corresponding to the states in A and columns corresponding to the states in $V \backslash A$. This time the matrix Q is obtained from P by deleting rows and columns corresponding to the states in A.

[31]I.e. c_{ij} is proportional to the effective resistance between node i and node j of the corresponding network, where a resistance $1/s_{ij}$ is assigned to each edge. See [161] for details.

algorithm, described in Sect. 3.1.5.6, to cluster the data. In [332] an adaptation of the FCM algorithm (discussed in Sect. 3.3) to the random walk distances is proposed. As its author state, this new algorithm outperforms other frequently used algorithms in data sets with different geometries. It should also be noted that while spectral clustering only uses k eigenvectors of Ψ, in this approach all the eigenvectors are exploited. Unfortunately, for large data sets the commute-time distance looses its discriminative properties, as shown in [484].

5.3.5 Total Variation Methods

Bresson et al. noted in [90] that using the concept of total variation it is possible to design algorithms that can obtain tighter relaxations than those used by spectral clustering. In brief, total variation techniques promote the formation of sharp indicator functions in the continuous relaxation. These functions equal one on a subset of the graph, zero elsewhere and exhibit a non-smooth jump between these two regions. In contrast to the relaxations provided by spectral clustering, total variation techniques therefore lead to quasi-discrete solutions that closely resemble the discrete solution of the original $\mathcal{NP}$-hard problem. In this section we give a sketch of these methods. We proceed this description by introducing an interesting extension of the notion of a graph Laplacian.

5.3.5.1 p-Laplacian and Its Properties

The properties (5.12a) and (5.18a) of the combinatorial and the normalized Laplacians can be generalized by introducing the operators $L^{[p]}$ and $\mathcal{L}^{[p]}$, $p > 1$ satisfying the condition

$$(L^{[p]}\mathbf{f})_i = \sum_{j=1}^{m} s_{ij} \rho^{[p]}(f_i - f_j) \tag{5.88a}$$

$$(\mathcal{L}^{[p]}\mathbf{f})_i = \frac{1}{\mathsf{d}_i} \sum_{j=1}^{m} s_{ij} \rho^{[p]}(f_i - f_j) \tag{5.88b}$$

where $\rho^{[p]}(x) = |x|^{p-1}\mathrm{sign}(x)$, and $\mathbf{f}$ is a signal on V. $L^{[p]}$ is said to be combinatorial p-Laplacian, and $\mathcal{L}^{[p]}$—normalized p-Laplacian [94, 337]. Obviously, if $p = 2$ both the p-Laplacians are identical with the already introduced Laplacians (5.8a) and (5.8b).

A real number $\lambda^{[p]}$ is said to be an eigenvalue of p-Laplacian, if there exists a real valued vector $\mathbf{v}$ such that

$$(L^{[p]}\mathbf{v})_i = \lambda^{[p]} \rho^{[p]}(v_i) \tag{5.89}$$

The vector $\mathbf{v}$ occuring in this equation is said to be the eigenvector of p-Laplacian corresponding to the eigenvalue $\lambda^{[p]}$. Bühler and Hein proved in [94] that $\mathbf{v} \in \mathbb{R}^m$ is an eigenvector of $L^{[p]}$ if it is a critical point of a generalized Rayleigh quotient

$$R^{[p]}(\mathbf{y}) = \frac{\langle \mathbf{y}, L^{[p]}\mathbf{y}\rangle}{\|\mathbf{y}\|_p^p} \tag{5.90}$$

To simplify the computations, these authors proved also, that the generalized Fiedler vector $\psi_2^{[p]}$ i.e. the p-eigenvector corresponding to $\lambda_2^{[p]}$ can be computed as

$$\psi_2^{[p]} = \underset{\mathbf{v}\in\mathbb{R}^n}{\arg\min} \frac{\mathbf{v}^T L^{[p]}\mathbf{v}}{\min_{c\in\mathbb{R}} \|\mathbf{v} - c\mathbf{e}\|_p^p} \tag{5.91}$$

It must be stressed that although p-Laplacian is, in general, non-linear operator, that is $L^{[p]}(\alpha\mathbf{y}) \neq \alpha L^{[p]}\mathbf{y}$, the functional $R^{[p]}$ is linear what simplifies the computation of eigenvectors.

Like in the "standard" case, the multiplicity of the first eigenvalue $\lambda_1^{[p]} = 0$ of the p-Laplacian $L^{[p]}$ is equal to the number k of connected components of the graph. Hence we say about p-Fiedler value which is positive only if the graph is connected. Other properties of the eigenvectors of p-Laplacian, as well as their usefulness in spectral clustering are discussed in [94, 337].

5.3.5.2 Sparsest Cuts

Apart of the cutting criteria (5.67) one can use two other balanced-cut criteria

$$RCC(Z, \bar{Z}) = \frac{\mathrm{cut}(Z, \bar{Z})}{\min(|Z|, |\bar{Z}|)} \tag{5.92a}$$

$$NCC(Z, \bar{Z}) = \frac{\mathrm{cut}(Z, \bar{Z})}{\min(\mathrm{vol}\, Z, \mathrm{vol}\, \bar{Z})} \tag{5.92b}$$

They are referred to, respectively, as the ratio Cheeger cut, or expansion of the cut $(Z, \bar{Z})$, and normalized Cheeger cut [115]. There holds the next relation between the ratio/normalized cut and the ratio/normalized Cheeger cut:

$$RCC(Z, \bar{Z}) \leq RCut(Z, \bar{Z}) \leq 2RCC(Z, \bar{Z})$$

$$NCC(Z, \bar{Z}) \leq NCut(Z, \bar{Z}) \leq 2NCC(Z, \bar{Z})$$

We already know that if $(Z_\tau, \bar{Z}_\tau)$ is the thresholded partition (described in Sect. 5.3.1.2) induced by the Fiedler vector ψ_2 then $R(\psi_2) = R^{[2]}(\psi_2) = RCut$ $(Z, \bar{Z})$, where $R^{[2]}(\psi_2)$ is the generalized Rayleigh quotient (5.90). Bühler and Hein proved in [94] that when the partition $(Z_\tau^{[p]}, \bar{Z}_\tau^{[p]})$ is induced by the p-Fiedler vector

$\psi_2^{[p]}$, see Sect. 5.3.5.1 for the definitions, then

$$\lim_{p \to 1} R^{[p]}(\psi_2^{[p]}) = RCC(Z_T^{[p]}, \bar{Z}_T^{[p]}) \tag{5.93}$$

i.e. in the limit as $p \to 1$ the cut found by thresholding the second eigenvector of the graph p-Laplacian converges to the optimal Cheeger cut. More precisely, if we denote $h_{RCC} = \min_{Z \in V} RCC(Z, \bar{Z})$, then

$$h_{RCC} \le RCC(Z_T^{[p]}, \bar{Z}_T^{[p]}) \le p \left(\max_{i \in V} d_i \right)^{\frac{p-1}{p}} h_{RCC}^{\frac{1}{p}}$$

i.e. by finding the solution of (5.91) for small p and thresholding it, we get a cut closely approximating h_{RCC}. Analogical property holds for the p-Fiedler vector of the normalized p-Laplacian.

Szlam and Bresson considered in [452] the case of $p = 1$. They noted that when $\mathbf{f}$ is the characteristic vector describing a partition into subsets Z and $\bar{Z}$, then

$$\|\mathbf{f} - m(\mathbf{f})\|_1 = \min \left(|Z|, |\bar{Z}| \right)$$

where $m(\mathbf{f})$ stands for the median of $\mathbf{f}$. Hence

$$\frac{\mathbf{f}^T L^{[1]} \mathbf{f}}{\|\mathbf{f} - m(\mathbf{f})\|_1} = 2 \frac{\text{cut}(Z, \bar{Z})}{\min(|Z|, |\bar{Z}|)} = 2RCC(Z, \bar{Z}) \tag{5.94}$$

and

$$\min_{\mathbf{f} \in \mathbb{R}^n} \frac{\mathbf{f}^T L^{[1]} \mathbf{f}}{\|\mathbf{f} - m(\mathbf{f})\|_1} \tag{5.95}$$

is a relaxation of the Cheeger cut problem, and what is more important, this relaxation is equivalent to the original problem, i.e. the minimum of (5.95) equals to h_{RCC}. This observation gave an inspiration for further studies of so-called 1-spectral clustering.[32]

Szlam and Bresson noted in [452] that the problem (5.95) is equivalent to the constrained minimization problem

$$\min_{\mathbf{v} \in \mathbb{R}^n} \frac{\|\mathbf{v}\|_{TV}}{\|\mathbf{v}\|_1}, \quad \text{s.t.} \ m(\mathbf{v}) = 0 \tag{5.96}$$

where $\|\mathbf{v}\|_{TV} = \sum_{i,j} s_{ij} |v(i) - v(j)|$ is the total variation norm. The authors show that their algorithm[33] performs very well, but they cannot prove that it converges.

---

[32] The "standard" spectral clustering can be termed 2-spectral clustering as it exploits 2-Laplacian, i.e. "standard" graph Laplacian. Consult the URL http://www.ml.uni-saarland.de/code/oneSpectralClustering/oneSpectralClustering.htm for more details, software and papers.

[33] Its implementation in MATLAB is provided by Szlam and can be found at http://www.sci.ccny.cuny.edu/~szlam/gtvc.tar.gz.

That is why Hein and Bühler proposed in [249] an alternative method inspired by the Inverse Power method used to determine the eigenvectors of a square matrix. This new algorithm is guaranteed to converge to a nonlinear eigenvector. Unfortunately, the algorithm performs well for bi-partitioning tasks only, and its recursive extensions yield rather strange results for multiclass clustering tasks.

An efficient generalization to the case of k-way partitioning—called Multiclass Total Variation (or MTV) clustering—is provided in [90] together with MATLAB implementation. First, these authors proposed new generalization of the Cheeger cut, namely

$$RCC(Z_1, \ldots, Z_k) = \sum_{i=1}^{k} \frac{\text{cut}(Z_i, \bar{Z}_i)}{\min \left(\alpha |Z_i|, |\bar{Z}_i| \right)} \tag{5.97}$$

where the parameter α controls the sizes of the sets Z_i in the partition. The authors of [90] propose to use $\alpha = k - 1$ to ensure that each of the groups approximately occupy the appropriate fraction $1/k$ of the total number of vertices. Second, they defined $m_\alpha(\mathbf{f})$ to be the α-median (quantile), i.e. the $(K+1)$-st maximal value in the range of $\mathbf{f}$, where $K = \lfloor (m/(\alpha + 1)) \rfloor$, and they introduced then norm

$$\|\mathbf{v}\|_{1,\alpha} = \sum_{i=1}^{n} |v(i)|_\alpha = \sum_{i=1}^{n} \begin{cases} \alpha v(i) & \text{if } v(i) \geq 0 \\ -v(i) & \text{otherwise} \end{cases}$$

Obviously, if $K = 2$, i.e. $\alpha = 1$, then $m_\alpha(v)$ is the standard median of the vector $\mathbf{v}$ and $\|\mathbf{v}\|_{1,\alpha} = \|\mathbf{v}\|_1$. Further, it is a simple exercise to verify, that if $\mathbf{f}_C$ is the characteristic vector of a subset Z of V, then

$$\frac{\|\mathbf{f}_Z\|_{TV}}{\|\mathbf{f} - m_\alpha(\mathbf{f}_Z)\|_{1,\alpha}} = \frac{2\text{cut}(Z, \bar{Z})}{\min \left(\alpha |Z|, |\bar{Z}| \right)}$$

Now, let $\mathcal{U}$ denotes the family of all partitions of the set V into k subsets, i.e. $\mathcal{U}$ contains all matrices $U = (\mathbf{u}_1, \ldots, \mathbf{u}_k) \in \{0, 1\}^{m \times k}$ such that $\mathbf{u}_1 + \cdots + \mathbf{u}_k = \mathbf{e}$. Then, the problem of finding the best partition with minimal value of the index (5.97) takes the form

$$\min_{(\mathbf{u}_1, \ldots, \mathbf{u}_k) \in \mathcal{U}} \sum_{i=1}^{n} \frac{\|\mathbf{u}_i\|_{TV}}{\|\mathbf{u}_i - m_\alpha(\mathbf{u}_i)\|_{1,\alpha}} \tag{5.98}$$

and its relaxation is formulated as follows

$$\min_{(\mathbf{f}_1, \ldots, \mathbf{f}_k) \in \mathcal{F}} \sum_{i=1}^{n} \frac{\|\mathbf{f}_i\|_{TV}}{\|\mathbf{f}_i - m_\alpha(\mathbf{f}_i)\|_{1,\alpha}} \tag{5.99}$$

where $\mathcal{F}$ is the family of all soft partitions of the set V into k subsets, i.e. $\mathcal{F}$ contains all matrices $F = (\mathbf{f}_1, \ldots, \mathbf{f}_k) \in [0, 1]^{m \times k}$ such that $\mathbf{f}_1 + \cdots + \mathbf{f}_k = \mathbf{e}$. The next two observations make the solution of (5.99) (relatively) easy accessible:

(i) The total variation terms give rise to quasi-indicator functions, i.e. the relaxed solutions $\mathbf{f}_1, \ldots, \mathbf{f}_k$ of (5.99) mostly take values near zero or one and exhibit a sharp, non-smooth transition between these two regions.

(ii) Both functions $\mathbf{f} \rightarrow \|\mathbf{f}\|_{TV}$ and $\mathbf{f} \rightarrow \|\mathbf{f} - m_\alpha(\mathbf{f})\|_{1,\alpha}$ are convex. The simplex constraint in (5.99) is also convex. Therefore solving (5.99) amounts to minimizing a sum of ratios of convex functions with convex constraints. This fact allows the authors to use machinery from convex analysis to develop an efficient algorithm for such problems.

More remarks on properties of graph-cut-based algorithms can be found in [466].

5.3.6 Out of Sample Spectral Clustering

The purpose of an out-of-sample extension is to enable clustering points that have not been part of the initial clustering process. Such an extension is helpful when not all data is available at the time of the initial clustering, or in applications where the data set is too large. There appears to be no simple way of treating out-of-sample data in spectral clustering based on the graph Laplacian. Alzate and Suykens show in [17] how to formulate an out-of-sample extension in the framework of weighted kernel PCA. In subsection below we briefly review the idea underlying kernel PCA, and in next subsection the clustering algorithm from [17] is described.

The basic idea behind this approach relies on the fact that for a population with well defined correlation matrix the "true" correlation matrix for a (very large, like in case of image data) may be approximated by a sufficiently big sample from the data set. If then the matrix X of size $m \times n$, in which rows represent centred data points describes the entire data set, and X' of size $m' \times n$, describes a random sample, then: (1) the similarity matrix XX^T and the covariance matrix $X^T X$ have the same set of dominating eigenvalues, and (2) covariance matrices $X^T X$ and $X'^T X'$ are identical up to proportionality factor (m'/m), so that their eigenvalues are proportional. We have already learnt that the Fiedler vector of a Laplacian is well suited for set bisection, and it is in turn related to the principal eigenvector of properly defined similarity matrix. So identifying the principal eigenvector of X' similarity matrix will bi-cluster the subsample X'. But given an SVD decomposition of X', we can easily (by matrix/vector multiplication) derive the principal eigenvector of X (without computing XX^T and its spectral analysis), and hence we get a straight forward bi-section of X. Partition into more clusters is a bit more complex. Note that SVD on centred data is in fact PCA. What the authors of [17] have essentially done is to (1) transfer this procedure from the actual data space into the feature space (in which the similarity matrix is more likely to be a correlation matrix) using the kernel method, to (2) exploit information beyond the principal eigenvector, but in the same simple form and to (3) apply the kernel trick so that we do not need to know the feature space and can work with kernel matrix without the need to perform first an embedding of the graph into an Euclidean space. Exactly for this reason we need to explain kernel PCA first.

5.3.6.1 Kernel PCA

Kernel principal component analysis, K-PCA for short, is an extension of classical (linear), principal component analysis (PCA for short). Recall, see e.g. [277], that the PCA was thought as a linear transformation that projects the data (assumed to be the random vectors sampled i.i.d. from a multivariate distribution $\mathcal{P}$) along the directions where there is the most variance. More precisely, if $\Sigma = (1/m) \sum_i \bar{\mathbf{x}}_i \bar{\mathbf{x}}_i^\mathsf{T} = (1/m)\bar{X}\bar{X}^\mathsf{T}$ is the covariance matrix of the centered data set, $\bar{X} = (\mathbf{x}_1 - \boldsymbol{\mu}, \ldots, \mathbf{x}_m - \boldsymbol{\mu})^\mathsf{T}$, where $\boldsymbol{\mu} = (1/m) \sum_i \mathbf{x}_i$, and $\mathbf{u}_1, \ldots, \mathbf{u}_m$ are the eigenvectors associated with the eigenvalues $\lambda_1 \geq \cdots \geq \lambda_m$ of the covariance matrix, then the projection $\pi(\mathbf{x})$ on these directions is

$$\pi_j(\mathbf{x}) = \mathbf{u}_j^\mathsf{T} \mathbf{x}_i, \ j = 1, \ldots, k \leq n$$

If the observations from $\bar{X}$ cannot be linearly separated in $k \leq n$ dimensions, we hope that it will be possible in $N \geq n$ dimensions, where $N \leq \infty$. That is, each data point $\mathbf{x}_i$ is transformed to a point $F(\mathbf{x}_i)$ in an N-dimensional feature space. Since N is very large, it is not practical to use the covariance matrix of size $N \times N$. Instead we operate with a kernel $\mathsf{K} \colon \mathbb{R}^n \times \mathbb{R}^n \to \mathbb{R}$ and the similarity between a pair of objects is defined as

$$s_{ij} = \mathsf{K}(\mathbf{x}_i, \mathbf{x}_j) = [F(\mathbf{x}_i)]^\mathsf{T} F(\mathbf{x}_i)$$

Under such a setting the steps of linear PCA in the feature space are as described by Schölkopf et al. [419]:

(a) Choose a kernel mapping K and compute the similarity matrix, S.
(b) Center the similarities

$$\begin{aligned}
\bar{s}_{ij} &= \left(F(\mathbf{x}_i) - \frac{1}{m}\sum_{\ell=1}^m F(\mathbf{x}_\ell)\right)^\mathsf{T}\left(F(\mathbf{x}_j) - \frac{1}{m}\sum_{\ell=1}^m F(\mathbf{x}_\ell)\right) \\
&= s_{ij} - \frac{1}{m}\sum_{\ell=1}^m s_{i\ell} - \frac{1}{m}\sum_{\ell=1}^m s_{\ell j} + \frac{1}{m^2}\sum_{\ell=1}^m\sum_{\ell'=1}^m s_{\ell\ell'}
\end{aligned}$$
(5.100)

Introducing the centering matrix

$$M = \mathbb{I} - \frac{1}{m}\mathbf{e}\mathbf{e}^\mathsf{T}$$

the Eq. (5.100) may be rewritten in matrix form as $\bar{S} = MSM$.

(c) Solve the eigenvalue problem

$$\bar{S}\mathbf{u}_i = \lambda_i m \mathbf{u}_i, \quad i = 1, \ldots, k \tag{5.101}$$

and stack the eigenvectors (ordered according to the eigenvalues sorted in descending order) into the columns of matrix $U_k = (\mathbf{u}_1, \ldots, \mathbf{u}_k)$.

(d) For each data point $\mathbf{x}$ compute its principal components in the feature space

$$\pi_j(\mathbf{x}) = \sum_{i=1}^{m} u_{ij} K(\mathbf{x}_i, \mathbf{x}), \quad j = 1, \ldots, k \tag{5.102}$$

Given a set of m_{new} observations $(\mathbf{x}_1^{new}, \ldots, \mathbf{x}_m^{new})$ sampled i.i.d. from the same distribution $\mathcal{P}$), the projection of these observations over the the eigenvectors $U_k = (\mathbf{u}_1, \ldots, \mathbf{u}_k)$ is

$$Z = (\mathbf{z}_1, \ldots, \mathbf{z}_k) = S^{new} U_k \tag{5.103}$$

where S^{new} is the $m_{new} \times m$ matrix with the elements $s_{ij}^{new} = K(\mathbf{x}_i^{new}, \mathbf{x}_j)$, $i = 1, \ldots, m_{new}, j = 1, \ldots, m$. The i-th column of matrix Z (called also score variable) represents projection of these new observations over i-th eigenvector.

Suykens et al. described in [451] a primal-dual support vector machine formulation to the problems of PCA and K-PCA. Below we present, after [17], the primal-dual formulation of the weighted kernel PCA. Let $W \in \mathbb{R}^{m \times m}$ be a symmetric positive definite weighting matrix.[34] Let $F = (F(\mathbf{x}_1), \ldots, F(\mathbf{x}_m))^T$ be the $m \times N$ feature matrix, and $\rho = (\rho_1, \ldots, \rho_m)^T$, where $\rho_i = \mathbf{u}^T F(\mathbf{x}_i)$, is the vector of projected variables. Then the weighted K-PCA can be represented by the following primal problem

$$\min_{\mathbf{u}, \rho} J(\mathbf{u}, \rho) = \frac{\gamma}{2m} \rho^T W \rho - \frac{1}{2} \mathbf{u}^T \mathbf{u}, \quad \text{s.t.} \quad \rho = F\mathbf{u} \tag{5.104}$$

where $\gamma \in \mathbb{R}_+$ is a regularization constant. The Lagrangian of (5.104) is

$$\mathcal{L}(\mathbf{u}, \rho, \alpha) = \frac{\gamma}{2} \rho^T W \rho - \frac{1}{2} \mathbf{u}^T \mathbf{u} - \alpha^T (\rho - F\mathbf{u})$$

and the conditions for optimality are

$$\frac{\partial \mathcal{L}}{\partial \mathbf{u}} = 0 \Rightarrow \mathbf{u} = F^T \alpha$$

$$\frac{\partial \mathcal{L}}{\partial \rho} = 0 \Rightarrow \alpha = (\gamma/m) W \rho$$

$$\frac{\partial \mathcal{L}}{\partial \alpha} = 0 \Rightarrow \rho = F\mathbf{u}$$

---

[34]Usually it is a diagonal matrix. If $W = \mathbb{I}$ the problem becomes unweighted kernel PCA.

Eliminating $\mathbf{u}$ and ρ we obtain the next eigenvalue problem

$$WS\alpha = \lambda\alpha \tag{5.105}$$

where α is an eigenvector of (possibly non symmetric but positive definite) matrix WS, $\lambda = m/\gamma$ is the corresponding eigenvalue, and S is the similarity matrix with the entries $s_{ij} = K(\mathbf{x}_i, \mathbf{x}_j)$.

Setting $W = D^{-1}$, where D is the degree matrix derived from S, we find that (5.105) reduces to the random walks Algorithm 5.5. In this case setting $\phi_\ell = D^{1/2}\alpha_\ell$, $\ell = 1, \ldots, n_k$, we obtain the NJW Algorithm 5.4. This establishes an interesting connection between weighted kernel PCA and spectral clustering.

5.3.6.2 Kernel Spectral Clustering

Alzate and Suykens described in [17] an extension of the weighted K-PCA which relies upon introducing the bias terms in the constraints of (5.104). The corresponding objective takes now the form

$$\min J(\mathbf{u}_\ell, \rho_\ell, b_\ell) = \frac{1}{2m}\sum_{\ell=1}^{n_k}\gamma_\ell\rho_\ell^{\mathsf{T}} W\rho_\ell - \frac{1}{2}\sum_{\ell=1}^{n_k}\mathbf{u}_\ell^{\mathsf{T}}\mathbf{u}_\ell \tag{5.106}$$
$$\text{s.t. } \rho_\ell = \mathbf{F}\mathbf{u}_\ell + b_\ell\mathbf{e}, \quad \ell = 1, \ldots, n_k$$

where $n_k \geq k-1$. The Lagrangian associated with this problem takes the form

$$\mathcal{L}(\mathbf{u}_\ell, \rho_\ell, b_\ell) = \frac{1}{2m}\sum_{\ell=1}^{n_k}\gamma_\ell\rho_\ell^{\mathsf{T}} W\rho_\ell - \frac{1}{2}\sum_{\ell=1}^{n_k}\mathbf{u}_\ell^{\mathsf{T}}\mathbf{u}_\ell$$
$$- \sum_{\ell=1}^{n_k}\alpha_\ell^{\mathsf{T}}(\rho_\ell - \mathbf{F}\mathbf{u}_\ell - b_\ell\mathbf{e})$$

where α_ℓ are the Lagrange multipliers. The optimality conditions are:

$$\frac{\partial \mathcal{L}}{\partial \mathbf{u}_\ell} = 0 \Rightarrow \mathbf{u}_\ell = \mathbf{F}^{\mathsf{T}}\alpha_\ell \tag{5.107a}$$

$$\frac{\partial \mathcal{L}}{\partial \rho_\ell} = 0 \Rightarrow \alpha_\ell = \frac{\gamma}{m}D^{-1}\rho_\ell \tag{5.107b}$$

$$\frac{\partial \mathcal{L}}{\partial b_\ell} = 0 \Rightarrow \mathbf{e}^{\mathsf{T}}\alpha_\ell \tag{5.107c}$$

$$\frac{\partial \mathcal{L}}{\partial \alpha_\ell} = 0 \Rightarrow \rho_\ell - \mathbf{F}\mathbf{u}_\ell - b_\ell\mathbf{e} = 0 \tag{5.107d}$$

Eliminating the primal variables $\mathbf{w}_\ell$, $\mathbf{e}_\ell$ and b_ℓ we obtain the dual problem represented by the following eigenvalue problem

$$WM_cS\alpha_\ell = \lambda_\ell\alpha_\ell, \quad \ell = 1, \ldots, n_k \tag{5.108}$$

where $\lambda_\ell = m/\gamma_\ell$, $S = \mathbf{FF}^\mathrm{T}$ is the similarity matrix with the elements $s_{ij} = \mathsf{K}(\mathbf{x}_i, \mathbf{x}_j)$, and

$$M_c = \mathbb{I} - \frac{1}{\sum_{i,j} w_{ij}}\mathbf{ee}^\mathrm{T}W$$

is the centering matrix. A reader should note, that if W is the identity matrix then M_c is identical with the centering matrix M introduced in previous subsection, and the procedure reduces to the kernel PCA. The *row* vector of bias terms become

$$\mathbf{b} = -\frac{1}{\sum_{i,j} w_{ij}}\mathbf{e}^\mathrm{T}WSA_{n_k} \tag{5.109}$$

where $A_{n_k} = (\alpha_1, \ldots, \alpha_{n_k})$.

Remark 5.3.3 As noted by Alzate and Suykens in [17], the effect of centering caused by the matrix M_c is that the eigenvector α_1 corresponding to the largest eigenvalue λ_1 in (5.108) already contains information about the clustering, i.e. $\mathbf{e}\alpha_1 = 0$—see Eq. (5.107c). Hence we can assume that $n_k = k - 1$. Furtermore, this feature is in contrast with the random walk Algorithm 5.5 where the dominating eigenvector is the constant vector. □

Combining the conditions (5.107a) and (5.107d) we state that the projection of data along ℓ-th eigenvector is $\rho_\ell = S\alpha_\ell + b_\ell\mathbf{e}$. Thus the n_k-dimensional projection of these data takes the form $R = (\rho_1, \ldots, \rho_\ell) = SA_{n_k} + \mathbf{eb}$. Given a set of m_{new} test points, $\mathbf{y}_1, \ldots, \mathbf{y}_{m_{new}}$, the projections over the eigenvectors being the columns of the matrix A_{n_k} are computed as

$$Z = S^{new}A_{n_k} + \mathbf{eb} \tag{5.110}$$

where S^{new} is the matrix of size $m_{new} \times m$ with the entries $s_{ij}^{new} = \mathsf{K}(\mathbf{y}_i, \mathbf{x}_j)$, $i = 1, \ldots, m_{new}$, $j = 1, \ldots, m$. The element $z_{i\ell}$ is said to be the score of i-th observation on the ℓ-th principal component.

To make the final assignment of new observations to already defied k groups the Error-Correcting Output Codes (ECOC) [154] approach was employed in [17]. In brief, an n_k bit error-correcting output code for a k-groups problem is represented by the codebook $\mathfrak{C}$ consisting of k codewords $\mathbf{c}_1, \ldots, \mathbf{c}_k$. Each codeword $\mathbf{c}_i$ is a unique binary string of length n_k. In the ideal case to distinguish among k groups it suffices to use codewords of length $n_k = k-1$. Using additional "error-correcting" bits, we can tolerate some error introduced by finite training sample, poor choice of input features, and flaws in the training algorithm. Such constructed classifier is more robust against the errors. With the codebook $\mathfrak{C}$ each new observation $\mathbf{y}_i$ described by the codeword $\mathbf{c}(\mathbf{y}_i)$ is assigned to the group $\ell(\mathbf{y}_i)$ such that

$$\ell(\mathbf{y}_i) = \underset{1 \le \ell \le k}{\arg\min} \, d_H(\mathfrak{c}(\mathbf{y}_i), \mathfrak{c}_\ell) \tag{5.111}$$

where d_H is the Hamming distance.

Alzate and Suykens suggested in [17] the next method of constructing the codebook. First, from the optimality condition (5.107b) it follows that

$$\text{sign}(\rho_l) = \text{sign}(\alpha_\ell), \ \ell = 1, \dots, n_k$$

Thus each row of the matrix $\text{sign}(A_{n_k})$ is the codeword of an observation from the training set X. As already mentioned these codewords are strings of length n_k over the alphabet $\{-1, 1\}$. The next step is to identify the unique codewords $\mathfrak{c}_1, \dots, \mathfrak{c}_{n'_k}$ together with the number of their occurrences in the set $\text{sign}(A_{n_k})$. Finally k codewords with highest occurrences are placed in the codebook. Each observation $\mathbf{x}_j$ from the training set X is assigned to the group $\ell(\mathbf{x}_j)$ determined according to the Eq. (5.111) with $\mathfrak{c}(\mathbf{x}_j) = \text{sign}(\alpha^j)$ where α^j is the j-th row of the matrix A_{n_k}. Similarly each observation $\mathbf{y}_i$ from the test set is assigned to the group $\ell(\mathbf{y}_i)$ determined according to the Eq. (5.111) with $\mathfrak{c}(\mathbf{y}_i) = \text{sign}(\mathbf{z}^j)$ where $\mathbf{z}^j$ is the j-th row of the scores matrix Z. This guarantees consistent group assignment of the observations from the training and testing set.

Algorithm 5.6 k-way kernel spectral algorithm [17]

Input: Training data $X = (\mathbf{x}_1, \dots, \mathbf{x}_m)^T$, testing data $Y = (\mathbf{y}_1, \mathbf{y}_{m_{test}})^T$, kernel K, number od groups k, $n_k \ge k-1$.

Output: Partition $\mathcal{C} = \{C_1, \dots, C_k\}$, codebook $\mathfrak{C} = \{\mathfrak{c}_1, \dots, \mathfrak{c}_k\}$, $\mathfrak{c}_k \in \{-1, 1\}^{n_k}$.

1: Compute the eigenvectors $\alpha_1, \dots, \alpha_{n_k}$ corresponding to the largest n_k eigenvalues of the matrix (WM_cS) where M_c is the centering matrix, and S is the training matrix with entries $s_{i,j} = \mathsf{K}(\mathbf{x}_i, \mathbf{x}_j)$.

2: Stack the eigenvectors α_i into columns of matrix A_{n_k} of size $m \times n_k$ and binarize its rows α^i, i.e. $\mathfrak{c}_i = \text{sign}(\alpha^i) \in \{-1, 1\}^{n_k}$ is the encoding vector for the training data point $\mathbf{x}_i$, $i = 1, \dots, m$.

3: Count the occurrences of the different encodings and choose the k encodings with largest occurrences to form the codebook $\mathfrak{C} = \{\mathfrak{c}_1, \dots, \mathfrak{c}_k\}$.

4: Assign each training data point $\mathbf{x}_i$ to the group $\ell(\mathbf{x}_i)$ using the Eq. (5.111).

5: Binarize the test data projections, i.e. compute $\mathfrak{c}(\mathbf{y}_j) = \text{sign}(\mathbf{z}^j)$ using the Eq. (5.111); here $\mathbf{z}^j$ is j-th row of the matrix Z defined in (5.110).

6: Assign each testing data point $\mathbf{y}_j$ to the group $\ell(\mathbf{y}_j)$ using the Eq. (5.111).

5.3.7 Incremental Spectral Clustering

Clustering algorithms discussed in previous section work in a "batch" mode, i.e. all data items are processed together at the same time. However, in many practical applications the links between nodes change in time, and/or the weights of such links are modified. Examples of such problems are, for instance, topic tracking, traffic jam predictions, clustering an incoming document streams, etc. In brief, incremental

Fig. 5.12 Eigenvalues of the
60×60 grid graph

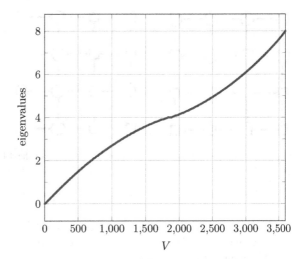

clustering is concerned with evolving data sets. More about this topic can be found
in [316, 376] and bibliography included in these papers.

5.3.8 Nodal Sets and Nodal Domains

Spectral clustering algorithms discussed in this chapter are not universal. In the
case of sparse graphs with uniform connectivity, there is no obvious optimal graph
partitioning solution, because of the lack of a meaningful eigengap—see e.g. Fig. 5.12
where 3600 eigenvalues of the 60×60 grid graph are depicted. This causes that the
task of choosing (in a fully unsupervised way) appropriate partition of the data set
becomes non-trivial.

Also, as noted in [426], the eigenvalues of any large semi-definite positive symmetric matrix are estimated only approximately; hence it is not easy to study the
eigenvalues' multiplicities (which play a crucial role in the analysis of the Laplacian
eigenvectors [76]) and that ordering the eigenvectors, needed to construct reliable
spectral embedding, based on these estimated eigenvalues is not reliable.

Fortunately, there exists a theoretical framework that extends the Fiedler's theorem to the other eigenvectors of a graph, namely, the discrete nodal domain theorem
formulated in [139]. It is a discrete version of the Courant's nodal domain theorem, which originally applies to the eigenfunctions of the Laplace-Beltrami operator
on continuous manifolds. To formulate this theorem we need few definitions. Let
$\Gamma = (V, E)$ be an undirected graph and ψ is an eigenvector corresponding to the
eigenvalue λ of its Laplacian. Then:

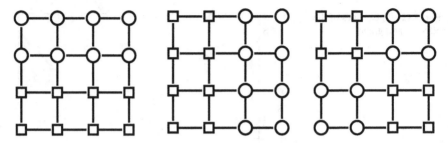

Fig. 5.13 Sign pattern of the eigenvectors ψ_2, ψ_3, ψ_4 of the 4-by-4 grid graph. Circle (square) denotes positive (negative) sign of the eigenvector at particular node. Note that here $\lambda_2 = \lambda_3 = 0.58579$

(1) The nodal set of the eigenvector ψ, $\mathfrak{N}(\psi)$, is the set of graph vertices for which ψ vanishes, i.e. $\mathfrak{N}(\psi) = \{v \in V : \psi(v) = 0\}$. Here $\psi(v)$ denotes the value of the eigenvector ψ at node v.

(2) A subset $S \subset V$ is a weak nodal domain of Γ induced by ψ if the subgraph $\Gamma(S)$ induced on Γ by S is a maximal connected component of either $\{v : \psi(v) > 0\}$ or $\{v : \psi(v) < 0\}$.

(3) A subset $S \subset V$ is a strong nodal domain of Γ induced by ψ if the subgraph $\Gamma(S)$ induced on Γ by S is a maximal connected component of either $\{v : \psi(v) \geq 0\}$ or $\{v : \psi(v) \leq 0\}$ and contains at least one node v such that $\psi(v) \neq 0$.

From these definitions it follows that the only possible nodal domain for λ_1 is Γ itself. Furthermore, as any eigenvector ψ_j, $j \geq 2$ has positive and negative entries, it has at least two components of different signs, i.e. any such eigenvector has at least two nodal domains. To illustrate these notions, consider the 4-by-4 grid graph. In this case $\mathfrak{N}(\psi_i) = \emptyset$ for all $j = 1, \ldots, 16$. Denoting by circle (square) the nodes with positive (negative) value of the eigenvector we obtain so-called sign pattern of this eigenvector. Figure 5.13 depicts the sign patterns of the eigenvectors corresponding to three eigenvalues λ_2, λ_3 and λ_4.

The number of weak and strong nodal domains is characterized by the next theorem

Theorem 5.3.1 (Discrete Nodal Domain Theorem [139]) *Let L be the Laplacian matrix[35] of a connected graph with m vertices. Then any eigenvector ψ_j corresponding to the j-th eigenvalue λ_j with multiplicity r has at most i weak nodal domains and $i + r - 1$ strong nodal domains.*   □

Nodal domains provide families of graph partitions, where there is one partitioning for each eigenvector of the Laplacian matrix [76]. An application of this fact to find a partitioning graph Γ is discussed e.g. in [247, 426].

---

[35] In fact instead of L we can consider any generalized Laplacian matrix [76, 139].

5.4 Local Methods

Spectral clustering allows extraction of clusters in datasets as well as cutting graphs into appropriate number of subgraphs. In this last case, apart from high computational complexity, the algorithms in question have at least two disadvantages. First, the spectral rounding not always results in the best cut—see [482] for an example. Second, in the case of large empirical graphs, unbalanced clusters are frequently returned [315].

For this reason, especially in the case of large graphs, a different approach has been proposed: starting with a small but consistent set of nodes, called *seed*, we search for the "natural neighbours" forming a "community" (or a collection of such communities) [25], where by the community[36] we understand a set of nodes $C \subset V$ of low conductance. All the nodes forming such a community communicate with the seed (or its subset) also included in this community. Low conductance means that only a small fraction of edges with one end in the community have the second end outside this community. In other words, the nodes belonging to the community C communicate with each other more often than with nodes outside C. Interestingly, the low value of the conductance $c(C)$ does not imply high intensity of connections between the nodes from C. Hence, the local methods can be used, first of all in graph clustering, but we can also try to apply such methods in the search for a set of objects sufficiently similar to a given item from the dataset. It must be stressed that we hope to identify the set C without examining too many vertices of the graph.

The first formal solution to the problem defined above was given by Spielnan and Teng in [439]. Their algorithm, called `Nibble`, allows to identify a set C together with the seed node $v \in V$ in time proportional to the size of this set.

The algorithms allowing to extract local communities are important for at least two reasons. First, they appear to be very effective, as they search only the "interesting" part of the graph (and the remaining subgraph, spanned over the nodes from the set $\overline{C}$ is ignored). Second, these algorithms allow for identifying small subsets existing inside very large graphs. Andersen et al. in a series of papers [22–25] proposed an interesting modification of `Nibble`, called `PageRank-Nibble`.

One of characteristic features of empirical graphs is small average distance between any two nodes,[37] which means a rapid increase (with distance) in the number of neighbours of the nodes forming a seed. This does not preclude the existence of communities that communicate with members of other communities only through a small number of connections (and therefore the cost of cutting such a community is small). It is usually assumed that the nodes forming a seed represent a unique community. Lang and Andersen proved in [25], that if there exists in a graph Γ, for a given seed, a well defined community, then their algorithm allows for extracting it. Let us repeat once again: the advantage of the local approach is that the time complexity of executing these procedures is proportional to the size of the generated

---

[36]This notion is discussed in next chapter.

[37]See e.g. Watts, D.J.; Strogatz, S.H. Collective dynamics of 'small-world' networks. *Nature* 393(6684), 1998, 409–410. https://doi.org/10.1038/30918.

clusters, and not—as in the case in conventional methods—to the size of the input graph, [23].

Below we present only a brief outline of the local clustering method; an interested reader is referred to the literature cited above. An application of this approach to local cluster discovery in protein networks is presented in [339]. It is worth noting that the approach outlined here is not the only possible way to determine the natural neighbourhood of some nodes of an empirical network. Methods developed in the context of social networks (discussed in next chapter) provide alternative solutions.

Remark 5.4.1 The following considerations apply to unweighted graphs. Thus we will use in this section the adjacency matrix A instead of the similarity matrix S. The off-diagonal elements of A take the values of 0 or 1. □

5.4.1 The **Nibble** Algorithm

The essence of the algorithm is to propose an order in which a walker will visit the nodes of an undirected graph $\Gamma = (V, E)$. At a first glance, it would seem that we should consider the nodes closest to the starting node $u^* \in V$. But most of empirical graphs, like social networks, or protein-protein networks, have small diameter, and such a solution becomes ineffective. Spielman and Teng proposed in [440] to order the nodes according to the probability of visiting the nodes during a short random walk starting from the node u^*.

Let A be the adjacency matrix of the graph Γ. If the graph contains self-loops, then $a_{ii} = z_i$ if there are $z_i \geq 1$ such loops in node i. Let D be the degree matrix and let

$$\widehat{P} = (AD^{-1} + \mathbb{I})/2 \tag{5.112}$$

be the *column* stochastic matrix, representing *lazy* random walk. In other words, if there are no self-loops and a random walker is in node u, then in the next time steps she/he will stay, with the probability $1/2$ in this node or, with the probability $1/(2 \cdot |N(u)|)$, she/he will move to a neighbour node from the set $N(u)$. Such a construction of the matrix $\widehat{P}$ guarantees that the random walk has the stationary distribution π (consult Sect. 5.2.5). Moreover, $\widehat{P}$ is diagonally dominant, i.e. $\widehat{p}_{ii} \geq \sum_{j \neq i} |\widehat{p}_{ij}|$ for each row i.

The **Nibble** algorithm refers to the property mentioned in Lemma 5.3.1, which states that if Z is a set with small conductance, then the probability of leaving it by the random walker is also small. Careful analysis, presented in [440], leads to a simple algorithm, which can be summarized as follows:

(a) Suppose that a random walk starts from a node $u^* \in V$. This implies the following starting probability distribution, defined on the set of states (nodes of the graph)

$$\mathbf{p}^{(0)}(v) = \chi_{u^*}(v) = \begin{cases} 1 \text{ if } v = u^* \\ 0 \text{ otherwise} \end{cases} \qquad (5.113)$$

(b) If, after the k-th step of random walk, the probability distribution over the set of states is $\mathbf{p}^{(k)}$, then in the next step it has the form $\mathbf{p}^{(k+1)} = \widehat{P}\mathbf{p}^{(k)}$.

(c) To improve the convergence of the algorithm, Spielman and Teng introduced censored walk, in which the original probability vector $\mathbf{p}^{(k+1)}$ is approximated by

$$[p^{(k+1)}]_\epsilon(v) = \begin{cases} p^{(k+1)}(v) \text{ if } p^{(k+1)}(v) \geq \epsilon d_v \\ 0 \qquad\qquad \text{otherwise} \end{cases} \qquad (5.114)$$

where ϵ is a small number.

From the formal standpoint, $[\mathbf{p}^{(k+1)}]_\epsilon$ is no longer a stochastic vector, but it was proved in [440], that such an approximation provides a correct cut, as well. What is important, the approximation allows significant reduction of the time of the computations.

(d) For a given vector $\mathbf{p}$ we define the ordering function $q : V \rightarrow \mathbb{R}$ of the form

$$q(v) = \frac{p(v)}{d_v} \qquad (5.115)$$

Let o stands for the ordering of the nodes $\{1, \ldots, m\}$ such that

$$q(v_{o(1)}) \geq q(v_{o(2)}) \geq \cdots \geq q(v_{o(m)})$$

and let $S_j^{\mathbf{p}}$ denotes the set of nodes of the form

$$S_j^{\mathbf{p}} = \{v_{o(1)}, \ldots, v_{o(j)}\} \qquad (5.116)$$

Algorithm 5.7 `Nibble`—algorithm for local identification of clusters [440]

1: *Input data*: Undirected graph $\Gamma = (V, E)$, starting node u^*, maximum value of conductance, ϕ, integer b.
2: declare maximal number of steps, k_{max} and compute the value ϵ.
3: set $r_0 = \chi_{u^*}$
4: **for** $k = 1, \ldots, k_{max}$ **do**
5:   $p^{(k)} = \widehat{P} r^{(k-1)}$
6:   $r^{(k)} = [p^{(k)}]_\epsilon$
7:   **if** $(\exists j \in V) : \left(\Phi(S_j^{p^{(k)}}) \leq \phi \right) \wedge \left(2^b \leq \text{vol}(S_j^{p^{(k)}}) \leq \frac{5}{6}\text{vol}(V) \right)$ **then**
8:     return the set $C = S_j^{p^{(k)}}$ and STOP
9:   **end if**
10: **end for**
11: return $C = \emptyset$

The essence of the algorithm represented by the pseudocode 5.7, is to determine the sequence of vectors

$$\mathbf{p}^{(k)} = \begin{cases} \chi_{u^*} & \text{if } k = 0 \\ \hat{P}\mathbf{r}^{(k-1)} & \text{otherwise} \end{cases} \tag{5.117}$$
$$\mathbf{r}^{(k)} = [\mathbf{p}^{(k)}]_\epsilon$$

while for every $k \in \{1, \ldots, k_{max}\}$ we check whether among the collection of the sets $S_j^{\mathbf{p}^{(k)}}$, there is one for which the conductance $\mathfrak{c}(S_j^{\mathbf{p}^{(k)}})$ does not exceed the value ϕ, and its volume is not too small and not too big.[38]

If C is a nonempty set $C \subset V$ returned by the Nibble, then:

(1) $\mathfrak{c}(C) \leq \phi$, hence C is a set with small conductance, and
(2) $\text{vol } C \leq \frac{5}{6}\text{vol}(V)$, i.e. C is not "too big", i.e. its volume is only a fraction of the total volume of the graph. Moreover, $\text{vol } C \geq 2^b$, i.e. C is not too small (nor too isolated) as the nodes in this set have not fewer than 2^b neighbours.

A reader interested in implementational details is referred to the original paper [440]. We should mention that, in agreement with its name, the algorithm can be used for clustering the nodes of the graph: after we obtain a nonempty set $C \subset V$, we delete it from V, together with the corresponding edges and repeat the whole procedure on the resulting diminished graph.

Although simple, the algorithm requires specification of many parameters. Moreover, the starting node u^* may lie outside the set C. The modification described in next section has no such disadvantages.

5.4.2 The `PageRank-Nibble` Algorithm

The `PageRank-Nibble` algorithm uses the so-called personalized PageRank vector instead of the stochastic vector $\mathbf{p}$ from the previous section. We will denote PageRank as $\mathfrak{p}(\mathbf{s}, \alpha)$, where $\alpha \in (0, 1)$, called dumping factor, is a parameter, and $\mathbf{s}$ is the vector of the form (5.113), or, more generally,

$$\mathbf{s}(v) = \begin{cases} 1/|S| & \text{if } v \in S \\ 0 & \text{otherwise} \end{cases} \tag{5.118}$$

where S is a subset of nodes. More information on PageRank is provided in Appendix C.

Instead of making k_{max} steps during a random walk, and verifying the stopping conditions at each step, in this algorithm the vector $\mathfrak{p}(\mathbf{s}, \alpha)$, is computed first, and next a family of sets $S_j^{\mathfrak{p}(\mathbf{s}, \alpha)}$, $j = 1, 2, \ldots$, defined in (5.116), is searched in order

---

[38]Spielman i Teng add one more condition, but, to save simplicity of presentation, we omit it.

to identify the set with minimal conductance. This set represents a natural cluster, containing the nodes specified in the vector **s**.

PageRank vector is nothing but a stationary distribution of a Markov chain characterized by the transition matrix

$$\check{P} = (1 - \alpha)\widehat{P} + \alpha\mathbf{s}\mathbf{e}^{\mathrm{T}} \tag{5.119}$$

where $\widehat{P}$ is the column-stochastic matrix defined in (5.112), **s** is the initial distribution of states, and $\alpha \in (0, 1)$ is the dumping factor: $\check{P} \to \widehat{P}$ if $\alpha \to 0$. The matrix $\check{P}$ describes a random walk, in which the walker decides, with probability $(1 - \alpha)$, on standard random walk, and with probability α she/he returns to a node specified in the set S (and thereafter continues the random walk). Thus, α is a probability of discouragement or irritation: the irritated walker starts a new random walk from a node randomly chosen from the set S.

In the standard definition [317], the vector **s** is the uniform probability distribution over the set of nodes V. Otherwise, we talk about personalized PageRank vector, since the distribution **s** favours certain nodes with respect to the others. In the sequel, we will consider personalized distributions (5.118); in particular, S may contain only one node. The vector **p** represents the stationary distribution. Hence, if $\check{P}$ has the form (5.119), then it follows from the equation $\mathbf{p} = \check{P}\mathbf{p}$ that this vector satisfies the condition (see Remark C.1.1)

$$\mathfrak{p}(\mathbf{s}, \alpha) = \alpha\mathbf{s} + (1 - \alpha)\check{P}\mathfrak{p}(\mathbf{s}, \alpha) \tag{5.120}$$

whence

$$\mathfrak{p}(\mathbf{s}, \alpha) = \alpha\left(\mathbb{I} - (1 - \alpha)\check{P}\right)^{-1}\mathbf{s} \tag{5.121}$$

Using the approximation $(\mathbb{I} - X)^{-1} = \sum_{j=0}^{\infty} X^{j}$, the above equation can be rewritten in the form[39]

$$\mathfrak{p}(\mathbf{s}, \alpha) = \alpha\mathbf{s} + \alpha\sum_{j=0}^{\infty}(1 - \alpha)^{j}\check{P}^{j}\mathbf{s} \tag{5.122}$$

Other methods of computing personalized PageRank $\mathfrak{p}(\mathbf{s}, \alpha)$ are mentioned in Appendix C.2.

Having the vector $\mathfrak{p}(\mathbf{s}, \alpha)$ we determine—like in the `Nibble` algorithm—the sets $S_j^{\mathfrak{p}(\mathbf{s},\alpha)}$, and we search for the set of minimal conductance. If the graph contains a well defined structure, the algorithm will find it—see [22, Sect. 6], where the full version of the algorithm, together with its properties, is described.

---

[39]Note that $\check{P}^j$ can be computed iteratively as $\check{P}\check{P}^{j-1}$, $j = 2, 3, \ldots$. Since $\check{P}$ is a sparse matrix, such a multiplication can be organized efficiently even in case of large graphs.

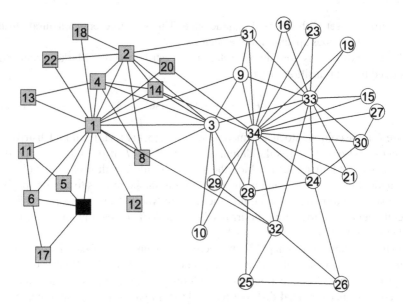

Fig. 5.14 Partition of the set of nodes of karate network into two groups. Black square denotes the starting node used to compute the personalized PageRank. Grey squares denote natural neighbors of the black square

Example 5.4.1 Consider the graph karate, representing social relationships among members of a karate club [525]—see Fig. 5.14. Black square shows the starting node #7. Grey squares denote members of the set S_{15}^p with minimal conductance. Left panel of Fig. 5.15 shows the values of the vector **q** (corresponding to the ordering function q from the previous section) for the original numbering of the nodes, and right panel—the values of conductance $\Phi(S_j)$. Note that the support of the vector **q** is whole set of nodes V. Note also the first local minimum on the right panel; it represents the set $\{5, 6, 7, 11, 17\}$, constituting a well defined micro-community. □

The local methods presented in this section may be of use in many situations. Some applications of this idea are mentioned below:

- *Data provenance.*[40] Systems for tracking the origin and transformation of historical data, which create and process huge graphs. There is no need, however, to examine the entire graph; it is important to select a subset of relevant nodes only.[41]

---

[40] See e.g. P. Buneman, S. Khanna, W.-C. Tan: Data provenance: Some basic issues. *Proc. of 20th Conf. Foundations of Software Technology and Theoretical Computer Science*, FST TCS 2000, New Delhi, India. Springer 2000, pp. 87–93.

[41] See P. Macko, D. Margo, M. Seltzer: Local clustering in provenance graphs. *Proc. of the 22nd ACM Int. Conf. on Information & Knowledge Management*. ACM, 2013, pp. 835–840.

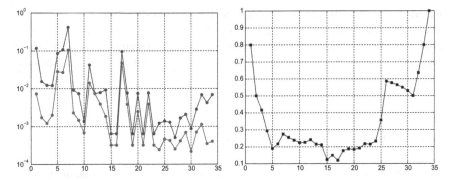

Fig. 5.15 The values of the ordering function q (left panel), and the values of conductance (right panel) computed for the `karate` network

- Macropol et al. showed in [339] a modification of the `PageRank-Nibble` algorithm. It allows for the detection of natural neighbours as well as for partition of the set of nodes into (possibly overlapping) groups.
- Determination of influential pages, i.e. an indication of the webpages affecting the rank of other pages, along with an indication of the intensity of the impact [20].
- The authors of [117] proposed a distance measure based on the order induced by the function $q(v)$. This allows for designing the advanced clustering algorithms.
- In [24] a method of identification of the densest part of a given graph is described. This idea can be used e.g. in bioinformatics to identify complex patterns in massive graphs.[42]
- Adaptation of the local approach to the analysis of directed graphs is proposed in [23]. This can be used e.g. in *topic mining*.[43] Discovering thematically related document groups in massive document collections can be viewed as identifying a subgraph of a huge graph. The subgraph represents a well defined topic. Local methods allow to solve the scalability problem successfully.

5.5 Large Datasets

Although spectral clustering is a very efficient method when applied to small or middle-sized datasets, it becomes almost impractical when applied to large data. This is mainly because the similarity matrix has m^2 entries and the process of computing the eigenvectors takes $O(m^3)$ time. Hence a number of modifications is required to make it feasible for large datasets. A most natural approach exploits random sampling

---

[42]B. Saha, et al. Dense subgraphs with restrictions and applications to gene annotation graphs. In *Research in Computational Molecular Biology*, Springer 2010, pp. 456–472.

[43]S.E.G. Villarreal, R.F. Brena: Topic mining based on graph local clustering. In: I. Batyrshin, G. Sidorov, eds. *Advances in Soft Computing*. LNCS 7095, Springer 2011, pp 201–212.

techniques. More elaborated approaches rely upon a supposition that if U, $\tilde{U} \in \mathbb{R}^{m \times k}$ are two matrices with orthogonal columns and $\|U - \tilde{U}\|_F \leq \epsilon$, for some $\epsilon > 0$, then clustering the rows of U and the rows of $\tilde{U}$ with the same method leads to the same partition of the set X [266]. Below we review some modifications of the standard spectral clustering.

5.5.1   Using a Sampling Technique

A most natural approach to coping with massive data seems to be an application of a sampling technique allowing to reduce original size of the data and then to interpolate obtained results on the whole data set. Below we briefly describe a "naïve" approach relying upon choosing a set of representatives from a large data set, and more advanced technique exploiting the Nyström technique.

5.5.1.1   Naïve Sampling Technique

Yan et al. [517] proposed a simple and fast procedure. Namely, they perform k-means algorithm on the data set with a large cluster number p, first. Then, the traditional spectral clustering is applied on the p cluster centres. Finally, each data point is assigned to the cluster as its nearest centre. This approach has been termed the k-means-based approximate spectral clustering (KASP). Its time complexity is $O(k^3 + kmt)$, where t is the number of iterations in k-means algorithms. As an alternative to k-means are so-called random projection (RP) trees [132], these authors propose also the (RASP) algorithm in which these trees are used instead of the k-means algorithm. The time complexity of this new algorithm is $O(k^3 + mh)$, where mh is the cost of building the h-level RP tree.

Shinnou and Sasaki [428], proposed a similar approach called Committees-based Spectral Clustering, CSC. Here the data set is partitioned, by means of k-means algorithm, into a large number p of clusters. But then the reduction relies upon removing those data points which are close (with pre-defined distance threshold) to the centres, called committees. Each removed set is replaced by a single "representative". Next, the standard spectral clustering is used to partition the set consisting of the committee "representatives" and the remaining data points. Finally, those removed data points are assigned to the cluster being their nearest centre. In the experiments, we named this approach.

Review of other methods is given in [97]. All these approaches use a sampling technique to select some key data points, which represent the remaining data points. In case of large datasets, these algorithms can only choose a limited number of data samples, usually determined by the available memory. This number is further reduced when the dataset has a large number of features. In effect, a lot of information of the detailed structure of the data is lost in the sampling step. Some improvements allowing to overcome this drawback are described in [494].

5.5.1.2 Nyström Extension

The Nyström method was originally intended as a technique for obtaining approximate numerical solutions to eigenfunction problems, for instance

$$\int_0^1 \mathsf{K}(x, y)u(y)dy = \lambda u(x)$$

where $\lambda \in \mathbb{R}$ and u is a real valued function, i.e. an eigenfunction of K. If $\{x_1, \ldots, x_r\} \subset [0, 1]$ is a set of evenly-spaced points, the numerical approximation of this equation is

$$\frac{1}{r} \sum_{j=1}^r \mathsf{K}(x, x_j)\hat{u}(x_j) = \lambda \hat{u}(x) \tag{5.123}$$

where $\hat{u}$ is an approximation of the true eigenfunction u. To solve this last problem, we set $x = x_i$ what leads to the system of equations

$$\frac{1}{r} \sum_{j=1}^r \mathsf{K}(x_i, x_j)\hat{u}(x_j) = \lambda \hat{u}(x_i), \quad i = 1, \ldots, r$$

It can be rewritten concisely in matrix form as

$$K\hat{\mathbf{u}} = \lambda \hat{\mathbf{u}}$$

where the elements of $r \times r$ matrix K are $k_{ij} = \mathsf{K}(x_i, x_j)$ and $\hat{\mathbf{u}} = (\hat{u}_1, \ldots, \hat{u}_r)^{\mathrm{T}}$ with $\hat{u}_i = \hat{u}(x_i)$. Obviously there are r (not necessarily unique) eigenpairs $(\lambda_i, \hat{u}_i)$ solving such an equation. To capture all r these eigenvectors we rewrite the above equation in the form

$$\hat{U} = K\hat{U}\Lambda^{-1}, \quad \hat{U} = (\hat{\mathbf{u}}_1, \ldots, \hat{\mathbf{u}}_r), \quad \Lambda = \mathrm{diag}(\lambda_1, \ldots, \lambda_r) \tag{5.124}$$

Suppose now that an $m \times m$ similarity matrix S is represented in the form

$$S = \begin{bmatrix} S_{11} & S_{12} \\ S_{21} & S_{22} \end{bmatrix} \tag{5.125}$$

where S_{11} is an $r \times r$ submarix of S with $r \ll m$, S_{12} is $r \times m - r$ submatrix, $S_{21} = S_{12}^{\mathrm{T}}$, and S_{22} is $m - r \times m - r$ submatrix of S. These submatrices are obtained e.g. by permuting rows and columns of the matrix S. Suppose next that the eigendecomposition of S_{11} is $U_{11}\Lambda_{11}U_{11}$. Then the Nyström's idea can be applied to approximate r eigenvectors of S by extending a $r \times r$ principal submatrix S_{11} of S. Namely, the extension (5.123) applied to the remaining $m - r$ values of the eigenvectors takes the form $\tilde{U}_{21} = S_{21}U_{11}\Lambda_{11}^{-1}$, where the rows $\mathbf{s}_{21}^\ell$, $\ell = 1, \ldots, m - r$, play the role of $\mathsf{K}(x, \cdot)$, and $x \in \{x_{r+1}, \ldots, x_m\}$. Hence the approximation $\tilde{U}$ of r eigenvectors of the matrix S is $\tilde{U} = (U_{11}^{\mathrm{T}} \, \tilde{U}_{21}^{\mathrm{T}})^{\mathrm{T}}$ and (see [185] for detailed derivation)

$$\tilde{S} = \tilde{U} \Lambda_{11} \tilde{U}^{\mathrm{T}} = \begin{bmatrix} S_{11} & S_{12} \\ S_{21} & S_{21} S_{11}^{-1} S_{12} \end{bmatrix} = \begin{bmatrix} S_{11} \\ S_{21} \end{bmatrix} S_{11}^{-1} [S_{11} \, S_{12}] \qquad (5.126)$$

what implies that

$$\| S - \tilde{S} \|_F = \| S_{22} - S_{21} S_{11}^{-1} S_{12} \|_F$$

Belabbas and Wolfe describe in [59] two strategies of ordering rows and columns of the matrix S in order to minimize the error $\| S - \tilde{S} \|_F$. Other strategies of selecting r columns from the set $\{1, \ldots, m\}$ are discussed in [312, 324].

It should be stressed that the columns of the matrix $\tilde{U}$ are not orthogonal. Two strategies of orthogonalizing the columns[44] are described in [185]. The time complexity of the whole procedure is $O(mr^2 + r^3)$, which for $r \ll m$ is significantly less than the $O(m^3)$.

5.5.2 Landmark-Based Spectral Clustering

Inspired by the achievements in sparse coding,[45] Chen and Cai proposed in [111] a scalable spectral clustering method termed *Landmark-based Spectral Clustering*, or LSC, for short. The algorithm starts from selecting $p \ll m$ representative data points, called landmarks. Then, according to sparse coding idea, the remaining data points are represented by the linear combinations of $r < p$ closest landmarks. The spectral embedding of the data can then be efficiently computed with the landmark-based representation. The proposed algorithm scales linearly with the problem size, and its time complexity is $O(mp^2 + p^3)$.

The essence of the algorithm relies upon representing the affinity matrix S in an economic form as follows:

$$S = Z^{\mathrm{T}} Z \qquad (5.127)$$

where $Z \in \mathbb{R}^{p \times m}$ is a sparse matrix. Similar idea was used in [431] in the context of information retrieval, where as Z the term-document matrix was used. In LSC a more general approach is applied. Assume namely that $\tilde{X} = (\mathbf{x}_1, \ldots, \mathbf{x}_m) \in \mathbb{R}^{n \times m}$ is the data matrix. It can be approximated by introducing two matrices $U \in \mathbb{R}^{n \times p}$ and $H \in \mathbb{R}^{p \times m}$ that minimize the error

$$\| \tilde{X} - U H \|_F$$

---

[44]They depend on whether S_{11} is positive definite.

[45]See e.g. H. Lee, et al. Efficient sparse coding algorithms. In *Proc. of the 19-th Int'l Conf. on Neural Information Processing Systems*, NIPS'06, pp. 801–808, 2006, MIT Press.

The matrix $\hat{X} = UH$ is said to be a nonnegative matrix factorization (NMF for brevity) of $\tilde{X}$, see e.g. [67] for a review of various factorizing algorithms. According to this paradigm, each basis vector (i.e. each column of U) can be regarded as a concept, and the matrix H (called also encoding) indicates that each data point is a linear combination of *all* the concepts, i.e.

$$\mathbf{x}_j \approx \sum_{i=1}^{p} \mathbf{u}_i h_{ij} = U\mathbf{h}_j$$

In LSC the basis vectors are treated as the landmark points of the data set. Sparse coding adds a sparsity constraint on Z. The resulting sparse matrix Z has at least two obvious advantages:

(a) Each data point is represented as a linear combination of a small number of basis vectors, i.e. the data points can be interpreted in a more clear way.
(b) Sparse matrix increases the efficiency of manipulating the affinity matrix (5.127).

The sparsification applied in [111] is realized in a simple way. Suppose for a moment that the landmarks U are fixed and $N_U(j, r)$ denotes the indices of $r < p$ landmarks nearest to $\mathbf{x}_j$. Define H to be a matrix with elements

$$h_{ij} = \begin{cases} \dfrac{K(\mathbf{u}_i, \mathbf{x}_j)}{\sum_{\ell \in N_U(j,r)} K(\mathbf{u}_\ell, \mathbf{x}_j)} & \text{if } \mathbf{u}_i \in N_U(j, r) \\ 0 & \text{otherwise} \end{cases}, \quad i = 1, \ldots, p, \, j = 1, \ldots, m \tag{5.128}$$

where $K(\mathbf{u}_i, \mathbf{x}_j)$ measures the degree of similarity between the landmark $\mathbf{u}_i$ and the observation $\mathbf{x}_j$; for instance the Gaussian kernel, $K(\mathbf{u}_i, \mathbf{x}_j) = \exp(-\|\mathbf{u}_i - \mathbf{x}_j\|/2\sigma^2)$, can be used. A reader should note that the representation (5.128) guarantees that each column of H has exactly $p - r$ zeroes, i.e. the parameter r controls the sparsity of H. Moreover, $\sum_i h_{ij} = 1$.

Denote next $C \in \mathbb{R}^{p \times p}$ the diagonal matrix with elements

$$c_{ii} = \sum_{j=1}^{m} h_{ij}, \quad i = 1, \ldots, p$$

and define Z as follows

$$Z = C^{-1/2} H \tag{5.129}$$

Then the similarity matrix $S \in \mathbb{R}^{m \times m}$ takes the form

$$S = Z^{\mathsf{T}} Z = H^{\mathsf{T}} C^{-1} H \tag{5.130}$$

i.e. its elements are of the form $s_{ij} = \mathbf{z}_i^{\mathsf{T}}\mathbf{z}_j$, where $\mathbf{z}_i$ stands for i-th column of the matrix Z. The matrix S is symmetric and the elements of each its row and column sum up to 1. Thus the degree matrix of S is the identity matrix. It means that both the combinatorial and the normalized Laplacian of the graph represented by S has the form $L = \mathcal{L} = \mathbb{I} - Z^{\mathsf{T}}Z$.

Suppose for a while that Z factorizes as $A\Sigma B^{\mathsf{T}}$, where $\Sigma = \mathrm{diag}(\sigma_1, \ldots, \sigma_p)$ is the diagonal matrix with singular values of Z indexed in descending order, the columns of A (resp. B) are left (resp. right) singular vectors of Z. Since $ZZ^{\mathsf{T}} = A\Sigma B^{\mathsf{T}}B\Sigma = A\Sigma^2 A^{\mathsf{T}}$ we state that the columns of A are the eigenvectors of the matrix ZZ^{T} and σ_i^2's are the eigenvalues of this matrix. The matrix ZZ^{T} has size $p \times p$, hence the time complexity of finding its eigenvectors is $O(p^3)$. Similarly one can verify that the columns of $B \in \mathbb{R}^{m \times p}$ are the eigenvectors of the matrix $S = Z^{\mathsf{T}}Z$. These eigenvectors can be easily computed as

$$B^{\mathsf{T}} = \Sigma^{-1}A^{\mathsf{T}}Z \qquad (5.131)$$

When using the NJW Algorithm 5.4, it suffices to form the matrix Y consisting of k row-normalized columns of the matrix B, and apply the k-means algorithm to this matrix. For reader's convenience we summarize the whole procedure in pseudocode 5.8.

Algorithm 5.8 Landmark-based Spectral Clustering [111]

1: *Inicjalizacja.* Data matrix $\tilde{X} = (\mathbf{x}_1, \ldots, \mathbf{x}_m)$, k—number of groups, $p \ll m$—number of landmarks, $r < p$—sparsity constraint.
2: Determine p landmarks stacked in columns of the matrix $U \in \mathbb{R}^{n \times p}$.
3: Determine the encodings H according to the Eq. (5.128).
4: Compute the matrix Z according to the Eq. (5.129).
5: Compute the eigenvectors of the matrix ZZ^{T} and determine the matrix B according to the Eq. (5.131).
6: Normalize rows of the matrix B. Denote B' resulting matrix.
7: Apply k-means to the data B'

The landmarks, i.e. columns of the matrix U can be selected randomly, or—in analogy with KASP algorithm—by using k means algorithm. The authors of [111] claimed that choosing the landmarks by random sampling results in a balance between accuracy and performance. Unfortunately, the weights defined in Eq. (5.128) are sensitive to the choice of the kernel $\mathbf{K}$, and in case of the Gaussian kernel—to the choice of the hyperparameter σ. A more elaborated strategy, relying upon constrained optimization of

$$\min_{Z \in \mathbb{R}^{p \times m}} \|X - UH\|_F^2$$
$$s.t. \ \mathbf{e}^{\mathsf{T}}\mathbf{h}_j = 1, h_{ij} \geq 0, \quad i = 1, \ldots, p, j = 1, \ldots, m$$

was proposed in [333].

Moreover, if the elements of matrix H are computed using the Eq. (5.128), the error $\|\tilde{X} - UH\|_F$ is rather high. Shao et al. proposed in [425] another method leading to the optimization problem

$$\min_{W,Z} \|X - WUH\|_F^2 + \lambda\|Z\|_F^2 \qquad (5.132)$$

where $W \in \mathbb{R}^{n \times n}$ is a linear transformation function, columns of $U \in \mathbb{R}^{n \times p}$ are landmarks, the elements of $H \in \mathbb{R}^{p \times m}$ are the encodings for the dataset, and λ is the regularizer. Again the set of landmarks is fixed at the beginning. The reader is referred to the original papers [333, 425] for details.

5.5.3 Randomized SVD

Halko, Martinsson and Tropp described in [226] a fast randomized algorithm which returns, among others, Singular Value Decomposition of a rectangular matrix $A \in \mathbb{R}^{m \times n}$. The emphasis was put on obtaining low-rank approximation of a given matrix. Note that if A is a square and symmetric matrix, e.g. Laplacian matrix, then its left singular vectors are identical to its eigenvectors, what justifies our interest in this algorithm. The entire algorithm consist of two stages:

(A) Compute an approximate basis for the range of the input matrix A, i.e. determine a matrix Q with orthogonal columns, and such that

$$A \approx QQ^{\mathsf{T}}A$$

The columns of Q span a low-dimensional subspace that captures the action of the matrix A. Thus the matrix Q should contain as few columns as possible, but the accuracy of approximation of the input matrix A is of primal importance.

(B) Use the matrix Q to compute the SVD factorization of A.

Stage (A) obeying steps (1)–(3) in the pseudocode 5.9 is executed efficiently with random sampling methods. Recall first (consult e.g. [354, Sect. 4.2]) that the range of a matrix A is a subspace $\mathcal{R}(A)$ of $\mathbb{R}^m$ generated by the range of the function $f(\mathbf{x}) = A\mathbf{x}$, i.e.:

$$\mathcal{R}(A) = \{A\mathbf{x} \colon \mathbf{x} \in \mathbb{R}^n\}$$

Instead of finding exact form of the basis of this subspace, we simply take a random sample of $k \leq \min(m, n)$ vectors $\boldsymbol{\omega}_1, \ldots, \boldsymbol{\omega}_k$ and we compute the vectors $\mathbf{y}_i = A\boldsymbol{\omega}_i$, $i = 1, \ldots, k$. These vectors are linearly independent, and thus they span the range of A. This is step (1) in the pseudocode 5.9. If the matrix A is disturbed, i.e. $A = B + E$, where B is a rank-k matrix containing the information we seek and E is a small perturbation, we generate $k + p$ samples with the hope that the enriched set of $\mathbf{y}_i$'s has a better chance of spanning the required subspace. Halko et al. suggest that setting $p = 5$ or $p = 10$ often gives quite good results, see [226, Sect. 4.2].

As the singular spectrum of the matrix A may decay slowly, step (2) of the pseudocode is performed. Observe namely, that the matrix $B = (AA^{\mathrm{T}})^q A$ has the same singular vectors as the input matrix A, but its singular values decay much more quickly:

$$\sigma_j(B) = [\sigma_j(A)]^{2q+1}$$

Thus the matrix $A\Omega$ is replaced by the matrix $(AA^{\mathrm{T}})^q A\Omega$, where usually $q = 1$ or $q = 2$. Step (3) returns the matrix $Q \in \mathbb{R}^{m \times (k+q)}$ with orthogonal columns.

Knowing the matrix Q it is easy to compute a standard factorization, e.g. SVD of A. To do so, construct a matrix $B = Q^{\mathrm{T}} A$ and compute its SVD decomposition: $B = U \Sigma V^{\mathrm{T}}$. Note that B is a small matrix of size $(k+q) \times n$ and its SVD can be computed efficiently. Finally, the matrix $U_A = QU$ contains $k+q$ dominating left singular vectors (or simply: eigenvectors, if A is square and symmetric) of the matrix A. The steps (4) and (5) in the pseudocode 5.9 represent this procedure.

Algorithm 5.9 Randomized SVD [226]

Input: Matrix $A \in \mathbb{R}^{m \times n}$, number of projection vectors k, exponent q, oversampling parameter p.
1: Generate a random Gaussian matrix $\Omega \in \mathbb{R}^{m \times (k+p)}$.
2: Form $Y = (AA^{\mathrm{T}})^q A\Omega$ by multiplying alternately with A and A^{T}.
3: Using QR decomposition factorize Y as $Y = QR$ where $Q \in \mathbb{R}^{m \times (k+p)}$ is the matrix with orthogonal columns.
4: Form $B = Q^{\mathrm{T}} A$ and compute its SVD factorization: $B = U \Sigma V^{\mathrm{T}}$.
5: Set $U_A = QU$.
Output: Approximate SVD of A as $A \approx (QU) \Sigma V^{\mathrm{T}}$.

When A is the normalized similarity matrix S used in step 2 of the NJW Algorithm 5.4, then the matrix Y from step 2 of the above algorithm has the form $Y = S^{2q+1}\Omega \in \mathbb{R}^{m \times (k+p)}$ and the matrix U_A contains an approximation of $k+q$ dominating eigenvectors of S. Hence, choosing first k columns of U_A we obtain an approximation of the matrix Φ_k occurring in step 3 of the NJW algorithm.

5.5.4 Incomplete Cholesky Decomposition

Recall that any positive-definite matrix $K \in \mathbb{R}^{m \times m}$ can be represented as the product of a lower triangular matrix $G \in \mathbb{R}^{m \times m}$ and its transpose, i.e. $K = GG^{\mathrm{T}}$. This representation is just the Cholesky decomposition [204] or [354]. The matrix G is said to be Cholesky triangle.

By incomplete Cholesky decomposition we understand a sparse approximation of the Cholesky factorization. Namely for a symmetric positive-definite matrix K we search for a matrix $C \in \mathbb{R}^{m \times r}$, $r \ll m$ such that

$$\|K - CC^{\mathrm{T}}\|_F \leq \epsilon \tag{5.133}$$

for some $\epsilon > 0$ [204]. In other words we search for such a matrix C that has some tractable sparsity structure and is "sufficiently close" to the Cholesky triangle G.

A very efficient algorithm for finding the matrix C has been proposed by Bach and Jordan in [38] (see also [190] for a pseudocode). Its time complexity is $O(mr^2)$, and—what is more important—in the case of Gaussian kernel we even do not need the full matrix K. What is really needed are the diagonal elements of K (in case of Gaussian kernel these elements are all equal to one), and other elements of this matrix that are needed to find C are computed on demand. Thus we do not have to waste time and space to compute and store the entries of K. Furthermore, if the eigenvalues of K decay quickly (what is true for Gaussian kernel), the value of ϵ is really small, and the number of columns r of the matrix C is also small [190]. Call `chol_gass` a procedure returning the matrix C for a given Gaussian similarity matrix S.[46] Its input parameters are: the entire data set X, width σ, and the tolerance, `tol`.

The Algorithm 5.10 for computing approximate values ot the dominant eigenvectors of the matrix S consists of four main steps. First an approximation $C \in \mathbb{R}^{m \times r}$ of the Cholesky triangle of the similarity matrix S is computed. The number r of columns of C depends on the value of `tol`: the greater `tol` value, the smaller r and poorer approximation quality ϵ. Note also that the diagonal elements of resulting matrix $\tilde{S} = CC^{\mathrm{T}}$ are close to 1 In step 2 the approximation of node degrees is computed, i.e. $D = \mathrm{diag}(Se)$ is approximated by $\tilde{D} = \mathrm{diag}(C(C^{\mathrm{T}}e))$. In step 3 the matrix $\tilde{D}^{-1/2}C$ is represented as the product of an orthogonal matrix $Q \in \mathbb{R}^{m \times r}$ and an upper triangular matrix $R \in \mathbb{R}^{r \times r}$.[47] With such a decomposition the matrix $\tilde{S} = \tilde{D}^{-1/2}\tilde{S}\tilde{D}^{-1/2}$ takes the form $\tilde{S} = QRR^{\mathrm{T}}Q^{\mathrm{T}}$. Suppose now that, as in step 4 of the algorithm, the matrix RR^{T} has the eigendecomposition $RR^{\mathrm{T}} = U\Lambda U^{\mathrm{T}}$. Since this matrix has size $r \times r$ we easily find its decomposition. Now the normalized similarity matrix $\tilde{S}$ can be represented as $\tilde{S} = (QU)\Lambda(QU)^{\mathrm{T}}$ what means that the columns of (QU) are eigenvectors of $\tilde{S}$.

Algorithm 5.10 Approximating eigenvectors of a normalized similarity matrix via incomplete Cholesky decomposition [316]

Input: Data matrix $X \in \mathbb{R}^{m \times n}$, parameter σ, and tollerance tol.
1: `C = chol_gauss(X',sigma,tol)`.
2: `D = C*sum(C,2)`.
3: `[Q,R] = qr(`$D^{-1/2}$`C,0)`
4: `[U,Lambda] = eig(RR')`
Output: Approximation `Q*U` of the dominating eigenvectors of matrix S

---

[46] Its MATLAB implementation can be found here: http://www.di.ens.fr/~fbach/kernel-ica/.

[47] In MATLAB the command `[Q,R] = qr(A,0)`, where the number of rows of `A` is larger than the number of columns, then an "economy" factorization is returned, omitting zeroes of `R` and the corresponding columns of `Q`.

An interesting application of the incomplete Cholesky decomposition in kernel spectral clustering (see Sect. 5.3.6.2) was proposed in [18].

5.5.5 Compressive Spectral Clustering

The Johnson-Lindenstrauss lemma[48] states that if $X = (\mathbf{x}_1, \ldots, \mathbf{x}_m)^\mathrm{T}$ is the matrix representing m data items, ϵ is a small number (tolerance), $\beta > 0$ and

$$d \geq d(m, \epsilon) = \left\lfloor \frac{4 + 2\beta}{\epsilon^2/2 - \epsilon^3/3} \log m \right\rfloor \tag{5.134}$$

then there exists a mapping $f : \mathbf{R}^n \to \mathbb{R}^d$ such that

$$(1 - \epsilon)\|\mathbf{x}_i - \mathbf{x}_j\|^2 \leq \|f(\mathbf{x}_i) - f(\mathbf{x}_j)\|^2 \leq (1 + \epsilon)\|\mathbf{x}_i - \mathbf{x}_j\|^2, \quad i, j = 1, \ldots, m \tag{5.135}$$

with probability at least $1 - m^{-\beta}$. In particular, if R is a $n \times d$ random matrix whose elements r_{ij} are independent Gaussian variables of zero mean value and variance $1/d$, the mapping f has the form $f(\mathbf{x}_i) = \mathbf{x}_i^\mathrm{T} R$. In other words, f maps the i-th row of X to the i-th row of the matrix XR. The integer $d(m, \epsilon)$ is said to be the Johnson-Lindenstraus dimension, or simply JL-dimension. Similarly, the lemma will be referred to as the JL-lemma.

Suppose now that, as in the NJW algorithm, Φ_k is the matrix whose columns are first k eigenvectors of the normalized Laplacian. Let δ_i be the i-th row of the identity matrix and $V = \mathrm{diag}(\nu_1^k, \ldots, \nu_m^k)$, where $\nu_i^k = (\sum_{j=1}^k \phi_{ij}^2)^{-1/2} = 1/\|\delta_i \Phi_k\|$. Then

$$\mathbf{y}_i^\mathrm{T} = \delta_i V \Phi_k \tag{5.136}$$

is the i-th row of the matrix Y constructed in step (4) of the NJW Algorithm 5.4.

According to the methods developed in the frames of graph signal processing [415, 432], given a continuous filter function h defined on $[0, 2]$, its associated graph filter operator $H \in \mathbb{R}^{\times m}$ is defined as

$$H = h(\Lambda') = \Phi\mathrm{diag}\big(h(\lambda_1'), \ldots, h(\lambda_m')\big)\Phi^\mathrm{T} \tag{5.137}$$

Particularly, if h is a step function

$$h_{\lambda_k}(\lambda) = \begin{cases} 1 \text{ if } \lambda \leq \lambda_k \\ 0 \text{ otherwise} \end{cases} \tag{5.138}$$

---

[48]See e.g.: D. Achlioptas. Database-friendly random projections: Johnson-Lindenstrauss with binary coins. *J. of Comput. and Syst. Sci.*, 66(4):671–687, 2003.

In such a case $H_k = \Phi_k \Phi_k^{\mathsf{T}}$. It is easy to note the following properties of the matrix H_k:

(a) H_k is a symmetric matrix of size $m \times m$.
(b) The spectrum of H_k, $\sigma(H_k)$, that is the set of eigenvalues of the matrix H_k is $\sigma(H_k) = \{0^{m-k}, 1^k\}$, i.e. it contains $m-k$ zeroes and k ones. Thus tr $(H_k) = k$.
(c) $H_k \mathbf{x} = (\mathbf{x}^{\mathsf{T}} H_k)^{\mathsf{T}}$ for any vector (signal) $\mathbf{x}$. Furthermore, $H_k \phi_j = \phi_j$ if $j \le k$ and $H_k \phi_j = \mathbf{0}$ otherwise.

The matrix H_k is said to be the low-pass filter. The signal $\mathbf{x}$ filtered by H_k is $H_k \mathbf{x}$, or equivalently, $\mathbf{x}^{\mathsf{T}} H_k$.

Let

$$\tilde{\mathbf{y}}_i^{\mathsf{T}} = \delta_i V H_k R$$

denotes i-th row of the matrix $V H_k R$, where R is the random matrix defined at the beginning of this section. Then

$$
\begin{aligned}
\|\tilde{\mathbf{y}}_i^{\mathsf{T}} - \tilde{\mathbf{y}}_j^{\mathsf{T}}\|^2 &= \|(\delta_i - \delta_j) V H_k R\|^2 = \|(\delta_i - \delta_j) V \Phi_k \Phi_k^{\mathsf{T}} R\|^2 \\
&= \|(\delta_i V \Phi_k - \delta_j V \Phi_k) \Phi_k^{\mathsf{T}} R\|^2 \\
&= \|(\mathbf{y}_i^{\mathsf{T}} - \mathbf{y}_j^{\mathsf{T}}) R'\|^2
\end{aligned}
\tag{5.139}
$$

where R' is a random matrix whose elements r'_{ij} are independent Gaussian variables of zero mean value and variance $1/d$ (because the columns of Φ_k are orthogonal, and the elements of each column sum up to 1). Thus we can apply the JL-lemma, and state that for a given tolerance ϵ and the constant $\beta > 0$

$$(1 - \epsilon)\|\mathbf{y}_i - \mathbf{y}_j\|^2 \le \|\tilde{\mathbf{y}}_i - \tilde{\mathbf{y}}_j\|^2 \le (1 + \epsilon)\|\mathbf{y}_i - \mathbf{y}_j\|^2 \tag{5.140}$$

holds with the probability not less than $1 - m^{-\beta}$ if $R \in \mathbb{R}^{m \times d}$ is a random matrix and d is not less than the JL-dimension $d(m, \epsilon)$.

The inequalities (5.140) imply that the distances between the points $\mathbf{y}_i$ and $\mathbf{y}_j$ may be computed by using new coordinates $\tilde{\mathbf{y}}_i$ and $\tilde{\mathbf{y}}_j$. Note that while $\mathbf{y}_i$ is a k-dimensional vector defined as in Step (4) of Algorithm 5.4, $\tilde{\mathbf{y}}_i^{\mathsf{T}}$ is the i-th row of the matrix $V K_k R$; it has $d > k$ elements. To fully exploit this fact, observe first that instead of computing the matrix V, we can use the formula

$$\tilde{\mathbf{y}}_i = \frac{\delta_i H_k R}{\|\delta_i H_k R\|} \tag{5.141}$$

Second, the ideal low-pass filter H_k can be approximated by using a polynomial approximation of the filter function

$$\tilde{h}(\lambda)_{\lambda_k} = \sum_{j=0}^{\infty} \alpha_j \lambda^j \approx \sum_{j=0}^{p} \alpha_j \lambda^j \tag{5.142}$$

where p is a sufficiently large number. For instance, a Tchebyshev approximation[49] can be applied. With such an approximation the low-pass filter is expressed as

$$\tilde{H}_k = \sum_{j=0}^{p} \alpha_j \mathcal{L}^j \tag{5.143}$$

and

$$\tilde{H}_k R = \sum_{j=0}^{p} \alpha_j \mathcal{L}^j R = \alpha_0 R + \alpha_1 \mathcal{L}R + \alpha_2 \mathcal{L}(\mathcal{L}R) + \cdots$$

The only problem is the determination of k-th lowest eigenvalue of the normalized Laplacian $\mathcal{L}$. Detailed description of a procedure for estimating λ_k is given [399, Sect. 4.3].

In summary, using the approximation $\tilde{H}_k$ of the ideal low-pass filter H_k allows to approximate the distances between the points in k-dimensional spectral space by the distances between the points in d-dimensional space with the coordinates computed as in (5.141). Tremblay et al. provide a detailed discussion of the quality of such an approximation [465, Sect. 4.2]. The advantage of this approach is that we do not need to compute the eigenvectors of $\mathcal{L}$.

In case of massive data sets the application of k-means algorithm at step (5) of the NJW algorithm is both memory and time consuming. Hence, Puy et al. proposed in [399] another approach: choose a random sample of $m' < m$ of feature vectors $\tilde{\mathbf{y}}$, apply the k-means algorithm to this reduced set to obtain the reduced indicator vectors $\mathbf{c}_\ell^r$, $\ell = 1, \ldots, k$, and interpolate them to obtain the indicator vectors $\tilde{\mathbf{c}}_\ell$ defined on the full set of nodes. Tremblay et al. suggest in [465] to choose $m' = 2k \log k$.

All these observations lead to a compressive spectral clustering, or CSC for short, presented in [465]. Its time complexity, when $m' = 2k \log k$ is

$$O\left(k^2 \log^2 k + pm(\log m + k)\right)$$

where p is the order of approximation (5.142). This suggests that the CSC is faster than NJW for massive data sets (large m) that are partitioned into large groups. To reduce the dimensionality of new representation the authors use $d = 4 \log m$ what is much smaller than the JL-dimension (5.134).

We summarize this algorithm in pseudocode 5.11.

---

[49]Concise description of Tchebyshev approximation can be found e.g. in W.H. Press, S.A. Teukolsky, W.T. Vettrrling, and B.P. Flannery: *Numerical Recipes in C. The Art of Scientific Computing*, 2-nd Edition, Cambridge University Press, 1992.

Algorithm 5.11 Compressive spectral clustering [465]

Input: Laplacian matrix $\mathcal{L}$, the number of clusters k, parameters: $m' = 2k \log k$, $d = 4 \log n$,
   $p = 50$
1: Estimate k-th eigenvalue λ_k of the Laplacian $\mathcal{L}$
2: Compute the polynomial approximation $\tilde{h}_{\lambda_k}$ of order p of the ideal low-pass filter h_{λ_k}
3: Generate d random Gaussian signals $\mathbf{r}_i \in \mathbb{R}^{m \times 1}$ of mean 0 and variance $1/d$ and stack them in
   columns of matrix R of size $m \times d$
4: Filter R with $\tilde{H}_k = \tilde{h}_{\lambda_k}(\mathcal{L})$ and define, for each node i, its feature vector according to the
   Eq. (5.141)
5: Choose randomly m' feature vectors $Y^r = (\tilde{\mathbf{y}}_{(1)}, \ldots, \tilde{\mathbf{y}}_{(m')})$.
6: Run k-means on the reduced dataset Y^r with the Euclidean distance $\tilde{d}_{ij} = \|\tilde{\mathbf{y}}_{(i)} - \tilde{\mathbf{y}}_{(j)}\|$. Denote
   $\mathbf{c}_\ell^r$, reduced indicator vectors assigning m' objects to $\ell = 1, \ldots, k$ groups.
7: Interpolate each reduced indicator vector $\mathbf{c}_\ell^r$ to obtain the k indicator vectors $\tilde{\mathbf{c}}_\ell^r \in \mathbb{R}^{m \times 1}$ on the
   full set of nodes.

Chapter 6
Community Discovery and Identification in Empirical Graphs

Abstract This chapter discusses a specific, yet quite important application area of cluster analysis that is of community detection in large empirical graphs. The chapter introduces a number of alternative understandings of the term *community*, which lead to a diversity of community detection algorithms. Then it presents specific object similarity measures as well as community quality measures (modularity and its derivatives) which require special adaptation or creation of new clustering algorithms to address community detection. In particular, we give some insights into Newman, Louvain, FastGreedy algorithms developed specially for community detection as well as other, like kernel k-means or spectral methods, which were adopted for this purpose. Finally issues of overlapping community and multi-layered community detection are discussed.

Classical examples of complex networks,[1] called also empirical graphs, include Internet [80], trust and social networks [168], information and citation networks,[2] collaboration networks,[3] grid-like networks,[4] organic chemical molecules[5] metabolic networks,[6] or protein interaction networks.[7]

---

[1] A primary distinction between a graph and a network is that the former is a method of representation of the latter, not necessarily reflecting all the aspects. We say that a network consists of nodes, that may be labelled with various attributes, and of relations between nodes, which may be of various types, and may possess attributes reflecting the presence, absence, degree or missing knowledge about a relation. A graph representing a network would consist of network nodes and would contain edges representing all or some of the types of relations between the nodes.

[2] like patent networks, journal citation networks.

[3] co-authorship networks, actor co-staging networks.

[4] road network, rail-way network, airport network, power supply network, telecommunication network.

[5] genes, proteins, enzymes etc. viewed as networks of atoms, with chemical bonds as links.

[6] Compare H. Jeong, B. Tombor, R. Albert, Z.N. Oltvai i A.-L. Barabasi, The large-scale organization of metabolic networks, *Nature*, 407:651–654, 2000.

[7] Consult M. Strong, Th.G. Graeber, M. Beeby, M. Pellegrini, M.J. Thompson, T.O. Yeates, D. Eisenberg. Visualization and interpretation of protein networks in Mycobacterium tuberculosis based on hierarchical clustering of genome-wide functional linkage maps. *Nucl. Acids Res.* 31(24):7099–7109, 2003.

© Springer International Publishing AG 2018   261
S. T. Wierzchoń and M. A. Kłopotek, *Modern Algorithms of Cluster Analysis*,
Studies in Big Data 34, https://doi.org/10.1007/978-3-319-69308-8_6

Initial approach to model such graphs was based on the concept of random graphs. However, it turned out quickly that complex networks are characterised by a series of features distinguishing them significantly from random graphs [13].

One of such important and widely known properties is the small world effect,[8] manifestation of which is a low link distance between the nodes of the network. Another one is the scale-free property which means that a large number of nodes communicates with a small set of other nodes while a tiny set of "centres" is connected with a large number of nodes. Mathematically this implies that the degree (d) distribution has the form $P(\text{d} = k) = ck^{-\gamma}$, where c is a normalising constant, and γ is a constant usually taking values from the interval (1, 3) [13]. A next significant feature of empirical networks is the transitivity: if two nodes $\mathbf{x}_i$, $\mathbf{x}_j$ have a common neighbour, $\mathbf{x}_k$, then with high probability they are neighbours of one another, that is $(\mathbf{x}_i, \mathbf{x}_j) \in E$. A measure of intensity of this effect is so-called clustering coefficient C. It is equal one for a complete graph ("everybody knows everybody"), and for empirical graphs it ranges between 0.1 and 0.5 [499].

Beside properties related to individual nodes or node pairs a significant feature of empirical graphs are so called motifs,[9] that is characteristic connections of 3–4 nodes. They emerge significantly more frequently than in random graphs [357]. Furthermore, the concentration of motifs in subgraphs is similar to their concentration in the whole empirical graph. This is in contrast to random graphs, where the concentration of motifs goes down with the increase of the size of the subgraph. Authors of [357] state that—like in statistical physics—the number of occurrences of each motif in the empirical graph is an extensive variable (that is it grows linearly with the growth of the size of the graph) while it is a non-extensive variable for random graphs. From the point of view of the information theory the motifs can be considered as elementary computational structures.[10] One can assume that motifs are a consequence of rules governing the growth of a given empirical graph.

Finally, the phenomenon was observed that in empirical graphs there exist groups of nodes communicating with one another more frequently than with nodes from other groups that is while they are globally sparse, nonetheless the empirical graphs are locally dense. Those dense regions of a graph constitute so-called communities (referred to also as modules) that is subsets of nodes connected more frequently to one another than to other nodes. Usually they serve as functional entities like inner organs of living organisms, [182]. This means that network properties within a

---

[8]On June 16th 1998 𝕿𝖍𝖊 𝕹𝖊𝖜 𝖄𝖔𝖗𝖐 𝕿𝖎𝖒𝖊𝖘 reported that "Mathematicians Prove That It's a Small World". The article is available at the address http://www.nytimes.com/library/national/science/061698sci-smallworld.html.

[9]We recommend the Web page http://www.weizmann.ac.il/mcb/UriAlon/networkMotifsMore.html, dealing with this topic. Notice also that there is an entire research area of frequent pattern mining, analogous to basket analysis, but taking into account links and their types in a graph, e.g. in pharmacology.

[10]Cf. S. Shen-Orr, R. Milo, S. Mangan & U. Alon, Network motifs in the transcriptional regulation network of *Escherichia coli. Nature Genetics*, 31:64–68 (2002).

concrete community[11] differ from those of the whole network. We devote this chapter to this phenomenon.

Research concerning communities concentrates around two topics: their discovery and identification. *Community discovery* is understood as extraction of all communities of a given graph $\Gamma = (V, E)$. *Community identification* consists in finding the community containing nodes from a given subset $S \subset V$. The publication [182] contains an extensive, 100 pages long, overview of actual methods and trends in community discovery. A concise introduction to this topic can be found in papers like [216, 369]. Further overview papers [131] and [180] may be recommended.

Clauset et al. describe in [122] a hierarchical technique for induction of the structure of an empirical network demonstrating, that the existence of this hierarchy explains the occurrence of the above-mentioned characteristic features of empirical graphs. Furthermore they show that once one knows the hierarchical structure, one can use it to predict missing links in partially discovered structures. They claim that hierarchy is a central constituting principle organising empirical networks.

An easy introduction to this topic can be found in the paper [51]. There exists a number of books devoted to the issues of empirical networks. The monograph of Newman [372] as well as [423] and [498] are good starting points.

6.1   The Concept of the Community

No generally accepted definition of a community seems to exist right now. Intuitively, its members would communicate more often within the community than with members of other communities. In the language of the graph theory we can say that a community is defined as such a set of nodes $W \subset V$ of the graph $\Gamma = (V, E)$ that the majority of neighbours of each node from W belongs to the very same set W. Even if the graph Γ is sparse, nonetheless the nodes W induce a locally dense subgraph of Γ. This qualitative statement, however, is expressed by different researches via different formulas.

Fortunato [182, III B] distinguishes three principal ways of defining a community: local, global, and similarity related ones. In [207] we find still another one - based on the membership degrees.

Independently of a particular community definition, one can identify physical constraints related to an "effective" community. It is claimed that if the persons constituting a community are expected to understand sufficiently not only their own

---

[11] A community in a network is usually understood as a cluster in a graph representing some aspect of the network. Therefore, community discovery is usually understood as graph clustering and algorithms of graph clustering are applied also for community detection.

bilateral relations with other members, but also the relations of each community member with the rest of the community, then the maximal size of the community must be limited. Usually this upper bound (called Dunbar's number) is assumed to range from 100 to 230, and the frequently suggested value amounts to 150.[12]

6.1.1 Local Definitions

Local definitions rely on the assumption that communities operate as stand-alone entities, so that one can concentrate on properties of respective subgraphs. So-called communities in a weak sense and in a strong sense can serve as sample definitions in this spirit [401]. A *community in a strong sense* is understood as such a set of nodes $C \subset V$ of the graph Γ that

$$\forall i \in C : \mathsf{d}_i^{in} > \mathsf{d}_i^{out} \tag{6.1}$$

where $\mathsf{d}_i^{in} = \sum_{j \in C} a_{ij}$, $\mathsf{d}_i^{out} = \sum_{j \notin C} a_{ij}$, while $a_{ij} \in \{0, 1\}$ indicates presence or absence of a link between nodes i and j. The above definition reflects the assumption that each node of the community communicates with other members more intensely than with nodes outside of the community. If we replace the condition (6.1) by

$$\sum_{i \in C} \mathsf{d}_i^{in} > \sum_{i \in C} \mathsf{d}_i^{out} \tag{6.2}$$

we obtain the definition of a *community in a weak sense*. Here it is only required that the total number of connections between the community members should be greater than the combined number of connections between nodes from the set C and the other nodes of the graph. For the purpose of extracting such communities Radicchi et al. [401] suggest a recursive division of the graph that is terminated as soon as a subsequent split leads to emergence of a group that is not a community in a strong or a weak sense.

Wang et al. [495] introduced a still more subtle distinction. They define so-called *community kernels* and *auxiliary communities*. For a given number of communities, kernels $\mathcal{K}_1, \ldots \mathcal{K}_k$ are disjoint subsets of nodes (not necessarily including all nodes of the network) such that

$$\forall i \in \mathcal{K}_j \forall l \notin \mathcal{K}_j : \mathsf{d}(i, \mathcal{K}_j) \ge \mathsf{d}(l, \mathcal{K}_j) \tag{6.3}$$

where $\mathsf{d}(i, S) = \sum_{j \in S} a_{ij}$. For each kernel $\mathcal{K}_j$ an auxiliary community $\mathcal{A}_j$ (disjoint from it) is defined such that

---

[12]Cf. e.g. A. Hernando, D. Villuendas, C. Vesperinas, M. Abad, A. Plastino. Unravelling the size distribution of social groups with information theory on complex networks. arXiv:0905.3704v3 [physics.soc-ph].

$$\forall i \in \mathcal{A}_j: \mathsf{d}(i, \mathcal{K}_j) \geq \mathsf{d}(i, \mathcal{K}_l), l \neq j \qquad (6.4)$$

$$\sum_{i \in \mathcal{A}_j} \mathsf{d}(i, \mathcal{K}_j) \geq \sum_{i \in \mathcal{K}_j} \mathsf{d}(i, \mathcal{K}_j) \qquad (6.5)$$

Kernels are apparently communities in the strong sense, while auxiliary communities concentrate around the "kernels", playing a kind of leader role.

6.1.2 Global Definitions

On the other hand global definitions assume that a graph with communities differs significantly from a random, graph. In this case a so-called referential model (called also *null model*), that is a random graph with parameters closest to those of the analysed graph, is used as a golden standard of a graph lacking communities. Divergence from this standard is deemed as an indication of the existence of a community. One of the most popular referential models was the model of Newman and Girvan [373] that is random graph $\Gamma_l = (V, E_l)$, edges of which are assigned randomly in such a way that the expected value of the degree of each node is identical with the degree of this node in the original graph Γ.

6.1.3 Node Similarity Based Definitions

The third group of definitions refers to the natural interpretation of a community as a set of objects with similar properties. It requires definition of a similarity measure between pair of nodes. Examples of such measures can be found in Sects. 6.2 and 6.4.4. This approach is mainly used in spectral clustering, described in Subsect. 6.4.5.

There is a significant difference between spectral clustering and the extraction of communities in a graph. In the first case the number of clusters is usually known in advance and we seek groups of approximately same size (e.g. in case of task planning for a distributed processing). In the second case the communities usually differ by size, their number is not known in advance, and, what is most important, the individual nodes can belong to multiple communities at the same time.

6.1.4 Probabilistic Labelling Based Definitions

Finally, we can imagine that community membership is provided by an explicit community label.

To make it not that simple, Gopalan and Blei [207] assume that instead the community membership is to some extent "fuzzy", that is each node belongs to each community (from a list of communities) with some probability. The actual network is a (stochastic) manifestation of these membership probabilities where the probability of a link is proportional to the sum of weighted products of community memberships for each of the community. The number of communities, given in advance, be k. Then a link $i - -j$ between nodes i and j manifests itself in the actual network with probability

$$p_{i--j} = \sum_{l=1}^{k} p_{il} p_{jl} \beta_l + \sum_{l=1}^{k} (1 - p_{il})(1 - p_{jl})\epsilon \qquad (6.6)$$

where β_l is a (high) value from the range (0,1) for each $l = 1, \ldots, k$, and ϵ is some low probability.

6.2 Structure-based Similarity in Complex Networks

Edges linking nodes in graphs express the close similarity of two individual nodes, based usually on their properties. However, for a number of applications, like link prediction or clustering, similarity measures based on structural properties also may be taken into account.

Three basic brands of structural similarity may be considered:[13]

- local similarity measures
- global similarity measures
- in-between similarity measures.

6.2.1 Local Measures

The *Common Neighbour* measure is based on the sets of neighbours of nodes. For a node x let $\mathcal{N}(x)$ denote its neighbours. Then Common Neighbour similarity can be expressed as:

$$s^{CN}(x, y) = |\mathcal{N}(x) \cap \mathcal{N}(y)|$$

If A is the adjacency matrix, then $s^{CN}(x, y) = (A^2)_{xy}$ that is it is the number of different paths of length 2 linking x and y. It turns out that this measure for collaboration networks is positively correlated with the probability of the scientists to work together in future.

The Common Neighbour measure reflects the direct relationship between nodes without taking into account a broader context of other nodes of the graph. The

---

[13] see e.g. L. Lü, T. Zhou: Link Prediction in Complex Networks: A Survey. CoRR 1010.0725

subsequent indexes try to eliminate this disadvantage, stressing, however different aspects of this context.

Salton index (referred to also as *cosine index*) is defined as

$$s^{Salton}(x, y) = \frac{|\mathcal{N}(x) \cap \mathcal{N}(y)|}{\sqrt{k_x \cdot k_y}}$$

where $k_x = |\mathcal{N}(x)|$ denotes the degree of node x. It equals Common Neighbour index divided by the geometric mean of degrees of nodes. It decreases the similarity between nodes if the common neighbours are a small fraction of all neighbours of the considered nodes.

Jaccard index is defined as

$$s^{Jaccard}(x, y) = \frac{|\mathcal{N}(x) \cap \mathcal{N}(y)|}{|\mathcal{N}(x) \cup \mathcal{N}(y)|}$$

and states how many common neighbours nodes have compared to the maximum possible number. It decreases the similarity between nodes in the same spirit as Salton index does, but is insensitive to the degree discrepancies between nodes x and y.

Sorensen index is defined as

$$s^{Sorensen}(x, y) = \frac{2|\mathcal{N}(x) \cap \mathcal{N}(y)|}{k_x + k_y}$$

Contrary to Salton index, the number of common neighbours is divided by the arithmetic and not the geometric mean of degrees of nodes x and y.

Hub Promoted index is defined as

$$s^{HPI}(x, y) = \frac{|\mathcal{N}(x) \cap \mathcal{N}(y)|}{\min\{k_x, k_y\}}$$

It equals common neighbour index divided by the smaller degree of nodes, promoting similarity of nodes of strongly differing degrees.

Hub Depressed index is defined as

$$s^{HDI}(x, y) = \frac{|\mathcal{N}(x) \cap \mathcal{N}(y)|}{\max\{k_x, k_y\}}$$

It equals Common Neighbour index divided by the larger degree of nodes. The underlying idea is that a node can be linked to a hub by chance, therefore such relations should be suppressed.

Leicht-Holme-Newman index is defined as

$$s^{LHM}(x, y) = \frac{|\mathcal{N}(x) \cap \mathcal{N}(y)|}{k_x \cdot k_y}$$

and is similar to Salton index, but it deprecates similarities between higher degree nodes.

Preferential Attachment index is defined as

$$s^{PA}(x, y) = k_x \cdot k_y$$

It expresses the idea of a network growth model in which higher degree nodes are more likely to be linked, so they are deemed to be more similar. The actual common neighbours are not taken into account.

Adamic-Adar index is defined as

$$s^{AA}(x, y) = \sum_{z \in \mathcal{N}_x \cap \mathcal{N}_y|} \frac{1}{\log k_z}$$

This index modifies the Common Neighbour measure by giving more weight to lower degree nodes. No dangling nodes are assumed.

Resource Allocation index is defined as

$$s^{RA}(x, y) = \sum_{z \in \mathcal{N}_x \cap \mathcal{N}_y|} \frac{1}{k_z}$$

This index is motivated by the idea of resource distribution. Assume that each node has one unit of a resource that it distributes uniformly to its neighbours. s^{RA} measures, how much a node obtains from its neighbours. Here, contrary to Salton and similar indexes, not the degrees of the nodes x, y play a role, but rather the degrees of their common neighbours. So the common neighbours are connecting x, y more strongly, if they have fewer links not directed to neither x nor y.

According to a number of studies, RA and AA indices seem to best express the similarity between nodes in terms of their dynamics.

6.2.2   Global Measures

The local measures can be obviously computed based on direct neighbours of the two nodes for which the similarity is sought so that the overall network structure has no impact on the perception of node similarity.

In contrast to that the global measures allow to view the similarity of two nodes in the context of the entire network.

Katz index counts all the paths connecting two nodes giving more weight to the shorter ones. Let $paths_{xy}^{<i>}$ denote the set of all paths of length i connecting x and y. Then

$$s^{Katz}(x, y) = \sum_{i=1}^{\infty} \beta^i |paths_{xy}^{<i>}| = \beta A_{xy} + \beta^2(A^2)_{xy} + \beta^3(A^3)_{xy} + \dots = ((I - \beta A)^{-1} - I)_{xy}$$

Here β is the user-defined so-called damping parameter, controlling the weight of path length. For the above sum to converge (and hence to make sense) β must be lower than the reciprocal of the largest eigenvalue of A.

The *Global Leicht-Holme-Newman index* defined as

$$s^{GLHN}(x, y) = \alpha(I - \beta A)_{xy}^{-1}$$

with two parameters α, β bears some formal resemblance to Katz index. It may be understood as promoting similarity if the neighbours are themselves similar. This is illustrated via the recursive formulation for the similarity matrix

$$S = \beta AS + \alpha I$$

Recall that $(A^i)_{xy}$ is the total number of paths of length i between x and y. The expected value of $(A^i)_{xy}$, $(E(A^i)_{xy})$, amounts to $\frac{k_x k_y}{2m}\lambda_1^{i-1}$ where m is the total number of edges, λ_1 is the largest eigenvalue of A.

Average Commute Time index between nodes x and y is defined as the inverse of the sum of average number of steps of a random walker required to reach y starting at x and average number of steps of a random walker required to reach x starting at y. The random walker goes to the neighbouring node from a given node with uniform probability.

Random Walk with Restart index is the sum of probabilities that a random walker starting at x reaches y and one starting at y reaches x. This time, however, the random walker with some fixed probability instead of going to a neighbouring node, jumps back to the starting point.

SimRank index is the number of steps after which two random walkers, respectively starting from nodes x and y, are expected to meet at any node.

Maximum Entropy SimRank index is the number of steps after which two random walkers, respectively starting from nodes x and y, are expected to meet at any node, given that we transform the graph adding weights in such a way that any infinitely long path through the graph has the same probability [325]. Note that in *SimRank*, due to the structure of the (unweighted) network some paths may prove more probable than the other.

Matrix Forest Index is defined as the ratio of the number of spanning rooted forests such that nodes x and y belong to the same tree rooted at x to all spanning rooted forests of the network. A forest is a set of disjoint trees. A spanning forest covers the whole graph. A rooted forest consists of rooted trees only (a rooted tree is a tree with a distinguished starting point, root).

6.2.3 Quasi-Local Indices

Local Path Index is an extension of the Common Neighbour measure by taking into account also paths of length 3 and is defined as $s^{LP}(x, y) = (A^2 + \beta A^3)_{xy}$ where β is a user defined parameter (when $\beta = 0$, then $s^{LP} = s^{CN}$). It can be generalised for broader neighbourhoods $s^{LP(n)}(x, y) = (A^2 + \beta A^3 + \cdots + \beta^{n-1} A^n)_{xy}$. With increasing n it will become similar to Katz index and Global Leicht-Holme-Newman index.

Local Random Walk index is the sum of probability that within a given number of steps t a random walker reaches y starting from x and of probability that within a given number of steps t a random walker reaches x starting from y

Superposed Random Walk is similar to the above but we allow that the endpoints are reached either within 1 step, or 2 steps, ..., or t steps.

6.3 Modularity—A Quality Measure of Division into Communities

Extraction of communities cuts essentially a graph into subgraphs. Hence we need to evaluate the quality of such a split. Minimisation of cut cost given by the formula (5.58) does not lead to good results as one does not seek to get a small number of connections with nodes outside of the community. Rather we seek such a cut that the number of such connections should be smaller than one in referential random graph [371]. Besides, the minimisation of *cut* leads to communities of low cardinality which induces the need of additional limitations on the group size. For this reason Newman and Girvan [373] introduced so-called modularity measure Q defined as follows:

$\mathcal{Q} =$(real percentage of edges within the communities)

$-$ (expected percentage of edges within same node groups if they were not communities)

$$(6.7)$$

A formal expression of the qualitative postulate saying that the number of connections between nodes constituting the community should differ from the expected number of random connections may be stated as follows:

$$Q(A, C) = \sum_{i=1}^{m} \sum_{j=1}^{m} (\frac{a_{ij}}{2m} - p_{ij}) \delta(c_i, c_j) \tag{6.8}$$

where p_{ij} denotes probability that two randomly selected nodes v_i, v_j are neighbours in the graph Γ with adjacency matrix $A = [a_{ij}]$. Symbol c_i, c_j denotes groups to which nodes i, j resp. are assigned under the partition C of the nodes V of the graph, and δ is the Kronecker symbol. In the referential random model the probability p_{ij} is of the form [372]

$$p_{ij} = \frac{d_i d_j}{(2m)^2} \tag{6.9}$$

Probabilities defined in this way fit the natural constraint $2m \sum_j p_{ij} = d_i$ validating that the sum of elements in each row of the so-called modularity matrix

$$B = A - (2m)P = A - \frac{dd^T}{2m} \tag{6.10}$$

is equal to zero. The matrix B is a kind of substitute for the Laplacian used in spectral optimisation. As the sum of elements of each row (as well as column) equals zero, the eigenvector $w = \alpha e$ corresponds to the eigenvalue $\beta = 0$.

The index Q takes values from the interval $[-0.5, 1]$. Hereby $Q = 0$ means that we have to do with a random graph that is all the nodes constitute in reality one community. Q lower than zero indicates that there are communities in the graph, but our guess on which nodes constitute them is worse than a random one. If Q approaches 1, then the underlying communities are strongly separated (do not communicate with each other) and our split into communities is correct.

Hence the search for communities means looking for such a partition C^* that maximises $Q(A, C)$:

$$C^* = \underset{C \in PART(V)}{\arg\min} \ Q(A, C)$$

Then quantity $Q_o(A)$ would be then called *optimal modularity*. Optimal modularity ranges from 0 to 1.[14]

Typical values of optimal modularity in empirical graphs range from 0.3 to 0.7. As Newman and Girvan [373] state, higher values are rarely observed. Same authors suggest that high values of Q_o correlate with a stricter split of a graph into subgraphs. This is confirmed by White and Smyth [501], who observed, that the index Q may be used both for determining the correct number k of clusters and of the content of the respective clusters.

Relations between objects corresponding to nodes of abstract network differ frequently by intensity, therefore depicting them in a binary way is a simplification. Therefore subsequently we will assume that we have to do with a weighed undirected graph[15] $\Gamma = (V, E, S)$, in which s_{ij} is the weight of the edge connecting nodes v_i and v_j. As Newman [368] notes, weighed graphs can be analysed using

---

[14]It is easily seen that isolated nodes do not impact the modularity value. It is known that a clustering with maximum modularity does not contain any cluster that consists of a single node with degree 1. Furthermore, there exists always a clustering with maximum modularity, in which each cluster is a connected subgraph. Clustering according to maximal modularity exhibits non-local behaviour on adding new nodes, and in particular it is sensitive to satellite nodes (attached leaf nodes). Generally, the size and structure of clusters in the optimal clustering are influenced by the total number of links in the graph. So structures observed locally in a subgraph may disappear upon embedding into a larger graph. See [88].

[15]We will discuss the case of directed graphs in point 6.5.

standard techniques in many cases through application of a simple mapping of a weighed graph onto one without weights. In the simplest case one can assume that:

$$a_{ij} = \begin{cases} 1 \text{ if } s_{ij} > 0 \\ 0 \text{ otherwise} \end{cases}$$

In particular the definition (6.8) can be rewritten as

$$Q = \frac{1}{2w} \sum_{i=1}^{m} \sum_{j=1}^{m} \left(s_{ij} - \frac{s_i s_j}{2w} \right) \delta(c_i, c_j) \tag{6.11}$$

whereby

$$s_i = \sum_{j=1}^{m} s_{ij} \tag{6.12}$$

and

$$2w = \sum_{i=1}^{m} s_i = \sum_{i=1}^{m} \sum_{j=1}^{m} s_{ij} \tag{6.13}$$

It is worth noting that Q is an additive function of the influences of individual nodes of the graph. To show this, let us denote with q_i the quantity [28]

$$q_i = \frac{1}{2w} \sum_{j=1}^{m} \left(s_{ij} - \frac{s_i s_j}{2w} \right) \delta(c_i, c_j), \quad i = 1, \ldots, m \tag{6.14}$$

Then $Q = \sum_{i=1}^{m} q_i$.

The formula (6.11) can be rephrased in a way proposed by [123] to reflect the requirement (6.7). Let

$$e_{ij} = \frac{1}{2w} \sum_{u=1}^{m} \sum_{v=1}^{m} s_{uv} \delta(c_v, i) \delta(c_v, j) \tag{6.15}$$

denote the empirical probability that ends of a randomly selected edge belong to ith and jth group resp. Let further

$$a_i = \frac{1}{2w} \sum_{u=1}^{m} s_u \delta(c_u, i) \tag{6.16}$$

denote probability that at least one end of a randomly chosen edge belongs to the i-th group.

Remembering that $\delta(c_u, c_v) = \sum_{g=1}^{k} \delta(c_u, g)\delta(c_v, g)$ and taking into account Eq. (6.9) we can transform the formula (6.11) to the form:

$$
\begin{aligned}
Q &= \frac{1}{2w} \sum_{u=1}^{m} \sum_{v=1}^{m} \left[s_{uv} - \frac{s_u s_v}{2w} \right] \sum_{g=1}^{k} \delta(c_u, g)\delta(c_v, g) \\
&= \sum_{g=1}^{k} \left[\frac{1}{2w} \sum_{u=1}^{m} \sum_{v=1}^{m} s_{uv}\delta(c_u, g)\delta(c_v, g) - \frac{1}{2w} \sum_{u=1}^{m} s_u\delta(c_u, g)\frac{1}{2w} \sum_{v=1}^{m} s_v\delta(c_v, g) \right] \\
&= \sum_{g=1}^{k} (e_{gg} - a_g^2)
\end{aligned}
$$

(6.17)

Here e_{gg} is the fraction of edges connecting nodes belonging to the community g, while a_g denotes the expected number of edges within the group g under assumption that nodes of the graph Γ have been randomly connected while maintaining the original node degrees in the graph.

Q, like many metrics, is not fault-free. Fortunato and Barthelémy observed so-called resolution limit resulting from the assumption that in the referential random model $p_{ij} \sim d_i d_j$, [183]. In large graphs the expected number of links between small groups is too small so that even a single link can paste both groups into a non-distinguishable entity. More precisely, if

$$
e_{gg} < \sqrt{w} - 1
$$

(6.18)

then the group g will not be extracted by an algorithm maximising Q. Arenas et al. proposed in [29] a method overcoming this imperfection.[16]

Modularity possesses also a very distinctive advantage: maximising Q we get as a by-product the number of groups occurring in the given network. Cost functions of classic clustering algorithms like $J(U, M)$ defined by (3.2) for k-means tend to reach their optimal (minimal) point when the set $\mathbf{X}$ is split into m one-element clusters. Spectral algorithms behave in a similar manner. But Q reaches its maximum for a certain $k \ll m$ (which is implied also by the assumption about the form of probabilities p_{ij}). This fact makes it an interesting object of research.

Evans and Lambiotte [172] draw attention to still another view of the modularity from (6.8), rewritten as

---

[16]Whether the resolution limit is an advantage or disadvantage, depends on the application. We may be not interested in having too many groups to consider and in this case resolution limit may prove to be a useful vehicle.

$$Q(A, C) = \sum_{i=1}^{m} \sum_{j=1}^{m} \left(\frac{a_{ij}}{2m} - \frac{d_i d_j}{(2m)^2} \right) \delta(c_i, c_j) \tag{6.19}$$

$$= \sum_{C \in \mathcal{C}} \sum_{i \in C} \sum_{j \in C} \left(\frac{a_{ij}}{d_j} \frac{d_j}{2m} - \frac{d_i}{2m} \cdot \frac{d_j}{2m} \right) \tag{6.20}$$

Consider now random walk in stationary conditions in an undirected network. The expression $\sum_{i \in C} \sum_{j \in C} \frac{a_{ij}}{d_j} \frac{d_j}{2m}$ describes the probability that a random walker for two consecutive steps stays in the cluster C. The expression $\sum_{i \in C} \sum_{j \in C} \frac{d_i}{2m} \cdot \frac{d_j}{2m}$ is the probability that two independent random walkers in one step will be in C. Note that the matrix $P = AD^{-1}$, expressing the transition probability between nodes has d as its (unnormalised) eigenvector for eigenvalue 1. So the same vector is also eigenvector of P^t being the transition matrix in t steps. And the matrix $P^t D$ is symmetric with jth row/column sum being jth node degree d_j. Hence one may generalise the notion of modularity to

$$Q(t, A, C) = \sum_{C \in \mathcal{C}} \sum_{i \in C} \sum_{j \in C} \left(\frac{(P^t)_{ij} d_j}{2m} - \frac{d_i}{2m} \cdot \frac{d_j}{2m} \right) \tag{6.21}$$

being the difference of probabilities of one random walker staying in the same cluster for $t + 1$ consecutive steps minus the probability of two independent random walkers "meeting" in the same cluster at one point in time (under stationary conditions).

Last not least let us draw the attention of the reader to an intriguing interpretation of the modularity coefficient in the next section.

6.3.1   Generalisations of the Concept of Modularity

The descriptive definition of a community, its qualitative character, inspires search for more formal foundations allowing to express the essence of community more precisely. Reichardt and Bornholdt proposed in [405] to form a community by attaching to it the nodes that communicate with one another and move to other communities those nodes that do not communicate with one another. If we denote with c a vector, elements c_i of which indicate the identifier of the community to which the ith node belongs, then this principle leads to the Hamiltonian

$$\mathcal{H}(\mathbf{c}) = - \sum_{i<j} \alpha_{ij} a_{ij} \delta(c_i, c_j) + \sum_{i<j} \beta_{ij} (1 - a_{ij}) \delta(c_i, c_j) \tag{6.22}$$

The first component sums up all nodes communicating with one another, and the second—nodes that do not communicate with one another in the graph $\Gamma = (V, E)$. Each pair of linked nodes (i, j) is rewarded by the weight α_{ij}, and each pair $(i, j) \notin E$

is assigned the punishment β_{ij} if only these two nodes are assigned to the same community.

Let us consider the problem of such an assignment of nodes to groups that this Hamiltonian reaches the minimum. Solution of this task requires first defining the coefficients α_{ij} and β_{ij}. Reichardt and Bornholdt noted that if the weights are chosen in such a way that the equilibrium condition below is met

$$\sum_{i<j} \alpha_{ij} a_{ij} = \sum_{i<j} \beta_{ij} (1 - a_{ij}) \tag{6.23}$$

then one can assume that $\alpha_{ij} = 1 - \gamma p_{ij}$ and $\beta_{ij} = \gamma p_{ij}$ for some γ, which reduces the number of Hamiltonian (6.22) parameters significantly. In order for the above condition to hold, we have to require $\gamma = 2|E| = 2m$. Under the above assumptions the Eq. (6.22) turns to a simpler form

$$\mathcal{H}(\mathbf{c}) = - \sum_{i<j} (a_{ij} - \gamma p_{ij}) \delta(c_i, c_j) \tag{6.24}$$

By defining p_{ij} in accordance with Eq. (6.9) we reduce the problem of minimising the Hamiltonian (6.22) to the task of maximising the modularity Q. This demonstrates that the problem of detecting the structure in an empirical graph can be treated as a search for configuration of spins of a spin glass with minimal energy.

In subsequent sections we will deal with various variants of modularity and will present selected methods of its optimisation. One shall keep in mind that optimisation of the Q index is a $\mathcal{NP}$-hard task [87].

6.3.2   Organised Modularity

Rossi and Villa-Vialaneix [408] propose a modification of Newman's modularity measure which would reflect not only the partition of the data points but also reveal a structure of partitions that may be presented in a plane. In its spirit it is similar to the concept of SOM (Self-Organising Maps). For purposes of visualisation, a graph is coarsely grained and closer (more similar) clusters are displayed closer to one another. A representation framework of the graph on the plane is assumed, for example a rectangle subdivided into squares being predefined placements for a predefined set of clusters. We have e.g. clusters $C_1, ..., C_N$ with geometric positions in this framework $x_1, ... x_N$. In order to enforce the distribution of the actual clusters of the graph on this plane, they redefine the modularity measure Q given by Eq. (6.8) as follows.

$$Q(A, C, M) = \sum_{i=1}^{m} \sum_{j=1}^{m} \left(\frac{a_{ij}}{2m} - p_{ij} \right) M(c_i, c_j) \tag{6.25}$$

where $M(c_i, c_j)$ is a function representing the geometrical layout, for example $M(c_i, c_j) = \exp(-\sigma \|x_i - x_j\|^2)$. In Newman's modularity, an edge connecting different clusters did not count, as Kronecker δ was used. Now M is used instead causing all the edges to count always, but with different weights, depending on how distant are the clusters placed on the plane.

6.3.3 Scaled Modularity

Arenas et al. [29] are concerned with the so-called resolution limit of the modularity. Their approach to handle this issue is to apply modularity search at "different topological scales". To facilitate such an approach they reweigh edges in a graph, so they in fact work with the modularity version for weighted graphs from Eq. (6.11). They investigate the fact that given the total sum of weight of edges equals w, then two identically structured groups of nodes each with the total sum of weights of edges inside the group s_c and connected by a single link will not be distinguished by Q maximisation, if $s_c < \sqrt{w/2} - 1$. They propose to cure such a situation by increasing the weights of each edge by a quantity such that the sum of weights of edges incidental with any node increases by r. This may be achieved via adding a self-loop at each node with this weight r. Now the total sum of weights amounts to $w + mr$ and the weight within a cluster to $s_c + n_c r$. Hence the condition turns to $s_c < \frac{1}{2} \left(\sqrt{2w + mr} - n_c r - 2 \right)$. As square root of r grows more slowly than r, this equation means that we can add such a weight that the groups will become separable by the algorithm. At lower values of r the modules will tend to be bigger, at higher—smaller. The maximum value of r that makes sense to be investigated is the one when each node becomes a module of its own. This is the smallest value where for each edge (i, j) $s_{ij} < \frac{(s_i + r)(s_j + r)}{2w + mr}$. r may be also negative leading to "superstructures" of the network. The lowest possible value that still makes sense is the one that leads to a network consisting of a single module. The authors of [29] show that it is approximately equal $r_{app} = -\frac{2w}{m}$. They investigate the persistence of modules while r changes.

6.3.4 Community Score

Pizzuti [394] proposes a modularity inspired, but a bit different measure of community tendency in a graph. He deals with undirected graphs. Hence the connection matrix A is symmetric. His new measure of community quality is built around the "volume" and the "density" of the community subgraph. The volume $\text{vol}(C)$ of the community C is understood here as double number of links within the cluster C. For a node i of the community C its connectivity $z_i(C)$ to C is understood as the number of links leading from i to any other node in C divided by cardinality of $C/\{i\}$. He defines the power connectivity of rank r of the community C as $Z(C, r) = \frac{\sum_{i \in C} z_i(C)^r}{|C|}$. The product $Qu(C) = Z(C, r) \text{vol}(C)$ is called community quality. The sum of $Qu(C)$ over all communities constituting the partition is called the Community Score CS. Maximal score is sought when looking for best split into communities. The higher r the more dense communities will be identified.

6.4 Community Discovery in Undirected Graphs

Community detection plays an important role in those areas where the relations between entities are modelled using graphs, e.g. in sociology, biology or in computer science. It is a difficult problem with apparently no general solutions. Difficulties are multiplied by the fact that in general it is not known in advance how many communities exist in a structure, while some structure elements may belong to several communities at once.

6.4.1 Clique Based Communities

Let $\Gamma = (V, E)$ be an empirical graph. A natural representative of a community in such a graph is a clique or near-clique.

Palla et al. [384] investigates communities that consist of overlapping k-cliques (cliques consisting of k nodes). For two nodes, in order to belong to the same community, there must exist a chain of overlapping k-cliques leading from one node to the other. The neighbouring cliques in the chain must have $k - 1$ nodes in common. For $k = 2$, this definition of a community reduces to that of a connected component of a graph. To detect such communities, they elaborated so-called "clique percolation method".

Kleinberg introduced the concept of a *hub* and *authority* in a directed graph, c.f. e.g. [198]. A node is called authority, if it is pointed at by many other ones. A hub is a node that points to many other nodes. A community is considered to be a set of "high" authorities together with hubs connecting them.

6.4.2 Optimisation of Modularity

Agarwal and Kempe [6] suggested that the problem of community identification should be treated as a problem of linear programming.

Let us denote with $\mathbf{x}_{ij}$ that variable that is equal 1 if the nodes i and j belong to the same community and is equal 0 otherwise. Then we obtain the program

$$\max \frac{1}{2m} \sum_{i=1}^{m} \sum_{j=1}^{m} b_{ij}(1 - x_{ij})$$

$$\text{subject to } x_{ik} \le x_{ij} + x_{jk}, \quad \forall(i, j, k)$$

$$x_{ij} \in \{0, 1\}$$

(6.26)

The first constraint results from the necessity to ensure transitivity: if nodes i, j belong to the same group and nodes j and k do so also, then also the nodes i and k must belong to the same group.

Solution to the zero-one problem is a $\mathcal{NP}$-had task, therefore the authors propose to relax it. On the other hand in [514] integer programming is proposed to solve the considered problem.

6.4.3 Greedy Algorithms Computing Modularity

Computing the value of optimal modularity is an $\mathcal{NP}$-hard task.

Therefore a number of algorithms have been developed for its computation, in particular greedy algorithms that are fast but without any guarantees of approaching the real value [88, Theorem 7]. So one can speak in fact about "optimal modularity according to algorithm XYZ".

There exist essentially two brands of modularity computation algorithms: top down and bottom up.

One of the first bottom-up approaches was proposed by Newman [366]. Initially each node constitutes its own community. Then the two communities are merged that yield the greatest increase of Q among candidate pairs (even if this "increase" is negative). It is proceeded until all nodes are merged to a single community. Then the maximum Q is selected among all the intermediate community structures.

Algorithm 6.1 Newman [366] algorithm of community detection

Input: Similarity matrix S.
Output: The number of clusters k and a clustering $C = \{C_1, \ldots, C_k\}$.
1: Create an initial clustering $C_0 = \{C_1, \ldots, C_m\}$ where each element lies in a separate cluster.
2: **for** $i = 1$ to $m - 1$ **do**
3:   Find pair of clusters $C_a, C_b \in C_{i-1}$ such that $Q(S, (C_{i-1} - \{C_a, C_b\}) \cup \{C_a \cup C_b\}))$ attains maximum value among all pairs of clusters in C_{i-1}.
4:   Create $C_i = (C_{i-1} - \{C_a, C_b\}) \cup \{C_a \cup C_b\}$.
5: **end for**
6: Find $i_{best} = \arg\max_i Q(S, C_i)$
7: **return** $C_{i_{best}}$.

Clauset et al. [123] modified this algorithm proposing so-called FastGreedy algorithm, which detects communities in logarithmic time with network size, and was demonstrated to be suitable for networks of 500,000 nodes. Essentially it concentrates on creating and maintaining a hierarchical structure which accelerates look-up of tables used in Newman's algorithm [366].

Wakita and Tsurumi [489] pointed at a weakness of FastGreedy algorithm as the hierarchical structure could degenerate to a linear one and so instead of logarithmic a linear complexity arises. Therefore they proposed another method of Q computation for networks of more than 5,000,000 nodes. Their modification aims at balancing the dendrogram of merged communities via so-called "consolidation ratios".

Algorithm 6.2 Louvain [77] algorithm of community detection

Input: Similarity matrix S.
Output: The number of clusters k and a clustering $C = \{C_1, \ldots, C_k\}$.
1: Create an initial level 0 set of sets of sets $C_0 = \{C_1, \ldots, C_m\}$ where for each node i $C_i = \{\{i\}\}$.
2: For a set of sets of sets $D = \cup_j\{\{\cup_l\{D_{jl}\}\}\}$ where D_{jl} is a set define $collapse(D) = \cup_j\{\{\cup_l D_{jl}\}\}$
3: For a set of sets $D = \cup_j\{D_j\}$ where D_j is a set define $lift(D) = \cup_j\{\{D_j\}\}$
4: $mergerpossible = TRUE$
5: $level = 0$
6: **while** $megerpossible$ **do**
7:    $mergerpossible = FALSE$
8:    $mergerpossibleatlevel = TRUE$
9:    **while** $megerpossibleatlevel$ **do**
10:       Find pair of sets $C_a, C_b \in C_{level}$ and a $C \in C_a$ such that $Q(S, collapse((C_{level} - \{C_a, C_b\}) \cup \{C_a - \{C\}, C_b \cup \{C\}\})$ attains maximum value among all pairs of clusters and elements of the first one in C_{level}.
11:       **if** $(Q(S, collapse((C_{level} - \{C_a, C_b\}) \cup \{C_a - \{C\}, C_b \cup \{C\}\}) < Q(S, collapse(C_{level})))$
          then
12:          $mergerpossibleatlevel = FALSE$
13:       **else**
14:          $mergerpossible = TRUE$
15:          $C_{level} := (C_{level} - \{C_a, C_b\}) \cup \{C_a - \{C\}, C_b \cup \{C\}\}$
16:       **end if**
17:    **end while**
18:    $C_{level+1} = lift(collapse(C_{level}))$
19:    $level = level + 1$
20: **end while**
21: Find $i_{best} = \arg\min_i Q(S, C_i)$
22: **return** $collapse(C_{level})$.

Another modification of Newman's approach was proposed by Blondel et al. [77], called frequently *Louvain algorithm*. It runs as follows: each node constitutes its own community at the beginning. Then individual nodes are moved between neighbouring communities (to increase the value of Q—only the biggest increase is permitted in each step) until a stopping criterion is reached (e.g. no positive increase of Q can be achieved.). We obtain in this way level 1 communities. We start operating on "level 1" communities performing very same operations, treating now the level 1 communities like previously single nodes. We form initial level 2 communities as level 1 communities, then move level-1 communities between level 2 communities until no more improvement of Q can be achieved. And so forth for next community levels until at a given level no merger is possible. Few levels only need to be considered. However, the first level is computationally expensive. The disadvantage of this algorithm are the computation burden for the first level.

Other variants, like [430], concentrate on pruning the clustered graph as efficiently as possible. Still another approach [109] adds a splitting phase to communities to get rid of the problem of greedy algorithms that get stuck in local minima.

An example of a top-down approach is the algorithm of Girvan and Newman [201]. They concentrate on removing of edges that for sure lie between communities. So for each edge they compute the edge betweenness (number of shortest paths between network nodes running through this edge), remove the edge with the highest betweenness, recompute the betweenness for the network and repeat the steps until no edge is left. In the process, the network is stepwise split into communities that are refined and for each set of communities obtained underway we can compute Q and identify the highest one.

Dinh and Thai [157], while considering constant factor approximations to modularity in scale-free networks, developed *Low-degree Following Algorithm* (LDF), applicable for scale-free undirected or directed networks with high exponents $\gamma (\geq 2)$ of the power distribution of node degrees (a node has the degree k with probability of $\frac{e^\alpha}{k^\gamma}$ for some normalising constant e^α). The algorithm divides the nodes into three subsets L (leaders), M (members) and O (orbiters) and the remaining ones that are unlabelled. Each orbiter is assigned a member that it follows. Each member follows a leader. Initially all nodes are unlabelled. A user-defined parameter d_0 is introduced being the upper bound for degrees of members and orbiters. The algorithm processes unlabelled nodes one by one. If the unlabelled node i is of degree at most d_0, if there exists a node $j \in N_i - M$, where N_i is the neighbourhood of i, such that j is a leader or unlabelled, then i is added to M and j to L, and i follows j. Otherwise all neighbours of i are members and i is inserted into O and is assigned to follow one arbitrary chosen elements of its neighbourhood. Finally, all members that follow the same leader together with orbiters following them constitute one community. Each node that was left unlabelled forms a community of its own. d_0 is either set arbitrarily, or is determined by setting d_0 to each possible value of node degree and choosing the d_0, that maximises the modularity. In case of $\gamma < 2$, the number of low degree vertices decreases radically. As a result, LDF fails to provide any performance guarantee as it cannot produce enough edges with both endpoints inside the same community. As a backup solution for such a case Dinh and Thai [157] suggest to use a bisectional approach.

6.4.4 Hierarchical Clustering

Hierarchical clustering is a classical method for extraction of clusters the elements of which are sufficiently similar. To construct the respective dendrogram, the agglomerative algorithm 2.1 is used. The most popular variant is the so-called single-linkage where edge is added to include into the community a node that is most similar to any node of the community.

However, the complete linkage method appears to be most appropriate here. A newly attached node needs to have the highest similarity to all the nodes of the community. It is assumed here that the communities are maximal graph cliques of the graph hence the method is rarely applicable. The first reason is that it is a hard task to detect communities. Here an algorithm developed by Bron and Kerbosch [91] is applied. The second reason is that the cliques are rarely unambiguous. Usually, a node belongs to several cliques so a decision on the membership needs to be made. Usually a node is attached to the largest community.

Another problem is the choice of an appropriate similarity measure. In sociology the principle of structural equivalence is applied: two nodes are structurally equivalent if they have identical neighbourhood. In practice, however, this condition is met only to certain degree. As a consequence Euclidean distance is frequently used (keeping in mind that this is not a similarity measure but rather a dissimilarity measure).

$$d_{ij} = \sqrt{\sum_{k \neq i,j} (a_{ik} - a_{jk})^2} \tag{6.27}$$

Alternatively one may exploit the Pearson's correlation coefficient

$$r_{ij} = \frac{\frac{1}{m} \sum_{k=1}^{m} (a_{ik} - \mu_i)(a_{jk} - \mu_j)}{\sigma_i \sigma_j} \tag{6.28}$$

where a_{ij} is an element of the neighbourhood matrix, and μ_i and σ_i^2 are mean and variance of the ith row resp., that is

$$\mu_i = \frac{1}{m} \sum_{j=1}^{m} a_{ij}, \quad \sigma_i^2 = \frac{1}{m} \sum_{j=1}^{m} (a_{ij} - \mu_j)^2$$

Positive r_{ij} values mean that nodes i and j can be assigned to the same group, and in case of $r_{ij} < 0$—to distinct groups. However a deeper analysis reveals that the value of the correlation coefficient depends on the network size and may vanish for large networks [160, 276].

One may use other association measures, borrowed from statistical mechanics, like correlation coefficient τ_b of Kendall, [287], or coefficient γ_g of Goodman-Kruskal.[17] Another simple similarity measure is a variant of the Jaccard coefficient

$$\omega_{ij} = \frac{|N(i) \cap N(j)|}{|N(i) \cup N(j)|} \tag{6.29}$$

---

[17]Goodman, Leo A.; Kruskal, William H. (1954)."Measures of Association for Cross Classifications". J. of the American Statistical Association 49 (268): 732–764.

where $N(i)$ denotes the set of neighbours of the node i.

Girvan and Newman [201] discuss two similarity measures applied in classical agglomerative algorithm. One is the (maximum) number of node-independent paths linking the compared nodes.[18] It is known from the flow network theory that the number of node-independent (resp. edge-independent) paths linking two nodes v_i and v_j is equal to the minimal number of nodes (resp. edges) that need to be removed from the graph in order to achieve that v_i and v_j belong to disjoint subgraphs.

Another method of determining the similarity consists in computing the number of all possible paths from v_i to v_j. These numbers are elements of the matrix

$$W = \sum_{l=0}^{\infty} (\alpha A)^l = (I - \alpha A)^{-1} \tag{6.30}$$

To guarantee that the above sum is finite, α must be lower than the reciprocal of the largest eigenvalue of the matrix A.

Regrettably, each of these measures yields good results only under some circumstances.

Girvan and Newman decided to identify those edges that connect various communities to the highest degree b. The method starts with the original graph from which nodes with the highest coefficient b are removed. Then the degrees b are computed again for the resulting graph and this procedure is repeated until there is no edge in the graph. Quick algorithms computing such measures are discussed in [86, 367].

An edge e is assigned the number of shortest paths linking any two nodes and containing this edge.

Other variants of applicable similarity measures were already listed in Sect. 6.2

6.4.5 Spectral Methods

Authors of [501] pioneered in application of spectral methods to modularity optimisation. Subsequent important results stem from Newman [370, 372]. Discovery of communities in very large networks was considered in papers [77, 489]. The paper [521] presents application of kernel methods.

6.4.5.1 Community Discovery as a Task of Spectral Optimisation

White and Smyth [501] approached the problem in the spirit of spectral relaxation (exploited in chapter 5), and proposed two algorithms of optimisation of the modularity Q in undirected weighted graphs. Their idea was to treat the maximisation of

---

[18]Two paths linking two nodes v_i and v_j are *node-independent* if their only common nodes are the starting and ending nodes of each path that is v_i and v_j. Similarly *edge-independent paths* can be defined.

the index Q as a quadratic assignment task. They replace the task with its continuous version and find an approximate solution. Then they select $k-1$ leading eigenvectors of a proper matrix and reach the final solution.

Assume that we split the set of nodes into k subset, that is $V = V_1 \cup \cdots \cup V_k$. Let us rewrite the formula (6.17) in the form

$$Q = \sum_{g=1}^{k} \left[\frac{A(V_g, V_g)}{V} - \left(\frac{A(V_g, V)}{V} \right)^2 \right] \tag{6.31}$$

where

$$A(V', V'') = \sum_{i \in V', j \in V''} s_{ij}$$

and $V = A(V, V) = \operatorname{vol} V$. Let further $C = [C_{ij}]_{m \times k}$ be a matrix representing the partition into k disjoint subsets, that is

$$C_{ij} = \begin{cases} 1 \text{ when } i\text{th node belongs to } j\text{the community} \\ 0 \text{ otherwise} \end{cases} \tag{6.32}$$

The mentioned authors demonstrate in [501], that the value Q can be computed as

$$\begin{aligned} Q &\propto \sum_{g=1}^{k} \left[C_g^T S C_g V - (\mathbf{d}^T C_g)^2 \right] \\ &= \sum_{g=1}^{k} \left[C_g^T S C_g V - C_g^T \mathbf{d}\mathbf{d}^T C_g \right] \end{aligned} \tag{6.33}$$

where C_g denotes gth column of the matrix C, S is the matrix representing the weights of connections, and $\mathbf{d}$ is the vector of generalized degrees with components $\mathbf{d}_i = \sum_j s_{ij}$. By denoting $\mathcal{D} = \mathbf{d}\mathbf{d}^T$ we give the above formula the form

$$\begin{aligned} Q &\propto V \cdot \operatorname{tr}(C^T S C) - \operatorname{tr}(C^T \mathcal{D} C) \\ &= \operatorname{tr}\left[C^T (V S - \mathcal{D}) C \right] \end{aligned} \tag{6.34}$$

The problem consists in maximisation (over all assignments C) of the above formula under the constraint $C^T C = M$, where M is the diagonal matrix in which m_{ii} represents the cardinality of the set V_i. This is an $\mathcal{NP}$-hard task, therefore White and Smyth replace the constraints $C_{ij} \in \{0, 1\}$ with the condition $C_{ij} \in \mathbb{R}$, and relax the optimisation task to $\operatorname{tr}\left[C^T (V S - \mathcal{D}) C \right] + (C^T C - M) \Lambda$. We take a derivative with respect to C of this expression and compare it to 0 (to get the optimum), which leads to the task:

$$L_Q S = S \Lambda \tag{6.35}$$

where $L_Q = \mathcal{D} - V S$ is called Q-Laplacian, and Λ is the diagonal matrix of Lagrange multipliers. It turns out that, with the increase of the cardinality m of the set V, L_Q

Fig. 6.1 Result of 10-fold application of the NJW algorithm for a neighbourhood matrix describing relations between jazz musician (*Jazz* network). The maximal value of $Q = 0.438591$ is obtained each time when splitting into $k = 4$ groups

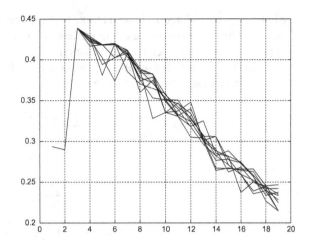

converges rapidly to the matrix S', the rows of which sum up to 1. In the paper [501] it was assumed that $S' = D^{-1}S$. The problem consists therefore in finding $k - 1$ leading eigenvectors of the matrix S', building of the matrix U, the columns of which are these eigenvectors and applying a quick clustering algorithm to obtain the proper partition. For this task, the algorithm of Ng, Jordan and Weiss (see pseudo-code 5.4) can be used (Fig. 6.1).

Example 6.4.1 To check the effectiveness of the method described here, the NJW algorithm was applied in the analysis of three benchmark data sets: *Jazz*, *metabolic* and *e-mails*. k-means was used for clustering the nodes, where k most distant points were chosen as initial centres.

For the data set *Jazz* a unique value $Q_{max} = 0.438591$, corresponding to a partition into 4 disjoint groups, was obtained.

For the data set *C.elegans* (metabolic) the situation is more complicated. Different (local) maxima were obtained for various numbers of clusters: $Q_{max} = 0.4109(k = 19), 0.4147(k = 14), 0.4097(k = 12), 0.4127(k = 14), 0.4078(k = 23), 0.4136(k = 14)$.

For the data set *e-mail* we obtained $Q_{max} = 0.5516$ for $k = 12$ and $k = 10$. A typical distribution of the values of Q as a function of k is presented in Fig. 6.2.

Problem: What is the reason for such a diversity of results? For example, in case of the data set *C.elegans* we had nearly identical input matrix (with variation of about 10^{-6}). Fluctuations of the results are stemming from the initialisation of the k-means algorithm and not from the input data. Interestingly, the values of modularity are identical for $k < k_{opt}$, and they diverge for $k > k_{opt}$, though this last observation does not hold for *C.elegans*.

An attempt to normalise the columns to the range $[-1, +1]$ worsens the results!!!   □

Zhang, Wang and Zhang [534] modified the formula (6.17) introducing the dependence of modularity on a membership function which divides the node set into k not

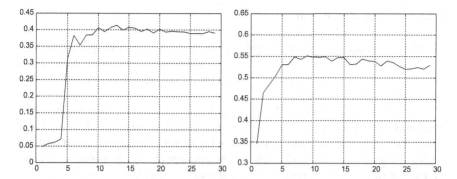

Fig. 6.2 Distribution of the values of the modularity Q as a function of the number of clusters for the sets *C.elegans* (to the left) and *e-mail* (to the right)

necessarily fuzzy subsets. Let U denote, like in Sect. 3.3, a matrix of dimensions $m \times k$, elements u_{ij} of which represent degree of membership of the ith node to the jth class. The new modularity function is of the form

$$\tilde{Q} = \sum_{g=1}^{k} \left[\frac{A(\tilde{V}_g, \tilde{V}_g)}{2w} - \left(\frac{A(\tilde{V}_g, V)}{2w} \right)^2 \right] \tag{6.36}$$

where

$$\tilde{V}_c = \{i \in V : u_{ic} > \lambda\}, \qquad \lambda \in (0, 1)$$

$$A(\tilde{V}_g, \tilde{V}_g) = \sum_{i,j \in \tilde{V}_g} [(u_{ig} + u_{jg})/2]s_{ij}$$

$$A(\tilde{V}_g, V) = A(\tilde{V}_g, \tilde{V}_g) + \sum_{i \in \tilde{V}_g, j \in V \setminus \tilde{V}_c} [(u_{ig} + (1 - u_{jg}))/2]s_{ij}$$

while the quantity $2w$ is defined via Eq. (6.12).

To determine the communities, the authors of [534] propose to use the NJW algorithm (see pseudo-code 5.4), that is the neighbourhood matrix A and the degree matrix D are determined, then the generalized eigen-problem $Ax = \mu Dx$ is solved. FCM algorithm from Sect. 3.3 is used to cluster objects with coordinates represented by the matrix Y from the Algorithm 5.4. In case of big data sets one of its modifications from Sect. 3.3.5 may be applied. The choice of the proper number of groups is made by checking for which value $g \in \{2, \ldots, k_{max}\}$, $k_{max} < m$, the index (6.36) reaches the maximal value.

6.4.5.2   Spectral Bisection

The classical spectral bisection algorithm from Sect. 5.3.1.2 is also applied for community detection. The split into communities is achieved via analysis of eigenvectors of the Laplacian $L = D - A$, where A is the neighbourhood matrix and D is the degree matrix of the graph Γ.

If Γ contains k disjoint communities, i.e. $V = C_1 \cup \cdots \cup C_k$ and no nodes of C_i communicates with any node from C_j, $i \neq j$, then Laplacian is a diagonal block square matrix, where each block represents Laplacian of a separate community. In such a situation algebraic multiplicity of the minimal (zero) eigenvalue λ_0 is equal k, and gth value λ_0^g corresponds to the eigenvector v_0^g with components $v_{0,i}^g = \alpha$, when ith node belongs to community c_g and $v_{0,i}^g = 0$ otherwise. Here α is a positive constant normalising the eigenvector.

If the communities communicate with each other, then there exists exactly one vector v_0 corresponding to the minimal eigenvalue λ_0 and $k-1$ positive eigenvalues. If exactly two communities exist in Γ, then they can be identified via computation of Fiedler vector.

As an example, consider the well-known karate club graph [525], representing relationships between club members. Fiedler value of the Laplacian representing this graph equals $\lambda_1 = 0.468525$. It is not very low (which may suggest existence of further subgroups), but the corresponding Fiedler vector quite well indicates the two groups into which this club was split in reality (cf. Fig. 6.3). One of these groups consists of nodes $\{1, 2, 4, 5, 6, 7, 8, 11, 12, 13, 14, 17, 18, 20, 22\}$, and the other one the remaining nodes from the overall set of 34 nodes. As Newman [369] noticed, only node 3 was wrongly classified. But one sees easily that its group membership is very low, hence the algorithm does not "insist" on membership of this node in any group.

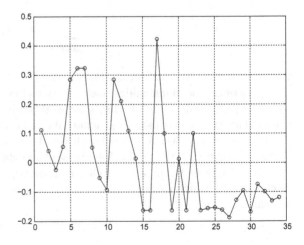

Fig. 6.3 Fiedler vector values for a graph representing friendships among members of a karate club [525]

The advantage of the spectral bisection is its simplicity and relative speed. A vice of this method is the split into two groups. Though one could divide each of the groups into further subgroups, but nonetheless the results aren't always satisfactory. And even if they are, there exists no clear criterion when to stop the procedure of subsequent partitions. This disadvantage is absent in Newman's method [370] presented in the next section.

6.4.5.3 Newman's Bisection Method Modification

Assume that we have to do with a graph having neighbourhood matrix A, and that we want to divide a set V into two disjoint subsets C and $\overline{C}$. Similarly as in section 5.3.1.2 the membership of nodes to a concrete set is represented by the vector χ with elements defined as in Subsect. 5.3.1, that is

$$\chi_i = \begin{cases} +1 \text{ if } \mathbf{x}_i \in C \\ -1 \text{ if } \mathbf{x}_i \in \overline{C} \end{cases} \quad i = 1, \ldots, m \tag{6.37}$$

Note that $\delta(g_i, g_j) = \frac{1}{2}(\chi_i \chi_j + 1)$. If we denote by P a matrix with elements p_{ij} given by the Eq. (6.9), then we can transform the definition (6.8) into a computationally convenient form

$$Q = \frac{1}{4|E|} \chi^{\mathsf{T}} (A - (2\mathrm{m})P)\chi = \frac{1}{4|E|} \chi^{\mathsf{T}} B \chi \tag{6.38}$$

where B means the modularity matrix (6.10).

By analogy to section 5.3.1 let us treat the vector χ as a linear combination of eigenvectors of the matrix B

$$\chi = \sum_{i=1}^{m} a_i \mathbf{u}_i$$

Then we can rewrite the Eq. (6.38) as

$$Q = \frac{1}{4|E|} \sum_{i=1}^{m} a_i^2 \beta_i \tag{6.39}$$

where β_i are eigenvalues corresponding to the eigenvectors $\mathbf{u}_i$.

Now, however, instead of minimising the cost of $cut(C, \overline{C})$, we are interested in maximising the index Q. If we arrange eigenvalues of the matrix B in decreasing order $\beta_0 \geq \beta_1 \geq \cdots \geq \beta_{m-1}$, then the requirement of maximising (6.39) leads to the rule

$$\chi_i = \begin{cases} +1 \text{ if } u_{0,i} \geq 0 \\ -1 \text{ if } u_{0,i} < 0 \end{cases}, \quad i = 1, \ldots, m \tag{6.40}$$

where $u_{0,i}$ denotes the ith component of the dominant eigenvector that is the vector corresponding the largest eigenvalue of the matrix B.

In case of dividing the graph into more than two subgroups, we introduce, following [370], the matrix S with elements

$$s_{ij} = \begin{cases} 1 \text{ if } i\text{th node belongs to } j\text{th community} \\ 0 \text{ otherwise} \end{cases} \tag{6.41}$$

The expression $\delta(c_i, c_j)$ from Eq. (6.8) is equivalent to

$$\delta(c_i, c_j) = \sum_{g=1}^{k} s_{ig} s_{jg}$$

Hence Eq. (6.8) can be rewritten as

$$Q = \frac{1}{2m} \sum_{i=1}^{m} \sum_{j=1}^{m} \sum_{g=1}^{k} b_{ij} s_{ig} s_{jg} = \frac{1}{2m} \text{tr}(S^{\mathsf{T}} B S) \tag{6.42}$$

where $\text{tr}(\cdot)$ means the matrix trace.

We know already that elements of the vector $\mathbf{u}_0$ indicate "membership degrees" of concrete nodes to a group. Newman [370] proposes the following heuristic to "improve" the split resulting from the simple bisection algorithm:

(a) Let Q denote the value of modularity obtained for the current partition.
(b) Move node v_i, $i = 1, \ldots, m$ to the opposite group and compute the value Q_i corresponding to the modified partition.
(c) If $Q_{i*} = \max_i Q_i > Q$, then modify the partition moving the node v_{i*} to the opposite group. Replace Q by Q_{i*}.
(d) Repeat steps (b) and (c) as long as an improvement of the current value of Q is possible under the constraint that each node can be moved to the opposite group *only once*.

This method of improvement of the current partition is called *fine-tuning* by Newman [370].

In order to decide whether or not the moving of nodes shall be stopped, the increase ΔQ of modularity is checked each time. Assume we want to split a group C_g consisting of n_g nodes. Let $B^{(g)}$ be a matrix with elements [370]

$$b_{ij}^{(g)} = b_{ij} - \delta_{ij} \left[d_i^{(g)} - d_i \frac{d_g}{2m} \right] \tag{6.43}$$

where $d_i^{(g)}$ denotes the degree of ith node in the subgraph g, and d_g is the sum of degrees d_i of nodes constituting this subgraph. In the end

$$\Delta Q = \chi^{\mathsf{T}} B^{(g)} \chi \tag{6.44}$$

Thus if we sort the nodes of the graph by decreasing value of the components $\mathbf{u}_0$, that is in such a way that $u_{0,1} \geq u_{0,2} \geq \cdots \geq u_{0,m}$, then the quality Q can be eventually improved by attempting to move to the other set the nodes with "undecided membership", that is ones which are assigned values $|u_{0,i}| \approx 0$. In this way we obtain bisectional spectral algorithm of community detection as described by the pseudo-code 6.3.

Algorithm 6.3 Bisectional spectral algorithm of community detection [370]

Input: Modularity matrix B.
Output: A cut $\{C, \overline{C}\}$ of the data set.
1: Determine the dominant eigenvector $\mathbf{u}_0$ of the modularity matrix B with elements
    $b_{ij} = a_{ij} - p_{ij}$
2: **for** $i = 1$ to m **do**
3:    **if** $(u_{0,i} \geq 0)$ **then**
4:        Assign node v_i to the set C
5:    **else**
6:        Assign the node v_i to the set $\overline{C}$
7:    **end if**
8: **end for**
9: Check whether the proposed partition can be improved by moving nodes v_i with low absolute
    values of $|u_{0,i}|$ to opposite groups.
10: **return** Cut $\{C, \overline{C}\}$.

6.4.5.4 Newman Method Implementation

Newman's method is particularly attractive from the computational point of view. First of all only the dominant eigenvector is computed in each step which can be done using power method in case of big data sets. Let us recall that this method consists in iterating the equation $\mathbf{x}^{(t+1)} = B\mathbf{x}^{(t)}$ until $\|\mathbf{x}^{(t+1)} - \mathbf{x}^{(t)}\|_1 \leq \epsilon$. Multiplication of the matrix B by a vector may be implemented as follows

$$B\mathbf{x} = S\mathbf{x} - \mathbf{d}\frac{\mathbf{d}^{\mathsf{T}}\mathbf{x}}{2m} \tag{6.45}$$

where S is a sparse matrix. Hence multiplication of $B\mathbf{x}$ is quick.

Additionally the dimension of the matrix S can be reduced by removing from the set all the leaf nodes (that is ones having only one neighbour) together with corresponding edges (called *hairs* in [28]). If the node l is a leaf, and j is his neighbour, then this operation is accompanied by modification of the weight s_{jj} according to the equation [28]

$$s_{jj} = s_{ll} + 2s_{il} \tag{6.46}$$

It is always permissible in case of graphs without loops. But if $s_{ll} > 0$, then its application is allowed only if the following inequality is matched [28]

$$s_{ll} \leq \frac{s_l^2}{2w} \tag{6.47}$$

where $s_l = \sum_j s_{lj}$, $2w = \sum_i \sum_j s_{ij}$. This procedure may reduce the number of nodes even by 40%—see Table 1 in [28].

Let us note further that the multiplication $\chi^T B \chi$ may be represented in the form of the sum

$$\chi^T B \chi = \sum_{i=1}^{m} \sum_{j=1}^{m} b_{ij} \chi_i \chi_j \tag{6.48}$$

This allows for a fast implementation of the method improving the partition. Namely, let χ^* denote a vector such that $\chi_i^* = \chi_i$ and $\chi_{i^*}^* = -\chi_{i^*}$ for a certain i^* (i.e. for a moment we move the node i^* to the opposite group). Then

$$
\begin{aligned}
\Delta Q &= \alpha \left[(\chi^*)^T B \chi^* - \chi^T B \chi \right] \\
&= -2\alpha * sign(\chi_{i^*}) \left(\sum_{j=1}^{m} \chi_{i^*} \chi_j (b_{i^*,j} + b_{j,i^*}) - 2b_{i^*,i^*} \right)
\end{aligned} \tag{6.49}
$$

where $\alpha = \frac{1}{4|E|}$. As we choose the node the shifting of which ensures the larger value of ΔQ, we can assume $\alpha = 1$.

Computation of Q for a partition into k classes is as fast as for two. In this case we use a proper substitution of the formula (6.17). Let $L = [l_{ij}]$ be the list of neighbours of graph nodes, that is the ith row contains $|d_i|$ entries, where l_{ij} indicates the index of jth neighbours of the node i. Let further x be a vector with elements x_i indicating the index of the cluster, to which ith node belongs, i.e. if $s_{ij} = 1$, then $x_i = j$. Then the diagonal elements of the matrix e from Eq. (6.17) are determined as described by the pseudo-code below (deg is the vector containing the values of d_i)

```
for (int i=0; i<m; i++) {
    for (int j=0; j<deg[i]; j++) {
        if (x[i]==x[L[i][j]]) {
            e[x[i]][x[i]]++;
        }
    }
}
```

The components of the vector a are computed as follows

```
for (int i=0; i<m; i++) {
```

```
        a[x}[i]] += deg[i];
}
```

Finally

```
for (int i=0; i<R; i++) {
    Q += e[i][i] - a[i]*a[i]/nnz;
}
```

where the symbol nnz denotes the number of elements stored in L. By dividing the value of Q by nnz we obtain the proper value of modularity (remember that nnz = $2w$).

6.4.5.5 Characterisation of Communities

In case of road maps, beside the information on towns and the connecting road, information about sizes of these towns (indicating their role in the region) are important. Similarly, in empirical graphs we seek not only information on the modules (groups of nodes) being equivalents of administrative units of areas depicted on the map, but also we want to know the role that the nodes play in this structure.

Hence finding modules (or communities) is only the first stage in the analysis of an empirical graph. Guimerà and Amaral [215] elaborated a method determining the role played by individual nodes in the graph. They assumed that nodes having similar function should be characterized by similar topological properties. Two basic properties have been investigated by them: within-module node degree and so-called participation coefficient. The participation coefficient is defined as follows

$$P_i = 1 - \sum_{g=1}^{R} \left(\frac{k_{ig}}{d_i}\right)^2 \tag{6.50}$$

where k_{ig} is the number of nodes of the group g, with which the node i communicates, d_i is its degree in the graph Γ, and R is the number of groups (modules).

To characterise the degree of the node within a group g, the *z-score* is used

$$z_i = \frac{k_{ig} - \widehat{k}_g}{\sigma_g} \tag{6.51}$$

where $\widehat{k}_g$ is the average number of edges linking a node within the group g with other nodes of the same group, and σ_g is the standard deviation of the quantity k_{ig}.

Thus, properties of each node are described by the pair (z_i, P_i). Guimerà and Amaral distinguished in [215] seven universal roles that nodes may play in a community. They notice that nodes playing the roles of *hubs* in a module have $z_i \geq 2.5$, while the remaining nodes $z_i < 2.5$. Based on the participation coefficient, they distinguished the roles of (a) ultra-peripheral node, with all its links within its module, $P_i \leq 0.05$, (b) peripheral node, with most links within its own module, $0.5 < P_i \leq 0.62$, (c) non-hub connector node, communicating with many nodes belonging to other modules, $0.62 < P_i \leq 0.8$, and (d) non-hub kinless node, communicating uniformly with all the modules, $P_i > 0.8$. The hubs are divided into (e) provincial hubs, communicating mainly with nodes of its own group, $P_i < 0.3$, (f) connector hubs, communicating also with nodes from most other groups, $0.3 < P_i \leq 0.75$, and (g) *kinless* hubs, that communicate to the same degree with nodes from their own group and from other groups, $P_i > 0.75$.

6.4.5.6 Balanced and Normalised Modularity

Bolla [79] pointed at the drawback of Newman' modularity consisting in finding communities of strongly varying sizes. Therefore she proposed two modifications of this measure along with providing spectral methods of their optimisation and revealing the relationship between Newman's modularity and the normalised cut and demonstrating relations with spectral properties (eigenvalues of proper matrices).

Balanced modularity, introduced by Bolla [79], penalizes clusters for having extreme sizes, compared to the weighted version of Newman modularity 6.11.

$$Q_b = \frac{1}{2w} \sum_{i=1}^{m} \sum_{j=1}^{m} \left(s_{ij} - \frac{s_i s_j}{2w} \right) \delta(c_i, c_j) |c_i|^{-1} \qquad (6.52)$$

Bolla assumes hereby that $\sum_{i=1}^{m} \sum_{j=1}^{m} s_{ij} = 2w = 1$ so that the above formula can be simplified by dropping the $2w$ factor.

Let $B = S - \mathbf{dd}^T/(2w)$ be the modularity matrix, being a negation of the so-called Q-Laplacian [501]

Bolla shows that the balanced modularity can be estimated from above by the sum of the k largest eigenvalues of this modularity matrix.

Normalised modularity, introduced also by Bolla [79], penalizes clusters for having extreme volumes.

$$Q_n = \frac{1}{2w} \sum_{i=1}^{m} \sum_{j=1}^{m} \left(s_{ij} - \frac{s_i s_j}{2w} \right) \delta(c_i, c_j) \mathrm{Vol}(c_i)^{-1} = -1 + \sum_{i=1}^{m} \left(\sum_{j=1}^{m} s_{ij} \delta(c_i, c_j) \right) \mathrm{Vol}(c_i)^{-1}$$
$$(6.53)$$

As visible from the last expression, maximising this quantity is equivalent to seek minimal normalised cut, if we fix the number of clusters to be equal some k. Therefore the same spectral method (that is applying the normalized Laplacian) can be used to find the maximum as is for the normalized cut problem.

Bolla introduces the normalized modularity matrix $B_D = D^{-1/2} B D^{-1/2}$, where D is the diagonal matrix in which the vector $\mathbf{d}$ constitutes this diagonal. She shows that the normalised modularity can be estimated from above by the sum of the k largest eigenvalues of this normalised modularity matrix.

For both balanced and normalised modularity she shows that using k-means algorithm for spectral representation the communities can be identified.

6.4.5.7 Kernel k-means Versus Community Detection via Normalised Cuts

The preceding Subsubsection 6.4.5.6 drew our attention to seemingly far-fetched relationship between optimisation of Newman's modularity and the normalised cuts of a graph. This brings us a step closer to a kind of unified view of the concept of a cluster. Interestingly, this unification can be moved a step further, to enclose the k-means clustering.

Dhillon et al. [149] draw attention to the formal similarities between the weighted kernel k-means objectives and the normalised cuts. The weighted kernel k-means objectives is defined as follows: minimise

$$J_2^\Phi(U) = \sum_{i=1}^m \min_{1 \le j \le k} w(\mathbf{x}_i) \| \Phi(\mathbf{x}_i) - \mu_j^\Phi \|^2 = \sum_{j=1}^k \sum_{i=1}^m u_{ij} w(\mathbf{x}_i) \| \Phi(\mathbf{x}_i) - \mu_j^\Phi \|^2$$

(6.54)

where u_{ij} indicates if the element i belongs to cluster j, and

$$\mu_j^\Phi = \frac{1}{\sum_{\mathbf{x}_i \in C_j} w(\mathbf{x}_i)} \sum_{\mathbf{x}_i \in C_j} w(\mathbf{x}_i) \Phi(\mathbf{x}_i)$$

(6.55)

Let us reformulate the goal of the weighted kernel k-means in terms of matrices. Let W be the diagonal matrix of weights of all data elements. W_j be the diagonal matrix of weights for cluster j, $\mathbf{e}$ be a (column) vector of ones. Let $s_j = \sum_{\mathbf{x}_i \in C_j} w(\mathbf{x}_i) = \mathbf{e}^T W_j \mathbf{e}$. Φ be the matrix of transforms of data points $\Phi = [\Phi(\mathbf{x}_1), \ldots, \Phi(\mathbf{x}_m)]$, and Φ_j be a submatrix related to data points from the cluster j. Under this notation

$$\mu_j^\Phi = \Phi_j \frac{W_j \mathbf{e}}{s_j}$$

The contribution of a single cluster j to $J_2^\Phi(U)$ can be expressed as

$$J_2^\Phi(U_j) = \sum_{i \in C_j} w(\mathbf{x}_i) \| \Phi(\mathbf{x}_i) - \mu_j^\Phi \|^2 \tag{6.56}$$

$$= \sum_{i \in C_j} w(\mathbf{x}_i) \| \Phi(\mathbf{x}_i) - \Phi_j \frac{W_j \mathbf{e}}{s_j} \|^2 \tag{6.57}$$

$$= \| (\Phi_j - \Phi_j \frac{W_j \mathbf{e} \mathbf{e}^T}{s_j}) W_j^{1/2} \|_F^2 \tag{6.58}$$

$$= \| \Phi_j W_j^{1/2} (I - \frac{W_j^{1/2} \mathbf{e} \mathbf{e}^T W_j^{1/2}}{s_j}) \|_F^2 \tag{6.59}$$

Recall the fact the for any matrix A we have $\|A\|_F^2 = tr(AA^T) = tr(A^T A)$.
Furthermore $\left(I - \frac{W_j^{1/2} ee^T W_j^{1/2}}{s_j}\right)$ is symmetrical, $\left(\frac{W_j^{1/2} ee^T W_j^{1/2}}{s_j}\right)^2 = \frac{W_j^{1/2} ee^T W_j^{1/2} W_j^{1/2} ee^T W_j^{1/2}}{s_j^2}$

$= \frac{W_j^{1/2} ee^T W_j ee^T W_j^{1/2}}{s_j^2} = \frac{W_j^{1/2} es_j W_j^{1/2}}{s_j^2} = \frac{W_j^{1/2} ew_j^{1/2}}{s_j}$, so that we get $\left(I - \frac{W_j^{1/2} ee^T W_j^{1/2}}{s_j}\right)^2 =$

$\left(I - \frac{W_j^{1/2} ee^T W_j^{1/2}}{s_j}\right)$. Hence

$$J_2^\Phi(U_j) \quad = tr\left(\Phi_j W_j^{1/2}\left(I - \frac{W_j^{1/2} ee^T W_j^{1/2}}{s_j}\right)^2 W_j^{1/2}\Phi_j^T\right) \tag{6.60}$$

$$= tr\left(\Phi_j W_j^{1/2}\left(I - \frac{W_j^{1/2} ee^T W_j^{1/2}}{s_j}\right) W_j^{1/2}\Phi_j^T\right) \tag{6.61}$$

$$= tr\left(\Phi_j W_j^{1/2} W_j^{1/2}\Phi_j^T - \Phi_j W_j^{1/2}\frac{W_j^{1/2} ee^T W_j^{1/2}}{s_j} W_j^{1/2}\Phi_j^T\right) \tag{6.62}$$

$$= tr\left(\Phi_j W_j^{1/2} W_j^{1/2}\Phi_j^T\right) - tr\left(\frac{\Phi_j W_j e}{\sqrt{s_j}} \cdot \frac{e^T W_j \Phi_j^T}{\sqrt{s_j}}\right) \tag{6.63}$$

$$= tr\left(W_j^{1/2}\Phi_j^T \Phi_j W_j^{1/2}\right) - tr\left(\frac{e^T W_j \Phi_j^T}{\sqrt{s_j}} \cdot \frac{\Phi_j W_j e}{\sqrt{s_j}}\right) \tag{6.64}$$

Let us arrange the matrix of transforms of data points in such a way that $\Phi = [\Phi_1, \ldots, \Phi_k]$, and let Y be a matrix with m rows and k columns such that the entry at $Y_{ij} = \sqrt{\frac{w_i}{s_j}}$ if element i belongs to cluster j, and $Y_{ij} = 0$ otherwise. Then

$$J_2^\Phi(U) = tr\left(W^{1/2}\Phi^T \Phi W^{1/2}\right) - tr\left(Y^T W^{1/2}\Phi^T \Phi W^{1/2} Y\right) \tag{6.65}$$

Note that Y is an orthonormal matrix ($Y^T Y = I$). As $tr\left(W^{1/2}\Phi^T \Phi W^{1/2}\right)$ is a constant, minimising $J_2^\Phi(U)$ means maximising the term $tr\left(Y^T W^{1/2}\Phi^T \Phi W^{1/2} Y\right)$. As $K = \Phi^T \Phi$ is the kernel matrix problem, we maximise

$$kern(Y) = tr\left(Y^T W^{1/2} K W^{1/2} Y\right).$$

Linear algebra tells us that if Y is allowed (as a relaxation) to be any orthonormal matrix, then the optimal value of Y are the k eigenvectors corresponding to the k largest eigenvalues of $Y^T W^{1/2} K W^{1/2} Y$, while the optimal value of the trace is the sum of the mentioned eigenvalues.

Let us now look at the normalised cut criterion for graph partitioning.

$$Ncut(U) = \sum_{j=1}^k \frac{\sum_{i=1}^m \sum_{l=1}^m u_{ij}(1 - u_{lj})s_{il}}{\sum_{i=1}^m \sum_{l=1}^m u_{ij}s_{il}} \tag{6.66}$$

where s_{ij} is some measure of similarity between elements i and j (in matrix S), u_{ij} indicates if element i belongs to cluster j (entry of matrix U). We seek to minimise the *Ncut* criterion for some fixed number of clusters k. Instead of minimising the number of "cut" edges in a graph, we can maximise the number of edges that are not cut.

$$Nuncut(U) = \sum_{j=1}^{k} \frac{\sum_{i=1}^{m} \sum_{l=1}^{m} u_{ij} u_{lj} s_{il}}{\sum_{i=1}^{m} \sum_{l=1}^{m} u_{ij} s_{il}} \tag{6.67}$$

The nominator represents the sum of similarities of edges lying within a cluster, while the denominator expresses the volume of the cluster that is the sum of similarity values of each cluster element for all edges incident with the node. If U_j represents the j-th column of matrix U, $D = S\mathbf{e}$, then the $Nuncut(U)$ can be expressed as

$$Nuncut(U) = \sum_{j=1}^{k} \frac{U_j^T S U_j}{U_j^T D U_j}$$

Let us introduce the scaled partition matrix $Z = U(U^T D U)^{-1/2}$. Then the normalised uncut can be expressed as

$$Nuncut(Z) = tr(Z^T S Z)$$

It is easily checked that $Z^T D Z = I$. By introducing $X = D^{-1/2} Z$ we obtain the property $X^T X = I$ and the normalised uncut will have the form

$$Nuncut(X) = tr(X^T D^{-1/2} S D^{-1/2} X)$$

Again upon relaxation of X to be any orthonormal matrix, the maximisation of $Nuncut(X)$ can be achieved by means of linear algebra eigenvector computations.

But the very interesting issue here is that the relaxed kernel k-means optimisation and normalised minimal cut optimisation bear strong formal resemblance. By substituting in $kern(Y)$ Y with X, K with S, W with D^{-1} we can reduce the task of seeking a normalised cut in a community graph to a kernel k-means clustering problem.

6.4.6 Bayesian Methods

Hofman and Wiggins [254] suggest to view the problem of a split of graph nodes into communities as a problem of estimating the most probable number of clusters given some graph generation model and given the generated graph.

The generation model runs as follows: First the number of clusters (communities) k is picked. We assume that a multinomial probability distribution π_k is associated with each k. From this distribution we draw a label for each element of the sample (of size m) independently. Then for each pair of elements if both elements got the same label previously, we assign with some probability p_w an edge, and if both elements got different labels, we assign an edge with a probability p_b, (intended to be smaller than p_w). Now we check under these circumstances, what is the probability of obtaining the actual graph Γ. We are interested now in such an estimation of the parameters p_w, p_b, k, π_k, that the probability of obtaining the actual graph is maximised. Hofman and Wiggins [254] propose an algorithm to find the optimal parameters under some simplifying assumptions. The algorithm converges towards a local minimum.

Reidy [406] proposes a measure of communitiveness and tries to redefine New-man's modularity from the information-theoretic point of view. He introduces two random variables L, C defined over the pairs of nodes. $L = 1$ if nodes are linked, 0 otherwise, $C = 1$ if nodes belong to the same cluster, 0 otherwise. He investigates the mutual information of both of them $I(L; C) = H(L) - H(L|C)$. With m nodes and m edges, we can calculate $H(L)$ as

$$H(L) = -\frac{2\mathrm{m}}{m(m-1)} \log_2 \frac{2\mathrm{m}}{m(m-1)}$$

If we denote by n_i the number of nodes in ith cluster, and with m_i the number of edges in ith cluster, then

$$H(L|C) = \left(\mathrm{m}_{int} \log_2 \mathrm{m}_{intfull} - \mathrm{m}_{intfull} \log_2 \mathrm{m}_{intfull} + (\mathrm{m}_{intfull} - \mathrm{m}_{int})\right.$$

$$\left. \log_2 \left(\mathrm{m}_{intfull} - \mathrm{m}_{int}\right) + \overline{\mathrm{m}}_{int} \log_2 \overline{\mathrm{m}}_{intfull} - \overline{\mathrm{m}}_{intfull} \log_2 \overline{\mathrm{m}}_{intfull}\right) / \binom{m}{2}$$

where (k being the number of clusters) $\mathrm{m}_{int} = \sum_{i=1}^{k} \mathrm{m}_i$, $\mathrm{m}_{intfull} = \sum_{i=1}^{k} \binom{n_i}{2}$, $\overline{\mathrm{m}}_{int} = \mathrm{m} - \mathrm{m}_{int}$, $\overline{\mathrm{m}}_{intfull} = \binom{m}{2} - \mathrm{m}_{intfull}$,

The idea behind this measure is that in case the communities do not have anything to do with linkages, then $H(L|C) = H(L)$. So the higher the mutual information, the better the clustering structure matches the link structure.

6.5 Discovering Communities in Oriented Graphs

Clustering of nodes in oriented graphs is a relatively new discipline, and the number of papers devoted to this topic is relatively limited. The most important are [261, 291, 321, 351] (LinkRank) and [99].

6.5.1 Newman Spectral Method

This is in fact a modification of the Newman algorithm presented in Sect. 6.4.5. Leicht and Newman [321] demonstrated that in case of directed graph the formula (6.38) still holds, i.e.

$$Q = \frac{1}{2m} \chi^T B \chi \qquad (6.68)$$

given, however, that we understand by B the matrix with elements

$$b_{ij} = a_{ij} - p_{ij} = a_{ij} - \frac{d_i^{in} d_i^{out}}{|E|} \qquad (6.69)$$

where d_i^{in} means the in-degree of the node v_i (that is the number of links pointing to this node), and deg_i^{in} is the out-degree of the node v_i. Note that the scaling factor is twice as big as before.

However, B is not a symmetric matrix any more. The simplest method of turning it into a symmetric one is to make the substitution $B \leftarrow (B + B^T)/2$, which finally leads to the index

$$Q = \frac{1}{4|E|} \chi^T (B + B^T) \chi \qquad (6.70)$$

The Algorithm 6.3 guarantees its maximization.

6.5.2 Zhou/Huang/ Schölkopf Method

Below we describe basic outline of the method presented in [538].

Let $\Gamma = (V, E, S)$ be a directed weighted graph, in which s_{ij} means the weight of the link $i \rightarrow j$. Let further

$$d^-(j) = \sum_{\substack{i \in V \\ i \rightarrow j}} s_{ij}, \qquad d^+(j) = \sum_{\substack{i \in V \\ j \rightarrow i}} s_{ji} \qquad (6.71)$$

mean out-degree and in-degree of a node $v \in V$, resp.

Let us assume for a moment that for each node $i \in V$ both these degrees are positive. One can construct then a Markov chain transition matrix P with elements

$$p_{ij} = \sum_{k \in V} \frac{s_{ki}}{d_i^-} \frac{s_{kj}}{d_j^+} = \sum_{k \in V} p_{ik}^- p_{kj}^+ \qquad (6.72)$$

There exists then a stationary distribution with values[19]

$$\pi_i = \frac{d_i^-}{\sum_{j\in V} d_j^-} = \frac{d_i^-}{\text{vol}\,(G)} \tag{6.73}$$

This equality results from

$$\begin{aligned}
\sum_{i\in V} \pi_i p_{ij} &= \sum_{i\in V} \frac{d_i^-}{\text{vol}\,(G)} \sum_{k\in V} \frac{s_{ki} s_{kj}}{d_i^- d_j^+} \\
&= \frac{1}{\text{vol}\,(G)} \sum_{k\in V} \frac{s_{kj}}{d_j^+} \sum_{i\in V} s_{ki} = \frac{d_j^-}{vol(G)} = \pi_j
\end{aligned} \tag{6.74}$$

In general case random walk with teleportation is applied. If $d_i^+ > 0$ then: (a) with probability $1 - \alpha$ a random node from the set $V\setminus\{i\}$ is selected, and with probability α one shall go to any neighbour of the node i. But if $d_i^+ = 0$, any node from the set $V\setminus\{i\}$ is selected.

One seeks a labelling function f assigning each node a label 1 or -1 in such a way that the following functional is minimised:

$$\Omega(f) = \arg\min_f 0.5 \sum_{(u,v)\in E} \pi(u) p(u,v) \left(\frac{f(u)}{\sqrt{\pi(u)}} - \frac{f(v)}{\sqrt{\pi(v)}} \right)^2 \tag{6.75}$$

where $p(u,v)$ is the amount of authority flowing from node u to node v (the probability that the random walker goes through the edge (u,v)).

6.5.3 Other Random Walk Approaches

Pons et al. [396] proposed an algorithm called Walktrap. It is based on the observation that random walks on a graph tend to get "trapped" in densely connected subgraphs, which represent communities.

They develop an algorithm essentially based on the Ward's method with a carefully chosen distance measure, reflecting the structural properties of the graph, derived from the stationary distribution approximated by random walks on the graph.

Meila and Shi [352] suggest to look for optimisation of their $NCut$ measure, reflecting the cut of flow of authority between communities, that is probability to step from one community to the other in a single step by a random walker starting at a stationary distribution. Let C and $\overline{C}$ be two complementary communities. Then

---

[19]The graph must be connected, not split into disjoint components.

$$NCut(C, \overline{C}) = \frac{Pr(C \to \overline{C})}{Pr(C)} + \frac{Pr(\overline{C} \to C)}{Pr(\overline{C})} \tag{6.76}$$

It turns out that this criterion can be approximated by the eigenvalues of the Laplacian matrix for directed networks [538], that is of the matrix

$$\mathfrak{L} = I - \frac{\Pi^{1/2} P^T \Pi^{-1/2} + \Pi^{-1/2} P \Pi^{1/2}}{2} \tag{6.77}$$

where Π is a matrix with stationary distribution on the diagonal, P is the matrix where P_{ij} means the conditional probability of moving to i given a random walker is at node j.

Capocci et al. [99], suggest that the underlying community structure can be revealed by looking at correlations of eigenvector components of the products of a "slightly modified" adjacency matrix AA^T and $A^T A$, where $A = P\Pi$, and where P, Π are defined as above.

Further methods of detecting communities in directed graphs are reviewed in [346].

6.6 Communities in Large Empirical Graphs

An intriguing property of communities in large empirical graphs has been observed by Leskovec *et al.* in [322]. Namely, they observed small (up to ≈ 100 nodes) communities that are loosely connected to the rest of the graph. On the other hand communities of size beyond ≈ 100 nodes turn into expander-like core of the network, and therefore do not look like proper communities. The quality of these communities (measured by conductance (see (5.79b)) or normalized cut metric, see eq. (5.67b)) decreases with increase of their size.

This may be interpreted as an empirical confirmation of the so-called Dunbar number limiting the size of a well-functioning community.

Their investigation method was essentially based on hierarchical bisectional clustering of the graphs. The bisection was performed via optimisation of the normalized cut (5.67b) metric. For the obtained clustering they created so-called network community profile plot. For each community size k they sought all the clusters of (approximately) that size and computed conductance (5.79b) for each of them, taking the minimum

$$\phi(k) = \min_{S \subset V; |S|=k} \phi(S)$$

6.7   Heuristics and Metaheuristics Applied
for Optimization of Modularity

As Caeri et al. [96] point out, a number of different heuristics were applied for search of optimal modularity. Beside the already mentioned agglomerative hier-archical clustering, techniques like simulated annealing, mean field annealing, genetic/evolutionary optimisation, extremal optimization, spectral clustering, linear programming, dynamical clustering, multilevel partitioning, contraction-dilation, multistep greedy search, quantum mechanics methods, just to mention a few, were applied. Let us discuss below some of them.

Caeri et al. [96] themselves propose a refinement, post-processing technique, that can be used to (locally) improve the results of a heuristic modularity optimisation algorithm. The post processing consists in alternating bisection of clusters and merger of them. The bisection is based on integer programming and yields the exact best bipartition.

Schuetz and Caflisch [421] on the other hand propose a modification of Newman's heuristic approach to formation of communities by merging multiple communities at one step. It is a kind of greedy search algorithm. While Newman merges in one step only the pair of communities that would increase the modularity most, they take l "levels" of highest increase of modularity of pairwise merge and merge all these pairs in decreasing order of modularity increase and increasing order of modularity value, excluding those pairs where one of pair elements was merged previously in the same step. In this way creation of a single large cluster is prohibited, diversity of clusters is ensured and their comparable quality is ensured.

Guimera and Amaral [215], propose a simulated annealing approach for increasing the modularity. *Simulated annealing* is an optimisation technique of a scoring function f, in which one travels through the space of possible solutions, considering one solution at a time. Upon attempt to move from the current solution s_c to the next solution s_n (usually chosen at random from some set of "neighbouring" solutions) one checks if the score of the next solution is higher than the current solution $(f(s_n) > f(s_c))$. If it is higher, then the next solution becomes the current solution and the process is repeated. If not, then a randomised decision is made. With probability $e^{-\alpha/T}$ the next solution becomes the current solution, and with the remaining probability the new solution is rejected and further search starts with the same current solution. α is a parameter and may depend on the score difference between the current and the next solution. T is a parameter called "temperature" that is systematically decreased during the search through the solution space. High values of T cause that with high probability a worse solution than the current solution will be accepted. So the exploitation of solution space is promoted. But with decrease of T the preference for better solutions is promoted so that finally one gets at a local optimum. This technique is applied by the authors of [215] as follows. Their solution space is the set of possible partitions on the network nodes into communities. Their scoring function is the Newman's modularity Q. They start with a more or less random partition. For a network with m nodes (and a user-defined factor b) at a given "temperature" T

they propose to perform bm^2 individual node move operations (moving a randomly selected node to a randomly selected community, just creating a new "neighbouring" solution) and bm merges/bisectional splits of clusters before decreasing the temperature by factor of 0.995. If a move/merger or split increases modularity, it is accepted. If not, it is accepted with a probability exponential in the negative quotient of the loss of modularity divided by the temperature.

Lehmann and Hansen [320] apply mean field annealing to the matrix B as defined by Eq. (6.10). *Mean field annealing* is a deterministic approximation to the already mentioned simulated annealing technique of solving optimization problems. It is based on the so-called mean field theory. The *mean field theory* substitutes the behavioural investigation of large complex stochastic models by a study of a much simpler model. The simplification relies on the observation that in a system with a big number of small individual components we do not need to study the impact of each component onto each other separately but rather we can substitute the influence of all the other components onto the component of interest by effect of the average representative of all the other components. Authors of [320] apply mean field annealing as follows: Let us introduce the community matrix S such that $S_{ic} = 1$ if node i belongs to cluster c and is zero otherwise. Under such notation $Q = \frac{\text{tr}(S^T BS)}{2m}$. In the mean field annealing approach we define variables μ_{ic} meaning the mean probability that i belongs to cluster c.

$$\mu_{ic} = \frac{\exp(\phi_{ic}/T)}{\sum_{c' \in C} \exp(\phi_{ic'}/T)}$$

where C is the set of clusters, T is the temperature parameter and $\phi_{ic} = \sum_j \frac{B_{ij}}{2m} \mu_{jc}$. For high temperatures, as expected μ_{ic} is identical for each cluster and each node. The computation is performed with decreasing temperature till one gets close to zero, where the decision on class membership is made. One has to assume the maximum number of clusters, but upon decision making the actual assignment may happen to a lower number of clusters.

Rossi and Villa-Vialaneix [408] suggest to apply so-called deterministic annealing for computing a slightly modified version of modularity, the so-called organised modularity (6.25). *Deterministic annealing* means an approach to solve the complex combinatorial problem of maximizing a function f defined over a large finite domain $\mathbb{M}$ by analysing Gibbs distribution obtained as the asymptotic regime of a classical simulated annealing. In this approach the function f is the function $Q(A, C, M)$, the domain $\mathbb{M}$ is the set of all cluster assignments (C, M) as defined in Eq. (6.25). They use the sampling distribution $P(C, M) = \frac{1}{z_P} \exp\left(\frac{Q(A,C,M)}{T}\right)$ with T being the temperature parameter and $\frac{1}{z_P}$ being a normalizing constant over all possible (C, M) combinations. As deterministic annealing prescribes, the authors compute the expectations of the (C, M) pair for a fixed temperature under Gibbs distribution and then observe the change of the expectations when temperature decreases. This leads them, via techniques from the mean field theory, to an algorithm as described in [408]. The basic complexity of their derivation lies in non-linearities of Q on its arguments.

Arenas et al. [29] apply so-called Tabu-search when seeking the optimal modularity (they apply it for the scaled modularity from Sect. 6.3.3). *Tabu search* is a search heuristic through a space of solutions in which one solution is considered at a time, then a move is made to another and so on. In order to avoid loops in this search process, a list of moves and/or solutions is created and maintained that cannot be performed/entered at a given moment. In [29] this technique is applied as follows: The current solution is a partition with the actual modularity value. The neighbourhood of the current solutions consists of solutions that can be obtained via a local "move" operator that consists in moving an element from one community to another or forming a separate community for it. The best neighbour (with highest Q) will be visited next unless on the Tabu list (as to escape from local optima). The Tabu list contains the nodes move of which caused improvement in the past (up to a number of steps ago) so for a number of steps the node on the list will not be moved unless its movement would lead to a better solution than the best found so far.

Duch and other [163] and [28] apply a technique called extremal optimization to find partition with highest Q. *Extremal optimization* algorithm aims at optimising a global target via improvements to extremal (worst) local components of the solution. Costs are assigned to individual components based on their contribution to the overall cost function. The components are assessed and ranked and weaker components are replaced by randomly selected ones. Duch et al. apply this technique via investigation of the contribution of a single node i to the modularity, defined as $q_i = \kappa_i - k_i p_i$, where κ_i is the number of links of the node i in its own community, k_i is the degree of the node and p_i is the probability, that a node would link to a node from the same community in case that the process of link insertion would be random. In this case $Q = \frac{1}{2m} \sum_{i \in V} q_i$ that is the global target function is a sum of node related components. They introduce the r_i contribution of a node relative to its degree: $r_i = \frac{q_i}{k_i} = \frac{\kappa_i}{k_i} - p_i$. They apply the following nested procedure to optimize Q. First they split randomly nodes into two communities. They iteratively pick the node with the lowest r_i and move it to the other community, and recalculate the r_i of many of the nodes as it changes. They stop when no improvement in Q can be achieved in a number of steps (usually equal to the number of nodes). Then they remove edges linking both communities and proceed with each one as above.

Tasgin and Bingo [455] use genetic algorithm to look for the optimal value of Q. *Genetic algorithm* is a population based optimization technique. One starts with a population of (alternative, randomly selected) solutions and searches the space of solutions via modifying the solutions of the population usually simultaneously using two types of operators: mutation and cross-over. As a result of these operations new solutions are added to the population. To keep a limited size of the population, operation of selection is applied to remove least promising candidate solutions (usually based on fitness that is the value of the cost function for the solution). To enable the mutation and cross-over, the solution needs to be encoded as a bit string, called chromosome. Mutation consists usually in flipping a randomly selected bit (or group of bits) in this string. Cross-over requires using two parent solutions to obtain a child solution. The child solution is obtained e.g. by taking the first n bits from the first parent and the rest from the second one. These bit operations may lead to a string that has no interpretation as the solution of the problem so various corrective operations are applied either after mutation and cross-over, as integral part of them or by ensuring a careful design of encoding the solutions. Tasgin and Bingo [455] use the modularity Q as fitness function. A chromosome is a vector of node assignments to modules (communities). While creating an element of the initial population of solutions, each node is assigned to a randomly selected community. But additionally randomly a subset of nodes is picked and for each node all its neighbours will be assigned to the community of the picked node. A population of such chromosomes is generated and then genetic algorithm operations are applied to them. Cross-over operation is performed as follows: two chromosomes are picked up at random, one being called source, the other destination. In the source chromosome a community is randomly selected. Then the community identifier of the nodes of this community is passed to the very same nodes of the destination chromosome. Mutation consists in random assignment of a node to a community of the network. Finally they use the clean-up operator. If for a randomly selected node of a chromosome the percentage of its links leading outside of its community exceeds some threshold, then this node along with its neighbours is assigned a new community identifier.

Pizzuti [394] also applies genetic computing, but to optimise a bit different measure of partition quality, the Community Score (see Sect. 6.3.4). A chromosome here is a vector $\mathbf{v}$ containing as many entries as there are nodes in the network. If $v[i] = j$, then both i and j are members of same community. The vector represents thus a forest (a set of trees) where each tree represents a separate community. Note that $v[i] = j$ does not need to mean that in the network there is a link between both. Based on this representation, they apply genetic operations: Random initialisation of a chromosome assigns any element of $\mathbf{v}$ a number from 1 to m (the number of nodes in the network). As such an initialisation may be meaningless, it is "repaired" afterwards: if $v[j] = j$ but there is no edge (i, j) in the network, then j is replaced by any neighbour of i. Note that in this way "safe chromosomes" are generated—if $v[i] = j$ then there exists a link (i, j). They apply so-called uniform cross-over. Given two parents $\mathbf{v}_1, \mathbf{v}_2$ they generate a binary random vector $\mathbf{u}$ of length m. The child chromosome $\mathbf{v}_c$ is obtained in such a way that $v_c[i] = v_1[i]$ if $u[i] = 1$ and $v_c[i] = v_2[i]$ otherwise. In this way from two "safe parents" also the child is "safe". A mutation consists in substituting $v[i]$ with a randomly selected neighbour of i in the network. Hence the mutated chromosome is also safe.

Many other techniques have been tried out for community detection, and also combined approaches have been investigated. Villa et al. [481] suggest combination of kernel methods or spectral methods with SOM (so called *SOM clustering*, with variants called kernel-SOM and spectral-SOM) which resemble the Algorithm 5.4 (on p. 220), in which the k-means algorithm was replaced with SOM. Both this paper and the already discussed [408] stress the visualisation aspect of community discovery. SOM was also exploited by Yang [518]. Tsomokos [470] reduces the process of community emergence to a time-continuous random walk through social network.

Niu et al. [377] explored a mixture of ideas comprising spectral clustering, kernel clustering, quantum mechanics and Bayesian Information Criterion (BIC).

The multiplicity of developments in the area prompts one to suspect that there is no solution of the type "one size fits all". And in fact one can summarise this section as follows:

- Tabu search is a quite precise heuristics for medium sized networks.
- Extremal optimisation is quite useful for larger networks as it balances quite well the precision and execution time.
- Spectral optimization is a very fast and useful heuristics for large networks. Its precision is low, however.

6.8 Overlapping Communities

Xie and Szymanski [510] proposed in 2012 so-called Speaker-listener Label Propagation Algorithm (SLPA) algorithm, called also GANXiS, which essentially extends the Label Propagation Algorithm (LPA) to detection of overlapping communities. Each node is equipped with a label vector. All nodes of the graph are initialised with unique labels, so that the label vector consists of one element. Then iteratively a listener is randomly selected and each of its neighbours "speaks" to it by sending him a label that the speaker randomly selects from its label vector with probability proportional to the label counts. The listener inserts one of received labels with the highest frequency into its label. If size limits of these vectors are imposed and exceeded, the least frequent labels are removed. After a user-defined number of iterations label probability distributions are computed for each node. If the probability of a label at a node falls below some level r, it is removed from its label vector. Nodes with common labels are grouped into communities. Nodes with multiple labels left will belong to several communities. If r is set above 0.5, communities will not overlap.

Another algorithm, Statistical Inference (SVINET) [207] developed by Gopalan and Blei uses a Bayesian mixed membership stochastic block model of overlapping communities in a graph. Here, knowledge of the number of communities in advance is assumed. The community model is as follows: Each node maintains a vector of community membership degrees, summing up to one (probability distribution of community membership). Out of this model instances of networks (graphs) can be generated, representing this community structure. A particular instance of a network representing the communities of this model is created as follows: We pick up two nodes i and j. According to their community membership distributions p_i, p_j, where p_{il} is the probability that node i belongs to community l, we pick randomly a community $z_{i/j}$ for i and a community $z_{j/i}$ for j. If both agree, a link between nodes i and j is inserted into the network with some high probability β_l, where

$l = z_{i/j} = z_{j/i}$. Otherwise (if the community indicators disagree), the link is inserted with some low probability ϵ. So the probability that there is a link between nodes i and j can be computed as

$$p_{i--j} = \sum_{l=1}^{k} p_{il} p_{jl} \beta_l + \sum_{l=1}^{k} (1 - p_{il})(1 - p_{jl})\epsilon \qquad (6.78)$$

Actually, all the quantities p_{il}, β_l, ϵ for all values of community idex l and node index i are hidden variables to be estimated from the observed network. An optimisation task is formulated so that p_{i--j} are high for nodes i, j connected by an edge, while they shall be low otherwise. To facilitate the search through a huge space of parameters, it is assumed that the probabilities stem from some families of probability distributions with low number of parameters. The task is still hard for large networks, therefore not the whole network is optimised at once, but rather samples of the nodes are taken (together with their links) to update some parts of the $p_{..}$ matrix and the $\beta_{.}$ vector.

The Top Graph Clusters (TopGC) algorithm by Macropol and Singh [340] probabilistically seeks top well-connected clique-like clusters in large graphs, scored highly according to the following cluster score measure

$$CS(C) = \sqrt{|C|} \frac{\sum_{i \in C} \sum_{j \in C} s_{ij}}{|C|(|C| - 1)} \qquad (6.79)$$

where C is the scored cluster, and s_{ij} is the similarity between nodes. In order to identify sets of nodes whose neighbourhoods are highly overlapping (like in cliques), they introduce a special hashing technique. They define a random neighbourhood of a node i as the node i itself plus a a subset of its neighbours where a neighbour j is randomly drawn proportionally to the similarity s_{ij} (They actually assume that s_{ij} is a conditional probability of including the neighbour j given we consider the neighbourhood of i). A number t of such random neighbourhoods is drawn, where t is the parameter of the algorithm. Then some r random permutations $\pi_1, \ldots, \pi_r$ of node identifiers are created. For the node i its hash code $h(i, e, b)$ for permutation π_b ($b = 1, \ldots, r$), and its random neighbourhood $N_{i,e}, e = 1, \ldots, t$ is the element of $N_{i,e}$ with the lowest rank under π_b. Let us pick up randomly a permutation π_b. The probability that two nodes i, j agree on the hashcode, that is that $h(i, e, b) = h(j, e, b)$ for a picked permutation π_b equals $\frac{|N_{i,e} \cap N_{j,e}|}{|N_{i,e} \cup N_{j,e}|}$ (Jaccard's index). If we pick additionally e randomly, then the probability that hash codes agree amounts to $\frac{\sum_{k \in N_i \cap N_j} s_{ik} s_{jk}}{|N_i \cup N_j|}$ If we pick l permutations randomly, and concatenate hash codes for each of them, then the probability of hash code agreement equals $\left(\frac{\sum_{k \in N_i \cap N_j} s_{ik} s_{jk}}{|N_i \cup N_j|} \right)^l$.

In a pre-pruning stage nodes together with their neighbourhood are evaluated by the cluster score measure in formula (6.79) and only those nodes are further considered for which this measure is high. For those selected the hashing is performed and those with agreeing hashes are put into one cluster/community.

In a comparative study [235] SLPA proved to perform best on a number of criteria, like precision, recall, F-measure, NMI. TopGC seemed to have the poorest performance.

6.8.1   Detecting Overlapping Communities via Edge Clustering

An interesting path of research on overlapping communities is to detect them via clustering not the nodes but rather the edges of a network graph (see e.g. [10, 172, 355, 540]). If a node coincides with edges from different edge communities, it has to belong to several communities.

Michoel and Nachtergael [355] suggest a procedure based on the intuitive criterion that a cluster is a part of the graph with high edge density. So the quality of a set C as a candidate for a cluster can be evaluated as

$$S(C) = \frac{\sum_{i,j \in C} a_{ij}}{|C|}$$

where a_{ij} is the element of adjacency matrix A expressing the connectedness of nodes i, j of the network.[20] Let χ_C denote a vector with components $\chi_{C,i} = 0$ if node $i \notin C$, and $\chi_{C,i} = |C|^{-1/2}$ if $i \in C$. Obviously $\chi_C^T \chi_C = 1$. and

$$S(C) = \chi_C^T A \chi_C \leq \max_{x \in \mathbb{R}^n, x \neq 0} \frac{x^T A x}{\|x\|^2} = \lambda_{max}$$

where λ_{max} is the largest eigenvalue of A. The Perron-Frobenius theorem ensures that if the graph represented by A is irreducible, then the dominant eigenvector $\mathbf{x}_{max}$, satisfying $A\mathbf{x}_{max} = \lambda_{max}\mathbf{x}_{max}$, is unique and has the property of being strictly positive ($x_{max,i} > 0$ for each i). So to find the "best" cluster one may seek such a set C for which the vector χ_C is as close as possible to $\mathbf{x}_{max}$, that is

$$\hat{C} = \arg\max_{C \subset V} \chi_C^T \mathbf{x}_{max} = \arg\max_{C \subset V} \frac{1}{|C|} \sum_{i \in C} x_{max,i}$$

This can be optimised in a straight forward manner: sort elements of $\mathbf{x}_{max}$ in decreasing order and try C taking k largest values of $\mathbf{x}_{max}$ for each $k = 1, \ldots, m$ finding the one that maximises $\frac{1}{|C|} \sum_{i \in C} x_{max,i}$. An even better solution C_{max}, with the same effort, can be found by trying out C corresponding to k largest values of $\mathbf{x}_{max}$ and looking for the one maximising $S(C)$. So they propose the algorithm

---

[20] Note the complementarity to the Ratio cut criterion from the criterion (5.67a).

1. Calculate the dominant eigenvector $\mathbf{x}_{max}$
2. Find the solution C_{max} and store the list of edges connecting them.
3. Remove all the edges between nodes of C_{max} from the graph and repeat the three steps till no edge is left in the graph.

Ahn et al. [10] propose to extract clusters of edges in a different way, using a hierarchical single-link algorithm. They define similarity between two edges (i, k) and (j, k) sharing a node k as $sim((i, k), (j, k)) = \frac{|N_+(j) \cap N_+(i)|}{|N_+(j) \cup N_+(i)|}$ where for a node i the $N_+(i)$ is the set consisting of i and its neighbours, and build the dendrogram of the edges (tiers are broken by simultaneous inclusion into the dendrogram). Subsequently the dendrogram is split creating a partition C of edges at some level, which is chosen to maximize the partition density

$$D_C = \frac{2}{m} \sum_{C_j \in C} m_j \frac{m_j - (m_j - 1)}{(m_j - 2)(m_j - 1)}$$

where m is the total number of edges, m_j is the number of edges in the edge cluster C_j, and m_j is the number of nodes in the cluster C_j. Again the edge clusters define the communities and the nodes incident with edges from various edge communities belong to overlapping communities.

Evans and Lambiotte [172] on the other hand propose an extension of the concept of Newman's modularity from nodes to edges, by the virtue of the concept of a random walker (see Eq. (6.19)). They consider two types of random walkers going from one edge to another (instead of one node to another node). One is to choose one of the edges incident with the current edge with uniform probability, the other to choose one of the edge ends first with probability 1/2 and then to choose uniformly one of the edges incident with that end. By maximising the respective modularity, they partition the set of edges, that represent the communities. A node incident with more than one edge cluster belongs to overlapping communities.

6.9 Quality of Communities Detection Algorithms

Rabbany et al. [400] drew attention to the fact that community mining algorithms, like clustering algorithms in general, require methods allowing to assess their quality.

As already mentioned in the context of general clustering problems in Sect. 4.3, there exist two brands of criteria for assessing clustering quality: external and internal ones. The external ones compare the obtained clustering against some benchmark clustering, provided e.g. by experts. Internal ones rely on some "aesthetic" criteria, reflecting the consistency of the clusters.

The external evaluation requires large benchmarks, which are not readily available, especially for purposes of a broad investigation and comparison of various algorithms. Therefore some benchmark data generators have been proposed in the past.

GN benchmark [400] is a synthetic network generator that produces graphs with 128 nodes, each with expected degree 16, divided into 4 communities. A link between a pair of nodes of the same community in generated with probability z_{in}, and between communities with probability $1 - z_{in}$. As this benchmark had serious and unrealistic limitations, (same expected number of links, same size of communities), another benchmark LFR [400] has been invented. It uses power law distributions of node degree and community size. Each node shares a fraction $1 - z$ of its links with the other nodes of its community and a fraction z with the other nodes of the same community.

For these generators and for benchmark sets the known external criteria from Sect. 4.4 are applicable,

As an internal criterion for a social network graph, a number of quality measures have been proposed, reflecting diverse ways of how a network may be looked at. As described in Sects. 6.5.2, 6.5.3, 6.2.2, one can view the communities from the perspective of keeping a flow inside it, thus measuring the number of edges cut by the split into clusters, or the time that a random walker remains within a community etc.

One popular measure of cluster quality to be used with community detection is conductance,[21] as discussed in the previous chapter. When looking for communities, it has however several disadvantages. For example a community encompassing several internally disconnected components may be scored higher than a similarly sized, better connected group of nodes. It is also distorted by overlapping communities: the edges leaving the community will penalize the score so that even separate communities may be merged together.

A different measure of score is to maximize how clique-like a community is[22] A limitation of such measures is that they do not give preference to larger in spite of the fact that a large clique is less likely to be formed by chance than a clique of size 3.

Another point of view is the compression aspect: how many bits of information are necessary to represent an instance of the network assuming a community structure. This approach is followed e.g. by Rosvall and Bergstrom in [409] in their infomap based compression method. They assume that a network will be summarized by a community assignment vector $\mathbf{a}$ and a community interconnection matrix M where the diagonal contains the number of links within each community, and off-diagonal elements M_{ij} tell how many links are between the communities i and j. The quality of a split into communities is the number of bits necessary to represent the model $(\mathbf{a}, M)$ plus the number of bits necessary to identify the concrete network among the ones covered by the model. Minimisation of this measure is aimed at.

The common choice for evaluation of community detection quality is the so-called modularity measure Q, as defined in Eq. (6.7). It reflects how much the detected structure differs from a random one. It exhibits, however, some drawbacks, which

---

[21] the ratio of the number of links within a cluster to the number of links leaving that cluster.

[22] One example is intra-cluster distance measure being the ratio between the number of edges in a cluster to the number of possible edges within that cluster. Another one is ratio association assigned to the set of communities, as the sum of the average link weights within each of communities.

become visible in particular for large networks, like WWW. First of all it requires the information on the structure of the entire network Furthermore, it has some resolution limitations so that it does overlook smaller structures, Finally, it takes into account existent links, but ignores the fact that elements are disconnected which may suggest they do not belong to the same community.

To overcome the need to known the structure of the entire large (undirected) graph Γ, Clauset [121] proposes so-called *local modularity R*. Let C be the community we want to assess. Let U be the set of nodes outside of C to which nodes of C link. B be the subset of nodes of C (called "boundary") which link to U. T be the sum of degrees of nodes belonging to B. Z be the number of links between U and B. Then R is defined as

$$R = \frac{T - Z}{T}$$

To handle the resolution limit, Arenas et al. [29] suggested the so-called *Scaled modularity*, described already in Sect. 6.3.3.

To take into account both connectivity and the disconnectivity, Chen et al. [108] propose so-called min-max modularity or MM modularity. Consider the graph $\Gamma = (V, E)$ and its complementary graph $\Gamma' = (V, E')$ where E' contains (all) edges absent in E. C be a clustering of nodes V. They define $Q_{max} = Q(A, C)$, $Q_{min} = Q(A', C)$ (Q from formula (6.8)), and

$$MM = Q_{max} - Q_{min}$$

where A is the neighbourhood matrix of Γ and A' is the neighbourhood matrix of Γ'.

Muff et al. [362] suggest still another modification of the Q modularity. They introduced so-called *localised modularity LQ*

$$LQ(A, C) = \frac{1}{2m} \sum_{i=1}^{m} \sum_{j=1}^{m} (a_{ij} - p_{i,ij}) \delta(c_i, c_j) \tag{6.80}$$

which differs essentially from formula (6.8) in the way how the link probability $p_{i,ij}$ is computed. For probability estimates only the cluster c_i and those clusters are considered with which the cluster c_i containing node i is linked. If each cluster is interlinked with any other, then LQ is identical with Q. Otherwise it can take values even bigger than 1.

Rabbany et al. [400] point at the fact that the above Newman's modularity based measures concentrated essentially around only one aspect of the communities, that is the randomness of links between and within communities. Let us add that the other ones, like cut edge count or flow based, though reflect other aspects, do not exhaust ways communities may be looked at.[23] As communities can be viewed as a special kind of clusters, Rabbany et al. argue that one should also apply other cluster

---

[23] As an analogy to density based clusters, we can think of cut edge count based clustering as seeking dense probability regions in the n-dimensional space, while Newman's modularity based clustering

quality criteria, like those mentioned already in Sect. 4.3. They point, however, at the specificity of communities which induces difficulties in direct application of these measures: The data points in the classical clustering lie in a feature space whereas members of communities are described by their links to other community members. This implies difficulties in defining distances (most quality criteria assume Euclidean distances) and definition of cluster gravity centres (which are obvious in Euclidean space, but not so in other spaces). Rabbany et al. suggest to consider the following distance measures

- *edge distance* $d_{ED}(i, j)$ as the inverse of edge weight between nodes i and j,
- *shortest path distance* $d_{SPD}(i, j)$ as the smallest sum of weights of edges constituting a path between i, j,
- *adjacency relation distance* $d_{ARD}(i, j)$, given the adjacency matrix A

$$d_{ARD}(i, j) = \sqrt{\sum_{k \neq i, j} (a_{ik} - a_{jk})^2}$$

- *neighbourhood overlap distance* $d_{NOD}(i, j)$, being the ration of unshared neighbours:

$$d_{NOD}(i, j) = 1 - \frac{|\mathcal{N}_i \cap \mathcal{N}_j|}{|\mathcal{N}_i \cup \mathcal{N}_j|}$$

where $\mathcal{N}_i$ is the set of nodes directly connected to i,
- *Pearson correlation distance* $d_{PCD}(i, j)$, computed as

$$d_{PCD}(i, j) = 1 - \frac{\sum_k (a_{ik} - \mu_i)(a_{jk} - \mu_j)}{m \sigma_i \sigma_j}$$

where m is the total number of nodes, $\mu_i = \sum_k \frac{a_{ik}}{m}$, $\sigma_i = \sqrt{\frac{1}{m} \sum_k (a_{ik} - \mu_i)^2}$, as examples.

Whenever the cluster C mean is referred to, they suggest to use one of the nodes, such that it is closest on average to all the other, instead.

$$\overline{C} = \arg \min_i \sum_{k \in C} d(i, k)$$

Under such assumptions, traditional cluster quality criteria may be accommodated quite conveniently. E.g. Dunn index from Eq. (4.9) or Davies-Bouldin index from Eq. (4.10) do not require formal change.

(Footnote 23 continued)
as a clustering of dependent dimensions, whereas the info-map type clustering as co-clustering of both; hence differences will emerge.

6.10 Communities in Multi-Layered Graphs

Applications related to analysis of social networks or mobile networks brought practical necessity to go beyond investigations of the plain existence of a relationship between objects and require differentiation between various aspects of human interactions, e.g. those between friends, co-workers, interest groups etc. (see [289]).

These aspects are modelled by multi-layered graphs. A multilayered graph $\Gamma^{[M]} = (\mathfrak{L}, \mathfrak{M})$ consists of a set $\mathfrak{L}$ of layers Γ_l being ordinary graphs $\Gamma_l = (V_l, E_l)$ and of a set $\mathfrak{M}$ of mappings M_h between sets of vertices V_l of the form $M_h : V_{h_1} \times V_{h_2} \rightarrow \{0, 1\}$. The layer graphs may be unweighted, or weighted ones. Layer can be defined directly as a structural information or indirectly as node attributes that connect nodes via calculation of a suitable (dis)similarity measure. In such a graph, one layer may correspond to relationships in a facebook-like network, the other in the blogosphere etc. As one user can have multiple accounts in these various media, the mappings will not be one-to-one.

An alternative representation to the multilayered graphs may be a typed graph or information network, being an ordinary directed graph in which each node belongs to some node type and each edge to some edge type.

The analysis of multilayered graphs may be performed in a number of ways, depending on the assumed layer interrelationship. The interrelationships between the layers can be viewed in various ways:

- there exists an underlying (hidden) split into communities and the various layers are manifestations of the same split communities (for example links by emails between scientists and co-authorships)
- the relationships are "nested" - one relationship by its nature is narrower than the other (e.g. family versus friends)
- the relationships are orthogonal to each other.

The type of dependencies between layers determines which methods of clustering are best suited for their analysis.

In the last case (orthogonality of relationships) each layer may be treated separately.

In case of nested relationships, researchers exploit to various extent the results of community detection process in one layer to drive the community detection in the other layer or in the entire graph. For example, one layer can be used to cluster the nodes on a finer level, while the other may produce a higher level of hierarchy out of the clusters formed by the first relationship. In case of layers derived from node attribute values, the layers may be redefined during clustering by giving more weight to node attributes that are shared within communities and are different between communities. Another possibility is to detect and remove "noisy" edges of one layer by exploiting information of the other layer.

The assumption of a common hidden community structure is most frequently exploited. The approaches can be divided into early integration, late fusion and linked co-clustering techniques[24]

Early integration approaches create a single graph out of the multiple layers. The simplest way is to form a set-theoretic sum of edges over all layers, or intersection of these sets. More sophisticated methods try to estimate the probability of existence of an edge between nodes given the occurrences in the various layers. After one graph is obtained, traditional community detection algorithms can be applied.

A more sophisticated and more rational approach would be to create the adjacency matrix as an average of adjacency matrices of the individual layers. As soon as one graph is formed, the traditional community detection algorithms may be applied.

The late fusion approaches like that of Tang et al.[25] aim at representing either the similarity matrix $A^{(l)}$ or the Laplacian $S^{(l)}$ of each layer Γ_l by an individual set of eigenvalues $\Lambda^{(l)}$ combined with a common matrix of (column) eigenvectors P ("common matrix factorisation") by minimising

$$\frac{1}{2}\sum_{l=1}^{\updownarrow}\|O^{(l)} - P\Lambda^{(l)}P^{\mathrm{T}}\|_F^2 + \frac{\alpha}{2}\left(\left(\sum_{l=1}^{\updownarrow}\|\Lambda^{(l)}\|_F^2\right) + \|P\|_F^2\right)$$

where $\updownarrow$ is the number of layers, $O^{(l)}$ is either the similarity matrix or its Laplacian, $\|.\|_F$ is the Frobenius norm.

Co-clustering methods try to combine the clustering of each layer into a single process. For example, the clustering objective (e.g. ncut) of the multilayered graph may be defined as a linear combination of objectives of each individual layer. Another approach would be to construct a unified distance measure based on the different layers. Still another approach seeks cross-graph quasi-cliques in a multi-layer graph. It is required that these quasi-cliques are frequent, coherent, and closed. The cross-graph quasi-clique has been defined as a set of vertices belonging to a quasi-clique (dense subgraph) that appears on at least some percentage of graph layers and must be the maximal set.

Liu et al.[26] propose to drive the search for communities by a multilayered modularity.

$$Q(\Gamma^{[M]}, C) = \sum_{l=1}^{\updownarrow}\frac{|E_l|}{\sum_{i=1}^{\updownarrow}|E_i|}Q(\Gamma_l, C)$$

[24] S. Paul, Y. Chen: Consistency of community detection in multi-layer networks using spectral and matrix factorization methods. https://arxiv.org/abs/1704.07353.

[25] W. Tang, Z. Lu, and I. S. Dhillon. Clustering with multiple graphs. In Proc. 9th Intern. Conf. on Data Mining, pages 1016–1021, Mianmi, Florida, Dec.2009; as well as Dong, P. Frossard, P. Vandergheynst, and N. Nefedov. Clustering with multi-layer graphs: A spectral perspective. IEEE Trans. on Signal Processing, 60(11):5820–5831, Dec. 2011.

[26] X. Liu, W. Liu, T. Murata, K. Wakita, A Framework for Community Detection in Heterogeneous Multi - Relational Networks, Advances in Complex Systems 17(6) (2014).

where the modularity Q for a single graph and a clustering C is the one defined in (6.8). Note that the optimal value of this formula is not a plain sum of optimal modularities of the individual layers because it refers to a common clustering of all the nodes with the additional constraint that the nodes related by an existent mapping have to belong to the same cluster of C.

6.11 Software and Data

A reader interested performing experiments in this area is first of all recommended to visit Web pages containing benchmark data in these areas like

- Stanford Large Network Dataset Collection: http://snap.stanford.edu/data/
- Alex Arenas' Web page (containing also programs) http://deim.urv.cat/~aarenas/data/welcome.htm
- various network dat sets http://www-personal.umich.edu/~mejn/netdata/
- LINQS project data—primarily used for link based classification http://www.cs.umd.edu/projects/linqs/projects/lbc/

A number of software packages is available on the Web that support community detection:

- software in MATLAB, supporting community detection, covering subgraph centrality, communicability and homogeneity—http://intersci.ss.uci.edu/wiki/index.php/Matlab_code_contributed_by_Scott_D._White and
- C++ and MATLAB programs for finding communities in large networks; program in C++ as well as in MATLAB http://sites.google.com/site/findcommunities/
- Tim Evans's Web page on community structure presentation software http://plato.tp.ph.ic.ac.uk/~time/networks/communitystructure.html
- Aaron Clauset, "Hierarchical Random Graphs" fitting procedures http://tuvalu.santafe.edu/~aaronc/hierarchy/
- Network Workbench (NWB): a project developing software for large-scale network analysis, modelling and visualisation; targeted application areas are biomedical sciences, social sciences and physics http://nwb.slis.indiana.edu/
- "Fast Modularity" method code, along with a discussion http://cs.unm.edu/~aaron/research/fastmodularity.htm
- MCL (Markov Cluster Algorithm)—a cluster algorithm for graphs http://micans.org/mcl/
- Java programs (jar, and test data) implementing CONGA, CliqueMod and COPRA algorithms by Steve Gregory http://www.cs.bris.ac.uk/~steve/networks/index.html
- WalkTrap—a random walk based clustering program for large networks by Pascal Pons, http://www-rp.lip6.fr/~latapy/PP/walktrap.html and http://www-rp.lip6.fr/~latapy/PP/walktrap.htm
- Clique Percolation Method: www.cfinder.org, [383]

It needs to be stressed that the scale of real graphs prompts the researchers to create parallel methods of community detection. An interested reader is recommended to study papers like

- Parallel Knowledge Community Detection Algorithm Research Based on MapReduce (Min Xu et al., 2011, Advanced Materials Research, 204–210, 1646)
- Yang, S., Wang, B., Zhao, H., & Wu, B. (2009). Efficient Dense Structure Mining Using MapReduce. 2009 IEEE International Conference on Data Mining Workshops, (v), 332–337

There are multiple visualisation tools serving the display of networks, in particular of large ones:

- software for drawing graphs http://www.dmoz.org/Science/Math/Combinatorics/ Software/Graph_Drawing/
- LaNet-vi (Large Network visualization) http://sourceforge.net/projects/lanet-vi/[27]
- statnet—Software tools for the analysis, simulation and visualization of network data, http://statnet.csde.washington.edu/

---

[27]Beiró et al. [57] proposed a further development of this package.

Chapter 7
Data Sets

Abstract In this chapter we present short characteristic of the data sets used through-out this book in illustrative examples.

We describe below the selected data sets used throughout this book.

The sets data3_2, data5_2, and data6_2 stem from the Web site http://www.isical.ac.in/~sanghami/data.html. These are 2-dimensional data sets. The first, consisting of 76 elements, breaks up into three clusters, the second one, containing 250 data points, splits into 5 clusters, and the third encompasses 300 points, forming six clusters. They are depicted in Fig. 7.1.

The set 2rings, depicted in Fig. 7.2a is a synthetic set consisting of 200 points. 100 points lie inside a circle of radius 1 and centre at (0,0); the remaining 100 points were distributed randomly in the outer ring. By adding 100 further points forming the next ring, surrounding the two former ones, the set 3rings, shown in Fig. 7.2b, was obtained.

The set 2spirals, visible in Fig. 7.2c, is a set comprising 190 points. Each subset with cardinality 95 lies on the arm of one of the spirals. This set is used in many papers.

We encourage the reader to experiment with other test data available for instance at the Web sites

- http://www.isical.ac.in/~sanghami/data.html,
- http://cs.joensuu.fi/sipu/datasets/,
- http://www.dr-fischer.org/pub/blockamp/index.html.

Many empirical data sets can be found at the Web site http://archive.ics.uci.edu/ml/. We exploit in this book the set iris.

Social networks of varying complexity are available at many sites, including

- http://deim.urv.cat/~aarenas/data/welcome.htm,
- https://sites.google.com/site/santofortunato/inthepress2,
- http://www-personal.umich.edu/~mejn/netdata/,
- http://www3.nd.edu/~networks/resources.htm.

© Springer International Publishing AG 2018   315
S. T. Wierzchoń and M. A. Kłopotek, *Modern Algorithms of Cluster Analysis*,
Studies in Big Data 34, https://doi.org/10.1007/978-3-319-69308-8_7

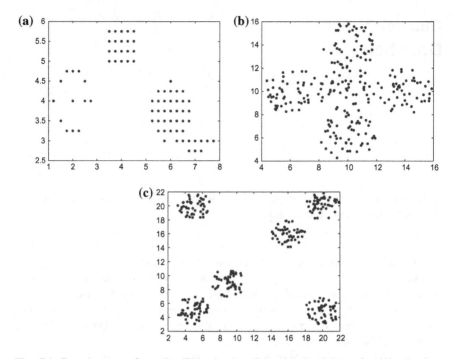

Fig. 7.1 Test data sets from the Web site http://www.isical.ac.in/~sanghami/data.html: **a** data3_2, **b** data5_2, **c** data6_2

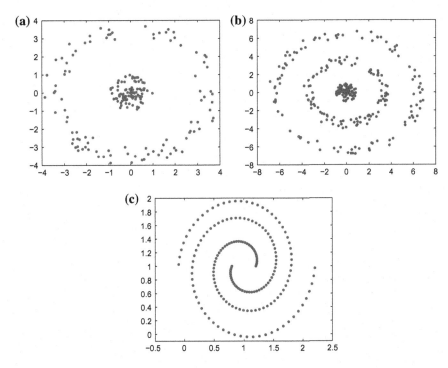

Fig. 7.2 Test data sets: **a** 2rings, **b** 3rings, **c** 2spirals

Appendix A
Justification of the FCM Algorithm

We present here a justification for the formulas (3.58) and (3.59). Recall that these formulas represent a minimum of the target function.

We formulate a Lagrangian in order to determine the elements u_{ij} of the assignment matrix

$$L(J_\alpha, \lambda) = \sum_{i=1}^{m} \sum_{j=1}^{k} u_{ij}^\alpha d^2(\mathbf{x}_i, \boldsymbol{\mu}_j) - \sum_{i=1}^{m} \lambda_i \left(\sum_{j=1}^{k} u_{ij} - 1 \right) \qquad (A.1)$$

We obtain the equation system, given below, by setting partial derivatives of the Lagrangian to zero.

$$\frac{\partial L}{\partial u_{ij}} = 0 \Leftrightarrow \alpha u_{ij}^{\alpha-1} d^2(\mathbf{x}_i, \boldsymbol{\mu}_j) = \lambda_i \ (a)$$

$$\frac{\partial L}{\partial \lambda_i} = 0 \Leftrightarrow \qquad \sum_{j=1}^{k} u_{ij} = 1 \ (b)$$

The equation (a) is equivalent to the following

$$u_{ij} = \left[\frac{\lambda_i}{\alpha d^2(\mathbf{x}_i, \boldsymbol{\mu}_j)} \right]^{\frac{1}{\alpha-1}}$$

Using (b), we can write

$$\left(\frac{\lambda_i}{\alpha} \right)^{\frac{1}{\alpha-1}} = \frac{1}{\sum_{j=1}^{k} d^{2/(1-\alpha)}(\mathbf{x}_i, \boldsymbol{\mu}_j)}$$

which implies the Eq. (3.61).

© Springer International Publishing AG 2018
S. T. Wierzchoń and M. A. Kłopotek, *Modern Algorithms of Cluster Analysis*,
Studies in Big Data 34, https://doi.org/10.1007/978-3-319-69308-8

In order to determine the components of the prototype, let us define

$$\gamma_j(\pmb{\mu}_j) = \sum_{i=1}^{m} u_{ij}^{\alpha} \|\mathbf{x}_i - \pmb{\mu}_j\|_A^2$$

Gradient of the function $\gamma_j(\pmb{\mu}_j)$ with respect to the components of the vector $\pmb{\mu}_j$ is of the form

$$\nabla\gamma_j(\pmb{\mu}_j) = \sum_{i=1}^{m} u_{ij}^{\alpha} \nabla \|\mathbf{x}_i - \pmb{\mu}_j\|_A^2$$

$$= -2A\left[\sum_{i=1}^{m} u_{ij}^{\alpha}(\mathbf{x}_i - \pmb{\mu}_j)\right] = 0$$

By definition of the norm, the matrix A is positive definite which guarantees the existence of the inverse matrix A^{-1}. Finally, we have

$$\pmb{\mu}_j = \frac{\sum_{i=1}^{m} u_{ij}^{\alpha} \mathbf{x}_i}{\sum_{i=1}^{m} u_{ij}^{\alpha}}$$

Of course, in order to check if the solution of the optimisation task, obtained in this way, is really a (local) minimum and not a saddle point, the Hessian of the function $J_\alpha(U, M)$ needs to be investigated.

Appendix B
Matrix Calculus

A detailed discussion of the topics presented here can be found in the books [354, 398, 403, 413, 464].

B.1 Vectors and Their Properties

We denote with the symbol $\mathbf{e}$ a vector having each component equal 1. We denote with $\mathbf{e}_j$ a vector identical with the jth column of a unit matrix.

Definition B.1.1 Let $\mathbf{x}, \mathbf{y}$ be n-dimensional vectors. They are

(a) orthogonal, which we denote $\mathbf{x} \perp \mathbf{y}$, if $\mathbf{x}^\mathsf{T}\mathbf{y} = 0$,
(b) orthonormal, if they are orthogonal vectors of unit length each. $\qquad\square$

If $V = \{\mathbf{v}_1, \ldots, \mathbf{v}_k\}$ is a set of linearly independent n-dimensional vectors, where $k \leq n$, then an orthogonal set $U = \{\mathbf{u}_1, \ldots, \mathbf{u}_k\}$ of vectors, spanning the same k-dimensional subspace of the space $\mathbf{R}^n$ as V does, is obtained by application of the Gram-Schmidt orthogonalisation procedure described e.g. in [354].

B.2 Matrices and Their Properties

Definition B.2.1 Let $A = [a_{ij}]_{m \times m}$ be a square matrix with real-valued entries, i.e. $A \in \mathbb{R}^{m \times m}$. We will call it

(a) non-negative (resp. positive), denoted by $A \geq 0$ (resp. $A > 0$), if all its elements are non-negative (resp. positive)
(b) diagonal, denoted by $A = \text{diag}(d_1, \ldots, d_m)$, if $a_{ij} = 0$ wherever $i \neq j$ and $a_{ii} = d_i$, $i = 1, \ldots, m$. If all diagonal elements are identical and equal to 1, $d_i = 1$, then A is called unit matrix and is denoted with the symbol $\mathbb{I}$.

© Springer International Publishing AG 2018
S. T. Wierzchoń and M. A. Kłopotek, *Modern Algorithms of Cluster Analysis*,
Studies in Big Data 34, https://doi.org/10.1007/978-3-319-69308-8

(c) symmetric, if $A = A^T$.
(d) orthogonal, if $A^T = A^{-1}$, that is, if $A^T A = A A^T = \mathbb{I}$.
(e) normal, if $A A^T = A^T A$.   □

If A is an orthogonal matrix then $|\det(A)| = 1$.

Definition B.2.2 A symmetric matrix A with dimensions $m \times m$ is called positive semidefinite, denoted by $A \succeq 0$, if for any non-zero vector $\mathbf{v} \in \mathbb{R}^m$ the following holds: $\mathbf{v}' A \mathbf{v} \geq 0$. If we can sharpen this relation ($\mathbf{v}' A \mathbf{v} > 0$ for each nonzero vector) then A is called positive definite, denoted with the symbol $A \succ 0$.
On the other hand, if for each non-zero vector $\mathbf{v}$ $\mathbf{v}' A \mathbf{v} \leq 0$ holds, then we call A negative semidefinite. And if $\mathbf{v}' A \mathbf{v} < 0$ holds, then we call A negative definite. □

In practice, the Sylvester theorem is used to decide on definiteness of quadratic forms. Let us denote with Δ_i, $i = 1, \ldots, m$ the leading principal sub-determinants (leading principal minors) of the matrix A, that is

$$\Delta_1(A) = a_{11}, \quad \Delta_2(A) = \begin{vmatrix} a_{11} & a_{12} \\ a_{21} & a_{22} \end{vmatrix}, \ldots, \quad \Delta_m(A) = \begin{vmatrix} a_{11} & \ldots & a_{1m} \\ \ldots & \ldots & \ldots \\ a_{m1} & \ldots & a_{mm} \end{vmatrix}$$

Matrix A is positive semidefinite if $\Delta_i(A) \geq 0$, $i = 1, \ldots, m$. If, furthermore, all the leading principal minors of the matrix are positive, then it is positive definite. In the next section, see Lemma B.3.5, we present further characterisations of positive semidefinite matrices. For completeness, let us state that if $(-1)^j \Delta_j(A) > 0$ then matrix A is negative definite, and if $(-1)^j \Delta_j(A) \geq 0$ then matrix A is negative semidefinite. It is easily seen that if A is a positive (semi)definite matrix then $B = -A$ is a negative (semi)definite matrix.

The Gram matrix is an important example of a positive semidefinite matrix.

Definition B.2.3 If $M = (\mathbf{m}_1, \ldots, \mathbf{m}_k)$ is a matrix of k column vectors of dimension n then $G = M^T M$ is called Gram matrix.   □

The above-defined matrix G is a matrix of dimensions $k \times k$ with elements $g_{ij} = \mathbf{m}_i^T \mathbf{m}_j = g_{ji}$. Its determinant is non-negative.

Definition B.2.4 If a matrix A with dimensions $m \times m$ can be represented in the form $A = B B^T$, where B is a non-negative matrix of dimensions $m \times n$, then A is called a completely positive matrix.[1] The minimal number of columns of B, i.e. n, ensuring the above factorisation of matrix A is called the factorisation index or cp-rank of the matrix A.   □

Definition B.2.5 The number

$$\text{tr}(A) = \sum_{i=1}^{m} a_{ii}$$

---

[1]See, e.g., C. Xu, Completely positive matrices, *Linear Algebra and its Applications*, 379(1), 2004, 319–327.

is called matrix trace. It has the following properties:

(a) $\text{tr}\,(A) = \text{tr}\,(A^{\mathrm{T}})$,
(b) $\text{tr}\,(A + B) = \text{tr}\,(A) + \text{tr}\,(B)$,
(c) $\text{tr}\,(ABC) = \text{tr}\,(BCA) = \text{tr}\,(CAB)$ □

Definition B.2.6 A non-negative matrix P is called a row-stochastic or right-stochastic matrix, if all elements of each row sum up to one. If the sum of all elements of each of its columns equals one then P is called column-stochastic or left-stochastic. If P is both column-stochastic and row-stochastic then we call it doubly stochastic matrix. □

If $A > 0$ is a symmetric matrix, then by alternating the normalising operators of rows and columns we get a doubly stochastic matrix P [433]. Saying it differently, there exists a diagonal matrix $D = \text{diag}(\mathbf{d})$ such that $P = DAD$. Components of the vector $\mathbf{d}$ are equal to $d_i = \sqrt{p_{ii}/a_{ii}}$. This method of computing the doubly stochastic matrix is called Sinkhorn-Knopp method. The paper [295] presents a quick algorithm for balancing a symmetric matrix A, that is, an algorithm transforming it into doubly stochastic matrix P, and presents a review of other related methods.

Definition B.2.7 A stochastic matrix is called

(a) stable, if all of its rows are identical,
(b) column-wise accessible, if its each column contains at least one positive element.

□

B.3 Eigenvalues and Eigenvectors

B.3.1 Basic Facts

A square matrix A of dimensions $m \times m$ possesses an eigen (or characteristic) value λ and an eigen (or characteristic) vector $\mathbf{w} \neq \mathbf{0}$ if

$$A\mathbf{w} = \lambda\mathbf{w} \tag{B.1}$$

The pair $(\lambda, \mathbf{w})$, satisfying the above conditions is called eigenpair.

Equation (B.1) can be rewritten in the equivalent form

$$\det(A - \lambda\mathbb{I}_m) = 0 \tag{B.2}$$

where $\mathbb{I}_m$ is a unit matrix of dimensions $m \times m$. Knowing that formula (B.2) is a m-degree polynomial we can state that the matrix A possesses m (not necessarily distinct) eigenpairs $(\lambda_i, \mathbf{w}_i)$. If all eigenvalues are distinct then we call them non-degenerate.

The set of all distinct eigenvalues

$$\sigma(A) = \{\lambda_1, \ldots, \lambda_{m'}\}, \ m' \leq m \tag{B.3}$$

defines the spectrum of the matrix A, and the quantity

$$\rho(A) = \max_{1 \leq j \leq m} |\lambda_j| \tag{B.4}$$

is called spectral radius of matrix A.

If $\lambda_1, \ldots, \lambda_m$ are eigenvalues of the matrix A of dimensions m, then

(a) $\sum_{i=1}^{m} \lambda_i = \mathrm{tr}(A)$,
(b) $\prod_{i=1}^{m} \lambda_i = \det(A)$,
(c) $\lambda_1^k, \ldots, \lambda_m^k$ are eigenvalues of the matrix A^k.

If $A = \mathrm{diag}(\mathbf{a})$ is a diagonal matrix with the diagonal identical by the vector $\mathbf{a}$, then its eigenvalues are elements of the vector $\mathbf{a}$, with the ith eigenpair of the form $(a_i, \mathbf{e}_i)$, i.e. $\lambda_i = a_i$, and $\mathbf{e}_i = (0, \ldots, 0, 1, 0, \ldots, 0)^{\mathrm{T}}$ is the ith column of the unit matrix.

Lemma B.3.1 *If $(\lambda_i, \mathbf{w}_i)$ is the ith eigenpair of the matrix A, then eigenpairs of the matrix $c_1 \mathbb{I} + c_2 A$ with c_1, c_2 being any values, for which $c_2 \neq 0$ holds, are of the form $(c_1 + c_2 \lambda_i, \mathbf{w}_i)$* $\qquad\qquad\square$

Let λ_{max} be the eigenvalue with the biggest module, i.e. $|\lambda_{max}| = \max_{i=1,\ldots,m} |\lambda_i|$ and let $\mathbf{w}_{max}$ be the corresponding eigenvector. The pair (λ_{max}, w_{max}) is called the principal eigenpair.

If we know the eigenvector $\mathbf{w}_i$ of a *symmetric* matrix A, then we can compute the corresponding eigenvalue λ_i from the equation

$$\lambda_i = \frac{\mathbf{w}_i^{\mathrm{T}} A \mathbf{w}_i}{\mathbf{w}_i^{\mathrm{T}} \mathbf{w}_i} = R(A, \mathbf{w}_i) \tag{B.5}$$

The quantity $R(A, \mathbf{x})$, defined by Eq. (B.5), where $\mathbf{x}$ is any non-zero vector, is called Rayleigh quotient. One can easily check that $R(A, c\mathbf{x}) = R(A, \mathbf{x})$ for any constant $c \neq 0$.

Theorem B.3.1 *Let S_k denote k-dimensional subspace of the space $\mathbb{R}^m$ and let $\mathbf{x} \perp S_k$ denote that $\mathbf{x}$ is a vector orthogonal to any vector $\mathbf{y} \in S_k$. Let $A \in \mathbb{R}^{m \times m}$ be a symmetric matrix with eigenvalues $\lambda_1 \leq \lambda_2 \leq \cdots \leq \lambda_m$. Then*

$$\lambda_k = \max_{S_k} \min_{0 \neq \mathbf{x} \perp S_k} R(A, \mathbf{x}) \tag{B.6}$$

$\qquad\qquad\square$

This is the so-called Courant-Fischer minimax theorem. It implies that

$$(a) \ \lambda_1 = \min_{\mathbf{x} \neq 0} \frac{\mathbf{x}^T A \mathbf{x}}{\mathbf{x}^T \mathbf{x}}, \quad \mathbf{w}_1 = \arg \min_{\mathbf{x} \neq 0} \frac{\mathbf{x}^T A \mathbf{x}}{\mathbf{x}^T \mathbf{x}}$$

$$(b) \ \lambda_2 = \min_{\substack{\mathbf{x} \neq 0 \\ \mathbf{x} \perp \mathbf{w}_1}} \frac{\mathbf{x}^T A \mathbf{x}}{\mathbf{x}^T \mathbf{x}}, \quad \mathbf{w}_2 = \arg \min_{\substack{\mathbf{x} \neq 0 \\ \mathbf{x} \perp \mathbf{w}_1}} \frac{\mathbf{x}^T A \mathbf{x}}{\mathbf{x}^T \mathbf{x}} \qquad (B.7)$$

. . . .

$$(c) \ \lambda_m = \max_{\mathbf{x} \neq 0} \frac{\mathbf{x}^T A \mathbf{x}}{\mathbf{x}^T \mathbf{x}}, \quad \mathbf{w}_m = \arg \max_{\mathbf{x} \neq 0} \frac{\mathbf{x}^T A \mathbf{x}}{\mathbf{x}^T \mathbf{x}}$$

In this way we obtain an estimate $\lambda_1 \leq R(A, \mathbf{x}) \leq \lambda_m$.

Lemma B.3.2 *If A is a symmetric square matrix, and $\mathbf{x} \neq 0$ is a vector, then the value μ, minimising the expression $\| A\mathbf{x} - \mu\mathbf{x} \|$, is equal to the Rayleigh quotient.* $\square$

Let us mention some important properties of the eigenvectors:

(i) If A is a Hermitian matrix (in particular, a symmetric one with real-valued elements) then all its eigenvalues are real numbers.
(ii) If $0 \in \sigma(A)$, then A is a singular matrix (the one with determinant equal zero)
(iii) A square matrix A with positive elements has exactly one real-valued principal eigenvalue and all the elements of the corresponding eigenvector are of the same sign (Perron-Frobenius theorem).
(iv) Eigenvectors of a normal matrix with non-degenerate eigenvalues form a complete set and are orthonormal. This means that they are versors spanning an m dimensional vector space.
    One applies Gram-Schmidt orthogonalisation in case of degenerate eigenvalues. In this way one can find a set of eigenvectors that form a complete set and are orthonormal.
(v) (Gershogorin theorem [354, Example 7.1.4]) Each eigenvalue λ of a square matrix A of dimension m fulfils at least one of the inequalities:

$$|\lambda - a_{ii}| \leq r_i = \sum_{\substack{j \neq i \\ 1 \leq j \leq m}} |a_{ij}|$$

i.e. λ lies inside at least one (complex) circle with a centre at the point a_{ii} and radius r_i being the sum of absolute values of non-diagonal elements of the ith row. $\square$

Definition B.3.1 Matrices A and B are called similar , which we denote $A \approx B$, if there exists a non-singular matrix X such that

$$A = XBX^{-1} \qquad (B.8)$$

The mapping transforming the matrix B into the matrix A is called the similarity mapping. $\square$

Lemma B.3.3 *If A and B are similar matrices then they have identical eigenvalues, and their eigenvectors satisfy the condition* $\mathbf{w}_B = X^{-1}\mathbf{w}_A$.

Proof: Let $(\lambda, \mathbf{w}_A)$ be an eigenpair of the matrix A, i.e. $XBX^{-1}\mathbf{w}_A = \lambda\mathbf{w}_A$ (because $A \approx B$). Let us multiply both sides of this equation by X^{-1}. Then we obtain $B(X^{-1}\mathbf{w}_A) = \lambda(X^{-1}\mathbf{w}_A)$, which implies the thesis.   □

Definition B.3.2 The square matrix A is called a diagonalisable matrix if and only if there exists a non-singular matrix X of the same dimension as A, such that $X^{-1}AX$ is a diagonal matrix. If, furthermore, X is an orthogonal matrix, i.e. $X^{-1} = X^{T}$, then A is called an orthogonally diagonalisable matrix.   □

Let $(\lambda_i, \mathbf{w}_i), i = 1, \ldots, m$ denote the set of eigenpairs of the matrix A. The matrix is diagnosable if its eigenvectors are linearly independent.[2] By substituting $X = (\mathbf{w}_1, \ldots, \mathbf{w}_m)$, i.e. by placing eigenvectors of the matrix A in the columns of the matrix X we obtain $X^{-1}AX = \text{diag}(\lambda_1, \ldots, \lambda_m) = \Lambda$.

Lemma B.3.4 *A square matrix A is a symmetric matrix if and only if it is orthogonally diagonalisable.*

Proof: Let us restrict our considerations to the simpler necessary condition. If A is an orthogonally diagonalisable matrix then there exists such an orthogonal matrix X and a diagonal matrix D that $A = XDX^{T}$. Hence $A^{T} = (XDX^{T})^{T} = A$ □

Let us consider again the positive semi-definite matrices that were introduced in the preceding section.

Lemma B.3.5 *The following conditions are equivalent for a symmetric matrix A:*

(a) $A \succeq 0$.
(b) *All eigenvalues of the matrix A are non-negative.*
(c) $A = C^{T}C$, where $C \in \mathbb{R}^{m \times n}$.

Proof: We will show that (a) implies (b), (b) implies (c), and (c) implies (a).
$(a) \Rightarrow (b)$: As A is a symmetric matrix, hence all, its eigenvalues are real numbers. Let $(\lambda, \mathbf{w})$ be an eigenpair bound by the equation $A\mathbf{w} = \lambda\mathbf{w}$. Let us perform left multiplication of both sides of this equation with the vector $\mathbf{w}^{T}$. We obtain $\mathbf{w}^{T}A\mathbf{w} = \lambda\mathbf{w}^{T}\mathbf{w}$. Both the left side of this equation and $\mathbf{w}^{T}\mathbf{w}$ are non-negative numbers, hence $\lambda \geq 0$.
$(b) \Rightarrow (c)$: Let $\Lambda = \text{diag}(\lambda_1, \ldots, \lambda_m)$ and let $W = (\mathbf{w}_1, \ldots, \mathbf{w}_m)$ be a matrix with columns being eigenvectors corresponding to eigenvalues $\lambda_1, \ldots, \lambda_m$. The Eq. (B.1) can be rewritten in matrix form as $AW = W\Lambda$. By performing right multiplication of this equation with matrix W^{T} we obtain $AWW^{T} = W\Lambda W^{T}$. But eigenvectors of a symmetric matrix A are orthonormal, hence $A = W\Lambda W^{T}$. As $\lambda_i \geq 0$, hence matrix Λ can be represented as $\Lambda^{1/2}(\Lambda^{1/2})^{T}$. Therefore

---

[2]Because then its rank is equal m, which is a necessary condition for the existence of the matrix inverse to the given matrix A.

$$A = W\Lambda^{1/2}(\Lambda^{1/2})^{\mathrm{T}}W^{\mathrm{T}} = W\Lambda^{1/2}(W\Lambda^{1/2})^{\mathrm{T}} = C^{\mathrm{T}}C$$

where $C^{\mathrm{T}} = W\Lambda^{1/2}$.

$(c) \Rightarrow (a)$: Because of $A = C^{\mathrm{T}}C$, we have, for any non-zero vector $\mathbf{v}$,

$$\mathbf{v}^{\mathrm{T}}A\mathbf{v} = \mathbf{v}^{\mathrm{T}}C^{\mathrm{T}}C\mathbf{v} \Leftrightarrow \mathbf{v}^{\mathrm{T}}A\mathbf{v} = \mathbf{y}^{\mathrm{T}}\mathbf{y} \Leftrightarrow \mathbf{v}^{\mathrm{T}}A\mathbf{v} \geq 0$$

where $\mathbf{y} = C\mathbf{v}$.   □

Let $A_k = [a_{ij}]$, $i, j \in \{1, \ldots, k\}$ denote a submatrix of the positive semi-definite matrix A. Then, for each value $1 \leq k \leq m$ and each non-zero vector $\mathbf{x} \in \mathbb{R}^k$, the following conditions hold : (a) $\mathbf{x}^{\mathrm{T}}A_k\mathbf{x} \geq 0$ and (b) $\det(A_k) \geq 0$.

B.3.2 Left- and Right-Hand Eigenvectors

The eigenvectors $\mathbf{w}$ of a matrix A that we have been talking about so far are called also right eigenvectors because they stand to the right of the matrix A in the defining Eq. (B.1). In an analogous way also the left eigenvectors can be defined.

$$\mathbf{u}^{\mathrm{T}}A = \lambda\mathbf{u}^{\mathrm{T}} \qquad\qquad (B.9)$$

By transposing this equation we see that the left eigenvector of the matrix A is the right eigenvector of the matrix A^{T}. It is obvious that the distinction between left and right eigenvectors would be pointless if A were symmetric, because $\mathbf{u} = \mathbf{w}$, that is—the left and right eigenvectors are identical. Hence, let us subsequently consider predominantly the non-symmetric or even non-normal matrices. Such a distinction is not necessary for eigenvalues, because due to Eq. (B.2) and due to $\det(A) = \det(A^{\mathrm{T}})$ we see that in both cases one obtains the same eigenvalue.

By reasoning like in the proof of the Lemma B.3.3 we infer

Lemma B.3.6 *If A, B are similar matrices, then they have identical eigenvalues and their left eigenvectors are linked by the interelation $\mathbf{u}_B = X\mathbf{u}_A$.*

Lemmas B.3.3 and B.3.6 imply that for similar matrices A and B with A, being symmetric, the left ($\mathbf{u}_B$) and right ($\mathbf{w}_B$) eigenvectors of the matrix B can be expressed in terms of the eigenvectors of $\mathbf{v}_A$ of the matrix A as follows:

$$\mathbf{u}_B = X\mathbf{v}_A, \quad \mathbf{w}_B = X^{-1}\mathbf{v}_A$$

where X is a matrix of coefficients.

Lemma B.3.7 *Let A be a matrix of dimension m possessing m distinct left and right eigenvectors*[3] $\mathbf{u}_i$, $\mathbf{w}_i$. *Then*

$$A = \sum_{i=1}^{m} \lambda_i \mathbf{w}_i \mathbf{u}_i^{\,T} \tag{B.10}$$

The above spectral representation implies the equation

$$A^n = \sum_{i=1}^{m} \lambda_i^n \mathbf{w}_i \mathbf{u}_i^{\,T} \tag{B.11}$$

In particular, if A is a symmetric matrix, we get

$$A^n = \sum_{i=1}^{m} \lambda_i^n \mathbf{v}_i \mathbf{v}_i^{\,T} \tag{B.12}$$

where $\mathbf{v} = \mathbf{u} = \mathbf{w}$.

B.3.3 Determining Eigenvalues and Eigenvectors

Quick algorithms for computation of eigenvectors are indispensable for the spectral clustering methods. Many monographs have been devoted to this topic, e.g. [41, 413, 464]. Projection methods seem to play a key role here. Chapter 9 of the monograph [204] is devoted to them.

Here below we present the simplest method allowing to find quickly the principal eigenvector and then discuss its application for determining eigenpairs of the Laplacians.

B.3.3.1   The Power Method

As stated previously, eigenvectors $\mathbf{x}_1, \ldots, \mathbf{x}_m$ of a diagonalisable matrix A are linearly independent. Therefore, they constitute a vector base in the space $\mathbb{R}^m$. Any vector $\mathbf{x}^{(0)} \in \mathbb{R}^m$ can be represented in the form

$$\mathbf{x}^{(0)} = c_1 \mathbf{x}_1 + \cdots + c_m \mathbf{x}_m$$

---

[3]Eigenvector normalization footnote: We deliberately postponed the issue of normalization of the eigenvectors till this point because it becomes only here clear why some choices are done. One should be aware that if a vector $\mathbf{w}$ is a right eigenvector of a matrix A, according to Eq. (B.1), then also any vector $c \cdot \mathbf{w}$ for any non zero scalar c is. Similarly, if a vector $\mathbf{u}$ is a left eigenvector of a matrix A, according to Eq. (B.9), then also any vector $c \cdot \mathbf{u}$ for any non zero scalar c is. To get rid of such an ambiguity, we assume throughout this book, if not stated otherwise, that the vectors are normalised, that is, we select that right eigenvector $\mathbf{w}$, for which $\mathbf{w}^T\mathbf{w} = 1$ and the left eigenvector $\mathbf{u}$ having the corresponding normalised right eigenvector $\mathbf{w}$, for which $\mathbf{u}^T\mathbf{w} = 1$. Note that the right and left eigenvectors are normalised differently.

where $c_1, \ldots, c_m$ are scalars. By multiplying the above equation by $A^k, k = 1, 2, \ldots,$ we get

$$A^k \mathbf{x}^{(0)} = c_1(A^k \mathbf{x}_1) + \cdots + c_m(A^k \mathbf{x}_m)$$

$$= c_1(\lambda_1^k \mathbf{x}_1) + \cdots + c_m(\lambda_m^k \mathbf{x}_m) = c_1 \lambda_1^k \left[\mathbf{x}_1 + \sum_{j=2}^{k} c_j \left(\frac{\lambda_j}{\lambda_1} \right)^k \mathbf{x}_j \right] \quad \text{(B.13)}$$

We exploit here the fact that $A\mathbf{x}_j = \lambda_j \mathbf{x}_j$, where λ_j denotes the eigenvalue corresponding to the eigenvector $\mathbf{x}_j$, and that whenever λ_j is an eigenvalue of the matrix A, then λ_j^k is an eigenvalue of the matrix A^k. If eigenvalues are sorted according to their decreasing module, then $|\lambda_j/\lambda_1| \leq 1$. So, if only $|\lambda_j/\lambda_1| < 1$ then

$$\lim_{k \to \infty} \frac{\mathbf{x}^{(k)}}{\lambda_1^k} = \lim_{k \to \infty} \frac{A^k \mathbf{x}^{(0)}}{\lambda_1^k} = c_1 \mathbf{x}_1$$

that is, the series $\{\mathbf{x}^{(k)}/\lambda_1^k\}$ converges with speed dependent on the quotient $|\lambda_2/\lambda_1|$, to the vector $c_1 \mathbf{x}_1$.

In practice, the successive approximations of the eigenvector are not computed from the equation $\mathbf{x}^{(k)} = A^k \mathbf{x}^{(0)}$, but rather iteratively, $\mathbf{x}^{(k)} = A\mathbf{x}^{(k-1)}, k = 1, \ldots$. In this way we do not need to compute successive powers of the matrix A. Instead, we multiply each time the matrix A with the vector $\mathbf{x}^{(k-1)}$, obtained in the preceding step. As $\|\mathbf{x}^{(k)}\| \to 0$ when $|\lambda_1| < 1$ (or $\|\mathbf{x}^{(k)}\| \to \infty$ when $|\lambda_1| > 1$), hence, to avoid overflow and underflow, the vector $\mathbf{x}^{(k)}$ is normalised. If we denote by $\mathbf{y}^{(k)}$ the product $A\mathbf{x}^{(k-1)}$ then $\mathbf{x}^{(k)} = \mathbf{y}^{(k)}/m(\mathbf{y})$, where $m(\mathbf{y})$ is the first element of the vector $\mathbf{y}$ with the largest module. If, for example, $\mathbf{y}^{(k)} = (1, -5, 2, 5)$, then $m(\mathbf{y}^{(k)}) = -5$. Thereby not only the division error is minimized in line 6 of pseudocode B.1, but also the total computational burden is minimized, see [354, p. 534]; [403]. The value $m(\mathbf{y}^{(k)}) \to \lambda_1$ when $k \to \infty$, so we get also a method to determine an approximation of the principal eigenvalue. We choose usually, as the stopping criterion, the criterion of minimal correction, i.e. $\|\mathbf{x}^{(k+1)} - \mathbf{x}^{(k)}\| < \epsilon$, or $|m(\mathbf{y}^{(k+1)}) - m(\mathbf{y}^{(k)})| \leq \epsilon$. One initialises the vector $\mathbf{x}^{(0)}$ randomly. Such an initialisation is particularly recommended if the main diagonal contains elements that are significantly larger than the other elements of the matrix A.

The method of computation of the principal eigenpair is presented as the pseudocode B.1.

Remark B.3.1 The weakness of the power method is that if the principal eigenvalue is a complex number, then the approximation series $\mathbf{x}^{(k)}$ is not convergent! $\qquad \square$

The power method returns the *principal* eigenpair, which means that the obtained approximation $\tilde{\lambda}$ can be a negative number. If our goal is to find the eigenvector corresponding to the maximal *positive* eigenvalue, then we proceed as follows: We construct a matrix $A' = A - \tilde{\lambda}\mathbb{I}$. Its eigenvalues are $\lambda_i' = \lambda_i - \tilde{\lambda} \geq 0$. One can verify

---

Algorithm B.1 Power method returning the principal eigenpair of the matrix A

1: $k = 0$.
2: Initialise the vector $\mathbf{x}^{(k)}$. {$\mathbf{x}^{(k)}$ cannot be orthogonal to the principal eigenvector. In practice, it is sufficient that all its components have the same sign. }
3: **while** not done **do**
4:    $\mathbf{y}^{(k+1)} = A x^{(k)}$
5:    $\beta^{(k+1)} = m(\mathbf{y}^{(k+1)})$
6:    $\mathbf{x}^{(k+1)} = \mathbf{y}^{(k+1)}/\beta^{(k+1)}$
7:    **if** $\|\mathbf{x}^{(k+1)} - \mathbf{x}^{(k)}\| \leq \epsilon$ **then**
8:       done = true;
9:    **end if**
10:   $k = k + 1$
11: **end while**
12: **return** eigenvalue $\lambda \approx \beta^{(k+1)}$ and the corresponding eigenvector $\mathbf{x}^{(k)}$

---

that if λ_i is the ith eigenvalue of the matrix A, and $\mathbf{x}_i$ is the corresponding eigenvector then

$$(A - \widetilde{\lambda}\mathbb{I})\mathbf{x}_i = \lambda_i\mathbf{x}_i - \widetilde{\lambda}\mathbf{x}_i$$

Hence, the matrices A and A' have identical eigenvectors, while the principal eigenvalue of the matrix A' corresponds to the maximal positive eigenvalue of the matrix A increased by $o -\widetilde{\lambda}$. So, by repeating the power method, this time for the matrix A', we find the eigenvector corresponding to the maximal positive eigenvalue of the matrix A. This approach is referred to as the "deflation method".

B.3.3.2   Determining the Eigenpairs of the Laplacian

In case of spectral cluster analysis, we are interested in finding $p < m$ eigenvectors, corresponding to the lowest non-trivial eigenvalues of the Laplacian $L = D - A$, where $A \in \mathbb{R}^{m \times m}$ is a symmetric matrix,[4] and D is a diagonal matrix with elements $d_{ii} = \sum_{j=1}^{m} a_{ij}$. An interesting solution to this problem was proposed by Koren, Carmel and Harel in [302]. The method is outlined in the pseudocode B.2. Originally, the method was developed for drawing graphs, which means that only the first and the second (positive) non-trivial Laplacian eigenvalues are sought.

Let us note, first, that if $(\lambda_i, \mathbf{v}_i)$ are eigenpairs of the matrix A, then $(g-\lambda_i, \mathbf{v}_i)$ are eigenpairs of the matrix $\widetilde{A} = g\mathbb{I} - A$. In particular, if A is a Laplacian, and $g \geq \max_{1 \leq i \leq m} |\lambda_i|$, then by applying the deflation method one can determine the eigenpairs corresponding to the lowest eigenvalues of the Laplacian. Gershogorin theorem is used to estimate the value of g—see property (iv) from p. 323, according to which the eigenvalues of a matrix A belong to the set sum of discs $\mathcal{K}_i$ in the complex plane.

---

[4]A is interpreted as a neighbourhood matrix (similarity or adjacency matrix), see Sect. 5.2.

$$\mathcal{K}_i = \{z \in \mathbb{C} : |z - a_{ii}| \le \sum_{j \neq i} |a_{ij}|\}, \qquad i = 1, \dots m$$

Eigenvalues of any Laplacian $L = [l_{ij}]$ are non-negative. Furthermore $l_{ii} = \sum_{j \neq i} |l_{ij}|$. Therefore, to estimate the largest eigenvalue, it is sufficient to compute the value

$$g = 2 \cdot \max_{1 \le j \le m} l_{ii} \tag{B.14}$$

$L = D - A$, hence $\mathrm{diag}(L) = \mathrm{diag}(D)$, which allows to accelerate the computations significantly. It is the step 2 of the Algorithm B.2. Let us sort the eigenvalues of a Laplacian increasingly, and let λ_i be the ith eigenvalue in this order. Then, values $\widehat{\lambda}_i = g - \lambda_i$ (decreasing with growing i) are the eigenvalues of the matrix $\widehat{L} = g\mathbb{I} - L$. This transformation is performed in step 3 of the Algorithm B.2.

The eigenvector, corresponding to the value $\widehat{\lambda}_1 = g$, is a normalised unit vector, that is—its components are equal $1/\sqrt{m}$. Therefore, one determines the eigenvectors corresponding to eigenvalues $\widehat{\lambda}_2 \ge \dots, \ge \widehat{\lambda}_p$—steps 4–19. The initial, random approximation of the ith eigenvector (step 5) is subject to Gram–Schmidt orthogonalisation, so that the resultant vector will be orthogonal to the already determined approximations of eigenvectors (step 10). Next one step of the power method is executed.

Let us look at the stopping condition of the loop **while–do**. Formally, one can iterate the loop till the product $\widehat{\mathbf{v}}\mathbf{v}$ exceeds the threshold $1 - \epsilon$. In practice, one observes stabilisation after a number of iterations, $\widehat{\mathbf{v}}^{\mathrm{T}}\mathbf{v} = const$. Therefore, the formal condition was replaced by a more natural condition $prev - next = 0$, where $next$ (resp. $prev$) is the current (resp. previous) value of the product $\widehat{\mathbf{v}}^{\mathrm{T}}\mathbf{v}$.

The only serious computational burden of the algorithm is determination of the product $\widehat{L}\mathbf{v} = \widehat{D}\mathbf{v} - A\mathbf{v}$. Notice, however, that $\widehat{D}\mathbf{v}$ is a vector with components of the form $(g - \mathsf{d}_i)v_i$, and A is a sparse matrix.

B.4   Norms of Vectors and Matrices

A tool allowing to measure the "size" of a vector $\mathbf{x} \in \mathbb{R}^m$ is its norm $\|\mathbf{x}\|_p$, defined as follows:

$$\|\mathbf{x}\|_p = \left(\sum_{i=1}^m |x_i|^p \right)^{1/p}, \quad p = 1, 2, \dots \tag{B.15}$$

If we set $p = 1$, then we obtain the so-called Manhattan norm (called also taxicab metric, rectilinear distance, city block distance, Manhattan distance, or Manhattan length), $\|\mathbf{x}\|_1 = \sum_i |x_i|$, while $\|\mathbf{x}\|_2$ is called Euclidean norm, or simply—vector

Algorithm B.2 Applying power method to determine the first $p \leq m$ eigenvectors of the Laplacian

---

Input: A—neighbourhood matrix, $V = [v_{ij}]_{m \times p}$—a matrix of randomly generated and normalised columns; elements of the first column are of the form $v_{i1} = 1/\sqrt{m}$, $i = 1, \ldots, m$.

1: $L = D - A$, where D is the degree matrix
2: $g = \max_{1 \leq i \leq m} \left(l_{ii} + \sum_{j \neq i} |l_{ij}| \right)$
3: $\widehat{L} = g\mathbb{I} - L$ {inversion of the order of the eigenvalues }
4: **for** $i = 2$ **to** p **do**
5: $\quad \widehat{\mathbf{v}}_i = V(:, i)$ {i^{th} column of the matrix V}
6: $\quad prev = 1, next = 0, done = \text{false}$
7: $\quad$ **while** (**not** done) **do**
8: $\quad\quad \mathbf{v} \leftarrow \widehat{\mathbf{v}}_i$
9: $\quad\quad$ **for** $j = 1$ **to** $i - 1$ **do**
10: $\quad\quad\quad \mathbf{v} = \mathbf{v} - (\mathbf{v}^T\mathbf{v}_j)\mathbf{v}_j$ { Gram-Schmidt orthogonalisation}
11: $\quad\quad$ **end for**
12: $\quad\quad \widehat{\mathbf{v}} = \widehat{L}\mathbf{v}$
13: $\quad\quad \widehat{\mathbf{v}} = \widehat{\mathbf{v}}/\|\widehat{\mathbf{v}}\|$
14: $\quad\quad next = \widehat{\mathbf{v}}^T\mathbf{v}$
15: $\quad\quad done = |prev - next| < \epsilon$
16: $\quad\quad prev = next$
17: $\quad$ **end while**
18: $\quad V(:, i) = \widehat{\mathbf{v}}$ {saving the vector $\widehat{\mathbf{v}}$ in the ith column of the matrix V}
19: **end for**

---

length. Finally, $p = \infty$ corresponds to the maximum norm, also known as supremum norm, sup norm, the Chebyshev norm, the infinity norm or the "uniform norm":

$$\|\mathbf{x}\|_\infty = \max_{1 \leq i \leq m} \mathbf{x}_i \tag{B.16}$$

The case of $p \in (0, 1)$ was investigated exhaustively by Aggarwal, Hinneburg and Keim in the paper [8], where they suggest its high usefulness.

Given a rectangular matrix $A \in \mathbb{R}^{n \times m}$, one defines either the Frobenius norm

$$\|A\|_F = \sqrt{\text{tr}\,(A^T A)} = \sqrt{\sum_{i=1}^{n}\sum_{j=1}^{m} a_{ij}^2} \tag{B.17}$$

or one uses a matrix norm induced by a vector norm

$$\|A\|_p = \sup_{\mathbf{x} \neq 0} \frac{\|Ax\|_p}{\|\mathbf{x}\|_p} = \sup_{\|\mathbf{x}\|_p = 1} \|Ax\|_p \tag{B.18}$$

In particular:

(i) $\|A\|_1 = \max_{1 \leq j \leq m} \sum_{i=1}^{n} |a_{ij}|$ is the maximum value of the sum of modules of column elements,

(ii) $\|A\|_2 = \sqrt{\lambda_{max}}$, where λ_{max} is the maximal eigenvalue of the matrix $A^{\mathrm{T}}A$; this norm is called spectral norm,

(iii) $\|A\|_\infty = \max_{1 \le i \le n} \sum_{j=1}^{m} |a_{ij}|$ is the maximum value of the sum of modules of row elements.

If A is a diagonal matrix, $A = \mathrm{diag}(a_1, \ldots, a_m)$, then

$$\|A\|_p = \max_{1 \le j \le m} |a_j|, \quad p = 1, 2, \ldots$$

More information on this subject can be found for example, in Chap. 5 of the monograph [354].

B.5 Kernel Trick and the Euclidean Distance Matrix

Kernel based clustering methods, as described in Sects. 2.5.7 and 3.1.5.5 exploit frequently the so-called kernel-trick, as described by formula (2.49), in that a kernel matrix K is used. Its elements $k_{ij} = \Phi(\mathbf{x}_i)^{\mathrm{T}}\Phi(\mathbf{x}_j) = \mathsf{K}(\mathbf{x}_i, \mathbf{x}_j)$ represent dot-products of coordinate vectors of data points $\mathbf{x}_i, \mathbf{x}_j$ transformed via the function Φ from the original data space to the feature space. The essence of the kernel trick is not to have the function Φ in an explicit form and to work with the entries k_{ij} instead.[5]

But the question should be raised what properties the kernel matrix should have in order to (1) be really a matrix of dot products and (2) to enable to recover function Φ at the data points from the kernel matrix.

These questions may seem to be pretty easy: Let Y be a matrix $Y = (\Phi(\mathbf{x}_1), \Phi(\mathbf{x}_2), \ldots, \Phi(\mathbf{x}_m))^T$. Then apparently $K = YY^T$. Hence for any non-zero vector $\mathbf{u}$, $\mathbf{u}^{\mathrm{T}}K\mathbf{u} = \mathbf{u}^{\mathrm{T}}YY^{\mathrm{T}}\mathbf{u} = (Y^{\mathrm{T}}\mathbf{u})^{\mathrm{T}}(Y^{\mathrm{T}}\mathbf{u}) = \mathbf{y}^{\mathrm{T}}\mathbf{y} \ge 0$ where $\mathbf{y} = Y^{\mathrm{T}}\mathbf{u}$ so K must be positive semidefinite. But a matrix is positive semidefinite iff all its eigenvalues are non-negative. Furthermore, all its eigenvectors are real numbers.

So to identify Φ at data points, one has to find all eigenvalues λ_l, $l = 1, \ldots, m$ and corresponding eigenvectors $\mathbf{v}_l$ of the matrix K. If all eigenvalues are hereby non-negative, then construct the matrix Y that has as columns the products $\sqrt{\lambda_l}\mathbf{v}_l$. Rows of this matrix (up to permutations) are the values of the function Φ at data points $1, \ldots, m$. It may be verified that kernel-k-means algorithm with the above K and ordinary k-means for Y would yield same results.

Closely related is the following issue: For algorithms like k-means, instead of the kernel matrix the distance matrix D between the objects may be available, being the Euclidean distance matrix in the feature space. We will call D Euclidean matrix. The question is now: (3) can we obtain the matrix K from such data?

A number of transformations yielding the required kernel matrix has been proposed. The answer to the third question seems to be easily derivable from the paper

---

[5]This section is based on: M.A. Kłopotek: On the Existence of Kernel Function for Kernel-Trick of k-Means. CoRR abs/1701.05335 (2017).

by Balaji et al. [43]. One should use the transformation[6]

$$K = -\frac{1}{2}\left(\mathbf{I} - \frac{\mathbf{11}^T}{m}\right)D_{sq}\left(\mathbf{I} - \frac{\mathbf{11}^T}{m}\right)$$

(where D_{sq} is a matrix containing as entries squared distances from D) a result going back to a paper by Schoenberg (1932).

A generally accepted proof of a more general transformation can be found in the paper by Gower [208, Theorem 2, p. 5], who generalises the above result of Schoenberg (1935) to

$$K = (\mathbf{I} - \mathbf{1s}^T)\left(-\frac{D_{sq}}{2}\right)(\mathbf{I} - \mathbf{s1}^T)$$

for an appropriate choice of $\mathbf{s}$.

Let us recall that a matrix $D \in \mathbb{R}^{m \times m}$ is an Euclidean distance matrix between points $1, \ldots, m$ if and only if there exists a matrix $X \in \mathbb{R}^{m \times n}$ rows of which $(\mathbf{x_1}^T, \ldots, \mathbf{x_m}^T)$ are coordinate vectors of these points in an n-dimensional Euclidean space and $d_{ij} = \sqrt{(\mathbf{x_i} - \mathbf{x_j})^T(\mathbf{x_i} - \mathbf{x_j})}$

Gower in [209] proposes:

Lemma B.5.1 *D is Euclidean iff the matrix $F = (\mathbf{I} - \mathbf{1s}^T)(-\frac{1}{2})D_{sq}(\mathbf{I} - \mathbf{s1}^T)$ is positive semidefinite for any vector $\mathbf{s}$ such that $\mathbf{s}^T\mathbf{1} = 1$.*

Let $D \in \mathbb{R}^{m \times m}$ be a matrix of Euclidean distances between objects. Let D_{sq} be a matrix of squared Euclidean distances d_{ij}^2 between objects with identifiers $1, \ldots, m$. This means that there must exist a matrix $X \in \mathbb{R}^{m \times n}$ for some n rows of which represent coordinates of these objects in an n-dimensional space. If $E = XX^T$ (E with dimensions $m \times m$), then $d_{ij}^2 = e_{ii} + e_{jj} - 2e_{ij}$. We will call the matrix X an embedding of D. A matrix can be called Euclidean if and only if an embedding exists.

As a rigid set of points in Euclidean space can be moved (shifted, rotated, flipped symmetrically[7]) without changing their relative distances, there may exist many other matrices Y rows of which represent coordinates of these same objects in the same n-dimensional space after some isomorphic transformation. Let us denote the set of all such embeddings $\mathcal{E}(D)$. And if a matrix $Y \in \mathcal{E}(D)$, then for the product $F = YY^T$ we have $d_{ij}^2 = f_{ii} + f_{jj} - 2f_{ij}$. We will say that $F \in \mathcal{E}_{dp}(D)$.

For an $F \in \mathcal{E}_{dp}(D)$ define a matrix $G = F + \frac{1}{2}D_{sq}$. Hence $F = G - \frac{1}{2}D_{sq}$. Obviously then

---

[6]To follow the convention of Gower and other authors, in the rest of this appendix we use $\mathbf{e}$ instead of e.

[7]Gower does not consider flipping.

$$d_{ij}^2 = f_{ii} + f_{jj} - 2f_{ij} \tag{B.19}$$

$$= (g_{ii} - \frac{1}{2}d_{ii}^2) + (g_{jj} - \frac{1}{2}d_{jj}^2) - 2(g_{ij} - \frac{1}{2}d_{ij}^2) \tag{B.20}$$

$$= g_{ii} + g_{jj} - 2g_{ij} + d_{ij}^2 \tag{B.21}$$

(as $d_{jj} = 0$ for all j). This implies that

$$0 = g_{ii} + g_{jj} - 2g_{ij} \tag{B.22}$$

that is

$$g_{ij} = \frac{g_{ii} + g_{jj}}{2} \tag{B.23}$$

So G is of the form

$$G = \mathbf{g1}^T + \mathbf{1g}^T \tag{B.24}$$

with components of $\mathbf{g} \in \mathbb{R}^m$ equal $g_i = \frac{1}{2}g_{ii}$.

Therefore, to find $F \in \mathcal{E}_{dp}(D)$ for an Euclidean matrix D we need only to consider matrices deviating from $-\frac{1}{2}D_{sq}$ by $\mathbf{g1}^T + \mathbf{1g}^T$ for some $\mathbf{g}$. Let us denote with $\mathcal{G}(D)$ the set of all matrices F such that $F = \mathbf{g1}^T + \mathbf{1g}^T - \frac{1}{2}D_{sq}$. So for each matrix F if $F \in \mathcal{E}_{dp}(D)$ then $F \in \mathcal{G}(D)$, but not vice versa.

For an $F \in \mathcal{G}(D)$ consider the matrix $F^* = (\mathbf{I} - \mathbf{1s}^T)F(\mathbf{I} - \mathbf{1s}^T)^T$. we obtain

$$F^* = (\mathbf{I} - \mathbf{1s}^T)F(\mathbf{I} - \mathbf{1s}^T)^T \tag{B.25}$$

$$= (\mathbf{I} - \mathbf{1s}^T)(\mathbf{1g}^T + \mathbf{g1}^T - \frac{1}{2}D_{sq})(\mathbf{I} - \mathbf{1s}^T)^T \tag{B.26}$$

$$= (\mathbf{I} - \mathbf{1s}^T)\mathbf{1g}^T(\mathbf{I} - \mathbf{1s}^T)^T + (\mathbf{I} - \mathbf{1s}^T)\mathbf{g1}^T(\mathbf{I} - \mathbf{1s}^T)^T - \frac{1}{2}(\mathbf{I} - \mathbf{1s}^T)D_{sq}(\mathbf{I} - \mathbf{1s}^T)^T \tag{B.27}$$

Let us investigate $(\mathbf{I} - \mathbf{1s}^T)\mathbf{1g}^T(\mathbf{I} - \mathbf{1s}^T)^T$:

$$(\mathbf{I} - \mathbf{1s}^T)\mathbf{1g}^T(\mathbf{I} - \mathbf{s1}^T) = \mathbf{1g}^T - \mathbf{1g}^T\mathbf{s1}^T - \mathbf{1s}^T\mathbf{1g}^T + \mathbf{1s}^T\mathbf{1g}^T\mathbf{s1}^T \tag{B.28}$$

Let us make the following choice (always possible) of $\mathbf{s}$ with respect to $\mathbf{g}$: $\mathbf{s}^T\mathbf{1} = 1$, $\mathbf{s}^T\mathbf{g} = 0$.

Then we obtain from the above equation

$$(\mathbf{I} - \mathbf{1s}^T)\mathbf{1g}^T(\mathbf{I} - \mathbf{s1}^T) = \mathbf{1g}^T - \mathbf{101}^T - \mathbf{1g}^T + \mathbf{1s}^T\mathbf{1} \cdot 0 \cdot \mathbf{1}^T = \mathbf{00}^T \tag{B.29}$$

By analogy

$$(\mathbf{I} - \mathbf{1s}^T)\mathbf{g1}^T(\mathbf{I} - \mathbf{1s}^T)^T = ((\mathbf{I} - \mathbf{1s}^T)\mathbf{1g}^T(\mathbf{I} - \mathbf{s1}^T))^T = \mathbf{00}^T \tag{B.30}$$

By substituting (B.29) and (B.30) into (B.27) we obtain

$$F^* = (\mathbf{I} - \mathbf{1s}^T)F(\mathbf{I} - \mathbf{1s}^T)^T = -\frac{1}{2}(\mathbf{I} - \mathbf{1s}^T)D_{sq}(\mathbf{I} - \mathbf{1s}^T)^T \qquad (B.31)$$

So for any $\mathbf{g}$, hence an $F \in \mathcal{G}(D)$ we can find an $\mathbf{s}$ such that: $(\mathbf{I} - \mathbf{1s}^T)F(\mathbf{I} - \mathbf{1s}^T)^T = -\frac{1}{2}(\mathbf{I} - \mathbf{1s}^T)D_{sq}(\mathbf{I} - \mathbf{1s}^T)^T$.

For any matrix $F = -\frac{1}{2}(\mathbf{I} - \mathbf{1s}^T)D_{sq}(\mathbf{I} - \mathbf{1s}^T)^T$ for some $\mathbf{s}$ with $\mathbf{1}^T\mathbf{s} = 1$ we say that F is in multiplicative form or $F \in \mathcal{M}(D)$.

If $F = YY^T$, that is F is decomposable, then also $F^* = (\mathbf{I} - \mathbf{1s}^T)YY^T(\mathbf{I} - \mathbf{1s}^T)^T = ((\mathbf{I} - \mathbf{1s}^T)Y)((\mathbf{I} - \mathbf{1s}^T)Y)^T = Y^*Y^{*T}$ is decomposable. But

$$Y^* = (\mathbf{I} - \mathbf{1s}^T)Y = Y - \mathbf{1s}^T Y = Y - \mathbf{1v}^T \qquad (B.32)$$

where $\mathbf{v} = Y^T\mathbf{s}$ is a shift vector by which the whole matrix Y is shifted to a new location in the Euclidean space. So the distances between objects computed from Y^* are the same as those from Y, hence if $F \in \mathcal{E}_{dp}(D)$, then $Y^* \in \mathcal{E}(D)$.

Therefore, to find a matrix $F \in \mathcal{E}_{dp}(D)$, yielding an embedding of D in the Euclidean n dimensional space we need only to consider matrices of the form $-\frac{1}{2}(\mathbf{I} - \mathbf{1s}^T)D_{sq}(\mathbf{I} - \mathbf{1s}^T)^T$, subject to the already stated constraint $\mathbf{s}^T\mathbf{1} = 1$, that is ones from $\mathcal{M}(D)$.

So we can conclude: If D is a matrix of Euclidean distances, then there must exist a positive semidefinite matrix $F = -\frac{1}{2}(\mathbf{I} - \mathbf{1s}^T)D_{sq}(\mathbf{I} - \mathbf{s1}^T)$ for some vector $\mathbf{s}$ such that $\mathbf{s}^T\mathbf{1} = 1$, $\det((\mathbf{I} - \mathbf{1s}^T)) = 0$ and $D_{sq}\mathbf{s} \neq \mathbf{0}$. So if D is an Euclidean distance matrix, then there exists an $F \in \mathcal{M}(D) \cap \mathcal{E}_{dp}(D)$.

Let us investigate other vectors $\mathbf{t}$ such that $\mathbf{t}^T\mathbf{1} = 1$,

Note that

$$(\mathbf{I} - \mathbf{1t}^T)(\mathbf{I} - \mathbf{1s}^T) = \mathbf{I} - \mathbf{1t}^T - \mathbf{1s}^T + \mathbf{1t}^T\mathbf{1s}^T \qquad (B.33)$$

$$= \mathbf{I} - \mathbf{1t}^T - \mathbf{1s}^T + \mathbf{1s}^T \qquad (B.34)$$

$$= \mathbf{I} - \mathbf{1t}^T \qquad (B.35)$$

Therefore, for a matrix $F \in \mathcal{M}(D)$

$$(\mathbf{I} - \mathbf{1t}^T)F(\mathbf{I} - \mathbf{1t}^T)^T = -\frac{1}{2}(\mathbf{I} - \mathbf{1t}^T)(\mathbf{I} - \mathbf{1s}^T)D_{sq}(\mathbf{I} - \mathbf{1s}^T)^T(\mathbf{I} - \mathbf{1t}^T)^T \quad (B.36)$$

$$= -\frac{1}{2}(\mathbf{I} - \mathbf{1t}^T)D_{sq}(\mathbf{I} - \mathbf{1t}^T)^T \qquad (B.37)$$

But if $F = YY^T \in \mathcal{E}_{dp}(D)$, then

$$F' = (\mathbf{I} - \mathbf{1t}^T)F(\mathbf{I} - \mathbf{1t}^T)^T \tag{B.38}$$

$$= (\mathbf{I} - \mathbf{1t}^T)YY^T(\mathbf{I} - \mathbf{1t}^T)^T \tag{B.39}$$

$$= (Y - \mathbf{1}(\mathbf{t}^T)Y)(Y - \mathbf{1}(\mathbf{t}^T)Y)^T \tag{B.40}$$

and hence each $-\frac{1}{2}(\mathbf{I} - \mathbf{1t}^T)D_{sq}(\mathbf{I} - \mathbf{1t}^T)^T$ is also in $\mathcal{E}_{dp}(D)$, though with a different placement (by a shift) in the coordinate systems of the embedded data points. So if one element of $\mathcal{M}(D)$ is in $\mathcal{E}_{dp}(D)$, then all of them are.

So we have established that: if D is an Euclidean distance matrix, this means there is matrix X such that rows are coordinates of objects i an Euclidean space with distances as in D, then there exists a decomposable matrix $F = YY^T \in \mathcal{E}_{dp}(D)$ which is in $\mathcal{G}(D)$, $\mathcal{E}_{dp}(D) \subset \mathcal{G}(D)$. For each matrix in $\mathcal{G}(D) \cap \mathcal{E}_{dp}(D)$ there exists a multiplicative form matrix in $\mathcal{M}(D) \cap \mathcal{E}_{dp}(D)$. But if it exists, all multiplicative forms are there: $\mathcal{M}(D) \subset \mathcal{E}_{dp}(D)$

In this way we have proven the only-if-part of Theorem B.5.1 of Gower.

Gower [208] makes the following remark: $F = (\mathbf{I} - \mathbf{1s}^T)(-\frac{1}{2}D_{sq})(\mathbf{I} - \mathbf{s1}^T)$ is to be positive semidefinite for Euclidean D. However, for non-zero vectors $\mathbf{u}$

$$\mathbf{u}^T F \mathbf{u} = -\frac{1}{2}\mathbf{u}^T(\mathbf{I} - \mathbf{1s}^T)D_{sq}(\mathbf{I} - \mathbf{1s}^T)^T\mathbf{u} = -\frac{1}{2}((\mathbf{I} - \mathbf{1s}^T)^T\mathbf{u})^T D_{sq}((\mathbf{I} - \mathbf{1s}^T)^T\mathbf{u})$$
$$\tag{B.41}$$

But D_{sq} is known to be not negative semidefinite, so that F would not be positive semidefinite in at least the following cases: $\det((\mathbf{I} - \mathbf{1s}^T)) \neq 0$ and $D_{sq}\mathbf{s} = \mathbf{0}$. So neither can hold.

As we can see from the first case above, F, given by $F = -0.5(\mathbf{I} - \mathbf{1s}^T)D_{sq}(\mathbf{I} - \mathbf{1s}^T)^T$ does not need to identify uniquely a matrix D, as $(\mathbf{I} - \mathbf{1s}^T)$ is not invertible. Though of course it identifies a D that is an Euclidean distance matrix.

Let us now demonstrate the missing part of Gower's proof that D is uniquely defined given a decomposable F.

So assume that for some D (of which we do not know if it is Euclidean, but is symmetric and with zero diagonal), $F = -0.5(\mathbf{I} - \mathbf{1s}^T)D_{sq}(\mathbf{I} - \mathbf{1s}^T)^T$ and F is decomposable that is $F = YY^T$. Let $\mathcal{D}(Y)$ be the distance matrix derived from Y (that is the distance matrix for which Y is an embedding). That means F is decomposable into properly distanced points with respect to $\mathcal{D}(Y)$. And F is be in additive form with respect to it, that is $F \in \mathcal{G} \cap \mathcal{D}(Y)$ Therefore there must exist some $\mathbf{s}'$ such that the $F' = -0.5(\mathbf{I} - \mathbf{1s}'^T)\mathcal{D}(Y)_{sq}(\mathbf{I} - \mathbf{s}'\mathbf{1}^T)$ as valid multiplicative form with respect to $\mathcal{D}(Y)$, and it holds that $F' = (\mathbf{I} - \mathbf{1s}'^T)F(\mathbf{I} - \mathbf{s}'\mathbf{1}^T)$. But recall that $(\mathbf{I} - \mathbf{1s}'^T)F(\mathbf{I} - \mathbf{s}'\mathbf{1}^T) = (\mathbf{I} - \mathbf{1s}'^T)(-0.5(\mathbf{I} - \mathbf{1s}^T)D_{sq}(\mathbf{I} - \mathbf{s1}^T))(\mathbf{I} - \mathbf{s}'\mathbf{1}^T) = -0.5((\mathbf{I} - \mathbf{1s}'^T)(\mathbf{I} - \mathbf{1s}^T))D_{sq}((\mathbf{I} - \mathbf{1s}'^T)(\mathbf{I} - \mathbf{1s}^T))^T = -0.5(\mathbf{I} - \mathbf{1s}'^T)D_{sq}(\mathbf{I} - \mathbf{s}'\mathbf{1}^T)$.

Hence $-0.5(\mathbf{I} - \mathbf{1s}'^T)D_{sq}(\mathbf{I} - \mathbf{s}'\mathbf{1}^T) = -0.5(\mathbf{I} - \mathbf{1s}'^T)\mathcal{D}(Y)_{sq}(\mathbf{I} - \mathbf{s}'\mathbf{1}^T)$.

So we need to demonstrate that for two two symmetric matrices with zero diagonals D, D' such that

$$-\frac{1}{2}(\mathbf{I} - \mathbf{1s}^T)D_{sq}(\mathbf{I} - \mathbf{s1}^T) = -\frac{1}{2}(\mathbf{I} - \mathbf{1s}^T)D'_{sq}(\mathbf{I} - \mathbf{s1}^T)$$

the equation $D = D''$ holds.

It is easy to see that

$$-\frac{1}{2}(\mathbf{I} - \mathbf{1s}^T)(D_{sq} - D'_{sq})(\mathbf{I} - \mathbf{s1}^T) = \mathbf{00}^T$$

Denote $\Delta = D_{sq} - D'_{sq}$.

$$(\mathbf{I} - \mathbf{1s}^T)\Delta(\mathbf{I} - \mathbf{s1}^T) = \mathbf{00}^T$$

$$\Delta - \mathbf{1s}^T\Delta - \Delta \mathbf{s1}^T + \mathbf{1s}^T\Delta \mathbf{s1}^T = \mathbf{00}^T$$

With $\overline{\Delta}$ denote the vector $\Delta \mathbf{s}$ and with c the scaler $\mathbf{s}^T \Delta \mathbf{s}$. So we have

$$\Delta - \mathbf{1}\overline{\Delta}^T - \overline{\Delta}\mathbf{1}^T + c\mathbf{11}^T = \mathbf{00}^T$$

So in the row i, column j of the above equation we have: $\delta_{ij} + c - \overline{\delta}_i - \overline{\delta}_j = 0$. Let us add cells ii and jj and subtract from them cells ij and ji. $\delta_{ii} + c - \overline{\delta}_i - \overline{\delta}_i + \delta_{jj} + c - \overline{\delta}_j - \overline{\delta}_j - \delta_{ij} - c + \overline{\delta}_i + \overline{\delta}_j - \delta_{ji} - c + \overline{\delta}_j + \overline{\delta}_i = \delta_{ii} + \delta_{jj} - \delta_{ij} - \delta_{ji} = 0$. But as the diagonals of D and D' are zeros, hence $\delta_{ii} = \delta_{jj} = 0$. So $-\delta_{ij} - \delta_{ji} = 0$. But $\delta_{ij} = \delta_{ji}$ because D, D' are symmetric. Hence $-2\delta_{ji} = 0$ so $\delta_{ji} = 0$. This means that $D = D'$.

This means that D and $\mathcal{D}(Y)$ are identical, hence that decomposition is sufficient to prove Euclidean space embedding and yields this embedding. This proves the if-part of Gower's Theorem B.5.1.

Appendix C
Personalized PageRank Vector

In case of undirected graphs, the PageRank vector corresponds to a stationary distribution of the random walk, described by a transfer matrix of a special form, see the Remark C.1.1 below. We are interested in a special case of such a random walk, which starts from a particular initial node. Knowledge of such stationary distributions allows to construct a stationary distribution for any stochastic initial vector. We discuss below some specific features of such a random walk. Review of recent developments and various forms of the PageRank can be found e.g. in [202].

C.1   Basic Notions and Interdependences

Let $\Gamma = (V, E)$ be an undirected and connected graph, and $P = AD^{-1}$ be a (column) stochastic transition matrix, describing random walk in this graph. As before, A is the neighbourhood matrix representing links in graph Γ, and D is the diagonal degree matrix. The PageRank vector is defined, as proposed by Page and Brin, in [379], as the vector $\rho(\mathbf{s}, \beta)$ being the solution to the equation

$$\rho(\mathbf{s}, \beta) = \beta\mathbf{s} + (1 - \beta)P\rho(\mathbf{s}, \beta) \qquad (C.1)$$

where $\beta \in (0, 1]$ is the so-called damping factor, and $\mathbf{s}$ is the starting vector. In the original formulation, $\mathbf{s}$ is a vector with all elements equal to $1/|V|$. If only some elements of this vector are positive, while the other ones are equal zero, then we talk about the *personalized* PageRank vector. The set $supp(\mathbf{s}) = \{v \in V : \mathbf{s}(v) > 0\}$ indicates the range of personalization. Subsequently, we will denote a personalised PageRank vector with the symbol $\mathfrak{p}(\mathbf{s}, \beta)$.

Lemma C.1.1 *Let $G(V, E)$ be a connected and aperiodic undirected graph and let $\mathbf{d} = (\mathbf{d}_1, \ldots, \mathbf{d}_m)^T$ denote the vector of degrees. Then the vector*

© Springer International Publishing AG 2018
S. T. Wierzchoń and M. A. Kłopotek, *Modern Algorithms of Cluster Analysis*,
Studies in Big Data 34, https://doi.org/10.1007/978-3-319-69308-8

$$\pi = \frac{\mathbf{d}}{\sum_i \mathsf{d}_i} = \frac{\mathbf{d}}{vol\,V} \tag{C.2}$$

represents a stationary distribution of a Markov chain with transfer matrix P.

Proof: $D^{-1}\mathbf{d} = \mathbf{e}$, hence

$$Pd = AD^{-1}\mathbf{d} = Ae = \mathbf{d}$$

i.e. $\mathbf{d}$ is a right eigenvector of the matrix P (note that the last equality is a consequence of the symmetry of the matrix A). By normalising this vector we obtain a stochastic vector π of the above-mentioned form. $\qquad\square$

Lemma C.1.2 *PageRank vector possesses the following properties:*

(a) $\rho(\mathbf{s}, 1) = \mathbf{s}$,
(b) *If π is a stationary distribution with components of the form (C.2) then $\rho(\pi, \beta) = \pi$ holds for any value of the parameter $\beta \in (0, 1]$,*
(c) $\|\rho(\mathbf{s}, \beta)\|_1 = 1$ *if only* $\|\mathbf{s}\|_1 = 1$.

$\qquad\square$

Remark C.1.1 Please pay attention that in the original formulation of Page and Brin Γ is a directed graph, which requires a more careful transformation of the connection matrix into a stochastic matrix. Furthermore, in order to ensure that the corresponding transfer matrix has a single principal value, the authors modify the stochastic matrix P to the form $P' = \beta\mathbf{se}^T + (1 - \beta)P$. If we denote with $\rho(\mathbf{s}, \beta)$ the eigenvector, corresponding to the principal eigenvalue (equal 1) of the matrix P', we can easily check that, in fact, it fulfills the conditions of the Eq. (C.1), i.e.

$$P'\rho(\mathbf{s}, \beta) = \beta\mathbf{s}(\mathbf{e}^T\rho(\mathbf{s}, \beta)) + (1 - \beta)P\rho(\mathbf{s}, \beta)$$

because $\mathbf{e}^T\rho(\mathbf{s}, \beta) = 1$.

Many authors assume that ρ and $\mathbf{s}$ are *row* vectors. In such a case the Eq. (C.1) is of the form

$$\rho(\mathbf{s}, \beta) = \beta\mathbf{s} + (1 - \beta)\rho(\mathbf{s}, \beta)P^T$$

i.e. the matrix P^T is defined as the product $D^{-1}A$. $\qquad\square$

Remark C.1.2 Let us stress that the symbols $\rho(\mathbf{s}, \beta)$ and $\mathfrak{p}(\mathbf{s}, \beta)$ denote the same solution of the equation system (C.1). We introduce them solely for the convenience of the reader. When we use the symbol $\rho(\mathbf{s}, \beta)$, we have in mind the global PageRank vector, that is, a solution obtained for the *positive* vector $\mathbf{s}$, whereas the symbol $\mathfrak{p}(\mathbf{s}, \beta)$ is intended to mean the personalised PageRank vector that is the solution of equation system (C.1) for the *nonnegative* starting vector $\mathbf{s}$. $\qquad\square$

The transfer matrix P, which appears in equation (C.1), was replaced in the paper [22] by the *lazy* random walk matrix of the form

$$\widehat{P} = \frac{1}{2}(I + AD^{-1}) \tag{C.3}$$

In such a case, the personalised PageRank vector is a vector $\mathfrak{p}(s, \alpha)$, for which the equation below holds:

$$\mathfrak{p}(s, \alpha) = \alpha s + (1 - \alpha)\widehat{P}\mathfrak{p}(s, \alpha) = \alpha s + (1 - \alpha)\frac{1}{2}(\mathbb{I} + P)\mathfrak{p}(s, \alpha) \qquad (C.4)$$

Lemma C.1.3 *If $\mathfrak{p}(s, \alpha; \widehat{P})$ is a solution of the equation system (C.4), and $\mathfrak{p}(s, \beta; P)$ is a solution of the system (C.1), and if $\beta = 2\alpha/(1 + \alpha)$, then $\mathfrak{p}(s, \alpha; \widehat{P}) = \mathfrak{p}(s, \beta; P)$.*

Proof: By transforming the equation (C.4), we obtain

$$\frac{1 + \alpha}{2}\mathfrak{p}(s, \alpha; \widehat{P}) = \alpha s + \frac{(1 - \alpha)}{2}P\mathfrak{p}(s, \alpha; \widehat{P})$$

By multiplying both sides of the above equation by $2/(1 + \alpha)$ and noticing that $(1 - \alpha)/(1 + \alpha) = 1 - \beta$, where $\beta = 2\alpha/(1 + \alpha)$, we obtain the thesis. $\qquad\qquad\Box$

The above lemma demonstrates that the Eqs. (C.1) and (C.4) possess identical solutions if we apply in both cases the same starting vector and the damping factors satisfy the condition $\beta = 2\alpha/(1 + \alpha)$, and matrices P and $\widehat{P}$ correspond to one another, according to Eq. (C.3). Furthermore, Eqs. (C.1) and (C.4), defining $\mathfrak{p}(s, \beta)$ and $\mathfrak{p}(s, \alpha)$, have exactly the same formal form. Hence, algebraic considerations, as well as algorithms derived for $\mathfrak{p}(s, \beta)$ and β, P, can be mapped by simple symbol substitution into $\mathfrak{p}(s, \alpha)$ and $\alpha, \widehat{P}$ and vice versa. Therefore, we will limit ourselves to seeking the distributions of $\mathfrak{p}$ for $\alpha, \widehat{P}$ only.

Lemma C.1.4 *A PageRank vector, personalised with respect to the starting vector s, is a sum of the geometric series:*

$$\mathfrak{p}(s, \alpha) = \alpha\sum_{t=0}^{\infty}(1 - \alpha)^t\widehat{P}^t s = \alpha s + \alpha\sum_{t=1}^{\infty}(1 - \alpha)^t\widehat{P}^t s \qquad (C.5)$$

Proof: Let us rewrite the Eq. (C.4) in the form

$$[\mathbb{I} - (1 - \alpha)\widehat{P}]\mathfrak{p}(s, \alpha) = \alpha s$$

This implies that $\mathfrak{p}(s, \alpha) = \alpha[\mathbb{I} - (1 - \alpha)\widehat{P}]^{-1}s$ if there exists the matrix $[\mathbb{I} - (1 - \alpha)\widehat{P}]^{-1}$. Theorem 1.7 in [270] leads to the conclusion that if X is a square matrix such that $X^n \to 0$ when $n \to \infty$, then the matrix $(\mathbb{I} - X)$ possesses an inverse matrix of the form

$$(\mathbb{I} - X)^{-1} = \sum_{t=0}^{\infty}X^t$$

Substituting the matrix $(1 - \alpha)\widehat{P}$ for X we can check that the condition of the theorem is fulfilled, hence the conclusion. $\qquad\qquad\Box$

Note that in the proof of the above lemma we represented the personalised PageRank vector in the form $\mathfrak{p}(\mathbf{s}, \alpha) = \alpha[\mathbb{I} - (1 - \alpha)\widehat{P}]^{-1}\mathbf{s}$.

By substituting

$$R_\alpha = \alpha\mathbb{I} + \alpha \sum_{t=1}^{\infty} (1 - \alpha)^t \widehat{P}^t \qquad (C.6)$$

we can rewrite the Eq. (C.5) in the form

$$\mathfrak{p}(\mathbf{s}, \alpha) = R_\alpha \mathbf{s} \qquad (C.7)$$

This equation makes apparent that the presonalised PageRank vector is a linear transformation of the vector $\mathbf{s}$, and R_α is the matrix of this transformation. Two important properties of the PageRank vector can be derived from it.

Lemma C.1.5 *The PageRank vector possesses the following properties:*

(a) *It is linear with respect to the starting vector, i.e. $\mathfrak{p}(\mathbf{s}_1 + \mathbf{s}_2, \alpha) = \mathfrak{p}(\mathbf{s}_1, \alpha) + \mathfrak{p}(\mathbf{s}_2, \alpha)$ for any vectors $\mathbf{s}_1, \mathbf{s}_2$. This, in turn, means that the standard PageRank vector $\rho(\mathbf{s}, \alpha)$ is a weighted average of personalised PageRank vectors $\mathfrak{p}(\chi_v, \alpha)$, $v \in V$. Here, the symbol χ_v denotes the characteristic function of the set $S = \{v\}$.*
(b) *The operator $\mathfrak{p}(\mathbf{s}, \alpha)$ commutes with the matrix $\widehat{P}$, i.e. $\widehat{P}\mathfrak{p}(\mathbf{s}, \alpha) = \mathfrak{p}(\widehat{P}\mathbf{s}, \alpha)$. The equality below is a consequence of this property*

$$\mathfrak{p}(\mathbf{s}, \alpha) = \alpha\mathbf{s} + (1 - \alpha)\mathfrak{p}(\widehat{P}s, \alpha) \qquad (C.8)$$

Proof: The property (a) results directly from the formula $\mathfrak{p}(\mathbf{s}, \alpha) = R_\alpha\mathbf{s}$. The property (b) results from a simple transformation

$$\widehat{P}\mathfrak{p}(\mathbf{s}, \alpha) = \alpha(\widehat{P}s) + (1 - \alpha)\widehat{P}(\widehat{P}s) = \mathfrak{p}(\widehat{P}s, \alpha)$$

Taking this into account, we transform the definition of the PageRank vector into the form

$$\mathfrak{p}(\mathbf{s}, \alpha) = \alpha\mathbf{s} + (1 - \alpha)\widehat{P}\mathfrak{p}(\mathbf{s}, \alpha) = \alpha\mathbf{s} + (1 - \alpha)\mathfrak{p}(\widehat{P}s, \alpha)$$

$\square$

C.2 Approximate Algorithm of Determining the Personalized PageRank Vector

The PageRank vector represents, in fact, a stationary distribution of the matrix $P' = (1 - \beta)P + \beta\mathbf{s}\mathbf{e}^{\mathsf{T}}$, see Remark C.1.1. If $\Gamma = (V, E)$ is a connected graph, then P is a regular matrix as in a finite number of steps one can pass from any node $u \in V$ to any other node $v \in V$. According to the theory of regular Markov chains, see e.g. [270], the stationary distribution of the matrix P' has the form

$$\mathfrak{p}_i(\mathbf{s}, \alpha) = \frac{\Delta_i}{\sum_{j=1}^{m} \Delta_i}, i = 1, \ldots, m \qquad (C.9)$$

where Δ_i is the algebraic complement of the ith element, lying on the main diagonal in the matrix $\mathbb{I} - P'$, and $\mathfrak{p}_i(\mathbf{s}, \alpha)$ is the ith element of the personalised vector $\mathfrak{p}(\mathbf{s}, \alpha)$.

Application of the above formula requires computation of m determinants of matrices of dimensions $(m - 1) \times (m - 1)$, which is quite expensive in practice. The Eq. (C.5) implies a conceptually simple method of determining the approximate value of the personalised PageRank vector. It is illustrated by the Algorithm C.1.

Algorithm C.1 Algorithm of determination of the approximation to the personalised PageRank vector on the basis of Eq. (C.5)

1: *Input parameters* : Starting vector $\mathbf{s}$, damping coefficient α, column-stochastic matrix P, precision ϵ.
2: $\mathfrak{p}_{old} = 0, \mathfrak{p}_{new} = 0, \mathbf{s}_1 = \mathbf{s}$.
3: $\beta = 1, \alpha_1 = 1 - \alpha$
4: $res = 1$
5: **while** $(res > \epsilon)$ **do**
6:   $\quad \mathbf{s}_1 = P\mathbf{s}_1$
7:   $\quad \beta = \alpha_1 \beta$
8:   $\quad \mathfrak{p}_{new} = \mathfrak{p}_{new} + \beta \mathbf{s}_1$
9:   $\quad res = \|\mathfrak{p}_{new} - \mathfrak{p}_{old}\|_1$
10: $\quad \mathfrak{p}_{old} = \mathfrak{p}_{new}$
11: **end while**
12: Return the vector $\mathfrak{p} = \alpha(\mathbf{s} + \mathfrak{p}_{new})$

A disadvantage of the algorithm is the necessity to perform multiple multiplications of a vector by a matrix. Taking into account that P is a sparse matrix, this algorithm can be applied to the middle sized graphs.

In case of large graphs, a simulation method, suggested in the paper [22], can prove to be much more convenient. The authors exploited there in an interesting way the idea of Berkhina, who proposed in [65] the following metaphor: let us place a portion of paint in the starting node v, which is spilled over the neighbouring nodes. In each step, a fraction α of the paint, present in a given node v dries out, half of the still wet paint remains in the node v, and the remaining part spills in equal proportions onto the neighbours of the node v. Let $\mathbf{p}^t$ be a vector with elements $p^t(v)$, representing the amount of dried paint at node $v \in V$ at the time point t, and let $\mathbf{r}^t$ be the vector having the elements $r^t(v)$ representing the amount of wet paint present at the time point t in the node v. The process of graph "colouring" is described by two equations

$$\begin{aligned} \mathbf{p}^{t+1} &= \mathbf{p}^t + \alpha \mathbf{r}^t \\ \mathbf{r}^{t+1} &= (1 - \alpha)\widehat{P}\mathbf{r}^t \end{aligned} \qquad (C.10)$$

where $\widehat{P}$ is a matrix of lazy random walk of the form (C.3), $\mathbf{p}^0 = \mathbf{0}$, and $\mathbf{r}^0 = \mathbf{e}_u$.

It is not difficult to see that the vector $\mathbf{p}^{t+1}$ assumes the form

$$\mathbf{p}^{t+1} = \alpha \sum_{k=0}^{t} \mathbf{r}^k = \alpha \sum_{k=0}^{t} (1-\alpha)^k \widehat{P}^k \mathbf{r}^0$$

By substituting $\mathbf{r}^0 = s$ and $t \to \infty$, we obtain the Eq. (C.5).

In the algorithm, presented in the paper [22], further simplifications were proposed, consisting in ignoring the time and in local perspective on the colouring process. The details of this approach are available in the cited paper. Let us only mention here that the essence of the algorithm is the creation of the so-called ϵ-appproximation of the vector $\mathfrak{p}(\mathbf{s}, \alpha)$, that is, a vector $\mathbf{p}$, for which the equation

$$\mathbf{p} + \mathfrak{p}(\mathbf{r}, \alpha) = \mathfrak{p}(\mathbf{s}, \alpha)$$

holds, where $\mathbf{s}$ is the starting vector, and $\mathbf{r}$ is a non-negative vector with components $r(v) < \epsilon \mathsf{d}(v)$, representing the amount of non-dried (wet) paint at nodes $v \in V$.

A further improvement to this algorithm was proposed by Chung and Zhao in [118]. Its time complexity was reduced to $O(\alpha m \log(1/\epsilon))$. The essence of this idea is illustrated by the pseudocode C.2.

Algorithm C.2 Fast algorithm of determining an ϵ-approximation of the personalised PageRank vector [118]

1: *Input parameters*: Undirected graph $\Gamma = (V, E)$, initial vector (distribution) $\mathbf{s}$, coefficient $\alpha \in$ (0, 1), precision ϵ.
2: Substitute $\mathbf{p} = 0, \mathbf{r} = \mathbf{s}, e = 1$.
3: **while** $(e > \epsilon)$ **do**
4: $\quad e = e/2$
5: $\quad p' = 0$
6: $\quad$ **while** $(\exists v \in V : r(v) \geq e\mathsf{d}(v))$ **do**
7: $\qquad$ Choose node u such that $r(u) \geq \epsilon \mathsf{d}(u)$
8: $\qquad p'(u) = p'(u) + \alpha r(u)$
9: $\qquad r(v) = r(v) + \frac{1-\alpha}{\mathsf{d}(u)} r(u), \; \forall v \in N(u)$
10: $\qquad r(u) = 0$
11: $\quad$ **end while**
12: $\quad \mathbf{p} = \mathbf{p} + \mathbf{p}'$
13: **end while**
14: Return the vector $\mathbf{p}$

Let us underline that the method of choosing the node influences only the speed measured in terms of the number of iterations within the **while** loop. When lazy random walk is applied, about 50% more iterations are needed to find an approximation with required precision. The abovementioned results were obtained for the random walk with the random walk matrix P, teleportation coefficient $\alpha = 0.15$ and precision

$\epsilon = 10^{-12}$. As comparisons show further on, the variant C.2 can be considered as the quickest one.

It turns out that in the graph Γ there are not many links between the nodes with high PageRank values and those with a low PageRank [21]. More precisely, if we sort the graph nodes by the decreasing value $p(\mathbf{s}, \alpha)$, and the kth node in this order is assigned a higher portion of probability than the node of rank $k(1 + \delta)$, then there exist few connections between nodes with ranks $\{1, \ldots, k\}$ and those with ranks from the set $\{k(1 + \delta) + 1, \ldots, |V|\}$.

Appendix D
Axiomatic Systems for Clustering

A considerable amount of research work has been devoted to understanding the essentials of clustering, as briefly discussed in Sect. 2.6.[8]

A number of axiomatic frameworks[9] have been devised in order to capture the nature of the clustering process, the most often cited being probably the Kleinberg's system [293].

In general, the axiomatic frameworks of clustering address either:

- the required properties of clustering functions, or
- the required properties of the values of a clustering quality function, or
- the required properties of the relation between the qualities of different partitions (ordering of partitions for a particular set of objects and a given similarity/dissimilarity function).

We will now briefly overview Kleinberg's axioms and some work done around their implications.

D.1 Kleinberg's Axioms

As a justification for his axiomatic system, Kleinberg [293] claims that a good clustering may only be a result of a reasonable method of clustering. His axioms are dealing with distance-based cluster analysis. He defines the clustering function as follows:

---

[8]This appendix is partially based on: R.A. Klopotek, M.A. Klopotek: On the Discrepancy Between Kleinberg's Clustering Axioms and k-Means Clustering Algorithm Behavior. CoRR abs/1702.04577 (2017).

[9]Axiomatic systems may be traced back to as early as 1973, when Wright proposed axioms of weighted clustering functions. This means that every domain object was attached a positive real-valued weight, that could be distributed among multiple clusters, like in later fuzzy systems. See: W.E. Wright. A formalization of cluster analysis. *Pattern Recognition*, **5**(3):273–282, 1973.

© Springer International Publishing AG 2018
S. T. Wierzchoń and M. A. Kłopotek, *Modern Algorithms of Cluster Analysis*,
Studies in Big Data 34, https://doi.org/10.1007/978-3-319-69308-8

Definition D.1.1 A function f is a *clustering function* if it takes as its argument a distance function d on the set S and returns a partition $\mathcal{C}$ of X. The sets in $\mathcal{C}$ will be called its *clusters*.   □

He postulated that some quite "natural" axioms need to be met, when we manipulate the distances between objects. These are:

Axiom D.1.1 The clustering method should allow to obtain any clustering of the objects (so-called *richness property*). More formally, if $Range(f)$ denotes the set of all partitions $\mathcal{C}$ such that $f(d) = \mathcal{C}$ for some distance function d then $Range(f)$ should be equal to the set of all partitions of X.

Axiom D.1.2 The method should deliver clusterings invariant with respect to distance scale (so-called (scale)-*invariance property*). Formally: for any distance function d and any $\alpha > 0$, $f(d) = f(\alpha \cdot d)$ should hold.

Axiom D.1.3 The method should deliver the same clustering if we move objects closer to cluster centres to which they are assigned (so-called *consistency property*). Formally, for a partition $\mathcal{C}$ of X, and any two distance functions d and d' on X such that (a) for all $i, j \in S$ belonging to the same cluster of $\mathcal{C}$, we have $d'(i, j) \leq d(i, j)$, and (b) for all $i, j \in S$ belonging to different clusters of $\mathcal{C}$, we have $d'(i, j) \geq d(i, j)$ (we will say that d' is a $\mathcal{C}$-transformation of d), the following must hold: if $f(d) = \mathcal{C}$ then $f(d') = \mathcal{C}$.

D.1.1 Formal Problems

Though the axioms may seem to be reasonable, Kleinberg demonstrated that they cannot be met all at once (but only pair-wise).

Theorem D.1.1 *No clustering function can have at the same time the properties of richness, invariance and consistency*

The axiom set is apparently not sound. The proof is achieved via contradiction. For example, take a set of $n + 2$ elements. The richness property implies that under two distinct distance functions d_1, d_2 the clustering function f may form two clusterings, respectively $\mathcal{C}_1, \mathcal{C}_2$, where the first n elements form one cluster and in $\mathcal{C}_1$ the remaining two elements are in one cluster, and in $\mathcal{C}_2$ they are in two separate clusters. By the invariance property, we can derive from d_2 the distance function d_4 such that no distance between the elements under d_4 is lower than the biggest distance under d_1. By the invariance property, we can derive from d_1 the distance function d_3 such that the distance between elements $n + 1, n + 2$ is bigger than under d_4. We have then $f(\{1, \ldots, n + 2\}; d_4) = \mathcal{C}_2$, $f(\{1, \ldots, n + 2\}; d_3) = \mathcal{C}_1$. Now let us apply the consistency axiom. From d_4 we derive the distance function d_6 such that for elements $1, \ldots, n$ the d_1 and d_6 are identical, the distance between $n + 1, n + 2$ is the same as in d_4, and the distances between any element of $1, \ldots, n$ and any of $n + 1, n + 2$

is some l that is bigger than any distances between any elements under $d_1, \ldots, d_4$. From d_3 we derive the distance function d_5 such that for elements $1, \ldots, n$ the d_1 and d_5 are identical, the distance between $n + 1, n + 2$ is the same as in d_4 and the distances between any element of $1, \ldots, n$ and any of $n + 1, n + 2$ is the same l as above. We have, then, $f(\{1, \ldots, n + 2\}; d_6) = C_2$, $f(\{1, \ldots, n + 2\}; d_5) = C_1$. But this means a contradiction, because by construction, d_5 and d_6 are identical.

At the same time, Kleinberg demonstrated that there exist algorithms that satisfy any pair of the above conditions. He uses for the purpose of this demonstration the versions of the well known statistical single-linkage procedure. The versions differ by the stopping condition:

- k-cluster stopping condition (which stops adding edges as soon as the linkage graph for the first time consists of k connected components)—not "rich"
- distance-r stopping condition (which adds edges of weight at most r only)—not scale-invariant
- scale-stopping condition (which adds edges of weight being at most some percentage of the largest distance between nodes)—not consistent

Notice that also k-median and k-means, as well as the MDL clustering do not fulfill the consistency axiom.

Note, however, that Ben-David [62] drew attention by an illustrative example (their Fig. 2), that consistency alone is a problematic property as it may give rise to new clusters at micro or macro-level.

D.1.2 Common Sense Problems

As there are practical limitations on the measurement precision, scale-invariance may not be appropriate for the whole range of scaling factors.

The consistency axiom precludes a quite natural phenomenon that under stretching of distances new clusters may be revealed.

Finally, we are really not interested in getting clusters of cardinality say equal one, so that the richness axiom is in fact not intuitive.

D.1.3 Clusterability Theory Concerns

A set of objects is not all we want to know when saying that we discovered a cluster. We want to see that objects belonging to different clusters differ substantially from one another, for example that the clusters are separated from one another by some space.

But this separating space is something what the majority of clustering methods do not take into account and about which Kleinberg's axiomatisation does not care.

Speaking differently, a clustering algorithm may provide us with clusters even when there are none in the data.

The very existence of clusters in the data, or, more precisely, the separation of clusters, was the subject of research on the so-called clusterability,[10] see e.g. [2, 3].

Definition D.1.2 [3] Clusterability is a function of the set $X \subset \mathbb{R}^m$ mapping it into the set of real numbers that specifies how a set X is clusterable.

These functions are constructed in such a way that the higher the value of the function the stronger is the evidence that high quality clusterings may occur.

Methods of cluster analysis, examined by Ackerman [3], on which we base our reflection in this section, are limited to the so-called centre based (or centric) clustering,

Definition D.1.3 We say that a partition is a *centric clustering* if each cluster is distinguished by the centre or several centres, where the distance of any element to the nearest centre of its own cluster is not greater than that to any other centre (of any other cluster).

Centric clusterings are a special case of results from the distance driven clustering functions (Definition D.1.1) so that Kleinberg's axioms should be applicable.

To express the inadequacy of data clustering, Ackerman and Ben-David introduce the concept of *loss function* for this class of clustering. An optimal clustering minimizes the loss function for a particular data set.

Ackerman introduced the following concepts:

Definition D.1.4 Two centric clusterings are ϵ-close, if you can create a set of disjoint pairs of centres from both clusterings such that the distance between the centres of each pair is less than ϵ.

Definition D.1.5 Data is ϵ, δ-clusterable if the loss function for any clustering, that is ϵ-close to an optimal clustering, is not greater than $(1 + \delta)$ times the value of the loss function of the optimal clustering (perturbational clusterability).

The study [3] proposed algorithms detecting such clusterability for $\epsilon = \frac{radius(X)}{\sqrt{l}}$, where l is the cardinality of subsets of X, which all have to be considered.

Let us recall two concepts on which clusterability evaluation is based.

Definition D.1.6 The *separation in the clustering* is the minimum distance between the elements of different classes. *Clustering diameter* is the maximum distance between elements of the the same class.

It is argued in [3] that there is at most one clustering such that the separation is larger than the diameter.

With these concepts, Ackerman introduces in [3] various kinds of clusterability.

---

[10]Instead of clusterability, that is, the issue of the very existence of clusters, the cluster validity, that is—their agreement with some prior knowledge about expected clusters, may be investigated, as discussed already in Chap. 4. Cluster validity was also studied in papers [119, 364].

Definition D.1.7 (Clusterability and related concepts)

- *Worst-pair-quality of a clustering* is the ratio of separation and diameter.
- *Worst-pair-clusterability* is the minimal worst-pair-quality over all centric clusterings of the set X.
- *k-separable clusterability* is the decrease in the loss function while passing from $k - 1$ clusters to k clusters.
- *Variance-clusterability* is the quotient of the variance between clusters to the variance within clusters.
- *Target-clusterability* is measured as the distance to the clustering defined manually.

It is claimed that in polynomial time, one can calculate only the worst-pair-clusterability.

Note, first of all, that all these clusterability measures are sensitive to outliers because (most) clustering methods do not allow for an element to be left unclustered.

But more interesting is the relationship to the Kleinberg's axioms. We would naively expect that if a clustering algorithm returns a clustering, then there exists an intrinsic clustering. Furthermore, one would expect that if there is an intrinsic clustering, then the clustering algorithm returns it. And it returns (approximately) the clustering that is there in the data.

But any algorithm, seeking clusters fitting the requirements of ϵ, δ-clusterability, will fail under distance-scaling, so it will not satisfy the Kleinberg's invariance-axiom.

On the other hand, a clustering algorithm, seeking to match the worst-pair-clusterability or variance-clusterability or target-clusterability criterion, will perform well under scaling.

The k-separable clusterability may or may not depend on scaling this being determined by the loss function.

The axiom of consistency seems not to constitute any obstacle in achieving any of the mentioned clusterability criteria. However, the Kleinberg's C-transformation may lead to emergence of new clusters according to at least the variance-based clusterability criterion.

The Kleinberg's axiom of richness is, however, hard to fulfil at least by the variance-based clusterability. This clusterability criterion puts preference on clusters with small numbers of elements, but if the variance is to be estimated with any basic scrutiny, the minimum number of three elements within each cluster is necessary, which precludes matching of the richness axiom.

D.1.4   Learnability Theory Concerns

Learnability theory [473] defines learnability as the possibility to generalize from the sample to the population. Its basic postulate is: the concept is learnable if it can be falsified. In order to learn, the algorithm must be able to conclude that the data do not belong to the space of concepts. If the algorithm is able to assign to each set of data a concept from the concept space, it does not learn anything.

In the light of the learnability theory

- the classical cluster analysis does not reveal any "natural clusters", but provides a mixture of the intrinsic structure of the data and of the structure induced by the clustering algorithm
- the classical cluster analysis is not a method of learning without supervision ("without a teacher")[11]
- the capability of a clustering algorithm to recreate the external ("manual") clustering is not a criterion for the correctness of the algorithm, if it has not learning abilities
- if the cluster analysis has something to contribute, it cannot satisfy the axioms of Kleinberg that is to

  - produce any clustering possible (via manipulation of distance)
  - produce invariant results while scaling distance
  - produce invariant results with respect to reduction of distance to cluster centre.

because these axioms give a much too big number of degrees of freedom for manipulation of the function of distance.

D.2   Cluster Quality Axiomatisation

Ackerman and Ben-David[12] propose to resolve the problem of Kleinberg's axiomatics by axiomatising not the clustering function, but rather cluster quality function. We base the following on their considerations.

Definition D.2.1 Let $\mathfrak{C}(X)$ be the set of all possible clusterings over the set of objects X, and let $\mathfrak{D}(X)$ be the set of all possible distance functions over the set of objects X.

A clustering-quality measure (CQM) $J : X \times \mathfrak{C}(X) \times \mathfrak{D}(X) \rightarrow \mathbb{R}^+ \cup \{0\}$ is a function that, given a data set (with a distance function) and its partition into clusters, returns a non-negative real number representing how strong or conclusive the clustering is.

Ackerman and Ben-David propose the following axioms:

Axiom D.2.1 (*Scale Invariance*) A quality measure J satisfies scale invariance if for every clustering C of (X, d),[13] and every positive β, $J(X, C, d) = J(X, C, \beta d)$.

---

[11]There is, in fact, always a teacher who provides an aesthetic criterion of what is "similar"/"different".

[12]M. Ackerman and S. Ben-David: Measures of Clustering Quality: A Working Set of Axioms for Clustering. In: D. Koller and D. Schuurmans and Y. Bengio and L. Bottou (eds): Advances in Neural Information Processing Systems 21 NIPS09, 2009, pp. 121–128 .

[13]X is the set of objects, d is a distance function.

Note that if we define a clustering function in such a way that it maximises the quality function, then the clustering function has also to be invariant.

Axiom D.2.2 (*Consistency*) A quality measure J satisfies consistency if for every clustering C over (X, d), whenever d' is a C-transformation of d, then $J(X, C, d') \geq J(X, C, d)$.

Note that if we define a clustering function in such a way that it maximises the quality function, then the clustering function *does not need to be consistent*.

Axiom D.2.3 (*Richness*) A quality measure m satisfies richness if for each non-trivial clustering C^* of X, there exists a distance function d over X such that $C^* = argmax_C\{J(X, C, d)\}$.

Note that if we define a clustering function in such a way that it maximises the quality function, then the clustering function has also to be rich.

Ackerman and Ben-David claim that

Theorem D.2.1 *Consistency, scale invariance, and richness for clustering-quality measures form a consistent set of requirements.*

and prove this claim by providing a quality measure that matches all these axioms.

The quality measure they propose here is the "Relative Margin". First, one needs to compute the ratio of distance of a data point to its closest centre to that to the second closest centre. Then, a representative set of a clustering is defined as a set K containing exactly one element from each cluster. One computes the average of the ratio, mentioned before, for each potential representative set. The minimum over these averages is the cluster quality function called Relative Margin. The lower the value of the Relative Margin, the higher the clustering quality. This does not match quite their axiomatisation because the axiomatisation assumed that increasing quality is related to increasing quality function value. So to satisfy the axiomatic system, axioms have to be inverted, so that quality increase is related to function decrease.

A disadvantage of this measure is that it ranks highly a clustering in which each element is a separate cluster.

D.3 Relaxations for Overcoming the Kleinberg's Problems

In the preceding section one way of clustering axiomatisation improvement was presented. It consisted in shifting axiomatisation from clustering function to clustering quality, which implicitly relaxed at least one Kleinberg's axiom (consistency).

Let us look now at other proposals, concerning now the explicit relaxation of the Kleinberg's axioms, as summarised by Ben-David.

Axiom D.3.1 (*Relaxation of Kleinberg's richness*) For any partition C of the set **X**, consisting of exactly k clusters, there exists such a distance function d that the clustering function $f(d)$ returns this partition C.

This relaxation allows for the algorithms splitting the data into a fixed number of clusters, like k-means. But it still leaves the open problem of a minimal number of elements in a cluster.

Axiom D.3.2 (*Local Consistency*) Let $C_1, \ldots, C_k$ be the clusters of $f(d)$. For every $\beta_0 \geq 1$ and positive $\beta_1, \ldots \beta_k \leq 1$, if d' is defined by: $d'(a, b) = \beta_i d(a, b)$ for a and b in C_i, $d'(a, b) = \beta_0 d(a, b)$ for a, b not in the same $f(d)$-cluster, then $f(d) = f(d')$.

This axiom does not guarantee that d' is in fact a distance, so that it is hard to satisfy.

Axiom D.3.3 (*Refinement Consistency*) is a modification of the consistency axiom obtained by replacing the requirement that $f(d) = f(d')$ with the requirement that one of $f(d)$, $f(d')$ is a refinement of the other.

Obviously, the replacement of the consistency requirement with refinement consistency breaks the impossibility proof of Kleinberg's axiom system. But there is a practical concern: assume a data set of points uniformly randomly distributed in a plane on line segments $((a, 0), (2a, 6a))$, $((a, 0), (2a, -6a))$, $((-a, 0), (-2a, 6a))$, $((-a, 0), (-2a, -6a))$. A k-means algorithm with $k = 2$ would create two clusters: (1) segments $((a, 0), (2a, 6a))$, $((a, 0), (2a, -6a))$, (2) $((-a, 0), (-2a, 6a))$, $((-a, 0), (-2a, -6a))$. But if cluster (2) is shrunk as follows: all points are rotated around $(-a, 0)$ so that they lie on the X axis to the left of $(-a, 0)$, then k-means would change class allocation of a part of points from this cluster, violating refinement consistency.

D.4 k-means Algorithm—Popular, but not a Clustering Algorithm?

The well-known k-means clustering algorithm seeks to minimize the function[14]

$$Q(\mathcal{C}) = \sum_{i=1}^{m} \sum_{j=1}^{k} u_{ij} \|\mathbf{x}_i - \boldsymbol{\mu}_j\|^2 = \sum_{j=1}^{k} \frac{1}{n_j} \sum_{\mathbf{x}_i, \mathbf{x}_l \in C_j} \|\mathbf{x}_i - \mathbf{x}_l\|^2 \qquad (\text{D.1})$$

for a dataset $\mathbf{X}$ under some partition $\mathcal{C}$ into the predefined number k of clusters, where u_{ij} is an indicator of the membership of data point $\mathbf{x}_i$ in the cluster C_j having the centre at $\boldsymbol{\mu}_j$. It is easily seen that it is scale-invariant, but one sees immediately

---

[14]The considerations would apply also to kernel k-means algorithm using the quality function

$$Q(\mathcal{C}) = \sum_{i=1}^{m} \sum_{j=1}^{k} u_{ij} \|\Phi(\mathbf{x}_i - \boldsymbol{\mu}_j^{\Phi})\|^2$$

where Φ is a non-linear mapping from the original space to the so-called feature space.

that it is not rich (only partitions with k clusters are considered), it has also been demonstrated by Kleinberg that it is not consistent. So the widely used algorithm violates in practice two of three Kleinberg's axioms. so that it cannot be considered to be a "clustering function". We perceive this to be at least counter-intuitive. Ben-David and Ackerman in [62] in Sect. 4.2., raised also similar concern from the perspective of what an axiomatic system should accomplish. They state that one would expect, for the axiomatised set of objects, a kind of soundness that is that most useful clustering algorithms would fit the axioms and completeness that is that apparent non-clustering algorithms would fail on at least one axiom. While Kleinberg's axioms explicitly address the distance-based clustering algorithms (and not e.g. density based ones), they fall apparently short of reaching this goal. In this Appendix we demonstrate that even for a narrower set of algorithms, ones over data embedded in Euclidean space, the axioms fail.

There exist a number of open questions on why it is so. Recall that in [476] it has been observed that Kleinberg's proof of Impossibility Theorem stops to be valid in case of graph clustering. This raises immediately the question of its validity in $\mathbb{R}^m$ Euclidean space. Note that Kleinberg did not bother about embedding the distance in such a space. So one may ask whether or not k-means does not fit Kleiberg's axioms because this is a peculiar property of k-means or because any algorithm embedded in Euclidean space would fail to fit.

Therefore we made an effort to identify and overcome at least some reasons for the difficulties connected with axiomatic understanding of research area of cluster analysis and hope that this may be a guidance for further generalisations to encompass if not all then at least a considerable part of the real-life algorithms. This Appendix investigates why the k-means algorithm violates the Kleinberg's axioms for clustering functions. We claim that the reason is a mismatch between informal intuitions and formal formulations of these axioms. We claim also that there is a way to reconcile k-means with Kleinberg's consistency requirement via introduction of centric consistency which is neither a subset nor superset of Kleinberg's consistency, but rather a k-means clustering model specific adaptation of the general idea of shrinking the cluster.

To substantiate our claim that there is a mismatch between informal intuitions and formal formulations of Kleinberg's axioms, we present a series of carefully constructed examples. First of all Kleinberg claims that his axioms can be fulfilled pair-wise. But we demonstrate that richness and scaling-invariance alone may lead to a contradiction for a special case. Further we provide example showing that the consistency and scale-invariance axioms, even ignoring richness axiom, yield a contradiction to the intuitions behind clustering. We demonstrate this by showing that these axioms may lead to moving clusters closer to one another instead of away from one another. If the clusters are moved closer, it does not wonder when a clustering algorithm puts them together. Furthermore we show that consistency alone leads to contradictions. We demonstrate that in practical settings of application of many algorithms, that is in a metric m-dimensional space where m is the number of features, it is impossible to contract a single cluster without moving the other ones and as a consequence running at risk of moving some clusters closer together. We show in particular that k-means version where we allow for k to range over a set, will change the optimal clustering k when Kleinberg's $\mathcal{C}$ operation (consistency operation) is applied.

Finally to substantiate our claim that there exists a way to reconcile the formulation of the axioms with their intended meaning and that under this reformulation the axioms stop to be contradictory and we can reconcile k-means with Kleinberg's consistency requirement we first introduce the concept of centric consistency which is neither a subset nor superset of Kleinberg's consistency, but rather a k-means clustering model specific adaptation of the general idea of shrinking the cluster. It relies simply in moving cluster elements towards its centre. Then we provide an example of a clustering function that fits the axioms of near-richness, scale-invariance and possesses the property of centric consistency, so that it is clear that they are not contradictory. And then we prove mathematically that k-means fits the centric-consistency property in that both local and global optima of its target function are retained. Additionally we prove that k-means is not Kleinberg-consistent even when we restrict ourselves to m-dimensional metric space

Let us stress, however, that the proposed reformulation is not sufficient to be raised to the status of a set of sound and complete axioms. Therefore, following a number of authors, e.g. [4] we will rather talk about properties that may or may not be fulfilled by a clustering algorithm, just characterising some features of k-means that should possibly be covered, and at least not neglected by future development of axiomatisation in the domain of cluster analysis.

We demonstrate in Sect. D.8 that the richness axiom denies common sense by itself. We clam that there is a clash between the intended meaning and formalism of the Kleinberg's axioms and that there exists a way to reconcile them. We propose a reformulation of the axioms in accordance with the intuitions and demonstrate that under this reformulation the axioms stop to be contradictory (Sect. D.9). and even a real-world algorithm like k-means conforms to this augmented axiomatic system (Sect. D.10).

We start with some remarks on the relationship between Kleinberg's axioms and k-means (Sect. D.5).

D.5   Kleinberg's Axioms and k-means—Conformance and Violations

Let us briefly discuss here the relationship of k-means algorithm to the already mentioned axiomatic systems, keeping in mind that we apply it in $\mathbb{R}^m$ Euclidean space.

Scale-invariance is fulfilled because k means qualifies objects into clusters based on relative distances to cluster centres and not their absolute values as may be easily seen from Eq. (D.1).[15]

---

[15] However, this quality function fails on the axiom of Function Scale Invariance, proposed in [62].

On the other hand richness, a property denial of which has nothing to do with distances, hence with embedding in an Euclidean space, as already known from mentioned publications, e.g. [526], is obviously violated because k-means returns only partitions into k clusters.

But what about its relaxation that is k-richness. Let us briefly show here that

Theorem D.5.1 *k-means algorithm is k-rich*

Proof: We proceed by constructing a data set for each required partition. Let us consider n data points arranged on a straight line and we want to split them into k clusters with a concrete partition cluster. For this purpose arrange the clusters on the line in non-increasing order of their cardinality. Each cluster shall occupy (uniformly) a unit length. The space between the clusters (distance between ith and (i+1)st cluster) should be set as follows: For $i = 1, \ldots, k - 1$ let $dce(j, i)$ denote the distance between the most extreme data points of clusters j and i, $cardc(j, i)$ shall denote the combined cardinality of clusters $j, j + 1, \ldots, i$. The distance between clusters i and $i + 1$ shall be then set to $2 * dce(1, i)\frac{cardc(1,i)+cardc(i+1,i+1)}{cardc(i+1,i+1)}$. In this case application of k-means algorithm will lead the desired split into clusters.

Let us stress here that there exist attempts to upgrade k-means algorithm to choose the proper k. The portion of variance explained by the clustering is used as quality criterion.[16] It is well known that increase of k increases the value of this criterion. The optimal k is deemed to be one when this increase stops to be "significant". The above construction could be extended to cover a range of k values to choose from. However, the full richness is not achievable because a split into two clusters will be better than keeping a single cluster, and the maximum is attained for this criterion if $k = n$. So either the clustering will be trivial or quite a large number of partitions will be excluded. However, even k-richness offers a large number of partitions to choose from.

Kleinberg himself proved via a bit artificial example (with unbalanced samples and an awkward distance function) that k-means algorithm with $k = 2$ is not consistent. Kleinberg's counter-example would require an embedding in a very high dimensional space, non-typical for k-means applications. Also k-means tends to produce rather balanced clusters, so Kleinberg's example could be deemed to be eccentric.

Let us illustrate by a more realistic example (balanced, in Euclidean space) that this is a real problem. Let A, B, C, D, E, F be points in three-dimensional space with coordinates: $A(1, 0, 0), B(33, 32, 0), C(33, -32, 0), D(-1, 0, 0), E(-33, 0, -32),$ $F(-33, 0, 32)$. Let $X_{AB}, X_{AC}, X_{DE}, X_{DF}$ be sets of say 1000 points randomly uniformly distributed over line segments (except for endpoints) AB, AC, DE, EF resp. Let $X = X_{AB} \cup X_{AC} \cup X_{DE} \cup X_{EF}$. k-means with $k = 2$ applied to X yields a partition $\{X_{AB} \cup X_{AC}, X_{DE} \cup X_{DF}\}$. But let us perform a $\mathcal{C}$ transformation consisting in rotating line segments AB, BC around the point A in the plane spread by the first two coordinates towards the first coordinate axis so that the angle between this

---

[16]Such a quality function would satisfy axiom of Function Scale Invariance, proposed in [62].

axis and AB' and AC' is say one degree. Now the k-means with $k = 2$ yields a different partition, splitting line segments AB' and AC'.[17] With this example not only consistency violation is shown, but also refinement-consistency violation.

D.6   Problems with Consistency in Euclidean Space

How does it happen that seemingly intuitive axioms lead to such a contradiction. We need to look more carefully at the consistency axiom in conjunction with scale-invariance. C-transform does not do what Kleinberg claimed it should that is describing a situation when moving elements from distinct clusters apart and elements within a cluster closer to one another.[18]

For example, let X consist of four elements e_1, e_2, e_3, e_4 and let a clustering function partition it into $\{e_1\}, \{e_2, e_3\}, \{e_4\}$ under some distance function d_1. One can easily construct a distance function d_2 being a C-transform of d_1 such that $d_2(e_2, e_3) = d_1(e_2, e_3)$ and $d_2(e_1, e_2) + d_2(e_2, e_3) = d_2(e_1, e_3)$ and $d_2(e_2, e_3) + d_2(e_3, e_4) = d_2(e_2, e_4)$ which implies that these points under d_2 can be embedded in the space $\mathbb{R}$ that is the straight line. Without restricting the generality (the qualitative illustration) assume that the coordinates of these points in this space are located at points $0, 0.4, 0.6, 1$ resp. Now assume we want to perform C-transformation of Kleinberg (obtaining the distance function d_3) i such a manner that the data points remain in $\mathbb{R}$ and move elements of the second set i.e. $\{e_2, e_3\}$ $(d_2(e_2, e_3) = 0.2)$ closer to one another so that $e_2 = (0.5), e_3 = (0.6)$ $(d_3(e_2, e_3) = 0.1)$. e_1 may then stay where it is but e_4 has to be shifted at least to (1.1) (under d_3 the clustering function shall yield same clustering). Now apply rescaling into the original interval that is multiply the coordinates (and hence the distances, yielding d_4) by $1/1.1$. e_1 stays at (0), $e_2 = (\frac{5}{11}), e_3 = \frac{6}{11}, e_4 = (1)$. e_3 is now closer to e_1 than before. We could have made the things still more drastic by transforming d_2 to d_3' in such a way that instead of e_4 going to (1.1), as under d_3, we set it at (2). In this case the rescaling would result in $e_1 = (0), e_2 = (0.25), e_3 = (0.3), e_4 = (1)$ (with the respective distances d_4') which means a drastic relocation of the second cluster towards the first— the distance between clusters decreases instead of increasing as claimed by Kleinberg. This is a big surprise. The C transform should have moved elements of a cluster closer together and further apart those from distinct clusters and rescaling should not disturb the proportions. It turned out to be the other way. So something is wrong

---

[17]In a test run with 100 restarts, in the first case we got clusters of equal sizes, with cluster centres at $(17,0,0)$ and $(-17,0,0)$, (between_SS / total_SS = 40%) whereas after rotation we got clusters of sizes 1800, 2200 with centres at $(26,0,0)$, $(-15,0,0)$ (between_SS / total_SS = 59%).

[18]Recall that the intuition behind clustering is to partition the data points in such a way that members of the same cluster are "close" to one another, that is their distance is low, and members of two different clusters are "distant" from one another, that is their distance is high. So it is intuitively obvious that moving elements from distinct clusters apart and elements within a cluster closer to one another should make a partition "look better".

either with the idea of scaling or of C-transformation. We shall be reluctant to claim the scaling, except for the practical case when scaling down leads to indescernability between points ue to measurement errors.

D.7   Counter-Intuitiveness of Consistency Axiom Alone

So we will consider counter-intuitiveness of consistency axiom. To illustrate it, recall first the fact that a large portion of known clustering algorithms uses data points embedded in an m dimensional feature space, usually $\mathbb{R}^m$ and the distance is the Euclidean distance therein. Now imagine that we want to perform a C-transform on a single cluster of a partition that is the C-transform shall provide distances compatible with the situation that only elements of a single cluster change position in the embedding space. Assume the cluster is an "internal" one that is for a point e in this cluster any hyperplane containing it has points from some other clusters on each side. Furthermore assume that other clusters contain together more than m data points, which should not be an untypical case. Here the problem starts. The position of e is determined by the distances from the elements of the other clusters in such a way that the increase of distance from one of them would necessarily decrease the distance to some other (except for strange configurations). Hence we can claim

Theorem D.7.1 *Under the above-mentioned circumstances it is impossible to perform C-transform reducing distances within a single cluster.*

So the C-transform enforces either adding a new dimension and moving the affected single cluster along it (which does not seem to be quite natural) or to change positions of elements in at least two clusters within the embedding space. Therefore vast majority of such algorithms does not meet not only the consistency but also inner consistency requirement.

Why not moving a second cluster is so problematic? Let us illustrate the difficulties with the original Kleinberg's consistency by looking at an application of the known k-means algorithm, with k being allowed to cover a range, not just a single value, to the two-dimensional data set visible in Fig. D.1.[19] This example is a mixture of data points sampled from 5 normal distributions. The k-means algorithm with $k = 5$, as expected, separates quite well the points from various distributions. As visible from the second column of Table D.1, in fact $k = 5$ does the best job in reducing the unexplained variance. Figure D.2 illustrates a result of a C-transform on the results of the former clustering. Visually we would tell that now we have two clusters. A look into the third column of the Table D.1 convinces that really $k = 2$ is the best choice for clustering these data with k-means algorithm. This of course contradicts Kleinberg's consistency axiom. And demonstrates the weakness of outer-consistency concept as well.

---

[19] Already Ben-David [62] indicated problems in this direction.

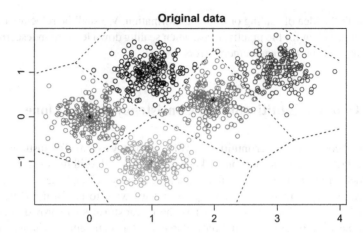

Fig. D.1 A mixture of 5 normal distributions as clustered by k-means algorithm (Voronoi diagram superimposed)

Table D.1 Variance explained (in percent) when applying k-means algorithm with $k = 2, \ldots, 6$ to data from Figs. D.1 (Original), D.2 (Kleinberg) and D.3 (Centric)

k	Original	Kleinberg	Centralised
2	54.3	98.0	54.9
3	72.2	99.17	74.3
4	83.5	99.4	86.0
5	90.2	99.7	92.9
6	91.0	99.7	93.6

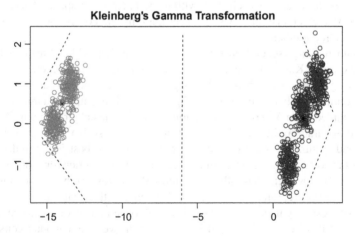

Fig. D.2 Data from Fig. D.1 after Kleinberg's C-transformation clustered by k-means algorithm into two groups

D.8 Problems of Richness Axiom

As already mentioned, richness or near-richness forces the introduction of "refinement-consistency" which is a too weak concept. But even if we allow for such a resolution of the contradiction in Kleinberg's framework, it still does not make it suitable for practical purposes. The most serious drawback of Kleinberg's axioms is the richness requirement.

But we may ask whether or not it is possible to have richness, that is for any partition there exists always a distance function that the clustering function will return this partition, and yet if we restrict ourselves to $\mathbb{R}^m$, the very same clustering function is not rich any more, or even it is not anti-chain.

Consider the following clustering function $f()$. If it takes a distance function $d()$ that takes on only two distinct values d_1 and d_2 such that $d_1 < 0.5d_2$ and for any three data points a, b, c if $d(a, b) = d_1, d(b, c) = d_1$ then $d(a, c) = d_1$, it creates clusters of points in such a way that a, b belong to the same cluster if and only if $d(a, b) = d_1$, and otherwise they belong to distinct clusters. If on the other hand $f()$ takes a distance function not exhibiting this property, it works like k-means. Obviously, function $f()$ is rich, but at the same time, if confined to $\mathbb{R}^m$, if $n > m + 1$ and $k \ll n$, then it is not rich—it is in fact k-rich, and hence not anti-chain.

Can we get around the problems all three Kleinberg's axioms in a similar way in $\mathbb{R}^m$? Regrettably,

Theorem D.8.1 *If C is a partition of $n > 2$ elements returned by a clustering function f under some distance function d, and f satisfies Consistency, then there exists a distance function d_E embedded in $\mathbb{R}^m$ for the same set of elements such that C is the partition of this set under d_E.*

The consequence of this theorem is of course that the constructs of contradiction of Kleinberg axioms are simply transposed from the domain of any distance functions to distance functions in $\mathbb{R}^m$.

Proof: To show the validity of the theorem, we will construct the appropriate distance function d_E by embedding in the $\mathbb{R}^m$. Let $dmax$ be the maximum distance between the considered elements under d. Let $C_1, \ldots C_k$ be all the clusters contained in C. For each cluster C_i we construct a ball B_i with radius r_i equal to $r_i = \frac{1}{2} \min_{x,y \in C_i, x \neq y} d(x, y)$. The ball B_1 will be located in the origin of the coordinate system. $B_{1...i}$ be the ball of containing all the balls $B_1 m \ldots, B_i$. Its centre be at $c_{1...i}$ and radius $r_{1..i}$. The ball B_i will be located on the surface of the ball with centre at $c_{1...i-1}$ and radius $r_{1...i-1} + dmax + r_i$. For each $i = 1, \ldots, k$ select distinct locations for elements of C_i within the ball B_i. The distance function d_E define as the Euclidean distances within $\mathbb{R}^m$ in these constructed locations.

Apparently, d_E is a C-transform of d, as distances between elements of C_i are smaller than or equal to $2r_i = \min_{x,y \in C_i, x \neq y} d(x, y)$, and the distances between elements of different balls exceed $dmax$. □

But richness is not only a problem in conjunction with scale-invariance and consistency, but rather it is a problem by itself.

It has to be stated first that richness is easy to achieve. Imagine the following "clustering function". You order nodes by average distance to other nodes, on tights on squared distance and so on, and if no sorting can be achieved, the unsortable points are set into one cluster. Then we create an enumeration of all clusters and map it onto unit line segment. Then we take the quotient of the lowest distance to the largest distance and state that this quotient mapped to that line segment identifies the optimal clustering of the points. Though the algorithm is simple in principle (and useless also), and meets axioms of richness and scale-invariance, we have a practical problem: As no other limitations are imposed, one has to check up to $\sum_{k=2}^{n} \frac{1}{k!} \sum_{j=1}^{k} (-1)^{k-j} \binom{k}{j} j^n$ possible partitions (Bell number) in order to verify which one of them is the best for a given distance function because there must exist at least one distance function suitable for each of them. This is prohibitive and cannot be done in reasonable time even if each check is polynomial (even linear) in the dimensions of the task (n).

Furthermore, most algorithms of cluster analysis are constructed in an incremental way. But this can be useless if the clustering quality function is designed in a very unfriendly way. For example as an XOR function over logical functions of class member distances and non-class member distances (e.g. being true if the distance rounded to an integer is odd between class members and divisible by a prime number for distances between class members and non-class members, or the same with respect to class centre or medoid).

Just have a look at sample data from Table D.2. A cluster quality function was invented along the above line and exact quality value was computed for partitioning first n points from this data set as illustrated in Table D.3. It turns out that the best partition for n points does not give any hint for the best partition for $n + 1$ points

Table D.2 Data points to be clustered using a ridiculous clustering quality function

id	x coordinate	y coordinate
1	4.022346	5.142886
2	3.745942	4.646777
3	4.442992	5.164956
4	3.616975	5.188107
5	3.807503	5.010183
6	4.169602	4.874328
7	3.557578	5.248182
8	3.876208	4.507264
9	4.102748	5.073515
10	3.895329	4.878176

Table D.3 Partition of the best quality (the lower the value the better) after including n first points from Table D.2

n	Quality	Partition
2	1270	{ 1, 2 }
3	1270	{ 1, 2 } { 3 }
4	823	{ 1, 3, 4 } { 2 }
5	315	{ 1, 4 } { 2, 3, 5 }
6	13	{ 1, 5 } { 2, 4, 6 } { 3 }
7	3	{ 1, 6 } { 2, 7 } { 3, 5 } { 4 }
8	2	{ 1, 2, 4, 5, 6, 8 } { 3 } { 7 }
9	1	{ 1, 2, 4, 5 } { 3, 8 } { 6, 9 } { 7 }
10	1	{ 1, 2, 3, 5, 9 } { 4, 6 } { 7, 10 } { 8 }

therefore each possible partition needs to be investigated in order to find the best one.[20]

Summarizing these examples, the learnability theory points at two basic weaknesses of the richness or even near-richness axioms. On the one hand the hypothesis space is too big for learning a clustering from a sample (it grows too quickly with the sample size). On the other hand an exhaustive search in this space is prohibitive so that some theoretical clustering functions do not make practical sense.

There is one more problem. If the clustering function can fit any data, we are practically unable to learn any structure of data space from data [294]. And this learning capability is necessary at least in the cases: either when the data may be only representatives of a larger population or the distances are measured with some measurement error (either systematic or random) or both. Note that we speak here about a much broader aspect than so-called cluster stability or cluster validity, pointed at by Luxburg [483, 485].

D.9   Correcting Formalisation of Kleinberg Axioms

It is obvious that richness axiom of Kleinberg needs to be replaced with a requirement of the space of hypotheses to be "large enough". For k-means algorithm it has been shown that k-richness is satisfied (and the space is still large, a Bell number of partitions to choose from). k means satisfies the scale-invariance axiom, so that only the consistency axiom needs to be adjusted to be more realistic.

Therefore a meaningful redefinition of Kleinberg's C-transform is urgently needed. It must not be annihilated by scaling and it must be executable.

---

[20]Strict separation [78] mentioned earlier is another kind of a weird cluster quality function, requiring visits to all the partitions.

Let us create for $\mathbb{R}$ a working definition of the C^* transform as follows: Distances in only one cluster X are changed by moving a point along the axis connecting it to cluster X centre reducing them within the cluster X by the same factor, the distances between any elements outside the cluster X are kept [as well as to the gravity centre of the cluster X][21]

Now consider the following one-dimensional clustering function: For a set of $n \geq 2$ points two elements belong to the same cluster if their distance is strictly lower than $\frac{1}{n+1}$ of the largest distance between the elements. When a, b belong to the same cluster and b, c belong to the same cluster, then a, c belong to the same cluster. As a consequence, the minimum distance between elements of distinct clusters is $\frac{1}{n+1}$ of the largest distance between the elements of X. It is easily seen that the weakened richness is fulfilled. The scale-invariance is granted by the relativity of inter-cluster distance. And the consistency under redefined C transform holds also. In this way all three axioms hold.

A generalization to an Euclidean space of higher dimensionality seems to be quite obvious if there are no ties on distances (the exist one pair of points the distance between which is unique and largest among distances[22]). We embed the points in the space, and then say that two points belong to the same cluster if the distance along each of the dimensions is lower than $\frac{1}{n+1}$ of the largest distance between the elements along the respective dimension. The distance is then understood as the maximum of distances along all dimensions.

Hence

Theorem D.9.1 *For each $n \geq 2$, there exists clustering function f that satisfies Scale-Invariance, near-Richness, and C^* based Consistency (which we term "centric-consistency"* [23]*).*

This way of resolving Kleinberg's contradictions differs from earlier approaches in that a realistic embedding into an $\mathbb{R}^m$ is considered and the distances are metric.

We created herewith the possibility of shrinking a single cluster without having to "move" the other ones. As pointed out, this was impossible under Kleinberg's C transform, that is under increase of all distances between objects from distinct clusters. In fact intuitively we do not want the objects to be more distant but rather the clusters. We proposed to keep the cluster centroid unchanged while decreasing distances between cluster elements proportionally, insisting that no distance of other

---

[21]Obviously, for any element outside the cluster X the distance to the closest element of X before the transform will not be smaller than its distance to the closest element of X after the transform. Note the shift of attention. We do not insist any longer that the distance to each element of other cluster is increased, rather only the distance to the cluster as a "whole" shall increase. This is by the way a stronger version of Kleinberg's local consistency which would be insufficient for our purposes.

[22]otherwise some tie breaking measures have to be taken that would break the any symmetry and allow to choose a unique direction.

[23]Any algorithm being consistent is also refinement-consistent. Any algorithm being inner-consistent is also consistent. Any algorithm being outer-consistent is also consistent. But there are no such subsumsions for the centric-consistency.

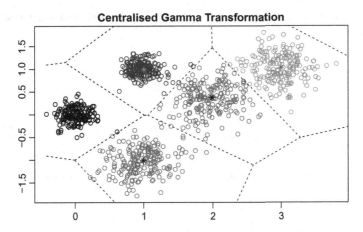

Fig. D.3 Data from Fig. D.1 after a corrected C-transformation, clustered by k-means algorithm into 5 groups

elements to the closest element of the shrunk cluster should decrease. This approach is pretty rigid. It assumes that we are capable to embed the objects into some Euclidean space so that the centroid has a meaning.

D.10 k-means Fitting Centric-Consistency Axiom

Our proposal of centric-consistency has a practical background. Kleinberg proved that k-means does not fit his consistency axiom. As shown experimentally in Table D.1, k-means algorithm behaves properly under C^* transformation. Figure D.3 illustrates a two-fold application of the C^* transform (same clusters affected as by C-transform in the preceding figure). As recognizable visually and by inspecting the forth column of Table D.1, here $k = 5$ is the best choice for k-means algorithm, so the centric-consistency axiom is followed.

Let us now demonstrate theoretically, that k-means algorithm really fits "in the limit" the centric-consistency axiom.

The k-means algorithm minimizes the sum[24] Q from Eq. (D.1). $V(C_j)$ be the sum of squares of distances of all objects of the cluster C_j from its gravity centre. Hence $Q(\mathcal{C}) = \sum_{j=1}^{k} \frac{1}{n_j} V(C_j)$. Consider moving a data point $\mathbf{x}^*$ from the cluster C_{j_0} to cluster C_{j_i} As demonstrated by [164], $V(C_{j_0} - \{\mathbf{x}^*\}) = V(C_{j_0}) - \frac{n_{j_0}}{n_{j_0}-1} \|\mathbf{x}^* - \boldsymbol{\mu}_{j_0}\|^2$ and $V(C_{j_i} \cup \{\mathbf{x}^*\}) = V(C_{j_i}) + \frac{n_{j_i}}{n_{j_i}+1} \|\mathbf{x}^* - \boldsymbol{\mu}_{j_i}\|^2$ So it pays off to move a point from

---

[24]We use here the symbol Q for the cluster quality function instead of J because Q does not fit axiomatic system for J—it is not scale-invariant and in case of consistency it changes in opposite direction, and with respect of richness we can only apply k-richness.

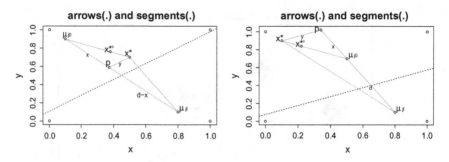

Fig. D.4 Impact of contraction towards cluster centre by a factor lambda—local optimum maintained

one cluster to another if $\frac{n_{j_0}}{n_{j_0}-1}\|\mathbf{x}^* - \boldsymbol{\mu}_{j_0}\|^2 > \frac{n_{j_l}}{n_{j_l}+1}\|\mathbf{x}^* - \boldsymbol{\mu}_{j_l}\|^2$. If we assume local optimality of $\mathcal{C}$, this obviously did not pay off. Now transform this data set to $\mathbf{X}'$ in that we transform elements of cluster C_{j_0} in such a way that it has now elements $\mathbf{x}_i' = \mathbf{x}_i + \lambda(\mathbf{x}_i - \boldsymbol{\mu}_{j_0})$ for some $0 < \lambda < 1$, see Fig. D.4. Now consider a partition $\mathcal{C}'$ of $\mathbf{X}'$ all clusters of which are the same as in $\mathcal{C}$ except for the transformed elements that form now a cluster C_{j_0}'. The question now is: does it pay off now to move a data point $\mathbf{x}'^*$ between the clusters? Consider the plane containing $\mathbf{x}^*, \boldsymbol{\mu}_{j_0}, \boldsymbol{\mu}_{j_l}$. Project orthogonally the point $\mathbf{x}^*$ onto the line $\boldsymbol{\mu}_{j_0}, \boldsymbol{\mu}_{j_l}$, giving a point $\mathbf{p}$. Either $\mathbf{p}$ lies between $\boldsymbol{\mu}_{j_0}, \boldsymbol{\mu}_{j_l}$ or $\boldsymbol{\mu}_{j_0}$ lies between $\mathbf{p}, \boldsymbol{\mu}_{j_l}$. Properties of k-means exclude other possibilities. Denote distances $y = \|\mathbf{x}^*, \mathbf{p}\|$, $x = \|\boldsymbol{\mu}_{j_0}, \mathbf{p}\|$, $d = \|\boldsymbol{\mu}_{j_0}, \boldsymbol{\mu}_{j_l}\|$ In the second case the condition that moving the point does not pay off means: $\frac{n_{j_0}}{n_{j_0}-1}(x^2 + y^2) \le \frac{n_{j_l}}{n_{j_l}+1}((d + x)^2 + y^2)$. If we multiply both sides with λ^2, we have: $\lambda^2\frac{n_{j_0}}{n_{j_0}-1}(x^2 + y^2) = \frac{n_{j_0}}{n_{j_0}-1}((\lambda x)^2 + (\lambda y)^2) \le \lambda^2\frac{n_{j_l}}{n_{j_l}+1}((d + x)^2 + y^2) = \frac{n_{j_l}}{n_{j_l}+1}(\lambda^2 d^2 + \lambda^2 2dx + \lambda^2 x^2 + \lambda^2 y^2) \le \frac{n_{j_l}}{n_{j_l}+1}(d^2 + 2d\lambda x + \lambda^2 x^2 + \lambda^2 y^2) = \frac{n_{j_l}}{n_{j_l}+1}((d + \lambda x)^2 + (\lambda y)^2)$ which means that it does not pay off to move the point $\mathbf{x}'^*$ between clusters either. Consider now the first case and assume that it pays off to move $\mathbf{x}'^*$. So we would have $\frac{n_{j_0}}{n_{j_0}-1}(x^2 + y^2) \le \frac{n_{j_l}}{n_{j_l}+1}((d - x)^2 + y^2)$ and at the same time $\frac{n_{j_0}}{n_{j_0}-1}\lambda^2(x^2 + y^2) > \frac{n_{j_l}}{n_{j_l}+1}((d - \lambda x)^2 + \lambda^2 y^2)$ Subtract now both sides: $\frac{n_{j_0}}{n_{j_0}-1}(x^2 + y^2) - \frac{n_{j_0}}{n_{j_0}-1}\lambda^2(x^2 + y^2) < \frac{n_{j_l}}{n_{j_l}+1}((d - x)^2 + y^2) - \frac{n_{j_l}}{n_{j_l}+1}((d - \lambda x)^2 + \lambda^2 y^2)$. This implies $\frac{n_{j_0}}{n_{j_0}-1}(1 - \lambda^2)(x^2 + y^2) < \frac{n_{j_l}}{n_{j_l}+1}((1 - \lambda^2)(x^2 + y^2) - 2d\lambda x)$. It is a contradiction because $\frac{n_{j_0}}{n_{j_0}-1}(1 - \lambda^2)(x^2 + y^2) > \frac{n_{j_l}}{n_{j_l}+1}(1 - \lambda^2)(x^2 + y^2) > \frac{n_{j_l}}{n_{j_l}+1}((1 - \lambda^2)(x^2 + y^2) - 2d\lambda x)$. So it does not pay off to move $\mathbf{x}'^*$, hence the partition $\mathcal{C}'$ remains locally optimal for the transformed data set. If the data have one stable optimum only like in case of "well separated" normally distributed k real clusters, then both turn to global optima.

However, it is possible to demonstrate that the newly defined transform preserves also the global optimum of k-means. Let us consider first the simple case of two clusters only (2-means). Let the optimal clustering for a given set of objects X consist of two clusters: T and Z. The subset T shall have its gravity centre at the origin of the coordinate system. The quality of this partition $Q(\{T, Z\}) = n_T Var(T) + n_Z Var(Z)$ where n_T, n_Z denote the cardinalities of T, Z and $Var(T)$, $Var(Z)$ their variances (averaged squared distances to gravity centre). We will prove by contradiction that by applying our $\mathcal{C}$ transform we get partition that will be still optimal for the transformed data points. We shall assume the contrary that is that we can transform the set T by some $1 > \lambda > 0$ to T' in such a way that optimum of 2-means clustering is not the partition $\{T', Z\}$ but another one, say $\{A' \cup D, B' \cup C\}$ where $Z = C \cup D$, A' and B' are transforms of sets A, B for which in turn $A \cup B = T$. It may be easily verified that

$$Q(\{A \cup B, C \cup D\}) = n_A Var(A) + n_A \mathbf{v}_A^2 + n_B Var(B) + n_B \mathbf{v}_B^2$$

$$+ n_C Var(C) + n_D Var(D) + \frac{n_C n_D}{n_C + n_D}(\mathbf{v}_C - \mathbf{v}_D)^2$$

while

$$Q(\{A \cup C, B \cup D\}) = n_A Var(A) + n_D Var(D) + + \frac{n_A n_D}{n_A + n_D}(\mathbf{v}_A - \mathbf{v}_D)^2$$

$$+ n_B Var(B) + n_C Var(C) + + \frac{n_B n_C}{n_B + n_C}(\mathbf{v}_B - \mathbf{v}_C)^2$$

and

$$Q(\{A' \cup B', C \cup D\}) = n_A \lambda^2 Var(A) + n_A \lambda^2 \mathbf{v}_A^2 + n_B \lambda^2 Var(B) + n_B \lambda^2 \mathbf{v}_B^2$$

$$+ n_C Var(C) + n_D Var(D) + \frac{n_C n_D}{n_C + n_D}(\mathbf{v}_C - \mathbf{v}_D)^2$$

while

$$Q(\{A' \cup C, B' \cup D\}) = n_A \lambda^2 Var(A) + n_D Var(D) + + \frac{n_A n_D}{n_A + n_D}(\lambda \mathbf{v}_A - \mathbf{v}_D)^2$$

$$+ n_B \lambda^2 Var(B) + n_C Var(C) + + \frac{n_B n_C}{n_B + n_C}(\lambda \mathbf{v}_B - \mathbf{v}_C)^2$$

The following must hold: $Q(\{A' \cup B', C \cup D\}) > Q(\{A' \cup D, B' \cup C\})$ and $Q(\{A \cup B, C \cup D\}) < Q(\{A \cup D, B \cup C\})$. Additionally also $Q(\{A \cup B, C \cup D\}) < Q(\{A \cup B \cup C, D\})$ and $Q(\{A \cup B, C \cup D\}) < Q(\{A \cup B \cup D, C\})$. These two latter inequalities imply:

$$\frac{n_C n_D}{n_C + n_D}(\mathbf{v}_C - \mathbf{v}_D)^2 < \frac{(n_A + n_B)n_C}{(n_A + n_B) + n_C}\mathbf{v}_C^2$$

and

$$\frac{n_C n_D}{n_C + n_D}(\mathbf{v}_C - \mathbf{v}_D)^2 < \frac{(n_A + n_B)n_D}{(n_A + n_B) + n_D}\mathbf{v}_D^2$$

Consider now an extreme contraction ($\lambda = 0$) yielding sets A'', B'' out of A, B. Then we have

$$Q(\{A'' \cup B'', C \cup D\}) - Q(\{A'' \cup C, B'' \cup D\})$$

$$= \frac{n_C n_D}{n_C + n_D}(\mathbf{v}_C - \mathbf{v}_D)^2 - \frac{n_A n_D}{n_A + n_D}\mathbf{v}_D^2 - \frac{n_B n_C}{n_B + n_C}\mathbf{v}_C^2$$

$$= \frac{n_C n_D}{n_C + n_D}(\mathbf{v}_C - \mathbf{v}_D)^2$$

$$-\frac{n_A n_D}{n_A + n_D}\frac{(n_A + n_B) + n_D}{(n_A + n_B)n_D}\frac{(n_A + n_B)n_D}{(n_A + n_B) + n_D}\mathbf{v}_D^2$$

$$-\frac{n_B n_C}{n_B + n_C}\frac{(n_A + n_B) + n_C}{(n_A + n_B)n_C}\frac{(n_A + n_B)n_C}{(n_A + n_B) + n_C}\mathbf{v}_C^2$$

$$= \frac{n_C n_D}{n_C + n_D}(\mathbf{v}_C - \mathbf{v}_D)^2$$

$$-\frac{n_A}{n_A + n_D}\frac{(n_A + n_B) + n_D}{(n_A + n_B)}\frac{(n_A + n_B)n_D}{(n_A + n_B) + n_D}\mathbf{v}_D^2$$

$$-\frac{n_B}{n_B + n_C}\frac{(n_A + n_B) + n_C}{(n_A + n_B)}\frac{(n_A + n_B)n_C}{(n_A + n_B) + n_C}\mathbf{v}_C^2$$

$$= \frac{n_C n_D}{n_C + n_D}(\mathbf{v}_C - \mathbf{v}_D)^2$$

$$-\frac{n_A}{n_A + n_B}(1 + \frac{n_B}{n_A + n_D})\frac{(n_A + n_B)n_D}{(n_A + n_B) + n_D}\mathbf{v}_D^2$$

$$-\frac{n_B}{n_A + n_B}(1 + \frac{n_A}{n_B + n_C})\frac{(n_A + n_B)n_C}{(n_A + n_B) + n_C}\mathbf{v}_C^2$$

$$< \frac{n_C n_D}{n_C + n_D}(\mathbf{v}_C - \mathbf{v}_D)^2$$

$$-\frac{n_A}{n_A + n_B}\frac{(n_A + n_B)n_D}{(n_A + n_B) + n_D}\mathbf{v}_D^2$$

$$-\frac{n_B}{n_A + n_B}\frac{(n_A + n_B)n_C}{(n_A + n_B) + n_C}\mathbf{v}_C^2 < 0$$

because the linear combination of two numbers that are bigger than a third yields another number bigger than this. Let us define a function

$$h(x) = n_A x^2 \mathbf{v}_A^2 + n_B x^2 \mathbf{v}_B^2 + \frac{n_C n_D}{n_C + n_D}(\mathbf{v}_C - \mathbf{v}_D)^2$$

$$-\frac{n_A n_D}{n_A + n_D}(x\mathbf{v}_A - \mathbf{v}_D)^2 - \frac{n_B n_C}{n_B + n_C}(x\mathbf{v}_B - \mathbf{v}_C)^2$$

It can be easily verified that $h(x)$ is a quadratic polynomial with a positive coefficient at x^2. Furthermore $h(1) = Q(\{A \cup B, C \cup D\}) - Q(\{A \cup C, B \cup D\}) < 0, h(\lambda) = Q(\{A' \cup B', C \cup D\}) - Q(\{A' \cup C, B' \cup D\}) > 0,\ h(0) = Q(\{A'' \cup B'', C \cup D\}) - Q(\{A'' \cup C, B'' \cup D\}) < 0$. But no quadratic polynomial with a positive coefficient at x^2 can be negative at the ends of an interval and positive in the middle. So we have the contradiction. This proves the thesis that the (globally) optimal 2-means clustering remains (globally) optimal after transformation.

Let us turn to the general case of k-means. Let the optimal clustering for a given set of objects X consist of two clusters: T and $Z_1, \ldots, Z_{k-1}$. The subset T shall have its gravity centre at the origin of the coordinate system. The quality of this partition

$$Q(\{T, Z_1, \ldots, Z_{k-1}\}) =$$

$$= n_T Var(T) + \sum_{i=1}^{k-1} n_{Z_i} Var(Z_i)$$

We will prove by contradiction that by applying our $\mathcal{C}$ transform we get partition that will be still optimal for the .... We shall assume the contrary that is that we can transform the set T by some $1 > \lambda > 0$ to T' in such a way that optimum of 2-means clustering is not the partition $\{T', Z\}$ but another one, say $\{T'_1 \cup Z_{1,1} \cup \ldots \cup Z_{k-1,1}, T'_2 \cup Z_{1,2} \cup \ldots \cup Z_{k-1,2} \ldots, T'_k \cup Z_{1,k} \cup \ldots \cup Z_{k-1,k}\}$ where $Z_i = \cup_{j=1}^k Z_{i,j}, T'_1, \ldots, T'_k$ are transforms of sets $T_1, \ldots, T_k$ for which in turn $\cup_{j=1}^k T_j = T$. It may be easily verified that

$$Q(\{T, Z_1, \ldots, Z_{k-1}\}) = \sum_{j=1}^k n_{T_j} Var(T_j) + \sum_{j=1}^k n_{T_j}\mathbf{v}_{T_j}^2$$

$$+ \sum_{i=1}^{k-1} n_{Z_i} Var(Z_i)$$

while (denoting $Z_{*,j} = \cup_{i=1}^{k-1} Z_{*,j}$)

$$Q(\{T_1 \cup Z_{*,1}, \ldots, T_k \cup Z_{*,k}\}) =$$

$$= \sum_{j=1}^{k} \left(n_{T_j} Var(T_j) + n_{Z_{*,j}} Var(Z_{*,j}) + + \frac{n_{T_j} n_{Z_{*,j}}}{n_{T_j} + n_{Z_{*,j}}} (\mathbf{v}_{T_j} - \mathbf{v}_{Z_{*,j}})^2 \right)$$

whereas

$$Q(\{T', Z_1, \ldots, Z_{k-1}\}) = \sum_{j=1}^{k} n_{T_j} \lambda^2 Var(T_j) + \sum_{j=1}^{k} n_{T_j} \lambda^2 \mathbf{v}_{T_j}^2$$

$$+ \sum_{i=1}^{k-1} n_{Z_i} Var(Z_i)$$

while

$$Q(\{T_1' \cup Z_{*,1}, \ldots, T_k' \cup Z_{*,k}\}) =$$

$$= \sum_{j=1}^{k} \left(n_{T_j} \lambda^2 Var(T_j) + n_{Z_{*,j}} Var(Z_{*,j}) + + \frac{n_{T_j} n_{Z_{*,j}}}{n_{T_j} + n_{Z_{*,j}}} (\lambda \mathbf{v}_{T_j} - \mathbf{v}_{Z_{*,j}})^2 \right)$$

The following must hold: $Q(\{T', Z_1, \ldots, Z_{k-1}\}) > Q(\{T_1' \cup Z_{*,1}, \ldots, T_k' \cup Z_{*,k}\})$ and $Q(\{T, Z_1, \ldots, Z_{k-1}\}) < Q(\{\{T_1 \cup Z_{*,1}, \ldots, Z_{*,k})$. Additionally also $Q(\{T, Z_1, \ldots, Z_{k-1}\}) < Q(\{\{T \cup Z_{*,1}, Z_{*,2}, \ldots, Z_{*,k})$ and $Q(\{T, Z_1, \ldots, Z_{k-1}\}) < Q(\{\{T \cup Z_{*,2}, Z_{*,1}, Z_{*,3}, \ldots, Z_{*,k})$ and …and $Q(\{T, Z_1, \ldots, Z_{k-1}\}) < Q(\{\{T \cup Z_{*,k}, Z_{*,1}, \ldots, Z_{*,k-1})$. These latter k inequalities imply that for $l = 1, \ldots, k$:

$$Q(\{T, Z_1, \ldots, Z_{k-1}\}) = n_T Var(T) + \sum_{j=1}^{k} n_{T_j} Var(T_j) + \sum_{j=1}^{k} n_{T_j} \mathbf{v}_{T_j}^2$$

$$+ \sum_{i=1}^{k-1} n_{Z_i} Var(Z_i) <$$

$$< Q(\{T \cup Z_{*,l}, Z_{*,1}, \ldots, Z_{*,l-1}, Z_{*,l+1} \ldots, Z_{*,k}\}) =$$

$$= n_T Var(T) + \sum_{j=1}^{k} n_{Z_{*,j}} Var(Z_{*,j}) + \frac{n_T n_{Z_{*,l}}}{n_T + n_{Z_{*,l}}} (\mathbf{v}_T - \mathbf{v}_{Z_{*,l}})^2$$

$$+ \sum_{i=1}^{k-1} n_{Z_i} Var(Z_i) <$$

$$< \sum_{j=1}^{k} n_{Z_{*,j}} Var(Z_{*,j}) + \frac{n_T n_{Z_{*,l}}}{n_T + n_{Z_{*,l}}} (\mathbf{v}_T - \mathbf{v}_{Z_{*,l}})^2$$

$$+ \sum_{i=1}^{k-1} n_{Z_i} Var(Z_i) - \sum_{j=1}^{k} n_{Z_{*,j}} Var(Z_{*,j}) <$$

$$< \frac{n_T n_{Z_{*,l}}}{n_T + n_{Z_{*,l}}} (\mathbf{v}_{Z_{*,l}})^2$$

Consider now an extreme contraction ($\lambda = 0$) yielding sets $T_j{}''$ out of T_j. Then we have

$$Q(\{T'', Z_1, \ldots, Z_{k-1}\}) - Q(\{T''_1 \cup Z_{*,1}, \ldots, T''_k \cup Z_{*,k}\})$$

$$= \sum_{i=1}^{k-1} n_{Z_i} Var(Z_i) - \sum_{j=1}^{k} \left(n_{Z_{*,j}} Var(Z_{*,j}) + \frac{n_{T_j} n_{Z_{*,j}}}{n_{T_j} + n_{Z_{*,j}}} (\mathbf{v}_{Z_{*,j}})^2 \right)$$

$$= \sum_{i=1}^{k-1} n_{Z_i} Var(Z_i) - \sum_{j=1}^{k} n_{Z_{*,j}} Var(Z_{*,j})$$

$$- \sum_{j=1}^{k} \frac{n_{T_j} n_{Z_{*,j}}}{n_{T_j} + n_{Z_{*,j}}} \frac{n_T + n_{Z_{*,j}}}{n_T n_{Z_{*,j}}} \frac{n_T n_{Z_{*,j}}}{n_T + n_{Z_{*,j}}} (\mathbf{v}_{Z_{*,j}})^2$$

$$= \sum_{i=1}^{k-1} n_{Z_i} Var(Z_i) - \sum_{j=1}^{k} n_{Z_{*,j}} Var(Z_{*,j})$$

$$- \sum_{j=1}^{k} \frac{n_{T_j}}{n_{T_j} + n_{Z_{*,j}}} \frac{n_T + n_{Z_{*,j}}}{n_T} \frac{n_T n_{Z_{*,j}}}{n_T + n_{Z_{*,j}}} (\mathbf{v}_{Z_{*,j}})^2$$

$$\leq \sum_{i=1}^{k-1} n_{Z_i} Var(Z_i) - \sum_{j=1}^{k} n_{Z_{*,j}} Var(Z_{*,j}) - \sum_{j=1}^{k} \frac{n_{T_j}}{n_T} \frac{n_T n_{Z_{*,j}}}{n_T + n_{Z_{*,j}}} (\mathbf{v}_{Z_{*,j}})^2 < 0$$

because the linear combination of numbers that are bigger than a third yields another number bigger than this. Let us define a function

$$g(x) = \sum_{j=1}^{k} n_{T_j} x^2 \mathbf{v}_{T_j}^2 + \sum_{i=1}^{k-1} n_{Z_i} Var(Z_i)$$

$$-\sum_{j=1}^{k}\left(n_{Z_{*,j}} Var(Z_{*,j}) + +\frac{n_{T_j} n_{Z_{*,j}}}{n_{T_j} + n_{Z_{*,j}}}(x\mathbf{v}_{T_j} - \mathbf{v}_{Z_{*,j}})^2\right)$$

It can be easily verified that $g(x)$ is a quadratic polynomial with a positive coefficient at x^2. Furthermore $g(1) = Q(\{T, Z_1, \ldots, z_{k-1}\}) - Q(\{T_1 \cup Z_{*,1}, \ldots, T_k \cup Z_{*,k}\}) < 0$, $g(\lambda) = Q(\{T', Z_1, \ldots, z_{k-1}\}) - Q(\{T_1' \cup Z_{*,1}, \ldots, T_k' \cup Z_{*,k}\}) > 0$, $g(0) = Q(\{T'', Z_1, \ldots, z_{k-1}\}) - Q(\{T''_1 \cup Z_{*,1}, \ldots, T''_k \cup Z_{*,k}\}) < 0$. But no quadratic polynomial with a positive coefficient at x^2 can be negative at the ends of an interval and positive in the middle. So we have the contradiction. This proves the thesis that the (globally) optimal k-means clustering remains (globally) optimal after transformation.

So summarizing the new $\mathcal{C}$ transformation preserves local and global optima of k-means for a fixed k. Therefore k-means algorithm is consistent under this transformation.

Hence

Theorem D.10.1 *k-means algorithm satisfies Scale-Invariance, k-Richness, and centric Consistency.*

Note that ($\mathcal{C}^*$ based) centric Consistency is not a specialization of Kleinberg's consistency as the requirement of increased distance between all elements of different clusters is not required in $\mathcal{C}^*$ based Consistency. Note also that the decrease of distance does not need to be equal for all elements as long as the gravity centre does not relocate. Also a limited rotation of the cluster may be allowed for. But we could strengthen centric-consistency to be in concordance with Kleinberg's consistency and under this strengthening k-means would of course still behave properly.

D.11 Moving Clusters

Kleinberg's consistency is split by some authors into the inner consistency (decreasing distances within the clusters) and the outer consistency (moving clusters away). We have just reformulated the inner consistency assumptions by Kleinvberg. Let us see what can be done with outer consistency.

As we have stated already, in the $\mathbb{R}^n$ it is actually impossible to move clusters in such a way as to increase distances to all the other elements of all the other clusters. However, we shall ask ourselves if we may possibly move away clusters as whole, via increasing the distance between cluster centres and not overlapping cluster regions, which, in case of k-means, represent Voronoi-regions.

Let us concentrate on the k-means case and let us look at two neighbouring clusters. The Voronoi regions are in fact polyhedrons, such that the "outer" polyhedrons (at least one of them) can be moved away from the rest without overlapping any other region.

So is such an operation on regions permissible without changing the cluster structure. A closer look at the issue tells us that it is not. As k-means terminates, the neighbouring clusters' polyhedrons touch each other via a hyperplane such that the straight line connecting centres of the clusters is orthogonal to this hyperplane. This causes that points on the one side of this hyperplane lie more closely to the one centre, and on the other to the other one. But if we move the clusters in such a way that both touch each other along the same hyperplane, then if it happens that some points within the first cluster will become closer to the centre of the other cluster and vice versa. So moving the clusters generally will change their structure (points switch clusters) unless the points lie actually not within the polyhedrons but rather within "paraboloids" with appropriate equations. Then moving along the border hyperplane will not change cluster membership (locally). But the intrinsic cluster borders are now "paraboloids". What would happen if we relocate the clusters allowing for touching along the "paraboloids"? The problem will occur again.

Hence the question can be raised: What shape should have the k-means clusters in order to be immune to movement of whole clusters?

Let us consider the problem of susceptibility to class membership change within a 2D plane containing the two cluster centres. Let the one cluster centre be located at a point (0,0) in this plane and the other at $(2x_0, 2y_0)$. Let further the border of the first cluster be characterised by a (symmetric) function $f(x)$ and le the shape of the border of the other one $g(x)$ be the same, but properly rotated: $g(x) = 2y_0 - f(x - 2x_0)$ so that the cluster centre is in the same. Let both have a touching point (we excluded already a straight line and want to have convex smooth borders). From the symmetry conditions one easily sees that the touching point must be (x_0, y_0). As this point lies on the surface of $f()$, $y_0 = f(x_0)$ must hold. For any point $(x, f(x))$ of the border of the first cluster with centre $(0, 0)$ the following must hold:

$$(x - 2x_0)^2 + (f(x) - 2f(x_0))^2 - x^2 - f^2(x) \geq 0$$

That is

$$-2x_0(2x - 2x_0) - 2f(x_0)(2f(x) - 2f(x_0)) \geq 0$$

$$-f(x_0)(f(x) - f(x_0)) \geq x_0(x - x_0)$$

Let us consider only positions of the centre of the second cluster below the X axis. In this case $f(x_0) < 0$. Further let us concentrate on x lower than x_0. We get

$$-\frac{f(x) - f(x_0)}{x - x_0} \geq \frac{x_0}{-f(x_0)}$$

In the limit, when x approaches x_0.

$$-f'(x_0) \geq \frac{x_0}{-f(x_0)}$$

Now turn to x greater than x_0. We get

$$-\frac{f(x) - f(x_0)}{x - x_0} \leq \frac{x_0}{-f(x_0)}$$

In the limit, when x approaches x_0.

$$-f'(x_0) \leq \frac{x_0}{-f(x_0)}$$

This implies

$$-f'(x_0) = \frac{-1}{\frac{f(x_0)}{x_0}}$$

Note that $\frac{f(x_0)}{x_0}$ is the directional tangent of the straight line connecting both cluster centres. As well as it is the directional tangent of the line connecting the centre of the first cluster to its surface. $f'(x_0)$ is the tangential of the borderline of the first cluster at the touching point of both clusters. The equation above means both are orthogonal. But this property implies that $f(x)$ must be definition (of a part) a circle centred at $(0, 0)$. As the same reasoning applies at any touching point of the clusters, a k-means cluster would have to be (hyper)ball-shaped in order to allow the movement of the clusters without elements switching cluster membership.

The tendency of k-means to recognize best ball-shaped clusters has been known long ago, but we are not aware of presenting such an argument for this tendency.

It has to be stated however that clusters, even if enclosed in a ball-shaped region, need to be separated sufficiently to be properly recognized. Let us consider, under which circumstances a cluster C_1 of radius r_1 containing n_1 elements would take over n_{21} elements (subcluster C_{21}) of a cluster C_1 of radius r_2 of cardinality n_2. Let $n_{22} = n_2 - n_{21}$ be the number of the remaining elements (subcluister C_{22} of the second cluster. Let the enclosing balls of both clusters be separated by the distance (gap) g. Let us consider the worst case that is that the centre of the C_{21} subcluster lies on a straight line segment connecting both cluster centres. The centre of the remaining C_{22} subcluster would lie on the same line but on the other side of the second cluster centre. Let r_{21}, r_{22} be distances of centres of n_{21} and n_{22} from the centre of the second cluster. The relations

$$n_{21} \cdot r_{22} = n_{22} \cdot r_{21}, \ r_{21} \leq r_2, \ r_{22} \leq r_2$$

must hold. Let denote with $SSC(C)$ the sum of squared distances of elements of the set C to the centre of this set.

So in order for the clusters to be stable

$$SSC(C_1) + SSC(C_2) \leq SSC(C_1 \cup C_{21}) + SSC(C_{22})$$

must hold. But

$$SSC(C_2) = SSC(C_{21}) + SSC(C_{22}) + n_{21} \cdot r_{21}^2 + n_{22} \cdot r_{22}^2$$

$$SSC(C_1 \cup C_{21}) = SSC(C_1) + SSC(C_{21}) + \frac{n_1 n_{21}}{n_1 + n_{21}}(r_1 + r_2 + g - r_{21})^2$$

Hence

$$SSC(C_1) + SSC(C_{21}) + SSC(C_{22}) + n_{21} \cdot r_{21}^2 + n_{22} \cdot r_{22}^2 \leq SSC(C_1)$$

$$+ SSC(C_{21}) + \frac{n_1 n_{21}}{n_1 + n_{21}}(r_1 + r_2 + g - r_{21})^2 + SSC(C_{22})$$

$$n_{21} \cdot r_{21}^2 + n_{22} \cdot r_{22}^2 \leq \frac{n_1 n_{21}}{n_1 + n_{21}}(r_1 + r_2 + g - r_{21})^2$$

$$\frac{n_{21} \cdot r_{21}^2 + n_{22} \cdot r_{22}^2}{\frac{n_1 n_{21}}{n_1 + n_{21}}} \leq (r_1 + r_2 + g - r_{21})^2$$

$$\sqrt{\frac{n_{21} \cdot r_{21}^2 + n_{22} \cdot r_{22}^2}{\frac{n_1 n_{21}}{n_1 + n_{21}}}} \leq r_1 + r_2 + g - r_{21}$$

$$\sqrt{\frac{n_{21} \cdot r_{21}^2 + n_{22} \cdot r_{22}^2}{\frac{n_1 n_{21}}{n_1 + n_{21}}}} - r_1 - r_2 + r_{21} \leq g$$

$$\sqrt{\frac{n_{21} \cdot r_{21}^2 + n_{21} \cdot r_{21} \cdot r_{22}}{\frac{1}{1/n_1 + 1/n_{21}}}} - r_1 - r_2 + r_{21} \leq g$$

$$\sqrt{(n_{21} \cdot r_{21}^2 + n_{21} \cdot r_{21} \cdot r_{22})(1/n_1 + 1/n_{21})} - r_1 - r_2 + r_{21} \leq g$$

$$\sqrt{(r_{21}^2 + r_{21} \cdot r_{22})(n_{21}/n_1 + 1)} - r_1 - r_2 + r_{21} \leq g$$

As $r_{22} = \frac{r_{21} n_{21}}{n_2 - n_{21}}$

$$\sqrt{(r_{21}^2 + r_{21} \cdot \frac{r_{21} n_{21}}{n_2 - n_{21}})(n_{21}/n_1 + 1)} - r_1 - r_2 + r_{21} \leq g$$

$$r_{21}\sqrt{(1 + \frac{n_{21}}{n_2 - n_{21}})(n_{21}/n_1 + 1)} - r_1 - r_2 + r_{21} \leq g$$

$$r_{21}\sqrt{\frac{n_2}{n_2 - n_{21}}\frac{n_1 + n_{21}}{n_1}} - r_1 - r_2 + r_{21} \leq g$$

$$r_{21}\sqrt{\frac{n_2}{n_1}\frac{n_1 + n_{21}}{n_2 - n_{21}}} - r_1 - r_2 + r_{21} \leq g$$

Let us consider the worst case when the elements to be taken over are at the "edge" of the cluster region ($r_{21} = r_2$). Then

$$r_2\sqrt{\frac{n_2}{n_1}\frac{n_1 + n_{21}}{n_2 - n_{21}}} - r_1 \leq g$$

The lower limit on g will grow with n_{21}, but $n_{21} \leq 0.5n_2$, because otherwise r_{22} would exceed r_2. Hence in the worst case

$$r_2\sqrt{\frac{n_2}{n_1}\frac{n_1 + n_2/2}{n_2/2}} - r_1 \leq g$$

$$r_2\sqrt{2(1 + 0.5n_2/n_1)} - r_1 \leq g$$

In case of clusters with equal sizes and equal radius this amounts to

$$g \geq r_1(\sqrt{3} - 1) \approx 0.7r_1$$

D.12 Learnability Oriented Axiomatisation

Let us now look at ways to formulate a learnability based axiomatic framework for clustering algorithms. The clustering algorithm is inevitably connected to the clusterings it can produce. So, first let us state the axioms necessary for clustering.[25]

Axiom D.12.1 (*Strong learnability*) Clustering must split the sample space (for each point in sample space we must be able to say to which cluster it belongs or if it belongs to the undecided space),

Axiom D.12.2 (*Weak learnability*) Clustering must be learnable from finite sample (the clustering to be discovered must belong to a class of clusterings such that the class is learnable in the sense of learnability theory),

---

[25]See: R. A. Kłopotek and M. A. Kłopotek: Fallacy of Kleinberg's Richness Axiom in Document Clustering. *Conference on Tools, Applications and Implementations of Methods For Determining Similarities Between Documents.* Warszawa, 22.4.2015 .

Axiom D.12.3 (*Separability*) Clusters in the clustering must be separable (for a sufficiently large sample it should be significantly more probable to have the closest neighbour from the same cluster than from a different one),

Axiom D.12.4 (*Enlightening*) Clustering must be enlightening (new information should be obtained via clustering compared to prior knowledge).

Therefore:

Axiom D.12.5 (*Learnability*) A clustering algorithm with high probability shall return a clustering if the intrinsic clustering of the data belongs to the class of clusterings for which the algorithm is designed and this clustering should be close to the intrinsic one,

Axiom D.12.6 (*Strong learnability*) A clustering algorithm with high probability shall return a failure information if the intrinsic clustering of the data does not belong to the class of clusterings for which the algorithm is designed or if there is no clustering behind the data,

Axiom D.12.7 (*Separability*) A clustering algorithm shall state with high reliability what is the separation between clusters,

Axiom D.12.8 (*Enlightening*) A clustering algorithm shall verify if there is a difference between the detected clustering and the prior knowledge of the sample space.

Note that these axioms are in opposition to axioms of Kleinberg. Learnability contradicts the richness axiom. Separability contradicts invariance. Enlightening opposes consistency.

D.13   Graph Clustering Axiomatisation

In the preceding sections we considered the case, when the clustered objects are placed in a space, where distances between objects may be defined. But a large portion of this book was devoted to clustering of graphs (Chap. 5).

Hence we cannot overlook axiomatic frameworks developed specially for them.

Van Laarhoven and Marchiori[26] developed an axiomatic framework, defining the desired properties of the clustering quality function, understood in a similar way as previously.

Their first axiom assumes "Permutation invariance", namely.

Axiom D.13.1 (*Permutation invariance*) The graph partition quality depends on node similarities but not on node labels.

The second axiom is the "Scale invariance" which means that

---

[26] see: T. van Laarhoven and E. Marchiori: Axioms for Graph Clustering Quality Functions, *Journal of Machine Learning Research* 15 (2014) 193–215.

Axiom D.13.2 (*Scale invariance*) Proportional increase/decrease of weights of edges (node similarities) does not change the partition quality.

Authors quoted request also the "Richness", that is

Axiom D.13.3 (*Richness*) For any partition of the sets of nodes there exists a graph, for which this partition is optimal with respect to the given quality function among all the partitions of the graph considered.

Further, they request the "Consistent improvement monotonicity", that is

Axiom D.13.4 (*Consistent improvement monotonicity*) If the edge weights within clusters are increased and between clusters they are decreased, then the quality function shall increase.

One sees immediately that these axioms parallel those from Sect. D.2—the difference is that we talk now about similarity measures and not distances and that some similarities could be equal zero. A similarity equal to zero breaks Kleinberg's contradiction proof, because scaling by the lowest and highest distances is necessary, and in this case the highest distance is infinite. However, there are two difficulties here. One is that the graph may have all edges with non-zero weights. In this case the claim of breaking Kleinberg's contradiction is not valid. Furthermore, the weight zero stems frequently from the fact of thresholding the similarity level between nodes. In this case the Scale invariance is violated, because an increase in similarity would introduce new edges, its decrease would remove the existing ones (crossing the threshold).

Van Laarhoven and Marchiori introduce, as well, an axiom for "Local consistency" (violated by the algorithms with a fixed number of clusters):

Axiom D.13.5 (*Local consistency*) For graphs agreeing on some subgraphs, if a change of clustering in the common subgraph increases the quality function in one graph, then the same change should cause an increase in the other.

Their final axiom is called "Continuity". That is

Axiom D.13.6 (*Continuity*) For each $\epsilon > 0$ there exists a $\delta > 0$ such that for two graphs differing by edge weights for each edge by only δ, the quality functions of the two graphs differ only by ϵ for each clustering C (of both graphs).

The same authors show that the popular graph partition quality measure called modularity[27] conforms only to four of their axioms (richness, continuity, scale invariance, permutation invariance) but violates monotonicity and locality.

Therefore, they propose a new quality function, called adaptive scale modularity,

$$Q_{M,C} = \sum_{c \in C} \left(\frac{w(c)}{M + Cvol(c)} - \left(\frac{vol(c)}{M + Cvol(c)} \right)^2 \right)$$

---

[27] See: Mark E. J. Newman and Michelle Girvan. Finding and evaluating community structure in networks. Phys. Rev. E, 69:026113, Feb 2004.

where $vol(c)$ is the volume of the cluster c, and $w(c)$ is the sum of weights of edges within the cluster c. This function satisfies all their axioms for $M = 0, C \geq 2$, and violates only scale invariance for $M > 0$.

For comparison, the original modularity was of the form

$$Q_{M,C} = \sum_{c \in C} \left(\frac{w(c)}{vol(G)} - \left(\frac{vol(c)}{vol(G)} \right)^2 \right)$$

where $vol(G)$ is the volume of the entire graph G.

It has been observed by Fortunato and Barthelemy[28] that modularity suffers from the so-called resolution limit. Consider a ring of cliques in which cliques are interconnected by a single link only. One would expect that an optimal clustering would contain exactly one clique in one cluster. However, optimality with respect to modularity does not behave in this way. Therefore, Fortunato and Barthelemy require that graph partition quality measure should be Resolution-limit-free. This means that: If a partition C of a graph G is optimal with respect to such a measure among all partitions of G, then for each subgraph G' of G the partition C' induced from C by G' should also be optimal among all partitions of G'.

Adaptive scale modularity does not fulfil this requirement but it does not suffer, nonetheless, from the resolution limit. Hence, the axiom of resolution-limit-freedom remains an open question for further investigations.

Let us make a further remark on modularity. It has a clear interpretation as a difference between the current link structure and a random one. This is in concordance with the enlightening axiom, mentioned before. The adaptive scale modularity lacks such an interpretation and one can suspect that it will promote big clusters (high volume compared to border, that is the sum of cluster node degrees being much larger than the number of edges linking the given cluster with other clusters). Hence, it seems that lack of contradictions in an axiomatic system is a necessary condition for the adequacy of this system, but it is not sufficient.

Furthermore, it turns out that finding a partition optimising the value of modularity is NP hard. Hence, in fact approximate, but fast algorithms are used instead, without even any guarantees of giving a solution within some bounds with respect to the optimal one.

Therefore, it seems reasonable to press on learnability rather than on richness in axiomatic systems.

This concluding remark would be conform with the opinion of Estivill-Castro, expressed in his position paper [29] that the lack of common definition of a cluster is the primary reason for the multitude of the clustering algorithms. The required properties of a clustering depend on the application. That is we have to define our notion of clustering and then prove that it is learnable from data using one's algorithm.

---

[28] See: Santo Fortunato and Marc Barthelemy. Resolution limit in community detection. Proc. Natl. Acad. Sci. USA, 104(1):36–41, 2007.

[29] Vladimir Estivill-Castro: Why So Many Clustering Algorithms: A Position Paper. SIGKDD Explor. Newsl., June 2002, Vol. 4, No. 1, 65–75.

Appendix E
Justification for the k-means++ Algorithm

The ultimate goal of the k-means algorithm is to minimise the partition cost function stated in formula (3.1)[30]

$$J(U, M) = \sum_{i=1}^{m} \sum_{j=1}^{k} u_{ij} \|\mathbf{x}_i - \boldsymbol{\mu}_j\|^2 \tag{E.1}$$

where u_{ij} is equal 1 if among all of cluster centres $\boldsymbol{\mu}_j$ is the closest to $\mathbf{x}_i$, and is 0 otherwise.

As already stated, Arthur and Vassilvitskii [32] prove the following property of their k-means++ algorithm.

Theorem E.0.1 *[32] The expected value of the partition cost, computed according to Eq. (3.1), is delimited in case of k-means++ algorithm by the inequality*

$$\mathbb{E}(J) \le 8(\ln k + 2) J_{opt} \tag{E.2}$$

where J_{opt} denotes the optimal value of the partition cost.

(For $k = 3$ this amounts to more than 24). Note that the proof in [32] refers to the initialisation step of the algorithm only so that subsequently we assume that k-means++ clustering consists of this initialisation only. In the full algorithm $\mathbb{E}(J)$ is lower than after initialisation step only as k-means can partially eliminate effects of poor initialisation.

The motivation for the invention of the k-means++ algorithm is, as [32] states, the poor performance of k-means for many real examples (i.e., $\frac{J}{J_{opt}}$ is unbounded even when m and k are fixed). However, k-means turns out to exhibit an appealing simplicity. Therefore it makes sense to try to improve its clustering quality. Arthur and Vassilvitskii created an algorithm with some goodness guarantees.

---

[30]This appendix is partially based on: M.A. Klopotek: On the Consistency of k-means++ algorithm. CoRR abs/1702.06120 (2017) .

© Springer International Publishing AG 2018
S. T. Wierzchoń and M. A. Kłopotek, *Modern Algorithms of Cluster Analysis*,
Studies in Big Data 34, https://doi.org/10.1007/978-3-319-69308-8

On the other hand Pollard [395] proved an interesting property of a hypothetical k-means$_{opt}$ algorithm that would produce a partition yielding J_{opt} for finite sample. He demonstrated the convergence of k-means$_{opt}$ algorithm to proper population clustering with increasing sample size. So one can state that such an algorithm would be strongly consistent. The problem of that algorithm is that it is NP-hard [395].

So let us investigate whether or not a realistic algorithm, k-means++, featured by reasonable complexity and reasonable expectation of closeness to optimality, is also strongly consistent.

Note that there existed and exists interest in strong consistency results concerning variants of k-means algorithm. However, like in [457] or [338], an abstract global optimiser is referred to rather than an actual algorithm performance.

In what follows we extend the results of Pollard.

Assume that we are drawing independent samples of size $m = k, k + 1, \ldots$ from some probability distribution P. Assume further, that the number of disjoint balls of radius ϵ_P for some $\epsilon_P > 0$ with non-zero P measure is much larger than k. Furthermore let

$$\int \|\mathbf{x}\|^{2(k+1)} P(d\mathbf{x})$$

be finite.

Given a sample consisting of m vectors $\mathbf{x}_i$, $i = 1, \ldots, m$, it will be convenient to consider an empirical probability distribution P_m assigning a probability mass of m^{-1} at each of the sample points $\mathbf{x}_i$. Let further M denote any set of up to k points (serving as an arbitrary set of cluster centres).

Let us introduce the function $J_{cp}(.,.)$, that for any set M of points in $\mathbb{R}^n$ and any probability distribution Q over $\mathbb{R}^n$, computes k-means quality function normalised over the probability distribution:

$$J_{cp}(M, Q) = \int D(\mathbf{x}, M)^2 Q(d\mathbf{x}) \qquad (E.3)$$

where $D(\mathbf{x}, M) = \min_{\boldsymbol{\mu} \in M} \|\mathbf{x} - \boldsymbol{\mu}\|$.

Then $J_{cp}(M, P_m)$ can be seen as a version of the function $J()$ from Eq. (3.1).

Note that for any sets M, M' such that $M \subset M'$ we have $J_{cp}(M, Q) \geq J_{cp}(M', Q)$.

Note also that if M has been obtained via the initialisation step of k-means++, then the elements of M can be considered as ordered, so we could use indexes $M_{[p:q]}$ meaning elements $\boldsymbol{\mu}_p, \ldots, \boldsymbol{\mu}_q$ from this sequence. As the cluster around $M_{[k]}$ will contain at least one element ($\boldsymbol{\mu}_k$ itself), we can be sure that $J_{cp}(M_{[1:k-1]}, P_m) > J_{cp}(M_{[1:k]}, P_m)$ as no element has a chance to be selected twice. Hence also $\mathbb{E}_M(J_{cp}(M_{[1:k-1]}, P_m)) > \mathbb{E}_M(J_{cp}(M_{[1:k]}, P_m))$.

If we assume that the distribution P is continuous, then for the expectation over all possible M obtained from initialisation procedure of k-means++, for any $j = 2, \ldots, k$ the relationship $\mathbb{E}_M(J_{cp}(M_{[1:j-1]}, P)) > \mathbb{E}_M(J_{cp}(M_{[1:j]}, P))$ holds. The reason is as follows: The last point, $\boldsymbol{\mu}_k$, was selected from a point of non-zero density, so that in ball of radius $\epsilon_b > 0$ such that ϵ_b is lower than $1/3$ of the distance of μ_j to each $\mu \in M_{[1:j-1]}$) around it, each point would be re-clustered to μ_j diminishing the overall within-cluster variance. As it holds for each set M, so it holds for the expected value too.

But our goal here is to show that if we pick the sets M according to k-means++ (initialization) algorithm, then $\mathbb{E}_M(J_{cp}(M, P_m)) \to_{m\to\infty} \mathbb{E}_M(J_{cp}(M, P))$, and, as a consequence, $\mathbb{E}_M(J_{cp}(M, P)) \le 8(\ln k + 2)J_{,opt}$, where $J_{,opt}$ is the optimal quality for the population.

We show this below.

Let $T_{m,k}$ denote the probability distribution for choosing the set M as cluster centres for a sample from P_m using k-means++ algorithm. From the k-means++ algorithm we know that the probability of choosing μ_j given $\mu \in M_{[1:j-1]}$ have already been picked amounts to $Prob(\mu_j|M_{[1:j-1]}) = \frac{D(\mu_j, M_{[1:j-1]})^2}{\int D(\mathbf{x}, M_{[1:j-1]})^2 P_m(d\mathbf{x})}$. whereas the point μ_1 is picked according to the probability distribution P_m. Obviously, the probability of selecting the set M amounts to:

$$Prob(M) = Prob(\mu_1) \cdot Prob(\mu_2|M_{[1]}) \cdot Prob(\mu_3|M_{[1:2]}) \cdots Prob(\mu_k|M_{[1:k-1]})$$

So the probability density $T_{m,k}$ may be expressed as

$$P_m(\mu_1) \cdot \prod_{j=2}^{k} \frac{D(\mu_j, M_{[1:j-1]})^2}{\int D(\mathbf{x}, M_{[1:j-1]})^2 P_m(d\mathbf{x})}$$

For P we get accordingly the T_k distribution.

Now assume we ignore with probability of at most $\delta > 0$ a *distant* part of the potential locations of cluster centres from T_k via a procedure described below. Denote with $\mathbf{M}_\delta$ the subdomain of the set of cluster centres left after ignoring these elements. Instead of expectation of

$$\mathbb{E}_M(J_{cp}(M, P_m)) = \int J_{cp}(M, P_m)T_{m,k}(dM)$$

we will be interested in

$$\mathbb{E}_{M,\delta}(J_{cp}(M, P_m)) = \int_{M\in\mathbf{M}_\delta} J_{cp}(M, P_m)T_{m,k}(dM)$$

and respectively for P instead of

$$\mathbb{E}_M(J_{cp}(M, P)) = \int J_{cp}(M, P)T_k(dM)$$

we want to consider

$$\mathbb{E}_{M,\delta}(J_{cp}(M, P)) = \int_{M\in\mathbf{M}_\delta} J_{cp}(M, P)T_k(dM)$$

$\mathbb{E}_M(J_{cp}(M, P))$ is apparently finite.

Let us concentrate on the difference $|\mathbb{E}_{M,\delta}(J_{cp}(M, P)) - \mathbb{E}_{M,\delta}(J_{cp}(M, P_m))|$ and show that it converges to 0 when m is increased for a fixed δ, that is for any ϵ and δ there exists such an $m_{\delta,\epsilon}$ that for any larger m almost surely (a.s.) this difference is lower than ϵ. Thus by decreasing δ and ϵ we find that the expectation on P_m converges to that on P.

The subdomain selection (ignoring δ share of probability mass of T_m) shall proceed as follows. Let $\delta_1 = \sqrt[k]{1 + \delta} - 1$.

Let us denote $J_{m,opt} = \min_M J_{cp}(M, P_m)$, $J_{.,opt} = \min_M J_{cp}(M, P)$, where M is a set of cardinality at most k. Let us denote $J_{m,opt,j} = \min_M J_{cp}(M, P_m)$, $J_{.,opt,j} = \min_M J_{cp}(M, P)$, where M be a set of cardinality at most j, $j = 1, \ldots, k$.

Let R_1 be a radius such that $\int_{\|\mathbf{x}\| > R_1} P(d\mathbf{x}) < \delta_1$. Further, let R_j for $j = 2, \ldots, k$ be such that $\int_{\|\mathbf{x}\| > R_j} \frac{(\|\mathbf{x}\| + R_1)^2}{J_{.,opt,j}} P(d\mathbf{x}) < \delta_1$.

We reject all M such that for any $j = 1, \ldots, k$ $\|\boldsymbol{\mu}_j\| > R_j$.

Let us show now that the rejected Ms constitute only δ of probability mass or less.

Consider the difference (for any $l = 2, \ldots, k$) between the probability mass before and after rejection.

$$\int_{\boldsymbol{\mu}_1} \cdots \int_{\boldsymbol{\mu}_l} \prod_{j=2}^{l} \frac{D(\boldsymbol{\mu}_j, M_{[1:j-1]})^2}{\int D(\mathbf{x}, M_{[1:j-1]})^2 P(d\mathbf{x})} P(d\boldsymbol{\mu}_l) \ldots P(d\boldsymbol{\mu}_1)$$

$$- \int_{\|\boldsymbol{\mu}_1\| \leq R_1} \cdots \int_{\|\boldsymbol{\mu}_l\| \leq R_l} \prod_{j=2}^{l} \frac{D(\boldsymbol{\mu}_j, M_{[1:j-1]})^2}{\int D(\mathbf{x}, M_{[1:j-1]})^2 P(d\mathbf{x})} P(d\boldsymbol{\mu}_l) \ldots P(d\boldsymbol{\mu}_1)$$

which is obviously non-negative.

Let us introduce the following notation: $\mathbf{r}$ be a vector of relations from $\{\leq, >\}^l$. So the above can be rewritten as

$$= \sum_{\mathbf{r} \in \{\leq, >\}^l} \int_{\|\boldsymbol{\mu}_1\| \mathbf{r}_1 R_1} \cdots \int_{\|\boldsymbol{\mu}_l\| \mathbf{r}_l R_l} \prod_{j=2}^{l} \frac{D(\boldsymbol{\mu}_j, M_{[1:j-1]})^2}{\int D(\mathbf{x}, M_{[1:j-1]})^2 P(d\mathbf{x})} P(d\boldsymbol{\mu}_l) \ldots P(d\boldsymbol{\mu}_1)$$

$$- \int_{\|\boldsymbol{\mu}_1\| \leq R_1} \cdots \int_{\|\boldsymbol{\mu}_l\| \leq R_l} \prod_{j=2}^{l} \frac{D(\boldsymbol{\mu}_j, M_{[1:j-1]})^2}{\int D(\mathbf{x}, M_{[1:j-1]})^2 P(d\mathbf{x})} P(d\boldsymbol{\mu}_l) \ldots P(d\boldsymbol{\mu}_1)$$

which is equivalent to

$$= \sum_{\mathbf{r} \in \{\leq, >\}^l - \{\leq\}^l} \int_{\|\boldsymbol{\mu}_1\| \mathbf{r}_1 R_1} \cdots \int_{\|\boldsymbol{\mu}_l\| \mathbf{r}_l R_l} \prod_{j=2}^{l} \frac{D(\boldsymbol{\mu}_j, M_{[1:j-1]})^2}{\int D(\mathbf{x}, M_{[1:j-1]})^2 P(d\mathbf{x})} P(d\boldsymbol{\mu}_l) \ldots P(d\boldsymbol{\mu}_1)$$

Consider that for any $j \geq 2$

$$\int_{\|\boldsymbol{\mu}_j\|\leq R_j} \frac{D(\boldsymbol{\mu}_j, M_{[1:j-1]})^2}{\int D(\mathbf{x}, M_{[1:j-1]})^2 P(dx)} P(d\boldsymbol{\mu}_j) \leq 1$$

On the other hand

$$\int_{\|\boldsymbol{\mu}_j\|> R_j} \frac{D(\boldsymbol{\mu}_j, M_{[1:j-1]})^2}{\int D(\mathbf{x}, M_{[1:j-1]})^2 P(dx)} P(d\boldsymbol{\mu}_j)$$

$$\leq \int_{\|\boldsymbol{\mu}_j\|> R_j} \frac{D(\boldsymbol{\mu}_j, M_{[1:j-1]})^2}{J_{.,opt,j-1}} P(d\boldsymbol{\mu}_j)$$

$$\leq \int_{\|\boldsymbol{\mu}_j\|> R_j} \frac{(\|\boldsymbol{\mu}_j\| + R_1)^2}{J_{.,opt,j-1}} P(d\boldsymbol{\mu}_j) \leq \delta_1$$

So the above sum is lower equal to:

$$\leq \sum_{\mathbf{r}\in\{\leq,>\}^l - \{\leq\}^l} \delta_1^{count(>,\mathbf{r})}$$

where $count(>, \mathbf{r})$ is the number of $>$ relations in the vector $\mathbf{r}$.

$$= (1 + \delta_1)^l - 1 \leq \delta$$

This confirms that upon the above restriction on R_j we reject at most δ of the mass of μ_l points under T_k.

So let us consider

$$|\mathbb{E}_{M,\delta}(J_{cp}(M, P)) - \mathbb{E}_{M,\delta}(J_{cp}(M, P_m))|$$

$$= \left| \int_{M\in\mathbf{M}_\delta} J_{cp}(M, P_m) T_{m,k}(dM) - \int_{M\in\mathbf{M}_\delta} J_{cp}(M, P) T_k(dM) \right|$$

$$= \left| \int_{M\in\mathbf{M}_\delta} J_{cp}(M, P_m) \prod_{j=2}^{k} \frac{D(\boldsymbol{\mu}_j, M_{[1:j-1]})^2}{\int D(\mathbf{x}, M_{[1:j-1]})^2 P_m(d\mathbf{x})} P_m(d\boldsymbol{\mu}_k) \ldots P_m(d\boldsymbol{\mu}_1) \right.$$

$$\left. - \int_{M\in\mathbf{M}_\delta} J_{cp}(M, P) \prod_{j=2}^{k} \frac{D(\boldsymbol{\mu}_j, M_{[1:j-1]})^2}{\int D(\mathbf{x}, M_{[1:j-1]})^2 P(d\mathbf{x})} P(d\boldsymbol{\mu}_k) \ldots P(d\boldsymbol{\mu}_1) \right|$$

$$= \left| \int_{M\in\mathbf{M}_\delta} J_{cp}(M, P_m) \prod_{j=2}^{k} \frac{D(\boldsymbol{\mu}_j, M_{[1:j-1]})^2}{J_{cp}(M_{[1:j-1]}, P_m)} P_m(d\boldsymbol{\mu}_k) \ldots P_m(d\boldsymbol{\mu}_1) \right.$$

$$\left. - \int_{M \in \mathbf{M}_\delta} J_{cp}(M, P) \prod_{j=2}^{k} \frac{D(\mu_j, M_{[1:j-1]})^2}{J_{cp}(M_{[1:j-1]}, P)} P(d\mu_k) \dots P(d\mu_1) \right|$$

For a sufficiently large m $\prod_{j=2}^{k} \frac{1}{J_{m,opt,j-1}}$ differs from $\prod_{j=2}^{k} \frac{1}{J_{.,opt,j-1}}$ by at most ϵ_{opt} (due to the result of Pollard). This difference can be made also that small between $\prod_{j=2}^{k} \frac{1}{J_{cp}(M_{[1:j-1]}, P_m)}$ and $\prod_{j=2}^{k} \frac{1}{J_{cp}(M_{[1:j-1]}, P)}$ by increasing m. The argument runs as follows: We can always find a radius R for which $\int_{\|\mathbf{x}\|>R} D(\mathbf{x}, M_{[1:j-1]})^2 P(d\mathbf{x})$ can be made as small as we want so that it is negligible for our purposes and within the limited domain of radius R with increase of m both $\int_{\|\mathbf{x}\|\leq R} D(\mathbf{x}, M_{[1:j-1]})^2 P_m(d\mathbf{x})$ and $\int_{\|\mathbf{x}\|\leq R} D(\mathbf{x}, M_{[1:j-1]})^2 P(d\mathbf{x})$ become as close as we want. Therefore

$$\leq \left| \int_{M \in \mathbf{M}_\delta} J_{cp}(M, P_m) \prod_{j=2}^{k} \frac{D(\mu_j, M_{[1:j-1]})^2 (1 + \pm\epsilon_{opt})}{J_{cp}(M_{[1:j-1]}, P)} P_m(d\mu_k) \dots P_m(d\mu_1) \right.$$

$$\left. - \int_{M \in \mathbf{M}_\delta} J_{cp}(M, P) \prod_{j=2}^{k} \frac{D(\mu_j, M_{[1:j-1]})^2}{J_{cp}(M_{[1:j-1]}, P)} P(d\mu_k) \dots P(d\mu_1) \right|$$

$$\leq \epsilon_{opt} \left| \int_{M \in \mathbf{M}_\delta} J_{cp}(M, P_m) \prod_{j=2}^{k} \frac{D(\mu_j, M_{[1:j-1]})^2}{J_{cp}(M_{[1:j-1]}, P)} P_m(d\mu_k) \dots P_m(d\mu_1) \right|$$

$$+ \left| \int_{M \in \mathbf{M}_\delta} J_{cp}(M, P_m) \prod_{j=2}^{k} \frac{D(\mu_j, M_{[1:j-1]})^2}{J_{cp}(M_{[1:j-1]}, P)} P_m(d\mu_k) \dots P_m(d\mu_1) \right.$$

$$\left. - \int_{M \in \mathbf{M}_\delta} J_{cp}(M, P) \prod_{j=2}^{k} \frac{D(\mu_j, M_{[1:j-1]})^2}{J_{cp}(M_{[1:j-1]}, P)} P(d\mu_k) \dots P(d\mu_1) \right|$$

The first addend, the product of an ϵ_{opt} and a finite quantity, can be set as low as needed by choosing a sufficiently low ϵ_{opt}.

Now following Pollard, we will decrease as much as required the second addend, which we will rewrite as.

$$\left| \int_{M \in \mathbf{M}_\delta} \int D(\mathbf{x}, M)^2 \prod_{j=2}^{k} \frac{D(\mu_j, M_{[1:j-1]})^2}{J_{cp}(M_{[1:j-1]}, P)} P_m(d\mathbf{x}) P_m(d\mu_k) \dots P_m(d\mu_1) \right.$$

$$\left. - \int_{M \in \mathbf{M}_\delta} \int D(\mathbf{x}, M)^2 \prod_{j=2}^{k} \frac{D(\mu_j, M_{[1:j-1]})^2}{J_{cp}(M_{[1:j-1]}, P)} P(d\mathbf{x}) P(d\mu_k) \dots P(d\mu_1) \right|$$

So select a finite set T_β of points from $\mathbf{M}_\delta$ such that each element of $\mathbf{M}_\delta$ lies within a distance of β from a point of T_β.

Let us define the function $g_M(x) = D(\mathbf{x}, M)^2 \prod_{j=2}^{k} \frac{D(\mu_j, M_{[1:j-1]})^2}{J_{cp}(M_{[1:j-1]}, P)}$.

Let us introduce also $\overline{D}(x, M)$ and $\underline{D}(x, M)$, $\overline{J}_{cp}(M, P)$ and $\underline{J}_{cp}(M, P)$, as follows: M^* be the set of elements from T_β such that $\|\mu_j - M^*[j]\|$ is less than β, and $|J_{cp}^{-1}(M, P) - J_{cp}^{-1}(M^*, P)|$ is less[31] than β. Then $\overline{D}(x, M) = D(x, M^*) + \beta$, $\underline{D}(x, M) = \max(D(x, M^*) - \beta, 0)$, $\overline{J}_{cp}(M, P) = J_{cp}^{-1}(M^*, P) + \beta$ and $\underline{J}_{cp}(M, P) = \max(J_{cp}^{-1}(M^*, P) - \beta, J_{max}^{-1})$, where J_{max} is the poorest initialisation within the domain M_δ. Now define the function

$$\underline{g}_M(x) = \underline{D}(\mathbf{x}, M)^2 \prod_{j=2}^{k} \underline{D}(\mu_j, M_{[1:j-1]})^2 \underline{J}_{cp}(M_{[1:j-1]}, P)$$

and

$$\overline{g}_M(x) = \overline{D}(\mathbf{x}, M)^2 \prod_{j=2}^{k} \overline{D}(\mu_j, M_{[1:j-1]})^2 \overline{J}_{cp}(M_{[1:j-1]}, P)$$

As $\|x, M^*[j]\| - \beta \le \|x - \mu_j\| \le \|x, M^*[j]\| + \beta$, it is easily seen that $\underline{D}(x, M) \le D(x, M) \le \overline{D}(x, M)$, similarly with J, and hence $\underline{g}_M(x) \le g_M(x) \le \overline{g}_M(x)$.

Therefore

$$\left| \int_{M \in \mathbf{M}_\delta} \int_{\mathbf{x}} g_M(x) P(dx) P(d\boldsymbol{\mu}_k) \dots P(d\boldsymbol{\mu}_1)) \right.$$

$$\left. - \int_{M \in \mathbf{M}_\delta} \int_{\mathbf{x}} g_M(x) P_m(dx) P_m(d\boldsymbol{\mu}_k) \dots P_m(d\boldsymbol{\mu}_1) \right|$$

is bounded from above by

---

[31] For a constraint domain of M this is always possible for the following reason. Let M^* and M^{**} be two sets of up to k distinct "corresponding" cluster centres. "Corresponding" shall mean here that μ_j^* and μ_j^{**} are much closer to each other than any other centre from $M^+ = M^* \cup M^{**}$. Let the distance between corresponding cluster centres be at most d. Recall that

$$J_{cp}(M, Q) = \int D(\mathbf{x}, M)^2 Q(d\mathbf{x}) = \int (\min_{\mu \in M} \|\mathbf{x} - \mu\|)^2 Q(d\mathbf{x})$$

This implies that $J_{cp}(M^*, Q) \ge J_{cp}(M^+, Q)$ and $J_{cp}(M^{**}, Q) \ge J_{cp}(M^+, Q)$. Note that $|\|\mathbf{x} - \mu^*\|^2 - \|\mathbf{x} - \mu^{**}\|^2|$
$= |(\|\mathbf{x} - \mu^*\| + \|\mathbf{x} - \mu^{**}\|)(\|\mathbf{x} - \mu^*\| - \|\mathbf{x} - \mu^{**}\|)| \le |(2\|\mathbf{x} - \mu^*\| + d) \cdot d|$ Hence $J_{cp}(M^*, Q) - J_{cp}(M^+, Q) \le d \int \min_{\mu^* \in M^*} (2\|\mathbf{x} - \mu^*\| + d) \cdot Q(d\mathbf{x})$ This difference can be kept as low as requested (but positive) by diminishing d. Analogical result is obtained for $J_{cp}(M^{**}, Q) - J_{cp}(M^+, Q)$. If we keep in mind that $|J_{cp}(M^*, Q) - J_{cp}(M^{**}, Q) \le J_{cp}(M^*, Q) - J_{cp}(M^+, Q) + J_{cp}(M^{**}, Q) - J_{cp}(M^+, Q)$ and that J_{cp} has always a positive minimum, then obviously we can keep the mentioned difference in a bounded area below β by an appropriately dense grid.

$$\int_{M\in\mathbf{M}_\delta}\int_{\mathbf{x}}(\overline{g}_M(x)-\underline{g}_M(x))P(dx)P(d\boldsymbol{\mu}_k)\ldots P(d\boldsymbol{\mu}_1)$$

$$+\max\left(\left|\int_{M\in\mathbf{M}_\delta}\int_{\mathbf{x}}\overline{g}_M(x)P(dx)P(d\boldsymbol{\mu}_k)\ldots P(d\boldsymbol{\mu}_1)\right.\right.$$

$$\left.-\int_{M\in\mathbf{M}_\delta}\int_{\mathbf{x}}\overline{g}_M(x)P_m(dx)P_m(d\boldsymbol{\mu}_k)\ldots P_m(d\boldsymbol{\mu}_1)\right|$$

$$\left|\int_{M\in\mathbf{M}_\delta}\int_{\mathbf{x}}\underline{g}_M(x)P(dx)P(d\boldsymbol{\mu}_k)\ldots P(d\boldsymbol{\mu}_1)\right.$$

$$\left.\left.-\int_{M\in\mathbf{M}_\delta}\int_{\mathbf{x}}\underline{g}_M(x)P_m(dx)P_m(d\boldsymbol{\mu}_k)\ldots P_m(d\boldsymbol{\mu}_1)\right|\right)$$

A sufficient increase on m will make the second addend as small as we like. We will show below that using an argument similar to Pollard, one can demonstrate that the first addend can be diminished as much as we like to by appropriate choice of β. Hence the restricted expectation can be shown to converge as claimed.

For any R

$$\int_{M\in\mathbf{M}_\delta}\int_{\mathbf{x}}(\overline{g}_M(x)-\underline{g}_M(x))P(dx)P(d\boldsymbol{\mu}_k)\ldots P(d\boldsymbol{\mu}_1)$$

$$=\int_{M\in\mathbf{M}_\delta}\int_{\|\mathbf{x}\|\le R}(\overline{g}_M(x)-\underline{g}_M(x))P(dx)P(d\boldsymbol{\mu}_k)\ldots P(d\boldsymbol{\mu}_1)$$

$$+\int_{M\in\mathbf{M}_\delta}\int_{\|\mathbf{x}\|> R}(\overline{g}_M(x)-\underline{g}_M(x))P(dx)P(d\boldsymbol{\mu}_k)\ldots P(d\boldsymbol{\mu}_1)$$

The second addend has the property:

$$\int_{M\in\mathbf{M}_\delta}\int_{\|\mathbf{x}\|> R}(\overline{g}_M(x)-\underline{g}_M(x))P(dx)P(d\boldsymbol{\mu}_k)\ldots P(d\boldsymbol{\mu}_1)$$

$$\le\int_{M\in\mathbf{M}_\delta}\int_{\|\mathbf{x}\|> R}\overline{g}_M(x)P(dx)P(d\boldsymbol{\mu}_k)\ldots P(d\boldsymbol{\mu}_1)$$

$$\le\int_{M\in\mathbf{M}_\delta}\int_{\|\mathbf{x}\|> R}(2R_k+2\sigma)^{2k+2}(\|x\|+R_k+2\sigma)^2P(dx)P(d\boldsymbol{\mu}_k)\ldots P(d\boldsymbol{\mu}_1)$$

which can be decreased as much as necessary by increasing R.

The first addend can be transformed

$$\int_{M \in \mathbf{M}_\delta} \int_{\|\mathbf{x}\| \leq R} (\overline{g}_M(x) - \underline{g}_M(x)) P(dx) P(d\boldsymbol{\mu}_k) \ldots P(d\boldsymbol{\mu}_1)$$

$$\leq \int_{M \in \mathbf{M}_\delta} \int_{\|\mathbf{x}\| \leq R} \sigma h_M(x) P(dx) P(d\boldsymbol{\mu}_k) \ldots P(d\boldsymbol{\mu}_1)$$

where $h_M(x)$ is a function of x and M. As both x and elements of M are of bounded length, $h_M(x)$ is limited from above. Hence this addend can be decreased to the desired value by appropriate choice of (small) σ.

Therefore the whole term can be decreased to a desired value (above zero of course), what we had to prove.

Hence our claim about consistency of k-means++ has been proven.

References

1. Z. Abbassi, V.S. Mirrokni, A recommender system based on local random walks and spectral methods, in *Proceedings of the 9th WebKDD and 1st SNA-KDD 2007 Workshop on Web Mining and Social Network Analysis*, San Jose, CA, USA, 12–17 Aug 2007 (ACM New York, NY, USA, 2007), pp. 102–108

2. M. Ackerman, S. Ben-David, *Which Data Sets are 'Clusterable'?—A Theoretical Study of Clusterability*. Preprint (D.R.C. School of Computer Science, University of Waterloo, 2008). http://www.ima.umn.edu/~iwen/REU/ability_submit.pdf

3. M. Ackerman, S. Ben-David, Clusterability: a theoretical study, in *Proceedings of the 12th International Conference on Artificial Intelligence and Statistics (AISTATS 2009), JMLR: W&CP 5*, 2009, pp. 1–8

4. M. Ackerman, S. Ben-David, D. Loker, Towards property-based classification of clustering paradigms, in *Advances in Neural Information Processing Systems 23*, ed. by J.D. Lafferty, C.K.I. Williams, J. Shawe-Taylor, R.S. Zemel, A. Culotta (Curran Associates, Inc., 2010), pp. 10–18

5. M. Ackerman, S. Dasgupta. Incremental clustering: the case for extra clusters, in *Advances in Neural Information Processing Systems 27: Annual Conference on Neural Information Processing Systems 2014, December 8–13, 2014*, Montreal, QC, Canada, 2014, pp. 307–315

6. G. Agarwal, D. Kempe, Modularity-maximizing graph communities via mathematical programming. Eur. Phys. J. B **66**(3), 409–418 (2008)

7. P.K. Agarwal, N.H. Mustafa, K-means projective clustering, in *Proceedings of the Twenty-third ACM SIGMOD-SIGACT-SIGART Symposium on Principles of Database Systems*, PODS '04, (New York, NY, USA, 2004). ACM, pp. 155–165

8. C.C. Aggarwal, A. Hinneburg, D.A. Keim, On the surprising behavior of distance metrics in high dimensional space, in *Proceedings of the 8th International Conference on Database Theory*, LNCS 1973 (Springer, Berlin, Heidelberg, 2001), pp. 420–430

9. ChC Aggarwal, ChK Reddy (eds.), *Data Clustering Algorithms and Applications* (CRC Data Mining and Knowledge Discovery Series (Chapman and Hall/CRC, Boca Raton, FL, 2013)

10. Y.Y. Ahn, J.P. Bagrow, S. Lehmann, Link communities reveal multiscale complexity in networks. Nature **466**, 761–764 (2010)

11. N. Ailon, R. Jaiswal, C. Monteleoni, Streaming k-means approximation, in *Advances in Neural Information Processing Systems 22*, ed. by Y. Bengio, D. Schuurmans, J.D. Lafferty, C.K.I. Williams, A. Culotta (Curran Associates, Inc., 2009), pp. 10–18

12. M.A. Ajzerman, E.M. Brawerman, L.I. Rozonoer, Theoretical foundations of the potential function method in pattern recognition learning. Automat. Rem. Contr. **25**, 821–837 (1964)

13. R. Albert, A.-L. Barabási, Statistical mechanics of complex networks. Rev. Mod. Phys. **74**(1), 47–97, 30 (2002)

14. D. Aldous, J.A. Fill, *Reversible Markov Chains and Random Walks on Graphs, 2002*. Unfinished monograph, recompiled 2014. https://www.stat.berkeley.edu/~aldous/RWG/book.html

© Springer International Publishing AG 2018
S. T. Wierzchoń and M. A. Kłopotek, *Modern Algorithms of Cluster Analysis*,
Studies in Big Data 34, https://doi.org/10.1007/978-3-319-69308-8

15. N. Alldrin, A. Smith, D. Turnbull, Clustering with EM and k-means. Technical report, Department of Computer Science (University of California, San Diego, La Jolla, CA 92037, 2003). http://neilalldrin.com/research/w03/cse253/project1.pdf

16. D. Aloise, A. Deshpande, P. Hansen, P. Popat, NP-hardness of Euclidean sum-of-squares clustering. *Cahiers du GERAD*, pp. G-2008-33 (2008)

17. C. Alzate, J. Suykens, Multiway spectral clustering with out-of-sample extensions through weighted kernel PCA. IEEE Trans. Pattern Anal. Mach. Intell. **32**(10), 335–347 (2010). https://doi.org/10.1109/TPAMI.2008.292

18. C. Alzate, J. Suykens, Sparse kernel spectral clustering models for large-scale data analysis. Neurocomputing **74**(9), 1382–1390 (2011). https://doi.org/10.1016/j.neucom.2011.01.001

19. M.R. Anderberg, *Cluster Analysis for Applications* (Academic Press, London, 1973)

20. R. Andersen, C. Borgs, J. Chayes, J. Hopcraft, V.S. Mirrokni, S.-H. Teng, Local computation of PageRank contributions. Internet Math. **5**(1–2), 23–45 (2008). https://doi.org/10.1080/15427951.2008.10129302

21. R. Andersen, F. Chung, Detecting sharp drops in PageRank and a simplified local partitioning algorithm, in *Proceedings of 4th International Conference on Theory and Applications of Models of Computation, TAMC 2007*, volume 4484 of LNCS, Shanghai, China, 22–25 May 2007. (Springer, 2007), pp. 1–13. https://doi.org/10.1007/978-3-540-72504-6_1

22. R. Andersen, F. Chung, K. Lang, Using PageRank to locally partition a graph. Internet Math. **4**(1), 35–64 (2007)

23. R. Andersen, F. Chung, K. Lang, Local partitioning for directed graphs using PageRank. Internet Math. **5**(1–2), 3–22 (2008)

24. R. Andersen, S.M. Cioabă, Spectral densest subgraph and independence number of a graph. J. Univ. Comput. Sci. **13**(11), 1501–1513 (2007). https://doi.org/10.3217/jucs-013-11-1501

25. R. Andersen, K.J. Lang, Communities from seed sets, in *Proceedings of the 15th International Conference on World Wide Web*, WWW'06, Edinburgh, Scotland, 23–26 May 2006 (ACM Press, New York, NY, 2006), pp. 223–232. https://doi.org/10.1145/1135777.1135814

26. M. Ankerst, M.M. Breunig, H.-P. Kriegel, J. Sander, OPTICS: Ordering points to identify the clustering structure, in *Proceedings of 1999 ACM-SIGMOD International Conference on Management of Data*, Philadelphia, PA, 1999. (ACM Press, 1999), pp. 49–60

27. A. Anthony, M. des Jardins, Open problems in relational data clustering, in *Proceedings of ICML Workshop on Open Problems in Statistical Relational Learning*, Pittsburgh, PA, 2006

28. A. Arenas, J. Duch, A. Fernández, S. Gómez, Size reduction of complex networks preserving modularity. New J. Phys. **9**, 176 (2007). https://doi.org/10.1088/1367-2630/9/6/176

29. A. Arenas, A. Fernández, S. Gómez, Analysis of the structure of complex networks at different resolution levels. New J. Phys. **10**, 053039 (2008). https://doi.org/10.1088/1367-2630/10/5/053039

30. D. Arthur, *Analyzing and improving local search: k-means and ICP*. Ph.D. thesis, Stanford University. Department of Computer Science, Stanford, CA, 2009

31. D. Arthur, B. Manthey, H. Röglin, Smoothed analysis of the k-means method. J. ACM **58**(5), 19 (2011). https://doi.org/10.1145/2027216.2027217

32. D. Arthur, S. Vassilvitskii, k-means++: the advantages of careful seeding, in *Proceedings of the Eighteenth Annual ACM-SIAM Symposium on Discrete Algorithms*, ed. by N. Bansal, K. Pruhs, C. Stein, SODA 2007, New Orleans, LA, USA, 7–9 Jan 2007. (SIAM, 2007), pp. 1027–1035

33. S. Arya, D.M. Mount, N.S. Netanyahu, R. Silverman, A.Y. Wu, An optimal algorithm for approximate nearest neighbor searching. J. ACM **45**(6), 891–8923 (1998). https://doi.org/10.1145/293347.293348

34. A. Asuncion, D.J. Newman, *UCI Machine Learning Repository* (University of California, Irvine, School of Information and Computer Sciences, 2007). http://www.ics.uci.edu/~mlearn/MLRepository.html

35. A. Azran, Z. Grahramani, Spectral methods for automatic multiscale data clustering, in *Proceedings of 2006 IEEE Computer Society Conference on Computer Vision and Pattern Recognition*, New York, NY, USA, 17–22 June 2006. (IEEE, 2006), pp. 190–197. https://doi.org/10.1109/CVPR.2006.289

36. R. Babuška, *Fuzzy Modeling for Control* (Kluwer Academic Publishers, Boston, USA, 1998)
37. R. Babuška, P.J. van der Veen, U. Kaymak, Improved covariance estimation for Gustafson-Kessel clustering, in *Proceedings of the 2002 IEEE International Conference on Fuzzy Systems, 2002. FUZZ-IEEE'02*, vol. 2, IEEE, 12–17 May 2002, pp. 1081–1085. https://doi.org/10.1109/FUZZ.2002.1006654
38. F.R. Bach, M.I. Jordan, Kernel independent component analysis. J. Mach. Learn. Res. **3**, 1–48 (2003). https://doi.org/10.1162/153244303768966085
39. R. Baeza-Yates, B. Ribeiro-Neto, *Modern Information Retrieval* (ACM Press/Addison-Wesley, New York, 1999)
40. B. Bahmani, B. Moseley, A. Vattani, R. Kumar, S. Vassilvitskii, Scalable k-means++. Proc. VLDB Endow. **5**(7), 622–633 (2012)
41. Z. Bai, J. Demmel, J. Dongarra, A. Ruhe, H. van der Vorst, *Templates for the Solution of Algebraic Eigenvalue Problems: A Practical Guide* (SIAM, Philadelphia, 2000)
42. E. Bair, R. Tibshirani, Semi-supervised methods to predict patient survival from gene expression data. *PLOS Biol.* **2**(4), e108 (2004). https://doi.org/10.1371/journal.pbio.0020108
43. R. Balaji, R.B. Bapat, On Euclidean distance matrices. Linear Algebra Appl. **424**(1), 108–117 (2007). https://doi.org/10.1016/j.laa.2006.05.013
44. M.F. Balcan, A. Blum, N. Srebro, A theory of learning with similarity functions. Mach. Learn. **72**(1–2), 89–112 (2008). https://doi.org/10.1007/s10994-008-5059-5
45. M.F. Balcan, A. Blum, S. Vempala, Clustering via similarity functions: theoretical foundations and algorithms. Unpublished, 2009. http://www.cs.cmu.edu/afs/cs.cmu.edu/Web/People/avrim/Papers/BBVclustering_full.pdf
46. P. Baldi, G.W. Hatfield, *DNA Microarrays and Gene Expression: From Experiments to Data Analysis and Modeling* (Cambridge University Press, Cambridge, UK, 2002)
47. G.H. Ball, D.J. Hall, ISODATA: a novel method of data analysis and pattern classification. Technical Report NTIS AD 699616 (Stanford Research Institute, Stanford, CA, 1965)
48. A. Banerjee, I. Dhillon, J. Ghosh, S. Merugu, D.S. Modha, A generalized maximum entropy approach to bregman co-clustering and matrix approximation. J. Mach. Learn. Res. **8**, 1919–1986 (2007)
49. A. Banerjee, C. Krumpelman, J. Ghosh, S. Basu, R.J. Mooney, Model-based overlapping clustering, in *Proceedings of the 11th ACM SIGKDD International Conference on Knowledge Discovery in Data Mining* (ACM New York, NY, USA, 2005), pp. 532–537. https://doi.org/10.1145/1081870.1081932
50. A. Banerjee, S. Merugu, I.S. Dhillon, J. Ghosh, Clustering with Bregman divergences. J. Mach. Learn. Res. **6**:1705–1749, 1 2005
51. A.-L. Barabási, E. Bonabeau, Scale-free networks. Sci. Am. (2003), pp. 50–59
52. W. Barbakh, C. Fyfe, Online clustering algorithms. Int. J. Neural Syst. **18**(3), 185–194 (2008). https://doi.org/10.1142/S0129065708001518
53. S.T. Barnard, A. Pothen, H.D. Simon, A spectral algorithm for envelope reduction of sparse matrices. Numer. Linear Algebra Appl. **2**(4), 317–334 (1995). https://doi.org/10.1002/nla.1680020402
54. M. Basseville, Distance measures for signal processing and pattern recognition. Sig. Proc. **18**(4), 349–369 (1989). https://doi.org/10.1016/0165-1684(89)90079-0
55. S. Basu, A. Banerjee, R. Mooney, Semi-supervised clustering by seeding, in *Proceedings of 19th International Conference on Machine Learning (ICML-2002)* (2002), pp. 19–26
56. S. Basu, A. Banerjee, R. Mooney, Active semi-supervision for pairwise constrained clustering, in *Proceedings of the SIAM International Conference on Data Mining, (SDM-2004)*, Lake Buena Vista, FL, 2004, pp. 333–344
57. M.G. Beiró, J.I. Alvarez-Hamelin, J.R. Busch, A low complexity visualization tool that helps to perform complex systems analysis. New J. Phys. **10**(12), 125003 (2008)
58. J. Bejarano, K. Bose, T. Brannan, A. Thomas, K. Adragni, N.K. Neerchal, Sampling within k-means algorithm to cluster large datasets. Technical Report HPCF-2011-12, High Performance Computing Facility (University of Maryland, Baltimore County, 2011)

59. M.-A. Belabbas, P.J. Wolfe, Spectral methods in machine learning: New strategies for very large datasets. Proc. Natl. Acad. Sci. USA **106**, 369–374 (2009)
60. M. Belkin, P. Niyogi, Laplacian eigenmaps for dimensionality reduction and data representation. Neur. Comput. **6**(15), 1373–1396 (2003)
61. M. Belkin, P. Niyogi, Towards a theoretical foundation for Laplacian-based manifold methods, in *Proceedings of the 18th Conference on Learning Theory (COLT)* ed. by P. Auer, R. Meir, LNCS 3559 (Springer, 2005), pp. 486–500. https://doi.org/10.1007/11503415_33
62. S. Ben-David, M. Ackerman, Measures of clustering quality: a working set of axioms for clustering, in *Advances in Neural Information Processing Systems 21* ed. by D. Koller, D. Schuurmans, Y. Bengio, L. Bottou (Curran Associates, Inc., 2009), pp. 121–128
63. A. Ben-Hur, D. Horn, H.T. Siegelmann, V. Vapnik, Support vector clustering. J. Mach. Learn. Res. **2**, 125–137 (2001)
64. A.M. Bensaid, L.O. Hall, J.C. Bezdek, L.P. Clarke, M.L. Silbiger, J.A. Arrington, R.F. Murtagh, Validity-guided (re)clustering with applications to image segmentation. IEEE Trans. Fuzzy Syst. **4**(2), 112–123 (1996)
65. P. Berkhin, Bookmark-coloring approach to personalized PageRank computing. Internet Math. **3**(1), 41–62 (2006). https://doi.org/10.1080/15427951.2006.10129116
66. P. Berkhin, A survey of clustering data mining techniques, in *Grouping Multidimensional Data*, ed. by J. Kogan, Ch. Nicholas, M. Teboulle (Springer, 2006), pp. 25–72
67. M.W. Berry, M. Browne, A.M. Langville, V.P. Pauca, R.J. Plemmons, Algorithms and applications for approximate nonnegative matrix factorization. Comput. Stat. Data Anal. **52**(1), 155–173 (2007)
68. M.W. Berry, S.T. Dumais, G.W. O'Brien, Using linear algebra for intelligent information retrieval. SIAM Rev. **37**(4), 573–595 (1995). https://doi.org/10.1137/1037127
69. J.C. Bezdek, A convergence theorem for the fuzzy ISODATA clustering algorithms. IEEE Trans. Pattern Anal. Mach. Intell. PAMI **2**(1), 1–8 (1980)
70. J.C. Bezdek, *Pattern Recognition with Fuzzy Objective Function Algorithms* (Plenum Press, New York and London, 1981)
71. J.C. Bezdek, R.J. Hathaway, Numerical convergence and interpretation of the fuzzy c-shells clustering algorithm. IEEE Trans. Neural Networks **3**(5), 787–793 (1992). https://doi.org/10.1109/72.159067
72. J.C. Bezdek, J. Keller, R. Krisnapuram, N.R. Pal, *Fuzzy Models and Algorithms for Pattern Recognition and Image Processing* (Kluwer Academic Publisher, Boston, London, 1999)
73. J.C. Bezdek, S.K. Pal, *Fuzzy Models for Pattern Recognition: Methods that Search for Structures in Data* (IEEE, New York, 1992)
74. N. Biggs, *Algebraic Graph Theory*, 2nd edn. (Cambridge University Press, 1993). https://doi.org/10.1017/CBO9780511608704
75. C.M. Bishop, *Neural Networks for Pattern Recognition* (Clarendon Press, London, 1995)
76. T. Biyikoglu, J. Leydold, P.F. Stadler, *Laplacian Eigenvectors, of Graphs: Perron-Frobenius and Faber-Krahn Type Theorems*, vol. 1915, Lecture Notes in Mathematics. (Springer, 2007). https://doi.org/10.1007/978-3-540-73510-6
77. V.D. Blondel, J.-L. Guillaume, R. Lambiotte, E. Lefebvre, Fast unfolding of communities in large networks. J. Stat. Mech., P10008 (2008). https://doi.org/10.1088/1742-5468/2008/10/P10008
78. A. Blum, Thoughts on Clustering. Essay for the 2009 NIPS Workshop "Clustering: Science or Art?" (2009)
79. M. Bolla, Penalized versions of the Newman-Girvan modularity and their relation to normalized cuts and k-means clustering. Phys. Rev. E Stat. Nonlin. Soft. Matter. Phys. **84**(1 Pt 2), 016108 (2011). Epub 2011(25) (2011)
80. A. Bonato, *A Course on the Web Graph*, vol. 89 (Graduate Studies in Mathematics (AMS, New York, 2008)
81. A. Bouchachia, W. Pedrycz, Data clustering with partial supervision. Data Min. Knowl. Disc. **12**, 47–78 (2006). https://doi.org/10.1007/s10618-005-0019-1

82. A. Bouchachia, W. Pedrycz, Enhancement of fuzzy clustering by mechanisms of partial supervision. Fuzzy Sets Syst. **157**, 1733–1759 (2006). https://doi.org/10.1016/j.fss.2006.02.015

83. Ch. Boutsidis, A. Zouzias, M.W. Mahoney, P. Drineas, Stochastic dimensionality reduction for k-means clustering. arXiv:1110.2897v1 [cs.DS], 13 Oct 2011

84. P.S. Bradley, K.P. Bennett, A. Demiriz, Constrained k-means clustering. Technical Report MSR-TR-2000-65, Microsoft Research (2000)

85. P.S. Bradley, O.L. Mangasarian, k-plane clustering. J. Global Optim. **16**(1), 23–32 (2000)

86. U. Brandes, A faster algorithm for betweenness centrality. J. Math. Sociol. **25**, 163–177 (2001)

87. U. Brandes, D. Delling, M. Gaertler, R. Göerke, M. Hoefer, Z. Nikoloski, D. Wagner, Maximizing modularity is hard. arXiv:physics/0608255v2 [physics.data-an], 30 Aug 2006

88. U. Brandes, D. Delling, M. Gaertler, R. Görke, M. Hoefer, Z. Nikoloski, D. Wagner, On finding graph clusterings with maximum modularity, in *Graph-Theoretic Concepts in Computer Science. 33rd International Workshop*, WG 2007, Dornburg, Germany, 21–23, 2007. Revised Papers (Springer, 2007), pp. 121–132

89. J.S. Breese, D. Heckerman, C. Kadie, Empirical analysis of predictive algorithms for collaborative filtering, in *Proceedings of 14th Conference on Uncertainty in Artificial Intelligence* (Morgan Kauffman, Madison, WI, 1998), pp. 43–52

90. X. Bresson, Th. Laurent, D. Uminsky, J. von Brecht, Multiclass total variation clustering, in *Advances in Neural Information Processing Systems 26*, ed. by C.J.C. Burges, L. Bottou, M. Welling, Z. Ghahramani, K.Q. Weinberger (Curran Associates, Inc., 2013), pp. 1421–1429. http://papers.nips.cc/paper/5097-multiclass-total-variation-clustering

91. C. Bron, J. Kerbosch, Finding all cliques of an undirected graph. Comm. ACM **16**, 575–577 (1973). https://doi.org/10.1145/362342.362367

92. J.-P. Brunet, P. Tamayo, T.R. Golub, J.P. Mesirov, Metagenes and molecular pattern discovery using matrix factorization. PNAS **101**(12), 4164–4169 (2004). https://doi.org/10.1073/pnas.0308531101

93. S. Bubeck, M. Meilă, U. von Luxburg, How the initialization affects the stability of the k-means algorithm. arXiv:0907.5494v1 [stat.ML], 31 July 2009

94. T. Bühler, M. Hein, Spectral clustering based on the graph p-Laplacian, in *Proceedings of the 26th Annual International Conference on Machine Learning, ICML'09* (ACM, New York, NY, USA, 2009). https://doi.org/10.1145/1553374.1553385

95. G.D. Cañas, T.A. Poggio, L. Rosasco. Learning manifolds with k-means and k-flats, in *Advances in Neural Information Processing Systems 25*, ed. by F. Pereira, C.J.C. Burges, L. Bottou, K.Q. Weinberger (Curran Associates, Inc., 2012), pp. 2474–2482

96. S. Cafieri, P. Hansen, L. Liberti, Improving heuristics for network modularity maximization using an exact algorithm, in *Matheuristics 2010: Third International Workshop on Model-based Metaheuristics* (Vienna, Austria, 28–30, 2010). http://www.lix.polytechnique.fr/~liberti/matheur-10.pdf

97. D. Cai, X. Chen, Large scale spectral clustering via landmark-based sparse representation. IEEE Trans. Cybern **45**(8), 1669–1680 (2014). https://doi.org/10.1109/TCYB.2014.2358564

98. R.J.G.B. Campello, D. Moulavi, J. Sander, Density-Based clustering based on hierarchical density estimates, in *Advances in Knowledge Discovery and Data Mining. PAKDD 2013*, ed. by J. Pei, V.S. Tseng, L. Cao, H. Motoda, G. Xu. LNCS, vol. 7819 (Springer, Berlin, Heidelberg, 2013), pp. 160–172. ISBN 978-3-642-37455-5

99. A. Capocci, V.D.P. Servedio, G. Caldarelli, F. Colaiori, Detecting communities in large networks. Physica A: Stat. Mech. Appl. **352**(2–4), 669–676 (2005). https://doi.org/10.1016/j.physa.2004.12.050

100. S.H. Cha, Comprehensive survey on distance/similarity measures between probability density functions. Int. J Math. Models Methods Appl. Sci. **4**(1), 300–307 (2007)

101. D. Chakrabarti, R. Kumar, A. Tomkins, Evolutionary clustering, in *Proceedings of the 12th ACM SIGKDD International Conference on Knowledge Discovery and Data Mining*, KDD '06 (ACM, New York, NY, USA, 2006), pp. 554–560

102. P.K. Chan, M.D.F. Schlag, J.Y. Zien, Spectral K-way ratio-cut partitioning and clustering. IEEE Trans. Comput. Aided Des. Integr. Circuits Syst. **13**(9), 1088–1096 (1994)

103. O. Chapelle, B. Schölkopf, A. Zien (eds.), *Semi-Supervised Learning* (MIT Press, Cambridge, MA, 2006). http://www.kyb.tuebingen.mpg.de/ssl-book

104. M. Charikar, C. Chekuri, T. Feder, R. Motwani, Incremental clustering and dynamic information retrieval, in *Proceedings of 29th Symposium on Theory of Computing* (ACM, New York, 1997), pp. 626–635

105. K. Chaudhuri, S. Dasgupta, Rates of convergence for the cluster tree, in *NIPS 2010 Conference*, ed. by J.D. Lafferty, C.K.I. Williams, J. Shawe-Taylor, R.S. Zemel, A. Culotta (Curran Associates, Inc., 2010), pp. 343–351

106. P. Cheeseman, J. Stutz, Bayesian classification (autoclass): theory and results, in *Advances in Knowledge Discovery and Data Mining*, ed. by U.M. Fayyad, G. Piatetsky-Shapiro, P. Smyth, R. Uthurusamy (AAAI Press/MIT Press, 1996), pp. 153–180

107. H. Chen, J. Peng, 0-1 semidefinite programming for graph-cut clustering: modelling and approximation, in *Data Mining and Mathematical Programming*, ed. by P.M. Pardalos, P. Hansen, CRM Proceedings & Lecture Notes (AMS, 2008), pp. 15–41

108. J. Chen, O.R. Zaiiane, R. Goebel, Detecting communities in social networks using max-min modularity, in *SIAM International Conference on Data Mining* (2009), pp. 978–989

109. M. Chen, K. Kuzmin, B.K. Szymanski, Community detection via maximization of modularity and its variants. IEEE Trans. Comput. Soc. Syst. **1**, 46–65 (2014). https://doi.org/10.1109/TCSS.2014.2307458

110. S. Chen, B. Ma, K. Zhang, On the similarity metric and the distance metric. Theor. Comput. Sci. **410**(24–25), 2365–2376 (2009). https://doi.org/10.1016/j.tcs.2009.02.023

111. X. Chen, D. Cai, Large scale spectral clustering with landmark-based representation, in *Proceedings of 25th AAAI Conference on Artificial Intelligence*, AAAI'11, San Francisco, CA, 7–11 Aug 2011, pp. 313–318. http://www.aaai.org/ocs/index.php/AAAI/AAAI11/paper/view/3484

112. Y. Chen, T. Georgiou, M. Pavon, A. Tannenbaum, Robust transport over networks. IEEE Trans. Autom. Control (99), 1–1, 2016. https://doi.org/10.1109/TAC.2016.2626796

113. T.W. Cheng, D.B. Goldof, L.O. Hall, Fast fuzzy clustering. Fuzzy Sets Syst. **93**(1), 49–56 (1998)

114. H. Choe, J. Jordan, On the optimal choice of parameters in a fuzzy c-means algorithm, in *Proceedings of First IEEE Conference on Fuzzy Systems* (San Diego, CA, 1992), pp. 349–353

115. F. Chung, *Spectral Graph Theory*, vol. 92 (American Mathematical Soc, Providence, RI, 1977)

116. F. Chung, Random walks and local cuts in graphs. Linear Algebra Appl. **423**(1), 22–32 (2007). https://doi.org/10.1016/j.laa.2006.07.018

117. F. Chung, A. Tsiatas, Finding and visualizing graph clusters using PageRank optimization, in *Algorithms and Models for the Web Graph. 7th International Workshop, WAW 2010. Stanford, CA, USA*, ed. by R. Kumar, D. Sivakumar, volume 6516 of LNCS (Springer, 13–14 Dec 2010), pp. 86–97. https://doi.org/10.1007/978-3-642-18009-5_9

118. F. Chung, W. Zhao, A sharp PageRank algorithm with applications to edge ranking and graph sparsification, in *Proceedings of 7th International Workshop on Algorithms and Models for the Web-Graph*, ed. by R. Kumar, D. Sivakumar, volume 6516 of WAW 2010 (Springer, Stanford, CA, USA, 13–14 Dec 2010), pp. 2–14. https://doi.org/10.1007/978-3-642-18009-5_2

119. A. Ciaramella, S. Cocozza, F. Iorio, G. Miele, F. Napolitano, M. Pinelli, G. Raiconi, R. Tagliaferri, Interactive data analysis and clustering of genomic data. Neural Networks **21**(2–3), 368–378 (2008). https://doi.org/10.1016/j.neunet.2007.12.026

120. K.L. Clarkson, Nearest-neighbor searching and metric space dimensions, in *Nearest-Neighbor Methods in Learning and Vision. Theory and Practice*, ed. by G. Shakhnarovich, T. Darrell, P. Indyk (MIT Press, 2006), pp. 26–71

121. A. Clauset, Finding local community structure in networks. Phys. Rev. E Stat. Nonlin. Soft Matter Phys. **72**, 026132 (2005)

122. A. Clauset, C. Moore, M.E.J. Newman, Hierarchical structure and the prediction of missing links in networks. Nature, **453**(7191), 98–101 (2008). https://doi.org/10.1038/nature06830

123. A. Clauset, M.E.J. Newman, C. Moore, Finding community structure in very large networks. Phys. Rev. E **70**(6), 066111 (2004). https://doi.org/10.1103/PhysRevE.70.066111

124. A. Coates, A.Y. Ng, Learning feature representations with k-means, in *Neural Networks: Tricks of the Trade*, 2nd edn. (2012), pp. 561–580

125. R. Coifman, S. Lafon, Diffusion maps. Appl. Comput. Harmonic Anal. **21**, 5–30 (2006). https://doi.org/10.1016/j.acha.2006.04.006

126. P. Corsini, B. Lazzerini, F. Marcelloni, A new fuzzy relational clustering algorithm based on the fuzzy c-means algorithm. Soft Comput. **9**, 439–447 (2005). https://doi.org/10.1007/s00500-004-0359-6

127. A. Criminisi, J. Shotton, Decision forests for classi cation, regression, density estimation, manifold learning and semi-supervised learning. Technical report tr-2011-114. Technical report, Microsoft Research, 2011

128. J.A. Cuesta-Albertos, A. Gordaliza, C. Matran, Trimmed k-means: an attempt to robustify quantizers. Ann. Stat. **25**(2), 553–576 (1997)

129. J. Czekanowski, *Zarys metod statystycznych w zastosowaniu do antropologii (An outline of statistical methods applied in anthropology)*, volume 5 of Travaux de la Société des Sciences de Varsovie. III - Classe des sciences mathématiques et naturelles. Towarzystwo Naukowe Warszawskie, Warszawa, 1913. reprint available at http://rcin.org.pl/dlibra/doccontent?id=29411&from=FBC

130. A. Damle, V. Minde, L. Ying, Robust and efficient multi-way spectral clustering. arXiv:1609.08251v2 [math.NA], 15 Apr 2016

131. L. Danon, J. Duch, A. Díaz-Guilera, A. Arenas, Community structure identification, in *Large Scale Structure and Dynamics of Complex Networks*, ed. by G. Caldarelli, A. Vespignani. volume 2 of Complex Systems and Interdisciplinary Science, Chapter 5 (World Scientific Publishing Co. Pte. Ltd., 2007), pp. 93–115

132. S. Dasgupta, Y. Freund, Random projection trees and low dimensional manifolds, in *Proceedings of the 40-th Annual ACM Symposium on Theory of Computing*, STOC '08. (ACM, New York, NY, USA, 2008), pp. 537–546. https://doi.org/10.1145/1374376.1374452

133. S. Dasgupta, A. Gupta, An elementary proof of a theorem of Johnson and Lindenstrauss. Random Struct. Algorithms **22**(1), 60–65 (2003). https://doi.org/10.1002/rsa.10073

134. S. Dasgupta, L. Schulman, A probabilistic analysis of EM for mixtures of separated, spherical Gaussians. J. Mach. Learn. Res. **8**, 203–226 (2007)

135. R.M. Dave, Fuzzy shell-clustering and applications to circle detection in digital images. Int. J. General Syst. **16**(4), 343–355 (1990). https://doi.org/10.1080/03081079008935087

136. R.N. Dave, Characterization and detection of noise in clustering. Pattern Recogn. Lett. **12**(11), 657–664 (1991). https://doi.org/10.1016/0167-8655(91)90002-4

137. R.N. Dave, Generalized fuzzy c-shells clustering and detection of circular and elliptical boundaries. Pattern Recogn. **25**(7), 713–721 (1992). https://doi.org/10.1016/0031-3203(92)90134-5

138. R.N. Dave, S. Sen, Robust fuzzy clustering of relational data. IEEE Trans. Fuzzy Syst. **10**(6), 713–727 (2001). https://doi.org/10.1109/TFUZZ.2002.805899

139. E.B. Davies, G.M.L. Gladwell, J. Leydold, P.F. Stadler, Discrete nodal domain theorems. Linear Algebra Appl. **336**(1–3), 51–60 (2001). https://doi.org/10.1016/S0024-3795(01)00313-5

140. J.V. de Oliveira, W. Pedrycz (eds.), *Advances in Fuzzy Clustering and its Applications* (Wiley, New York, 2007)

141. G. De Soete, J.D. Carroll, K-means clustering in a low-dimensional euclidean space, in *New Approaches in Data Analysis*, ed. by E. Diday, et al. (Springer, Heidelberg, 1994), pp. 212–219

142. J.-C. Delvenne, A.-S. Libert, Centrality measures and thermodynamic formalism for complex networks. Phys. Rev. E **83**, 046117 (2011). https://doi.org/10.1103/PhysRevE.83.046117

143. A. Demiriz, K. Bennett, M.J. Embrechts, Semi-supervised clustering using genetic algorithms, in *Artificial Neural Networks in Engineering (ANNIE-99)* (ASME Press, 1999), pp. 809–814

144. A.P. Dempster, N.M. Laird, D.B. Rubin, Maximum likelihood from incomplete data via the EM algorithm. J. Roy. Stat. Soc., Series B **39**, 1–38 (1967)

145. K. Devarajan, Nonnegative matrix factorization: an analytical and interpretive tool in computational biology. PLoS Comput. Biol. **4**(7), e1000029 (2008). https://doi.org/10.1371/journal.pcbi.1000029
146. I.S. Dhillon, J. Fan, Y. Guan, Efficient clustering of very large document collections, in *Data Mining for Scientific and Engineering Applications*, ed. R. Grossman, G. Kamath, R. Naburu (Kluwer Academic Publishers, 2001)
147. I.S. Dhillon, Y. Guan, J. Kogan, Iterative clustering of high dimensional text data augmented by local search, in *Proceedings of IEEE International Conference on Data Mining* (Maebashi City, Japan, 2002), pp. 131–138
148. I.S. Dhillon, Y. Guan, J. Kogan, Refining clusters in high dimensional text data, in *2-nd SIAM ICDM, Workshop on Clustering High Dimensional Data* (Arlington, VA, 2002). http://citeseer.ist.psu.edu/dhillon02refining.html
149. I.S. Dhillon, Y. Guan, B. Kulis, Kernel k-means, spectral clustering and normalized cuts. In *KDD'04, August 22–25, 2004*, Seattle, Washington, USA, 2004
150. I.S. Dhillon, Y. Guan, B. Kulis. Kernel k-means: spectral clustering and normalized cuts, in *Proceedings of the 10-th ACM SIGKDD International Conference on Knowledge Discovery and Data Mining*, KDD'04 (New York, NY, USA, 2004), pp. 551–556. https://doi.org/10.1145/1014052.1014118
151. I.S. Dhillon, Y. Guan, B. Kulis, A unified view of kernel k-means, spectral clustering and graph cuts. Technical Report TR-04-25 (University of Texas, Department of Computer Science, 2005). http://www.cs.utexas.edu/ftp/pub/techreports/tr04-25.pdf
152. I.S. Dhillon, D.S. Modha, Concept decompositions for large sparse text data using clustering. Mach. Learn. **42**, 143–175 (2001). https://doi.org/10.1023/A:1007612920971
153. I.S. Dhillon, S. Sra, Generalized nonnegative matrix approximations with Bregman divergences, in *Proceedings of the Neural Information Processing Systems Conference*, NIPS 2005 (Vancouver, Canada, 2005), pp. 283–290
154. T.G. Dietterich, G. Bakiri, Solving multiclass learning problems via error-correcting output codes. J. Artif. Intell. Res. **2**(1), 263–286 (1995). https://doi.org/10.1613/jair.105
155. E. Dimitriadou, S. Dolničar, A. Weingessel, An examination of indexes for determining the number of clusters in binary data sets. Psychometrica **67**(1), 137–159 (2002). https://doi.org/10.1007/BF02294713
156. C. Ding, X. He, H. Zha, M. Gu, H. Simon, A min-max cut algorithm for graph partitioning and data clustering. In *Proceedings of IEEE International Conference on Data Mining*, ICDM 2001 (San Jose, CA , USA, 2001), pp. 107–114. https://doi.org/10.1109/ICDM.2001.989507
157. T.N. Dinh, M.T. Thai, Community detection in scale-free networks: approximation algorithms for maximizing modularity. IEEE J. Sel. Areas Commun. **31**(6), 997–1006 (2013). https://doi.org/10.1109/JSAC.2013.130602
158. C. Domeniconi, D. Gunopulos, S. Ma, B. Yan, M. Al-Razgan, D. Papadopoulos, Locally adaptive metrics for clustering high dimensional data. Data Min. Knowl. Disc. **14**(1), 63–97 (2007)
159. W.E. Donath, A.J. Hoffman, Lower bounds for the partitioning of graphs. IBM J. Res. Develop. **17**, 420–425 (1973)
160. S.N. Dorogovtsev, A.L. Ferreira, A.V. Goltsev, J.F.F. Mendes, Zero Pearson coefficient for strongly correlated growing trees. Phys. Rev. E **81**, 031135 (2010). https://doi.org/10.1103/PhysRevE.81.031135
161. P.G. Doyle, J.L. Snell, Random walks and electric networks. ArXiv Mathematics e-prints, math/0001057, 2000
162. L. Duana, L. Xub, F. Guoc, J. Leea, B. Yana, A local-density based spatial clustering algorithm with noise. Inf. Syst. **32**(7), 978–986 (2007). https://doi.org/10.1016/j.is.2006.10.006
163. J. Duch, A. Arenas, Community detection in complex networks using extremal optimization. Phys. Rev. E **72**(2), 027104 (2005). https://doi.org/10.1103/PhysRevE.72.027104
164. R.O. Duda, P.E. Hart, G. Stork, *Pattern Classification*, 2nd edn. (Wiley, New York, 2000)
165. D. Dueck, Clustering by affinity propagation. Ph.D. thesis, Department of Electrical & Computer Engineering, University of Toronto, 2009. http://www.cs.columbia.edu/~delbert/docs/DDueck-thesis_small.pdf

166. J.C. Dunn, A fuzzy relative of the Isodata process and its use in detecting compact well-separated clusters. J. Cyber. **3**(3), 32–57 (1974)
167. M. Dyer, A. Frieze, A simple heuristic for the p-center problem. Oper. Res. Lett. **3**(6), 285–288 (1985)
168. D. Easley, J. Kleinberg, *Networks, Crowds, and Markets: Reasoning About a Highly Connected World* (Cambridge University Press, Cambridge, 2010)
169. C. Elkan, Using the triangle inequality to accelerate k-means, in *Proceedings of the 20th International Conference on Machine Learning*, ICML 2003, ed. by T. Fawcett, N. Mishra, Washington, DC, 21-24 2003. AAAI Press pp. 147–153
170. A. Ester, H.-P. Kriegel, J. Sander, M. Wimmer, X. Xu, Incremental clustering for mining in a data warehousing environment, in *Proceedings of the 2nd International Conference on Knowledge Discovery and Data Mining*, ed. E. Simoudis, J. Han, U.M. Fayyad (Morgan Kaufmann, New York, 1998), pp. 323–333
171. A. Ester, H.-P. Kriegel, J. Sander, X. Xu, A density-based algorithm for discovering clusters in large databases with noise, in *Proceedings of the 24th VLDB Conference on Knowledge Discovery and Data Mining* (AAAI Press, Portland, OR, 1996), pp. 226–231
172. T.S. Evans, R. Lambiotte, Line graphs, link partitions, and overlapping communities. Phys. Rev. E **80**, 016105 (2009)
173. B.S. Everitt, *Cluster Analysis*. Halsted Press (1993)
174. M. Fiedler, Algebraic connectivity of graphs. Czechoslovak Math. J. **23**(98), 298–305 (1973)
175. M. Fiedler, A property of eigenvectors of nonnegative symmetric matrices and its application to graph theory. Czechoslovak Math. J. **25**, 619–672 (1975)
176. M. Filippone, F. Camastra, F. Masulli, S. Rovetta, A survey on spectral and kernel methods for clustering. Pattern Recogn. **41**(1), 176–190 (2008). https://doi.org/10.1016/j.patcog.2007.05.018
177. M. Filippone, F. Masulli, S. Rovetta, Applying the possibilistic c-means algorithm in kernel-induced spaces. IEEE Trans. Fuzzy Syst. **18**(3), 572–584 (2010). https://doi.org/10.1109/TFUZZ.2010.2043440
178. I. Fischer, J. Poland, Amplifying the block matrix structure for spectral clustering, in *Proceedings of the 14th Annual Machine Learning Conference of Belgium and the Netherlands*, ed. by M. van Otterlo, M. Poel, A. Nijholt (2005), pp. 21–28. http://www.dr-fischer.org/pub/blockamp/index.html
179. P. Fjällström. Algorithms for graph partitioning: a survey. Linköping Electron. Art. Comput. Inform. Sci. **3**(10) (1998)
180. G.W. Flake, K. Tsioutsiouliklis, L. Zhukov, Methods for mining Web communities: bibliometric, spectral, and flow, in *Web dynamics: adapting to change in content, size, topology and use*, ed. by M. Levene, A. Poulovassilis (Springer, Berlin, 2004), pp. 45–68
181. E. Forgy, Cluster analysis of multivariate data: efficiency versus interpretability of classification. Biometrics **21**, 768–780 (1965)
182. S. Fortunato, Community detection in graphs. Phys. Rep. **486**(3–5), 75–174 (2010)
183. S. Fortunato, M. Barthélemy, Resolution limit in community detection. Proc. Nat. Acad. Sci. USA **104**(1), 36–41 (2007). https://doi.org/10.1073/pnas.0605965104
184. F. Fouss, A. Pirotte, J.-M. Renders, M. Saerens, Random-walk computation of similarities between nodes of a graph with application to collaborative recommendation. IEEE Trans. Knowl. Data. Eng. **19**(3), 355–369 (2007). https://doi.org/10.1109/TKDE.2007.46
185. C. Fowlkes, S. Belongie, F. Chung, J. Malik, Spectral grouping using the Nyström method. IEEE Trans. Pattern Anal. Mach. Intell. **26**(2), 214–225 (2004). https://doi.org/10.1109/TPAMI.2004.1262185
186. C. Fraley, A. Raftery, MCLUST: software for model-based cluster and discriminant analysis. Technical Report 342 (Department of Statistics, University of Washington, 1999)
187. A.L.N. Fred, A.K. Jain, Data clustering using evidence accumulation. In *Proceedings of the 16th International Conference on Pattern Recognition*, volume 4 of ICPR 2002, pp. 276–280. https://doi.org/10.1109/ICPR.2002.1047450, 11–15 Aug. 2002

188. A.L.N. Fred, A.K. Jain, Robust data clustering, in *Proceedings of IEEE Computer Society Conference on Computer Vision and Pattern Recognition*, volume 2 of CVPR 2003, pp. 128–136, Madison, Wisconsin, USA, 16–22 June 2003

189. A.L.N. Fred, A.K. Jain, Combining multiple clusterings using evidence accumulation. IEEE Trans. Pattern Anal. Mach. Intell. **27**(6), 835–850 (2005). https://doi.org/10.1109/TPAMI.2005.113

190. K. Frederix, M. Van Barel, Sparse spectral clustering method based on the incomplete Cholesky decomposition. J. Comput. Appl. Math. **237**(1), 145–161 (2013). https://doi.org/10.1016/j.cam.2012.07.019

191. B.J. Frey, D. Dueck, Clustering by passing messages between data points. Science **315**, 972–976 (2007). https://doi.org/10.1126/science.1136800

192. B. Fritzke, Some competitive learning methods. Draft (Institute for Neural Computation, Ruhr-Universität Bochum, Germany, 1997). http://www.ki.inf.tu-dresden.de/~fritzke/JavaPaper/

193. Y. Fukuyama, M. Sugeno, A new method of choosing the number of clusters for the fuzzy *c*-means method, in *Proceedings of 5th Fuzzy System Symptoms* (1989), pp. 247–250 (in Japanese)

194. C. Fyfe, J. Corchado, A comparison of kernel methods for instantiating case based reasoning systems. Adv. Eng. Inform. **16**(3), 165–178 (2002)

195. S. Galbraith, J.A. Daniel, B. Vissel, A study of clustered data and approaches to its analysis. J. Neurosci. **30**(32), 10601–10608 (2010)

196. L. Galluccio, O. Michel, P. Comon, A. Hero, Graph based *k*-means clustering. Sig. Process. **92**(9), 1970–1984 (2012)

197. I. Gath, A.B. Geva, Unsupervised optimal fuzzy clustering. IEEE Trans. Pattern Anal. Mach. Intell. **11**(7), 773–781 (1989). https://doi.org/10.1109/34.192473

198. D. Gibson, J. Kleinberg, P. Raghavan, Inferring Web communities from link topology, in *Proceedings of the 9th ACM Conference on Hypertext and Hypermedia*, Pittsburgh, PA, June 20–24 (ACM, New York, 1998), pp. 225–234. https://doi.org/10.1145/276627.276652

199. E. Giné, V. Koltchinskii, Empirical graph Laplacian approximation of Laplace-Beltrami operators: large sample results, in *Proceedings of the 4-th International Conference on High Dimensional Probability*, ed. by E. Giné, V. Koltchinskii, W. Li, J. Zinn. Institute of Mathematical Statistics, Lecture Notes—Monograph Series, vol. 51. IMS (Beachwood, Ohio, USA, 2006), pp. 238–259. https://doi.org/10.1214/074921706000000888

200. A. Gionis, H. Mannila, P. Tsaparas, Clustering aggregation. ACM Trans. Knowl. Discov. Data **1**(1), (2007). https://doi.org/10.1145/1217299.1217303

201. M. Girvan, M.E.J. Newman, Community structure in social and biological networks. Proc. Natl. Acad. Sci. USA **99**(12), 7821–7826 (2002). https://doi.org/10.1073/pnas.122653799

202. D.F. Gleich, PageRank beyond the Web. SIAM Rev. **57**(3), 321–363 (2015). https://doi.org/10.1137/140976649

203. X. Golay, S. Kollias, G. Stoll, D. Meier, A. Valavanis, P. Boesiger, A new correlation-based fuzzy logic clustering algorithm for FMRI. Magn. Reson. Med. **40**(2), 249–260 (1998). https://doi.org/10.1002/mrm.1910400211

204. G.H. Golub, C.F. Van Loan, *Matrix Computations*, 3rd edn. (Johns Hopkins University Press, Baltimore, MD, 1996)

205. R.C. Gonzalez, R.E. Woods, *Digital Image Processing (NJ*, 3rd edn. (Prentice Hall, Upper Saddle River, 2008)

206. T. Gonzalez, Clustering to minimize the maximum intercluster distance. Theoret. Comput. Sci. **38**, 293–306 (1985). https://doi.org/10.1016/0304-3975(85)90224-5

207. P.K. Gopalan, D.M. Blei, Efficient discovery of overlapping communities in massive networks. Proc. Natl. Acad. Sci. **110**(36), 14534–14539 (2013). https://doi.org/10.1073/pnas.1221839110

208. J.C. Gower, Euclidean distance geometry. Math. Scientist **7**, 1–14 (1982)

209. J.C. Gower, P. Legendre, Metric and euclidean properties of dissimilarity coefficients. J. Classif. **3**(1), 5–48 (1986). https://doi.org/10.1007/BF01896809

210. P. Grabusts, A. Borisov, Using grid-clustering methods in data classification, in *2002 International Conference on Parallel Computing in Electrical Engineering (PARELEC 2002)*, 22–25 Sept. 2002, Warsaw, Poland (2002), p. 425

211. L. Grady, Random walks for image segmentation. IEEE Trans. Pattern Anal. Mach. Intell. **28**(11), 1768–1783 (2006). https://doi.org/10.1109/TPAMI.2006.233

212. L. Grady, J.R. Polimeni, *Discrete Calculus: Applied Analysis on Graphs for Computational Science* (Springer, 2010). https://doi.org/10.1007/978-1-84996-290-2

213. P.L. Graham, P. Hell, On the history of the minimum spanning tree problem. IEEE Ann. Hist. Comput. **7**(1), 43–57 (1985). https://doi.org/10.1109/MAHC.1985.10011

214. D. Graves, W. Pedrycz, Kernel-based fuzzy clustering and fuzzy clustering: a comparative experimental study. Fuzzy Sets Syst. **161**(4), 522–543 (2010). https://doi.org/10.1016/j.fss.2009.10.021

215. R. Guimerà, L.A.N. Amaral, Functional cartography of complex metabolic networks. Nature **433**, 895–900 (2005). https://doi.org/10.1038/nature03288

216. N. Gulbahce, S. Lehmann, The art of community detection. BioEssays **30**(10), 934–938 (2008)

217. G. Gupta, J. Ghosh, Bregman bubble clustering: A robust framework for mining dense clusters. ACM Trans. Knowl. Discov. Data **2**(2), 8:1–8:49 (2008). https://doi.org/10.1007/978-3-642-23166-7_7

218. E.E. Gustafson, W.C. Kessel, *Fuzzy clustering with a fuzzy covariance matrix, in IEEE CDC* (CA, San Diego, 1979), pp. 761–766. https://doi.org/10.1109/CDC.1978.268028

219. A. Guttman, R-trees: a dynamic index structure for spatial searching, in *Proceedings of Third ACM SIGMOD International Conference on Management of Data*, SIGMOD '84 (1984), pp. 47–57

220. I. Guyon, A. Elisseeff, An introduction to variable and feature selection. J. Mach. Learn. Res. **3**, 1157–1182 (2003)

221. I. Guyon, S. Gunn, M. Nikravesh, L.A. Zadeh (eds.), *Feature Extraction. Foundations and Applications*. Studies in Fuzziness and Soft Computing, vol. 207 (Springer, 2006)

222. L. Hagen, A. Kahng, A new approach to effective circuit clustering, in *Proceedings of the IEEE/ACM International Conference on Computer-Aided Design*, ICCAD'92 (IEEE Computer Society Press, 1992), pp. 422–427

223. L. Hagen, A. Kahng, New spectral methods for ratio cut partitioning and clustering. IEEE Trans. Comput. Aided Des. **11**(9), 1074–1085 (1992). https://doi.org/10.1109/43.159993

224. M. Halkidi, Y. Batistakis, M. Vazirgiannis, On clustering validation techniques. Intell. Inf. Syst. **17**(2–3), 107–145 (2001). https://doi.org/10.1023/A:1012801612483

225. M. Halkidi, M. Vazirgiannis, Clustering validity assignment: finding the optimal partitioning of a data set, in *Proceedings of ICDM Conference* (2001)

226. N. Halko, P.G. Martinsson, J.A. Tropp, Finding structure with randomness: probabilistic algorithms for constructing approximate matrix decompositions. SIAM Rev. **53**(2), 217–288 (2011). https://doi.org/10.1137/090771806

227. K.M. Hall, An r-dimensional quadratic placement algorithm. Manage. Sci. **17**(3), 219–229 (1970). https://doi.org/10.1287/mnsc.17.3.219

228. L.O. Hall, I.B. Ozyurt, J.C. Bezdek, Clustering with a genetically optimized approach. IEEE Trans. Evol. Comput. **3**(2), 103–112 (1999). https://doi.org/10.1109/4235.771164

229. G. Hamerly, Learnig structure and concepts in data using data clustering. Ph.D. thesis, University of California, San Diego, 2003

230. G. Hamerly, C. Elkan, Alternatives to the k-means algorithm that find better clusterings, in *Proceedings of the ACM Conference on Information and Knowledge Management*, CIKM-2002 (2002), pp. 600–607

231. J. Handl, J. Knowles, D.B. Kell, Computational cluster validation in post-genomic data analysis. Bioinformatics **21**(15), 3201–3212 (2005). https://doi.org/10.1093/bioinformatics/bti517

232. W.-C. Hang, On using principal components before separating a mixture of two multivariate normal distributions. J. Roy. Stat. Soc. Ser. C (Appl. Stat.) **32**(3), 267–275 (1983)

233. P. Hansen, B. Jaumard, Cluster analysis and mathematical programming. Math. Program. **79**(1–3), 191–215 (1997). https://doi.org/10.1007/BF02614317

234. P. Hansen, N. Mladenović, j-Means: a new local search heuristic for minimum sum-of-squares clustering. Pattern Recogn. **34**, 405–413 (2002). https://doi.org/10.1016/s0031-3203(99)00216-2

235. S. Harenberg, G. Bello, L. Gjeltema, S. Ranshous, J. Harlalka, R. Seay, K. Padmanabhan, N. Samatova, Community detection in large-scale networks: a survey and empirical evaluation. WIREs Comput. Stat. **6**(6), 426–439 (2014). https://doi.org/10.1002/WICS.1319

236. J.A. Hartigan, Consistency of single linkage for high-density clusters. J. Am. Stat. Assoc. **76**(374), 388–394 (1981)

237. T. Hastie, R. Tibshirani, J. Friedman, *The Elements of Statistical Learning. Data Mining, Inference, and Prediction*. Springer Series in Statistics, 2nd edn. (Springer, Berlin, Heidelberg, New York, 2009)

238. T. Hastie, R. Tibshirani, M. Wainwright, *Statistical Learning with Sparsity: The Lasso and Generalizations* (Chapman and Hall/CRC 2015, 2015)

239. R.J. Hathaway, J.C. Bezdek, Recent convergence results for the fuzzy c-means clustering algorithm. J. Classif. **5**, 237–247 (1988). https://doi.org/10.1007/BF01897166

240. R.J. Hathaway, J.C. Bezdek, Optimization of clustering criteria by reformulation. IEEE Trans. Fuzzy Syst. **3**, 241–245 (1995). https://doi.org/10.1109/91.388178

241. R.J. Hathaway, J.C. Bezdek, Generalized fuzzy c-means clustering strategies using L_p norm distances. IEEE Trans. Fuzzy Syst. **8**(5), 576–582 (2000). https://doi.org/10.1109/91.873580

242. R.J. Hathaway, J.C. Bezdek, J.W. Davenport, On relational data versions of c-means algorithms. Pattern Recogn. Lett. **17**, 607–612 (1996). https://doi.org/10.1016/0167-8665(96)00025-6

243. R.J. Hathaway, J.C. Bezdek, W. Tucker, An improved convergence theorem for the fuzzy c-means clustering algorithm, in *The Analysis of Fuzzy Information*, vol. 3, ed. by J.C. Bezdek (CRC Press, Boca Raton, 1986), pp. 1–10

244. R.J. Hathaway, J.W. Davenport, J.C. Bezdek, Relational duals of the c-means clustering algorithms. Pattern Recon. **22**, 205–212 (1989). https://doi.org/10.1016/0031-3203(89)90066-6

245. D. Haussler, Convolution kernels on discrete structure. Technical Report UCSC-CRL-99-10 (Deptment of Computer Science, University of California at Santa Cruz, Santa Cruz, CA, USA, 1999)

246. T.C. Havens, R. Chitta, A.K. Jain, R. Jin, Speedup of fuzzy and possibilistic kernel c-means for large-scale clustering, in *IEEE International Conference on Fuzzy Systems*, Grand Hyatt Taipei, Taipei, Taiwan, 27–30 June 2011. (IEEE, 2011), pp. 463–470. https://doi.org/10.1109/FUZZY.2011.6007618

247. B. He, L. Gu, X.-D. Zhang, Nodal domain partition and the number of communities in networks. J. Stat. Mech. Theory Exp. **2012**(02), P02012 (2012)

248. M. Hein, J.-Y. Audibert, U. von Luxburg, From graphs to manifolds—weak and strong pointwise consistency of graph Laplacians, in *Proceedings of the 18th Conference on Learning Theory (COLT)*, ed. by P. Auer, R. Meir, LNCS 3559 (Springer, 2005), pp. 470–485. https://doi.org/10.1007/11503415_32

249. M. Hein, T. Bühler, An inverse power method for nonlinear eigenproblems with applications in 1-spectral clustering and sparse PCA, in *Advances in Neural Information Processing Systems 23*, NPIS 2010 (2010), pp. 847–855. http://www.ml.uni-saarland.de/code/oneSpectralClustering/oneSpectralClustering.htm

250. C. Hennig, What are the true clusters. Pattern Recogn. Lett. **64**, 53–62 (2015). https://doi.org/10.1016/j.patrec.2015.04.009

251. D.J. Higham, G. Kalna, M. Kibble, Spectral clustering and its use in bioinformatics. J. Comput. Appl. Math. **204**(1), 25–37 (2007). https://doi.org/10.1016/j.cam.2006.04.026

252. A. Hinneburg, E. Hinneburg, D.A. Keim, An efficient approach to clustering in large multimedia databases with noise, in *Proceedings of the 4th International Conference on Knowledge Discovery and Datamining*, KDD 1998 (AAAI Press, New York, NY, 1998), pp. 58–65

253. D.S. Hochbaum, D.B. Shmoys, A best possible heuristic for the k-center problem. Math. Oper. Res. **10**(2), 180–184 (1985). https://doi.org/10.1287/moor.10.2.180

254. J.M. Hofman, C.H. Wiggins, Bayesian approach to network modularity. Phys. Rev. Lett. **100**(25), 258701 (2008). https://doi.org/10.1403/PhysRevLett.100.258701

255. F. Höppner, F. Klawonn, R. Kruse, T. Runkler, *Fuzzy Cluster Anal* (Wiley, Chichester, England, 1999)

256. P. Hore, L.O. Hall, D.B. Goldgof, Single pass fuzzy *c* means, in *IEEE International Conference on Fuzzy Systems, FUZZ-IEEE*, Imperial College, London, UK, 2007. IEEE. http://www.csee.usf.edu/~hall/papers/singlepass.pdf

257. P. Hore, L.O. Hall, D.B. Goldgof, A scalable framework for cluster ensembles. Pattern Recogn. **42**(5), 676–688 (2009). https://doi.org/10.1016/j.patcog.2008.09.027

258. J. Hu, B.K. Ray, M. Singh, Statistical methods for automated generation of service engagement staffing plans. IBM J. Res. Dev. **51**(3), 281–293 (2007). https://doi.org/10.1147/rd.513.0281

259. A. Huang, Similarity measures for text document clustering. In *Proceedings of New Zealand Computer Science Research Student Conference*, ed. by J. Holland, A. Nicholas, D. Brignoli, NZCSRSC 2008 (Christchurch, New Zealand, 14–18 April 2008), pp. 49–56. http://nzcsrsc08.canterbury.ac.nz/site/digital-proceedings

260. J. Huang, F. Nie, H. Huang, Spectral rotation versus *k*-means in spectral clustering, in *Proceedings of 27th (AAAI) Conference on Artificial Intelligence*, ed. by M. desJardins, M.L. Littman, 14–18 July 2013, Bellevue, Washington, USA (AAAI Press, 2013). http://www.aaai.org/ocs/index.php/AAAI/AAAI13/paper/view/6462

261. J. Huang, T. Zhu, D. Schuurmans, Web communities identification from random walks, in *Knowledge Discovery in Databases: PKDD 2006*, ed. by J. Fürnkranz, T. Scheffer, M. Spiliopoulou. LNCS, vol. 4213 (Springer, Berlin, Heidelberg, 2006), pp. 187–198. https://doi.org/10.1007/11871637_21

262. J.Z. Huang, M.K. Ng, H. Rong, Z. Li, Automated variable weighting in *k*-means type clustering. IEEE Trans. Pattern Anal. Mach. Intell. **27**(5), 657–668 (2005). https://doi.org/10.1109/TPAMI.2005.95

263. Z. Huang, Extensions to the *k*-means algorithm for clustering large data sets with categorical values. Data Min. Knowl. Disc. **2**(3), 283–304 (1998). https://doi.org/10.1023/A:1009769707641

264. Z. Huang, A. Zhou, G. Zhang, Non-negative matrix factorization: a short survey on methods and applications, in *Computational Intelligence and Intelligent Systems*, ed. by Z. Li, X. Li, Y. Liu, Z. Cai. CCIS, vol. 316 (Springer, Berlin, Heidelberg, 2012), pp. 331–340

265. L. Hubert, Ph. Arabie, Comparing partitions. J. Classif. **2**(1), 193–218 (1985). https://doi.org/10.1007/BF01908075

266. B. Hunter, Th. Strohmer, Performance analysis of spectral clustering on compressed, incomplete and inaccurate measurements. arXiv:1011.0997 [math.NA] (2010)

267. T. Huntsberger, P. Ajjimarangsee, Parallel self-organizing feature maps for unsupervised pattern recognition. Int. J. Gen. Syst. **16**(4), 357–372 (1989)

268. H.M. Hussain, K. Benkrid, A. Ebrahim, A.T. Erdogan, H. Seker, Novel dynamic partial reconfiguration implementation of *k*-means clustering on FPGAs: comparative results with GPPs and GPUs. Int. J. Reconfig. Comp. **135926** (2012). https://doi.org/10.1155/2012/135926

269. M. Inaba, N. Katoh, H. Imai, Applications of weighted Voronoi diagrams and randomization to variance-based *k*-clustering: (extended abstract), in *Proceedings of the Tenth Annual Symposium on Computational Geometry*, SCG '94 (ACM, New York, NY, USA, 1994), pp. 332–339

270. M. Iosifescu, Finite Markov Process and Their Applications (Dover Books on Mathematics, 2007)

271. M.A. Ismail, S.Z. Selim, Fuzzy *c*-means optimality of solutions and effective termination of the algorithm. Pattern Recognit. **19**(6), 481–485 (1986)

272. A. Jain, Data clustering: 50 years beyond *k*-means. Pattern Recognit. Lett. **31**, 651–666 (2010). https://doi.org/10.1016/j.patrec.2009.09.011

273. A. Jain, R. Dubes, *Algorithms for Clustering Data* (Prentice Hall, Upper Saddle River, NJ, 1988)

274. A. Jain, M. Murty, P. Flynn, Data clustering: a review. ACM Comput. Surv. **31**, 264–323 (1999). https://doi.org/10.1145/331499.331504

275. H. Jia, S. Ding, X. Xu, R. Nie, The latest research progress on spectral clustering. Neural Comput. Applic. **24**(7–8), 1477–1486 (2014). https://doi.org/10.1007/s00521-013-1439-2

276. J. Menche, A. Valleriani, R. Lipowsky, Asymptotic properties of degree-correlated scale-free networks. Phys. Rev. E **81**(4), 046103 (2010). https://doi.org/10.1103/PhysRevE.81.046103

277. I.T. Jolliffe, *Principal Component Analysis*, 2nd edn. Springer Series in Statistics (Springer, NY, 2002). https://doi.org/10.1007/b98835

278. R. Kannan, H. Salmasian, S. Vempala, The spectral method for general mixture models. SIAM J. Comput. **38**(3), 1141–1156 (2008). https://doi.org/10.1137/S0097539704445925

279. R. Kannan, S. Vempala, A. Vetta, On clusterings: good, bad and spectral. J. ACM **51**(3), 497–515 (2004). https://doi.org/10.1145/990308.990313

280. T. Kanungo, D.M. Mount, N.S. Netanyahu, Ch.D. Piatko, A.Y. Wu, R. Silverman, A local search approximation algorithm for k-means clustering. Comput. Geom. **28**(2–3), 89–112 (2004). https://doi.org/10.1016/j.comgeo.2004.03.003

281. T. Kanungo, N.S. Netanyahu, ChD Piatko, R. Silverman, A.Y. Wu, An efficient k-means clustering algorithm: analysis and implementation. IEEE Trans. Pattern Anal. Mach. Intell. **24**(7), 881–892 (2002). https://doi.org/10.1109/TPAMI.2002.1017616

282. B. Karrer, E. Levina, and M.E.J. Newman. Robustness of community structure in networks. Phys. Rev. E, **77**(4), 046119, 2008. https://doi.org/10.1103/PhysRevE.77.046119

283. I. Katsavounidis, C.-C.J. Kuo, Z. Zhang, A new initialization technique for generalized Lloyd iteration. IEEE Signal Process. Lett. **1**(10), 144–146 (1994). https://doi.org/10.1109/97.329844

284. L. Kaufman, P. Rousseeuw, *Finding Groups in Data: An Introduction to Cluster Analysis* (Wiley, New York, 1990)

285. M. Kearns, Y. Mansour, A.Y. Ng, An information-theoretic analysis of hard and soft assignment methods for clusterig, in *Proceedings 13-th Conference Uncertainty in Artificial Intelligence* (Morgan Kaufmann, Madison, WI, 1997)

286. J.A. Kelner, J.R. Lee, G.R. Price, S.-H. Teng, Metric uniformization and spectral bounds for graphs. Geom. Funct. Anal. **21**(5), 1117–1143 (2011). https://doi.org/10.1007/s00039-011-0132-9

287. M. Kendall, J.D. Gibbons, *Rank Correlation Methods* (Edward Arnold, London, 1990)

288. D.W. Kim, K.Y. Lee, D. Lee, K.H. Lee, Evaluation of the performance of clustering algorithms in kernel-induced feature space. Pattern Recogn. **38**(4), 607–611 (2005). https://doi.org/10.1016/j.patcog.2004.09.006

289. J. Kim, J.G. Lee, Community detection in multi-layer graphs: a survey. SIGMOD Rec. **44**(3), 37–48 (2015). https://doi.org/10.1145/2854006.2854013

290. T. Kim, J.C. Bezdek, R.J. Hathaway, Optimality tests for fixed points of the fuzzy c-means algorithm. Pattern Recogn. **21**, 651–663 (1988). https://doi.org/10.1016/0031-3203(88)90037-4

291. Y. Kim, S.-W. Son, H. Jeong, Finding communities in directed networks. Phys. Rev. E **81**(1), 016103–11 (2010). https://doi.org/10.1103/PhysRevE.81.016103

292. D. Klein, S.D. Kamvar, C.D. Manning, From instance-level constraints to space-level constraints: making the most of prior knowledge in data clustering, in *Proceedings of the Nineteenth International Conference on Machine Learning*, ICML '02, San Francisco, CA, USA (Morgan Kaufmann Publishers Inc., 2002), pp. 307–314

293. J. Kleinberg, An impossibility theorem for clustering, in *Proceedings of NIPS 2002* (2002), pp. 446–453. http://books.nips.cc/papers/files/nips15/LT17.pdf

294. M.A. Kłopotek, On the phenomenon of flattening 'flexible prediction' concept hierarchy, in *Fundamentals of Artificial Intelligence Research. International Workshop*, ed. by P. Jorrand, J. Kelemen. Lecture Notes, in Artificial Intelligence, vol. 535, Smolenice, Czech-Slovakia, 8–13 Sept. 1991 (Springer, Berlin Heidelberg, New York, 1991), pp. 99–111

295. P.A. Knight, D. Ruiz, A fast algorithm for matrix balancing, in *Web Information Retrieval and Linear Algebra Algorithms*, ed. by A. Frommer, M.W. Mahoney, D.B. Szyld, number 07071

in Dagstuhl Seminar Proceedings. Internationales Begegnungs- und Forschungszentrum für Informatik (IBFI), Schloss Dagstuhl, Germany, 2007. http://drops.dagstuhl.de/opus/volltexte/2007/1073

296. P.A. Knight, The Sinkhorn-Knopp algorithm. Convergence and applications. SIMAX **30**(1), 261–275 (2008). https://doi.org/10.1137/060659624

297. J. Kogan, *Introduction to Clustering Large and High-Dimensional Data* (Cambridge University Press, Cambridge, 2007)

298. J.F. Kolen, T. Hutcheson, Reducing the time complexity of the fuzzy C-means algorithm. IEEE Trans. Fuzzy Syst. **10**(2), 263–267 (2002). https://doi.org/10.1109/91.995126

299. Y. Kondo, M. Salibian-Barrera, R. Zamar, A robust and sparse k-means clustering algorithm (2012). http://arxiv.org/pdf/1201.6082.pdf

300. D. Kong, C. Ding, Maximum consistency preferential random walks, in *Proceedings of the 2012 European Conference on Machine Learning and Knowledge Discovery in Databases— Volume Part II, ECML PKDD'12*, Berlin, Heidelberg (Springer, 2012), pp. 339–354

301. Y. Koren, R. Bell, Ch. Volinsky, Matrix factorization techniques for recommender systems. Computer **42**(8), 30–37 (2009). https://doi.org/10.1109/MC.2009.263

302. Y. Koren, L. Carmel, D. Harel, ACE: a fast multiscale eigenvectors computation for drawing huge graphs, in *Proceedings of IEEE Symposium on Information Visualization, INFOVIS 2002*, Boston, MA, USA (IEEE Computer Society, Washington, DC, USA, 2002), pp. 137–144. https://doi.org/10.1109/INFVIS.2002.1173159

303. J. Koronacki, J. Ćwik, *Statystyczne systemy uczące się*, 2nd edn. (EXIT, Warszawa, 2008). ISBN 978-83-60434-56-7

304. H.-P. Kriegel, P. Kröger, A. Zimek, Clustering high-dimensional data: a survey on subspace clustering, pattern-based clustering, and correlation clustering. ACM Trans. Knowl. Discov. Data (TKDD) **3**(1), 1 (2009). https://doi.org/10.1145/1497577.1497578

305. R. Krishnapuram, A. Joshi, O. Nasraoui, L. Yi, Low-complexity fuzzy relational clustering algorithms for Web mining. IEEE Trans. Fuzzy Syst. **9**(4), 595–608 (2001). https://doi.org/10.1109/91.940971

306. R. Krishnapuram, J.M. Keller, A possibilistic approach to clustering. IEEE Trans. Fuzzy Syst. **2**(2), 98–110 (1993). https://doi.org/10.1109/91.227387

307. R. Kruse, Ch. Döring, M.-J. Lesot, Fundamentals of fuzzy clustering, in *Advances in Fuzzy Clustering and its Applications, Chapter 1*, ed. by J.V. de Oliveira, W. Pedrycz (Wiley, 2007), pp. 3–30. https://doi.org/10.1002/9780470061190.ch1

308. B. Kulis, A.C. Surendran, J.C. Platt, Fast low-rank semidefinite programming for embedding and clustering, in *Proceedings of 11th International Conference on Artificial Intelligence and Statistics, AISTAT 2007*, ed. by M. Meila, X. Shen, S. Juan, P. Rico, 21–24 Mar 2007, pp. 235–242. http://jmlr.csail.mit.edu/proceedings/papers/v2/kulis07a/kulis07a.pdf

309. A. Kumar, Y. Sabharwal, S. Sen, A simple linear time $(1 + \varepsilon)$-approximation algorithm for k-means clustering in any dimensions, in *Proceedings of the 45th Annual IEEE Symposium on Foundations of Computer Science, FOCS'04* (IEEE Computer Society, 2004), pp. 454–462

310. M. Kumar, J.B. Orlin, Scale-invariant clustering with minimum volume ellipsoids. Comput. Oper. Res. **35**(4), 1017–1029 (2006). https://doi.org/10.1016/j.cor.2006.07.001

311. R. Kumar, Recommendation systems: a probabilistic analysis. J. Comput. Syst. Sci. **63**(1), 42–61 (2001). https://doi.org/10.1006/jcss.2001.1757

312. S. Kumar, M. Mohri, A. Talwalkar, Sampling methods for the Nyström method. J. Mach. Learn. Res. **13**(1), 981–1006 (2012)

313. L.I. Kuncheva, D.P. Vetrov, Evaluation of stability of k-means cluster ensembles with respect to random initialization. IEEE Trans. Pattern Anal. Mach. Intell. **28**(11), 1798–1808 (2006). https://doi.org/10.1109/TPAMI.2006.226

314. A. Lancichinetti, S. Fortunato, J. Kertész, Detecting the overlapping and hierarchical community structure of complex networks. New J. Phys. **11**, 033015 (2009). https://doi.org/10.1088/1367-2630/11/3/033015

315. K. Lang, Fixing two weaknesses of the spectral method, in *Advances in Neural Information Processing Systems 18*, ed. by Y. Weiss, B. Schölkopf, J. Platt (MIT Press, Cambridge, MA, 2006), pp. 715–722

316. R. Langone, M. van Barel, J.A.K. Suykens, Efficient evolutionary spectral clustering. Pattern Recogn. Lett. **84**, 78–84 (2016). https://doi.org/10.1016/j.patrec.2016.08.012
317. A.N. Langville, C.D. Meyer, *Google's PageRank and Beyond: The Science of Search Engine Rankings* (Princeton University Press, Princeton, NJ, USA, 2006)
318. L. De Lathauwer, B. De Moor, J. Vandewalle, A multilinear singular value decomposition. SIAM J. Matrix Anal. Appl. **21**(4), 1253–1278 (2000). https://doi.org/10.1137/S08954479896305696
319. D. Lee, H.S. Seung, Learning the parts of objects by nonnegative matrix factorization. *Nature*, **401**(6755), 788–791 (1999). https://doi.org/10.1038/44565
320. S. Lehmann, L.K. Hansen, Deterministic modularity optimization. Eur. Phys. J. B **60**(1), 83–88 (2007). https://doi.org/10.1140/epjb/e2007-00313-2
321. E.A. Leicht, M.E.J. Newman, Community structure in directed networks. Phys. Rev. Lett. **10**(11), 118703–4 (2008). https://doi.org/10.1103/PhysRevLett.100.118703
322. J. Leskovec, K.J. Lang, A. Dasgupta, M.W. Mahoney, Community structure in large networks: natural cluster sizes and the absence of large well-defined clusters. arXiv:0810.1355v1 [cs.DS], 8 Mar 2008
323. D.A. Levin, Y. Peres, E.L. Wilmer, *Markov Chains and Mixing Times* (AMS, 2009)
324. M. Li, W. Bi, J.T. Kwok, B.-L. Lu, Large-scale Nyström kernel matrix approximation using randomized SVD. IEEE Trans. Neural Netw. Learn. Syst. **26**(1), 152–164 (2015). https://doi.org/10.1109/TNNLS.2014.2359798
325. R.-H. Li, J.X. Yu, J. Liu. Link prediction: the power of maximal entropy random walk, in *Proceedings of the 20th ACM International Conference on Information and Knowledge Management, CIKM '11* (2011), pp. 1147–1156. https://doi.org/10.1145/2063576.2063741
326. T. Li, Clustering based on matrix approximation: a unifying view. Knowl. Inf. Syst. **17**(1), 1–15 (2008). https://doi.org/10.1007/s10115-007-0116-0
327. T. Li, Ch. Ding. Non-negative matrix factorizations for clustering: a survey, in *Data Clustering: Algorithms and Applications*, ed. by C.C. Aggarwal, C.K. Reddy. Data Mining and Knowledge Discovery Series, Chapter 7. (Chapman & Hall/CRC, 2013), pp. 149–176
328. T.W. Liao, Clustering of time series data—a survey. Pattern Recogn. **38**(11), 1857–1874 (2005). https://doi.org/10.1016/j.patcog.2005.01.025
329. Y. Lifshits, The homepage of nearest neighbors and similarity search, 2004–2007. http://simsearch.yury.name/tutorial.html
330. Y. Linde, A. Buzo, R. Gray, An algorithm for vector quantizer design. IEEE Trans. Commun. **28**(1), 84–95 (1980). https://doi.org/10.1109/TCOM.1980.1094577
331. H. Liu, H. Motoda (eds.), *Computational Methods of Feature Selection* (CRC Data Mining and Knowledge Discovery Series. Chapman and Hall/CRC, Bocca Raton, FL, 2007)
332. S. Liu, A. Matzavinos, S. Sethuraman, Random walk distances in data clustering and applications. Adv. Data Anal. Classif. **7**, 83–108 (2013). https://doi.org/10.1007/s11634-013-0125-7
333. W. Liu, J. He, S.-F. Chang, Large graph construction for scalable semi-supervised learning, in *Proceedings of 27th International Conference on Machine Learning*, ed. by J. Fürnkranz, Th. Joachims, ICML (Omnipress, 2010), pp. 679–686
334. S.P. Lloyd, Least squares quantization in PCM. IEEE Trans. Inf. Theory **28**(2), 129–137 (1982). https://doi.org/10.1109/TIT.1982.1056489
335. H. Lodhi, C. Saunders, J. Shwe-Taylor, N. Cristianini, C. Watkins, Text classification using string kernels. J. Mach. Learn. Res. **2**, 419–444 (2002). https://doi.org/10.1162/153244302760200687
336. L. Lovász, Random walks on graphs: a survey, in *Combinatorics, Paul Erdös is Eighty*, vol. 2, ed. by D. Miklós, V.T. Sós, T. Szönyi (János Bolyai Mathematical Society, Keszthely (Hungary), 1993), pp. 1–46
337. D. Luo, H. Huang, Ch. Ding, F. Nie, On the eigenvectors of p-Laplacian. Mach. Learn. **81**(1), 37–51 (2010). https://doi.org/10.1007/s10994-010-5201-z
338. J. MacQueen, Some methods for classification and analysis of multivariate observations, in *Proceedings of Fifth Berkeley Symposium on Mathematical Statistics and Probability*, vol. 1 (University of California Press, 1967), pp. 281–297

339. K. Macropol, T. Can, A.K. Singh, RRW: repeated random walks on genome-scale protein networks for local cluster discovery. BMC Bioinf. **10**(1), 283–292 (2009). https://doi.org/10. 1186/1471-2105-10-283

340. K. Macropol, A.K. Singh, Scalable discovery of best clusters on large graphs, in *Proceedings of VLDB Endowment 3(2010)*, 693–702 (2010)

341. R. De Maesschalck, D. Jouan-Rimbaud, D.L. Massart, The Mahalanobis distance. Chemometr. Intell. Lab. Syst. **50**, 1–18 (2000). https://doi.org/10.1016/S0169-7439(99)00047-7

342. M.W. Mahoney, Randomized algorithms for matrices and data. Found. Trends Mach. Learn. **3**(2), 123–224 (2011). https://doi.org/10.1561/2200000035

343. M.W. Mahoney, Lecture notes on spectral graph methods. arXiv:1608.04845v1 [cs.DS], 17 Aug 2016

344. M. Maier, U. von Luxburg, M. Hein, How the result of graph clustering methods depends on the construction of the graph. ESAIM: Probab. Stat. **17**, 370–418 (2013). https://doi.org/10. 1051/ps/2012001

345. R. Maitra, A.D. Peterson, A.P. Ghosh, A systematic evaluation of different methods for initializing the K-means clustering algorithm. Preprint (2010). http://www.public.iastate.edu/ ~apghosh/files/IEEEclust2.pdf

346. F.D. Malliaros, M. Vazirgiannis, Clustering and community detection in directed networks: A survey. Phys. Rep. **533**(4), 95–142 (2013). https://doi.org/10.1016/j.physrep.2013.08.002

347. Ch.D. Manning, P. Raghavan, H. Schütze, *Introduction to Information Retrieval* (Cambridge University Press, 1 Apr 2009)

348. G.J. McLachlan, T. Krishnan, *The EM Algorithm and Extensions* (Wiley, 1997)

349. M. Meilă, Comparing clusterings–an information based distance. J. Multivar. Anal. **98**(5), 873–895 (2007). https://doi.org/10.1016/j.jmva.2006.11.013

350. M. Meilă, D. Heckerman, An experimental comparison of model-base clustering methods. Mach. Learn. **42**(1/2), 9–29 (2001). https://doi.org/10.1023/A:1007648401407

351. M. Meilă, W. Pentney, Clustering by weighted cuts in directed graphs, in *Proceedings of the 7th SIAM International Conference on Data Mining*, Radisson University Hotel, Minneapolis, MN, 26–28 Apr 2007

352. M. Meilă, J. Shi, A random walks view on spectral segmentation, in *8th International Workshop on AI and Statistics, AISTATS 2001*. JMLR.org, Hyatt Hotel, Key West, Florida, USA, 4–7 Jan 2001. http://www.gatsby.ucl.ac.uk/aistats/aistats2001/files/meila177.ps

353. M.E.S. Mendes, L. Sacks, Dynamic knowledge representation for e-learning applications, in *Enhancing the Power of the Internet, vol. 139, Studies in Fuzziness and Soft Computing*, ed. by M. Nikravesh, B. Azvine, R. Yager, L.A. Zadeh (Springer, Berlin, Heidelberg, Berlin Heidelberg New York, 2004), pp. 259–282

354. C.D. Meyer, *Matrix Analysis and Applied Linear Algebra* (SIAM, Philadelphia, 2000)

355. T. Michoel, B. Nachtergaele, Alignment and integration of complex networks by hypergraph-based spectral clustering. Phys. Rev. E-Stat. Nonlin. Soft Matter Phys. **86**(5), 056111 (2012)

356. G. Milligan, M. Cooper, An examination of procedures for determining the number of clusters in data sets. Psychometrica **50**(2), 159–179 (1985). https://doi.org/10.1007/BF02294245

357. R. Milo, S. Shen-Orr, S. Itzkovitz, N. Kashtan, D. Chklovskii, U. Alon, Network motifs: simple building blocks of complex networks. Science **298**(5594), 824–827 (2002). https:// doi.org/10.1126/science.298.5594.824

358. B. Mohar, The Laplacian spectrum of graphs, in *Proceedings of the Sixth Quadrennial International Conference on the Theory and Applications of Graphs*, vol. 2 (Wiley, New York, Kalamazoo, MI, 1991), pp. 871–898

359. B. Mohar, Some applications of Laplace eigenvalues of graphs, in *Graph Symmetry: Algebraic Methods and Applications*, volume 497 of *NATO ASI Ser. C*, ed. by G. Hahn, G. Sabidussi (Kluwer, 1997), pp. 225–275

360. A. Moore, Very fast EM-based mixture model clustering using multiresolution kd-trees, in *Proceedings of the 1998 Conference on Advances in Neural Information Processing Systems II*, ed. by M. Kearns, D. Cohn (Morgan Kaufman, 1998), pp. 543–549

361. A. Moore, The anchors hierarchy: using the triangle inequality to survive high-dimensional data, in *Proceedings of the 12th Conference on Uncertainty in Artificial Intelligence* (AAAI Press, 2000), pp. 397–405

362. S. Muff, F. Rao, A. Caflisch, Local modularity measure for network clusterizations. Phys. Rev. E **72**, 056107 (2005). https://doi.org/10.1103/PhysRevE.72.056107

363. M. Muja, D.G. Lowe, Scalable nearest neighbor algorithms for high dimensional data. IEEE Trans. Pattern Anal. Mach. Intell. **36**(11), 2227–2240 (2014). https://doi.org/10.1109/TPAMI.2014.2321376

364. F. Napolitano, G. Raiconi, R. Tagliaferri, A. Ciaramella, A. Staiano, G. Miele, Clustering and visualization approaches for human cell cycle gene expression data analysis. Int. J. Approximate Reasoning **47**(1), 70–84 (2008). https://doi.org/10.1016/j.ijar.2007.03.013

365. O. Nasraoui, M. Soliman, E. Saka, A. Badia, R. Germain, A Web usage mining framework for mining evolving user profiles in dynamic websites. IEEE Trans. Knowl. Data Eng. **20**(2), 202–215 (2008). https://doi.org/10.1109/TKDE.2007.190667

366. M.E.J. Newman, Fast algorithm for detecting community structure in networks. Phys. Rev. E **69**, 066133 (2004). https://doi.org/10.1103/PhysRevE.69.066133

367. M.E.J. Newman, Scientific collaboration networks: II. Shortest paths, weighted networks, and centrality. Phys. Rev. E **64**(1), 016132 (2001). https://doi.org/10.1103/PhysRevE.64.016132

368. M.E.J. Newman, Analysis of weighted networks. Phys. Rev. E **70**(5), 056131 (2004). https://doi.org/10.1103/PhysRevE.70.056131

369. M.E.J. Newman, Detecting community structure in networks. Eur. Phys. J. B **38**, 321–330 (2004). https://doi.org/10.1140/epjb/e2004-00124-y

370. M.E.J. Newman, Finding community structure in networks using the eigenvectors of matrices. Phys. Rev. E **74**(3), 036104 (2006). https://doi.org/10.1103/PhysRevE.74.036104

371. M.E.J. Newman, Modularity and community structure in networks. Proc. Natl. Acad. Sci. USA **103**(23), 8577–8582 (2006). https://doi.org/10.1073/pnas.0601602103

372. M.E.J. Newman, *Networks: An Introduction* (Oxford University Press, Oxford, 2010)

373. M.E.J. Newman, M. Girvan, Finding and evaluating community structure in networks. Phys. Rev. E **69**(2), 026113 (2004). https://doi.org/10.1103/PhysRevE.69.026113

374. A. Ng, M. Jordan, Y. Weiss, On spectral clustering: analysis and an algorithm, in *Advances in Neural Information Processing Systems*, vol. 14, ed. by T. Dietterich, S. Becker, Z. Ghahramani (MIT Press, 2002), pp. 849–856

375. R. Ng, J. Han, Efficient and effective clustering methods for spatial data mining. In *Proceedings of 20-th Conference on VLDB* (Santiago, Chile, 1994), pp. 144–155

376. H. Ning, W. Xu, Y. Chi, Y. Gong, T. Huang, Incremental spectral clustering by efficiently updating the eigen-system. Pattern Recogn. **43**(1), 113–127 (2010). https://doi.org/10.1016/j.patcog.2009.06.001

377. Y.Q. Niu, B.Q. Hua, W. Zhang, M. Wang, Detecting the community structure in complex networks based on quantum mechanics. Phys. A: Stat. Mech. Appl. **387**(24), 6215–6224 (2008). https://doi.org/10.1016/j.physa.2008.07.008

378. R. Ostrovsky, Y. Rabani, L.J. Schulman, C. Swamy. The effectiveness of Lloyd-type methods for the k-means problem, in *Proceedings of the 47th Annual IEEE Symposium on Foundations of Computer Science*, FOCS'06 (Berkeley, CA, 2006), pp. 165–176

379. L. Page, S. Brin, R. Motwani, T. Winograd, The PageRank citation ranking: Bringing order to the Web. Technical Report (Stanford Digital Library Technologies Project, 1998. http://citeseer.ist.psu.edu/page98pagerank.html

380. N.R. Pal, J.C. Bezdek, On cluster validity for the fuzzy c-means model. IEEE Trans. Fuzzy Syst. **3**(3), 370–379 (1995). https://doi.org/10.1109/91.413225

381. N.R. Pal, J.C. Bezdek, E.C.-K. Tsao, Generalized clustering networks and Kohonen's self-organizing scheme. IEEE Trans. Neural Networks **4**(4), 549–557 (1993). https://doi.org/10.1109/72.238310

382. N.R. Pal, K. Pal, J.M. Keller, J.C. Bezdek, A possibilistic fuzzy c-means clustering algorithm. IEEE Trans. Fuzzy Syst. **13**(4), 517–530 (2005). https://doi.org/10.1109/TFUZZ.2004.840099

383. G. Palla, A.-L. Barabási, T. Vicsek, Quantifying social group evolution. Nature **446**, 664–667 (2007). https://doi.org/10.1038/nature05670

384. G. Palla, I. Derényi, I. Farkas, T. Vicsek, Uncovering the overlapping community structure of complex networks in nature and society. Nature **435**, 814–818 (2005)

385. H.-S. Park, C.-H. Jun, A simple and fast algorithm for k-medoids clustering. Expert Syst. Appl. **36**(2), 3336–3341 (2009). https://doi.org/10.1016/j.eswa.2008.01.039

386. L. Parsons, E. Haque, H. Liu, Subspace clustering for high dimensional data: a review. ACM SIGKDD Explor. Newsletter **6**(1), 90–105 (2004). https://doi.org/10.1145/1007730.1007731

387. W. Pedrycz, *Knowledge-Based Clustering* (Wiley, Hoboken, NJ, 2005)

388. W. Pedrycz, A. Amato, V. Di Lecce, V. Piuri, Fuzzy clustering with partial supervision in organization and classification of digital images. IEEE Trans. Fuzzy Syst. **16**, 1008–1026 (2008). https://doi.org/10.1109/TFUZZ.2008.917287

389. J.M. Pena, J.A. Lozano, P. Larrañaga, An empirical comparison of four initialization methods for the k-means algorithm. Pattern Recogn. Lett. **20**, 1027–1040 (1999). https://doi.org/10.1016/S0167-8655(99)00069-0

390. J. Peng, Y. Wei, Approximating K-means-type clustering via semidefinite programming. SIAM J. Optimization **18**(1), 186–205 (2007). https://doi.org/10.1137/050641983

391. A. Pérez-Suárez, J.F. Martinéz-Trinidad, J.A. Carrasco-Ochoa, J.E. Medina-Pagola, A new overlapping clustering algorithm based on graph theory, in *Advances in Artificial Intelligence*, vol. 7629, LNCS (Springer, 2013), pp. 61–72

392. F. Pernkopf, D. Bouchaffra, Genetic-based EM algorithm for learning Gaussian mixture models. IEEE Trans. Pattern Anal. Mach. Intell. **27**(8):1344–1348. https://doi.org/10.1109/TPAMI.2005.162

393. T. Pisanski, J. Shawe-Taylor, Characterising graph drawing with eigenvectors. J. Chem. Inf. Comput. Sci. **40**(3), 567–571 (2000). https://doi.org/10.1021/ci9900938

394. C. Pizzuti, GA-Net: a genetic algorithm for community detection in social networks, in *Proceedings of the 10th International Conference on Parallel Problem Solving from Nature*, volume 5199 of *LNCS*, pp. 1081–1090, Dortmund, Germany, September 13–17, 2008 (Springer, Berlin, 2008)

395. D. Pollard, Strong consistency of k-means clustering. Ann. Statist. **9**(1), 135–140 (1981)

396. P. Pons, M. Latapy, Computing communities in large networks using random walks. J. Graph Algorithms Appl. **10**(2), 191–218 (2006). https://doi.org/10.1007/11569596_31

397. F.P. Preparata, M.I. Shamos, *Computational Geometry* (Springer, Berlin, Heidelberg, New York, 1990)

398. W.H. Press, S.A. Teukolsky, W.T. Vetterling, B.P. Flannery, *Numerical Recipes: The Art of Scientific Computing*, 3rd edn. (Cambridge University Press, Cambridge, New York, 1992)

399. G. Puy, N. Tremblay, R. Gribonval, P. Vandergheynst, Random sampling of bandlimited signals on graphs. arXiv:1511.05118v2 [cs.SI], 20 May 2016

400. R. Rabbany, M. Takaffoli, J. Fagnan, O.R. Zaane, R.J.G.B. Campello, Relative validity criteria for community mining algorithms, in *Proceedings of Advances in Social Network Analysis and Mining ASONAM 2012* (IEEE Computer Society, 2012), pp. 258–265

401. F. Radicchi, C. Castellano, F. Cecconi, V. Loreto, D. Parisi, Defining and identifying communities in networks. Proc. Natl. Acad. Sci. USA **101**(9), 2658–2663 (2004). https://doi.org/10.1073/pnas.0400054101

402. A. Rajaraman, J.D. Ullman, *Mining of Massive Data Sets* (Stanford University, 2010). http://infolab.stanford.edu/~ullman/mmds.html

403. A. Ralston, A First Course in Numerical Analysis (McGraw-Hill, New York, 1965)

404. R.A. Redner, H.F. Walker, Mixture densities, maximum likelihood and the EM algorithm. SIAM Rev. **26**(2), 195–239 (1984)

405. J. Reichardt, S. Bornholdt, When are networks truly modular? Physica D: Nonlinear Phenom. **224**(1–2), 20–26 (2006). https://doi.org/10.1016/j.physd.2006.09.009

406. L.C. Reidy, An information-theoretic approach to finding community structure in networks. B.Sc. thesis. Trinity College, Dublin, 2009

407. J.L. Rodgers, W.A. Nicewander. Thirteen ways to look at the correlation coefficient. Am. Stat. **42**(1), 59–66 (1988). http://www.jstor.org/stable/2685263

408. F. Rossi, N. Villa-Vialaneix, Optimizing an organized modularity measure for topographic graph clustering: a deterministic annealing approach. Neurocomputing **73**(7–9), 1142–1163 (2010). https://doi.org/10.1016/j.neucom.2009.11.023

409. M. Rosvall, C.T. Bergstrom, An information-theoretic framework for resolving community structure in complex networks. Proc. Natl. Acad. Sci. **104**(18), 7327 (2007). https://doi.org/10.1073/pnas.0611034104

410. M. Roubens, Pattern classification problems and fuzzy sets. Fuzzy Sets Syst. **1**, 239–253 (1978). https://doi.org/10.1016/0165-0114(78)90016-7

411. S. Roweis, L. Saul, Nonlinear dimensionality reduction by locally linear embedding. Science **290**(5500), 2323–2326 (2000). https://doi.org/10.1126/science.290.5500.2323

412. E.H. Ruspini, A new approach to clustering. Inf. Control **15**, 22–32 (1969). https://doi.org/10.1016/S0019-9958(69)90591-9

413. Y. Saad, *Numerical Methods for Large Eigenvalue Problems*, 2nd edn. (SIAM, Philadelphia, 2011)

414. J. Sander, A. Ester, H.-P. Kriegel, X. Xu, Density-based clustering in spatial databases: the algorithm GDBSCAN and its applications. Data Min. Knowl. Disc. **2**(2), 169–194 (1998). https://doi.org/10.1023/A:1009745219419

415. A. Sandryhaila, J.M.F. Moura, Big data analysis with signal processing on graphs: Representation and processing of massive data sets with irregular structure. IEEE Signal Process. Mag. **31**(5), 80–90 (2014). https://doi.org/10.1109/MSP.2014.2329213

416. B.M. Sarwar, G. Karypis, J. Konstan, J. Riedl, Recommender systems for large-scale e-commerce: scalable neighborhood formation using clustering, in *Proceedings of the Fifth International Conference on Computer and Information Technology*, ICCIT 2002 (200)

417. L.K. Saul, S.T. Roweis, Think globally, fit locally: unsupervised learning of low dimensional manifolds. J. Mach. Learn. Res. **4**, 119–155 (2003). https://doi.org/10.1162/153244304322972667

418. S.E. Schaeffer, Graph clustering. Comput. Sci. Rev. **1**(1), 27–64 (2007). https://doi.org/10.1016/j.cosrev.2007.05.001

419. B. Schölkopf, A. Smola, K.-R. Müller, Nonlinear component analysis as a kernel eigenvalue problem. Neural Comput. **10**(5), 1299–1319 (1998). https://doi.org/10.1162/089976698300017467

420. M. Schonlau, The clustergram: a graph for visualizing hierarchical and non-hierarchical cluster analyses. Stata J. **2**(4), 391–402 (2002)

421. Ph Schuetz, A. Caflisch, Efficient modularity optimization by multistep greedy algorithm and vertex mover refinement. Phys. Rev. E **77**(4), 046112 (2008). https://doi.org/10.1103/PhysRevE.77.046112

422. V. Schwämmle, O.N. Jensen, A simple and fast method to determine the parameters for fuzzy *c*-means cluster validation. arxiv:1004.1307v1 [q-bio.qm], Department of Biochemistry and Molecular Biology, University of Southern Denmark, DK-5230 Odense M,Denmark (2010)

423. J. Scott, *Social Network Analysis: A Handbook*, 2nd edn. (Sage, London, 2000)

424. S.Z. Selim, M.A. Ismail, *k*-means-type algorithms: a generalized convergence theorem and characterization of local optimality. IEEE. Trans. Pattern Anal. Mach. Intell. **6**, 81–87 (1984). https://doi.org/10.1109/TPAMI.1984.4767478

425. M. Shao, S. Li, Z. Ding, Y. Fu, Deep linear coding for fast graph clustering, in *Proceedings of the 24th International Conference on Artificial Intelligence*, ed. by Q. Yang, M. Wooldridge, IJCAI'15 (AAAI Press, 2015), pp. 3798–3804

426. A. Sharma, Representation, Segmentation and Matching of 3D Visual Shapes using Graph Laplacian and Heat-Kernel. PhD thesis, Université de Grenoble, 29 Oct 2012

427. J. Shi, J. Malik, Normalized cuts and image segmentation. IEEE Trans. Pattern Anal. Mach. Intell. **22**(8), 888–905 (2000). https://doi.org/10.1109/34.868688

428. H. Shinnou, M. Sasaki, Spectral clustering for a large data set by reducing the similarity matrix size, in *Proceedings of the 6th International Conference on Language Resources and*

Evaluation (European Language Resources Association (ELRA), 2008). http://www.lrec-conf.org/proceedings/lrec2008/summaries/62.html

429. R. Shioda, L. Tunçel, Clustering via minimum volume ellipsoids. J. Comput. Optim. Appl. **37**(3), 247–295 (2007). https://doi.org/10.1007/s10589-007-9024-1

430. H. Shiokawa, Y. Fujiwara, M. Onizuka, Fast algorithm for modularity-based graph clustering, in *AAAI* 13 ed. by M. desJardins, M.L. Littman. (AAAI Press, 2013). https://www.aaai.org/ocs/index.php/AAAI13/paper/download/6188

431. L. Shu, A. Chen, M. Xiong, W. Meng, Efficient spectral neighborhood blocking for entity resolution, in *Proceedings of IEEE International Conference on Data Engineering*, ICDE 2011, Hannover, Germany, 2011, pp. 1067–1078. https://research.google.com/pubs/archive/36940.pdf

432. D.I. Shuman, S.K. Narang, P. Frossard, A. Ortega, P. Vandergheynst, The emerging field of signal processing on graphs: Extending high-dimensional data analysis to networks and other irregular domains. IEEE Signal Process. Mag. **30**(3), 83–98 (2013). https://doi.org/10.1109/MSP.2012.2235192

433. R. Sinkhorn, A relationship between arbitrary positive matrices and doubly stochastic matrices. Ann. Math. Stat. **35**, 876–879 (1964)

434. P.H. Sneath, R.R. Sokal, *Numerical Taxonomy* (Freeman, San Francisco, 1973)

435. S. Sorgun, Bounds for the largest Laplacian eigenvalue of weighted graphs. Intl. J. Comb., Article ID **520610**, (2013). https://doi.org/10.1155/2013/520610

436. H. Späth, *Cluster Analysis Algorithms for Data Reduction and Classification of Objects* (Ellis Harwood, Chichester, 1980)

437. D. Spielman, Spectral graph theory, in *Combinatorial Scientific Computing*, Chapter 18, ed. by U. Naumann, O. Schenk. Chapman & Hall/CRC Computational Science, 25 Jan 2012, pp. 495–524. http://www.cs.yale.edu/~spielman/PAPERS/SGTChapter.pdf

438. D.A. Spielman, Algorithms, graph theory, and linear equations in Laplacian matrices, in *Proceedings of the International Congress of Mathematicians, (ICM 2010)*, vol. 4, ed. by R. Bhatia, A. Pal, G. Rangarajan, V. Srinivas, M. Vanninathan, Hyderabad, India, 19–27 Aug 2010 (World Scientific Publishing, 2010), pp. 2698–2722

439. D.A. Spielman, S.-H. Teng, Spectral partitioning works: planar graphs and finite element meshes. Linear Algebra Appl. **421**(2–3), 284–305 (2007). https://doi.org/10.1016/j.laa.2006.07.020

440. D.A. Spielman, S.-H. Teng, A local clustering algorithm for massive graphs and its application to nearly-linear time graph partitioning. arXiv:0809.3232v1 [cs.DS], 18 Sept 2008

441. N. Srebro, G. Shakhnarovich, S. Roweis, When is clustering hard? PASCAL Workshop on Statistics and Optimization of Clustering Workshop, 2005. http://ttic.uchicago.edu/~nati/Publications/SrebroEtalPASCAL05.pdf

442. P.F. Stadler, Landscapes and their correlation functions. J. Math. Chem. **20**(1), 1–45 (1996). https://doi.org/10.1007/BF01165154

443. M. Steinbach, G. Karypis, V. Kumar, A comparison of document clustering techniques, in *Proceedings of KDD Workshop on Text Mining, Proceedings of the 6th International Conference on Knowledge discovery and Data Mining*, Boston, MA, (2000). http://citeseerx.ist.psu.edu/viewdoc/summary?doi=10.1.1.34.1505

444. H. Steinhaus, Sur la division des corp materiels en parties. Bull. Acad. Polon. Sci. **4**(12), 801–804 (1956)

445. S. Still, W. Bialek, L. Bottou, Geometric clustering using the information bottleneck method, in *Advances in Neural Information Processing Systems 16* (2003), pp. 1165–1172

446. H. Strange, R. Zwiggelaar, *Open Problems in Spectral Dimensionality Reduction* (Springer, 2014). https://doi.org/10.1007/978-3-319-03943-5

447. A. Strehl, J. Ghosh, Cluster ensembles—a knowledge reuse framework for combining multiple partitions. J. Mach. Learn. Res. **3**, 583–617 (2002). https://doi.org/10.1162/153244303321897735

448. T. Su, J.G. Dy, In search of deterministic methods for initializing k-means and Gaussian mixture clustering. Intell. Data Anal. **11**(4), 319–338 (2007)

449. C.A. Sugar, G.M. James, Finding the number of clusters in a data set: An information theoretic approach. J. Am. Stat. Assoc. **98**(463), 750–763 (2003). https://doi.org/10.1198/016214503000000666
450. P. Sun, R.M. Freund, Computation of minimum-volume covering ellipsoids. Oper. Res. **52**(5), 690–706 (2004). https://doi.org/10.1287/opre.1040.0115
451. J.A.K. Suykens, T. Van Gestel, J. Vandewalle, B. De Moor, A support vector machine formulation to PCA analysis and its kernel version. IEEE Trans. Neural Networks **14**(2), 447–450 (2003). https://doi.org/10.1109/TNN.2003.809414
452. A. Szlam, X. Bresson, Total variation and Cheeger cuts, in *Proceedings of the 27th International Conference on Machine Learning (ICML-10)* (Haifa, Israel, 2010), pp. 1039–1046
453. P.-N. Tan, M. Steinbach, V. Kumar, *Introduction to Data Mining* (Addison Wesley Longman, 2006)
454. J. Tanha, M. van Someren, H. Afsarmanesh, Semi-supervised self-training for decision tree classifiers. Int. J. Mach. Learn. Cybern., 1–16 (2015). https://doi.org/10.1007/S13042-015-0328-7
455. M. Tasgin, H. Bingol, Community detection in complex networks using genetic algorithm. arXiv:cond-mat/0604419v1 [cond-mat.dis-nn], 18 Apr 2006
456. J.B. Tenenbaum, V. de Silva, J.C. Langford, A global geometric framework for nonlinear dimensionality reduction. Science **290**(5500), 2319–2323 (2000). https://doi.org/10.1126/science.290.5500.2319
457. Y. Terada, Strong consistency of reduced k-means clustering (2014)
458. K. Thangavel, N.K. Visalakshi, Ensemble based distributed k-harmonic means clustering. Int. J. Recent Trends Eng. **2**(1), 125–129 (2009)
459. R. Tibshirani, G. Walther, T. Hastie, Estimating the number of clusters in a data set via the gap statistic. J. Royal Stat. Soc.: Ser. B (Stat. Methodol.) **63**(2), 411–423 (2001). https://doi.org/10.1111/1467-9868.00293
460. M.E. Timmerman, E. Ceulemans, K. Roover, K. Leeuwen, Subspace k-means clustering. Behav. Res. Methods **45**(4), 1011–1023 (2013)
461. N. Tishby, F.C. Pereira, W. Bialek, The information bottleneck method, in *Proceedings of 37th Allerton Conference on Communication, Control and Computing* (1999), pp. 368–377
462. D.A. Tolliver, G.L. Miller, Graph partitioning by spectral rounding: applications in image segmentation and clustering, in *IEEE Computer Society Conference on Computer Vision and Pattern Recognition*, vol. 1, ed. by A. Fitzgibbon, C.J. Taylor, Y. LeCun. IEEE Computer Soc., 17–22 Jun 2006, pp. 1053–1060. https://doi.org/10.1109/CVPR.2006.129
463. J.T. Tou, R.C. Gonzalez, *Pattern Recognition Principles* (Addison-Wesely Pub. Co., 1974)
464. L.N. Trefethen, D. Bau III, *Numerical Linear Algebra* (SIAM, Philadelphia, 1997)
465. N. Tremblay, G. Puy, R. Gribonval, P. Vandergheynst, Compressive spectral clustering, in *Proceedings of the 33rd International Conference on Machine Learning*, ed. by M.F. Balcan, K.Q. Weinberger. JMLR Workshop and Conference Proceedings, vol. 48 (2016). http://jmlr.org/proceedings/papers/v48/tremblay16.pdf
466. N.C. Trillos, D. Slepčev, J. von Brecht, Th. Laurent, X. Bresson, Consistency of cheeger and ratio graph cuts. arXiv preprint arXiv:1411.6590 (2014)
467. M. Trosset, Visualizing correlation. J. Comput. Graph. Stat. **14**(1), 1–19 (2005). https://doi.org/10.1198/106186005X27004
468. E.C.-K. Tsao, J.C. Bezdek, N.R. Pal, Fuzzy Kohonen clustering networks. Pattern Recogn. **27**(5), 757–764 (1994). https://doi.org/10.1109/FUZZY.1992.258797
469. P. Tseng, Nearest q-flat to m points. J. Optim. Theory Appl. **105**(1), 249–252 (2000). https://doi.org/10.1023/A:1004678431677
470. D.I. Tsomokos, Community detection in complex networks with quantum random walks. arXiv:1012.2405v1 [quant-ph], 10 Dec 2010
471. E. Tu, L. Cao, J. Yang, N. Kasabov, A novel graph-based k-means for nonlinear manifold clustering and representative selection. Neurocomputing **143**, 109–122 (2014). https://doi.org/10.1016/j.neucom.2014.05.067

472. J.K. Uhlmann, Satisfying general proximity/similarity queries with metric trees. Info. Process. Lett. **40**, 175–179 (1991). https://doi.org/10.1016/0020-0190(91)90074-R
473. L.G. Valiant, A theory of the learnable. Commun. ACM **27**(11), 1134–1142 (1984)
474. S. van Dongen. MCL—a cluster algorithm for graph. https://micans.org/mcl/
475. S. van Dongen, Performance criteria for graph clustering and Markov cluster experiments. Technical Report INS-R0012, Centre for Mathematics and Computer Science, Amsterdam, The Netherlands, 31 May 2000. http://www.cwi.nl/ftp/CWIreports/INS/INS-R0012.pdf
476. T. van Laarhoven, E. Marchiori, Axioms for graph clustering quality functions. J. Mach. Learn. Res. **15**, 193–215 (2014)
477. V.N. Vapnik, *The Nature of Statistical Learning Theory* (Springer, New York, 1995)
478. V.N. Vapnik, *Statistical Learning Theory* (Wiley-Interscience, New York, 1998)
479. S. Vempala, G. Wang, A spectral algorithm for learning mixture models. J. Comput. Syst. Sci. **68**(4), 841–860 (2004). https://doi.org/10.1016/j.jcss.2003.11.008
480. D. Verma, M. Meilă, A comparison of spectral clustering algorithms. Technical Report UW-CSE-03-05-01, University of Washington, 2003. http://www.stat.washington.edu/spectral/papers/UW-CSE-03-05-01.pdf
481. N. Villa, F. Rossi, Q.-D. Truong, Mining a medieval social network by kernel SOM and related methods, in *Proceedings of MASHS 2008 (Modèeles et Apprentissage en Sciences Humaines et Sociales)*, Créeteil, France (2008)
482. U. von Luxburg, A tutorial on spectral clustering. Stat. Comput. **17**(4), 395–416 (2007). https://doi.org/10.1007/s11222-007-9033-z
483. U. von Luxburg, Clustering stability: an overview. Found. Trends Mach. Learn. **2**(3), 235–274 (2009). https://doi.org/10.1561/2200000008
484. U. von Luxburg, A. Radl, M. Hein, Hitting and commute times in large random neighborhood graphs. J. Mach. Learn. Res. **15**, 1751–1798 (2014)
485. U. von Luxburg, R. C Williamson, I. Guyon, Clustering: Science or art? Proc. of ICML Workshop on Unsupervised and Transfer Leaving, ed. by I. Guyon, G. Dror, V. Lemaire, G. Taylor, D. Silver, 02 Jul 2012, pp. 65–79. https://doi.org/proceedings.mlr.press/v27/luxburg12a.html
486. S. Wagner, D. Wagner, Comparing clustering. Interner Bericht. Fakultät für Informatik, Universität Karlsruhe, 12 Jan 2007. http://digbib.ubka.uni-karlsruhe.de/volltexte/1000011477
487. K. Wagstaff, C. Cardie, Clustering with instance-level constraints, in *Proceedings of the Seventeenth International Conference on Machine Learning*, ICML '00, San Francisco, CA, USA, 2000. (Morgan Kaufmann Publishers Inc., 2000), pp. 1103–1110
488. K. Wagstaff, C. Cardie, S. Rogers, S. Schroedl, Constrained k-means clustering with background knowledge, in *Proceedings of the 18th International Conference on Machine Learning*, Williamstown, MA, USA, 28 Jun–1 July 2001, pp. 577–584
489. K. Wakita, T. Tsurumi, Finding community structure in mega-scale social networks. arXiv:cs/0702048v1 [cs.CY], 8 Feb 2007
490. D.L. Wallace, Comment. J. Am. Stat. Assoc. **78**(383), 569–576 (1983). https://doi.org/10.1080/01621459.1983.10478009
491. C. Walles, D. Dowe, Intristic classification by MML—the Snob program, in *Proceedings of 7th Australian Joint Conference on Artificial Intelligence*, Armidale, Australia (World Scientific Publishing Co., 1994), pp. 37–44
492. F. Wang, P. Li, A.Ch. König, M. Wan, Improving clustering by learning a bi-stochastic data similarity matrix, Knowl. Inf. Syst. **32**(2), 351–382 (2012). https://doi.org/10.1007/s10115-011-0433-1
493. F. Wang, Ch. Zhang, Label propagation through linear neighborhoods. IEEE Trans. Knowl. Data Eng. **20**(1), 55–67 (2008). https://doi.org/10.1109/TKDE.2007.190672
494. L. Wang, Ch. Leckie, K. Ramamohanarao, J. Bezdek, Approximate spectral clustering, in *Advances in Knowledge Discovery and Data Mining. Proceedings of 13th Pacific-Asia Conference, PAKDD 2009 Bangkok, Thailand, 27–30*, ed. by T. Theeramunkong, B. Kijsirikul, N. Cercone, T.-B. Ho. LNCS, vol. 5476 (2009), pp. 134–146. https://doi.org/10.1007/978-3-642-01307-2_15

495. L. Wang, T. Lou, J. Tang, J.E. Hopcroft, Detecting community kernels in large social networks, in *Proceedings of the 2011 IEEE 11th International Conference on Data Mining*, ICDM '11, Washington, DC, USA, 2011 (IEEE Computer Society, 2011), pp. 784–793

496. W. Wang, J. Yang, R.R. Muntz, Sting: a statistical information grid approach to spatial data mining, in *Proceedings of the 23rd International Conference on Very Large Data Bases*, VLDB '97, San Francisco, CA, USA (Morgan Kaufmann Publishers Inc., 1997), pp. 186–195

497. Y.-X. Wang, Y.-J. Zhang, Nonnegative matrix factorization: a comprehensive review. IEEE Trans. Knowl. Data Eng. **25**(6), 1336–1353 (2013)

498. S. Wasserman, K. Faust, *Social Network Analysis* (Cambridge University Press, Cambridge, 1994)

499. D.J. Watts, S. Strogatz, Collective dynamics of "small-world" networks. Nature **393**(6684), 440–442 (1998). https://doi.org/10.1038/30918

500. W. Wei, J.M. Mendel, Optimality tests for the fuzzy c-means algorithm. Pattern Recogn. **27**(11), 1567–1573 (1994). https://doi.org/10.1016/0031-3203(88)90037-4

501. S. White, P. Smyth, A spectral approach to finding communities in graphs, in *Proceedings of the 5th SIAM International Conference on Data Mining*, ed. by H. Kargupta, J. Srivastava, C. Kamath, A. Goodman, SIAM, Philadelphia (2005). http://www.datalab.uci.edu/papers/siam_graph_clustering.pdf

502. S.T. Wierzchoń, M.A. Kłopotek, *Algorithms of Cluster Analysis*, vol. 3 (Warszawa, Monograph Series (Institute of Computer Science PAS, 2015)

503. M.P. Windham, Numerical classification of proximity data with assignment measures. J. Classif. **2**, 157–172 (1985). https://doi.org/10.1007/BF01908073

504. D.M. Witten, R. Tibshirani, A framework for feature selection in clustering. J. Am. Stat. Assoc. **105**(490), 713–726 (2010). https://doi.org/10.1198/jasa.2010.tm09415

505. I.H. Witten, E. Frank, *Data Mining: Practical Machine Learning Tools and Techniques. Second Edition* (Morgan Kaufmann, 2005)

506. F. Wu, B.A. Huberman, Finding communities in linear time: a physics approach. Eur. Phys. J. B **38**(2), 331–338 (2004). https://doi.org/10.1140/epjb/e2004-00125-x

507. J. Wu, J. Chen, H. Xiong, M. Xie, External validation measures for k-means clustering: A data distribution perspective. Expert Syst. Appl. **36**(3), 6050–6061 (2009). https://doi.org/10.1016/j.eswa.2008.06.093

508. K.L. Wu, M.S. Yang, Alternative c-means clustering algorithms. Pattern Recogn. **35**(10), 2267–2278 (2002). https://doi.org/10.1016/s0031-3203(01)00197-2

509. X. Wu, V. Kumar, J.R. Quinlan, J. Ghosh, Q. Yang, H. Motoda, G.J. McLachlan, A. Ng, B. Liu, P.S. Yu, Z.-H. Zhou, M. Steinbach, D.J. Hand, D. Steinberg, Top 10 algorithms in data mining. Knowl. Inf. Syst. **14**(1), 1–37 (2007). https://doi.org/10.1007/s10115-007-0114-2

510. J. Xie, B.K. Szymanski, Towards linear time overlapping community detection in social networks, in *PAKDD*. Proceedings of the 16th Pacific-Asia Conference on Advances in Knowledge Discovery and Date Mining, May 29–June 01 2012. https://doi.org/10.1007/978-3-642-30220-6_3

511. X.L. Xie, G. Beni, A validity measure for fuzzy clustering. IEEE. Trans. Pattern Anal. Mach. Intell. **13**(8), 841–847 (1991). https://doi.org/10.1109/34.85677

512. E.P. Xing, M.I. Jordan, On semidefinite relaxation for normalized k-cut and connections to spectral clustering. Technical Report UCB/CSD-3-1265, Computer Science Division (EECS), University of California, Berkeley, CA 94720 (2003)

513. E.P. Xing, A.Y. Ng, M.I. Jordan, S. Russell, Distance metric learning with application to clustering with side-information, in *Advances in Neural Information Processing Systems* (2003), pp. 521–528

514. G. Xu, S. Tsoka, L.G. Papageorgiou, Finding community structures in complex networks using mixed integer optimisation. Eur. Phys. J. B—Condens. Matter Complex Syst. **60**(2), 231–239 (2007). https://doi.org/10.1140/epjb/e2007-00331-0

515. R. Xu, D. Wunsch II, Survey of clustering algorithms. IEEE Trans. Neural Networks **16**(3), 645–678 (2005). https://doi.org/10.1109/TNN.2005.845141

516. W. Xu, X. Liu, Y. Gong, Document clustering based on non-negative matrix factorization, in *Proceedings of the 26th Annual International ACM SIGIR Conference on Research and Development in Informaion Retrieval*, New York, NY, USA, 2003 (ACM Press, 2003), pp. 267–273

517. D. Yan, L. Huang, M.I. Jordan, Fast approximate spectral clustering, in *Proceedings of the 15th ACM SIGKDD Int'l Conference on Knowledge Discovery and Data Mining*, KDD'09, ACM, New York, NY, USA (2009), pp. 907–916. https://doi.org/10.1145/1557019.1557118

518. B. Yang, Self-organizing network evolving model for mining network community structure, in *Advanced Data Mining and Applications*, ed. by X. Li, O.R. Zaïane, Z. Li. LNCS, vol. 4093 (Springer, 2006), pp. 404–415. https://doi.org/10.1007/11811305_45

519. F. Yang, T. Sun, Ch. Zhang, An efficient hybrid data clustering method based on K-harmonic means and particle swarm optimization. Expert Syst. Appl. **36**, 9847–9852 (2009). https://doi.org/10.1016/j.eswa.2009.02.003

520. M.-S. Yang, A survey of fuzzy clustering. Math. Comput. Model. **18**(11), 1–16 (1993). https://doi.org/10.1016/0895-7177(93)90202-A

521. L. Yen, F. Fouss, Ch. Decaestecker, P. Francq, M. Saerens, Graph nodes clustering with the sigmoid commute-time kernel: A comparative study. Data Knowl. Eng. **68**(3), 338–361 (2009). https://doi.org/10.1016/j.datak.2008.10.006

522. J. Yu, Q. Cheng, H. Huang, Analysis of the weighting exponent in the FCM. IEEE. Trans. Syst. Man Cybern. B, Cybern. **34**(1), 634–638 (2004). https://doi.org/10.1109/TSMCB.2003. 810951

523. J. Yu, M.S. Yang, Optimality test for generalized FCM and its application to parameter selection. IEEE. Trans. Fuzzy Syst. **13**(1), 164–176 (2005). https://doi.org/10.1109/FUZZ.2004. 836065

524. S.X. Yu, J. Shi, Multiclass spectral clustering, in *Proceedings of the 9th IEEE International Conference on Computer Vision*. ICCV '03, vol. 2, Washington, DC, USA, 2003. (IEEE Computer Society, 2003), pp. 313–319

525. W.W. Zachary, An information flow model for conflict and fission in small groups. J. Anthropol. Res. **33**, 452–473 (1977)

526. R.B. Zadeh, S. Ben-David, A uniqueness theorem for clustering, in *Proceedings of the Twenty-Fifth Conference on Uncertainty in Artificial Intelligence*, UAI '09, Arlington, Virginia, United States, 2009 (AUAI Press, 2009), pp. 639–646

527. R. Zass, A. Shashua, A unifying approach to hard and probabilistic clustering, in *Proceedings of 10th IEEE International Conference on Computer Vision*. ICCV 2005, vol. 1, Beijing, China, 17–21 Oct 2005, pp. 294–301

528. L. Zelnik-Menor, P. Perona, Self-tuning spectral clustering, in *Advances in Neural Information Processing Systems*, NIPS 17 (MIT Press, 2004), pp. 1601–1608

529. H. Zha, X. He, Ch. Ding, M. Gu, H. Simon, Spectral relaxation for k-means clustering, in *Advances in Neural Information Processing Systems 14* ed. by T.G. Dietterich, S. Becker, Z. Ghahramani. NIPS, vol. 14 (MIT Press, 2001), pp. 1057–1064

530. B. Zhang, Generalized k-harmonic means—boosting in unsupervised learning. Technical Report HPL-2000-137 (Hewlett-Packard Labs, 2000). http://www.hpl.hp.com/techreports/ 2000/HPL-2000-137.html

531. B. Zhang, Dependence of clustering algorithm performance on clustered-ness of data. Technical Report HPL-2001-91 (Hewlett-Packard Labs, 2001)

532. D. Zhang, S. Chen, Clustering incomplete data using kernel-based fuzzy c-means algorithm. Neural Process. Lett. **18**(3), 155–162 (2003). https://doi.org/10.1023/B:NEPL.0000011135. 19145.1b

533. R. Zhang, A.I. Rudnicky, A large scale clustering scheme for kernel k-means, in *Proceedings of the 16th International Conference on Pattern Recognition*. ICPR'02, vol. 4, Washington, DC, USA, 11–15 Aug 2002. (IEEE Computer Society, 2002), pp. 40289–40292

534. S. Zhang, R.-S. Wang, X.-S. Zhang, Identification of overlapping community structure in complex networks using fuzzy c-means clustering. Physica A: Stat. Mech. Appl. **374**(1), 483–490 (2007). https://doi.org/10.1016/j.physa.2006.07.023

535. Q. Zhao, Cluster validity in clustering methods. Ph.D. thesis, University of Eastern Finland, 2012. http://epublications.uef.fi/pub/urn_isbn_978-952-61-0841-4/urn_isbn_978-952-61-0841-4.pdf

536. W. Zhao, H. Ma, Q. He, Parallel k-means clustering based on MapReduce, in *Proceedings of the 1st International Conference on Cloud Computing*, ed. by M.G. Jaatun, G. Zhao, C. Rong. LNCS, vol. 5931 (Springer-Verlag Berlin, Heidelberg, 2009), pp. 674–679. https://doi.org/10.1007/978-3-642-10665-1_71

537. S. Zhong, Efficient online spherical k-means clustering, in *Proceeding of IEEE International Joint Conference Neural Networks*, IJCNN 2005, Montreal, Canada, 31 July–4 Aug 2004 (IEEE Computer Society, 2004), pp. 3180–3185

538. D. Zhou, J. Huang, B. Schölkopf, Learning from labeled and unlabeled data on a directed graph, in *Proceedings of the 22nd International Conference on Machine Learning*, ICML '05, Bonn, Germany, 7–11 Aug 2005 (ACM, New York, 2005), pp. 1036–1043. https://doi.org/10.1145/1102351.1102482

539. D. Zhou, B. Schölkopf, A regularization framework for learning from graph data, in *ICML Workshop on Statistical Relational Learning and Its Connections to Other Fields*, vol. 15 (2004), pp. 132–137. http://www.cs.umd.edu/projects/srl2004/srl2004_complete.pdf

540. M. Zhou, Infinite edge partition models for overlapping community detection and link prediction, in *Proceedings of the Eighteenth International Conference on Artificial Intelligence and Statistics, AISTATS 2015*, San Diego, California, USA, 9–12 May 2015 (2015)

541. S. Zhou, J.Q. Gan, Mercer kernel, fuzzy c-means algorithm, and prototypes of clusters, in *Intelligent Data Engineering and Automated Learning—IDEAL 2004*, ed. by Z.R. Yang, H. Yin, R.M. Everson. LNCS, vol. 3177 (Springer, 2004), pp. 613–618

542. X. Zhu, Z. Ghahramani, J. Lafferty, Semi-supervised learning using Gaussian fields and harmonic functions, in *Proceedings of the Twentieth International Conference on Machine Learning (ICML-2003)*, vol. 3, ed. by T. Fawcett, N. Mishra, Washington, DC, 2003 (AAAI Press, 2003), pp. 912–919

543. X. Zhu, J.D. Lafferty, Z. Ghahramani, Semi-supervised learning: From Gaussian fields to Gaussian processes. Technical Report CMU-CS-03-175 (School of Computer Science, Carnegie Mellon University, Pittsburgh, PA 15213, 18 Aug 2003)

544. R. Zhang, A. Rudnicky, A new data selection principle for semi—supervised incremental learning (2006), in Proceedings of 18 International Conference on Pattern Recognition, 20–24 Aug 2006. https://doi.org/10.1109/ICPR.2006.115

Index

© Springer International Publishing AG 2018

S. T. Wierzchoń and M. A. Kłopotek, *Modern Algorithms of Cluster Analysis*,
Studies in Big Data 34, https://doi.org/10.1007/978-3-319-69308-8

Printed in the United States
By Bookmasters